Handbook of Squirrel Monkey Research

Handbook of Squirrel Monkey Research

Edited by
Leonard A. Rosenblum
State University of New York Downstate Medical Center
Brooklyn, New York
and
Christopher L. Coe
Stanford University School of Medicine
Stanford, California

Plenum Press • New York and London

Library of Congress Cataloging in Publication Data

Main entry under title:

Handbook of squirrel monkey research.

Includes bibliographies and index.
1. Squirrel monkeys. I. Rosenblum, Leonard A. II. Coe, Christopher L.
QL737.P925H36 1985 599.8′2 84-24859
ISBN 978-1-4757-0814-1 ISBN 978-1-4757-0812-7 (eBook)
DOI 10.1007/978-1-4757-0812-7

Softcover reprint of the hardcover 1st edition 1985
A Division of Plenum Publishing Corporation
233 Spring Street, New York, N.Y. 10013

Contributors

Christian R. Abee, Department of Comparative Medicine, University of South Alabama College of Medicine, Mobile, Alabama 36688

Eleanor R. Adair, John B. Pierce Foundation and Yale University, New Haven, Connecticut 06519

Lynne M. Ausman, School of Nutrition, Tufts University, Medford, Massachusetts 02155, and Department of Nutrition, Harvard School of Public Health, Boston, Massachusetts 02115

John D. Baldwin, Department of Sociology, University of California at Santa Barbara, Santa Barbara, California 93106

James E. Barrett, Department of Psychiatry, Uniformed Services University of the Health Sciences, School of Medicine, Bethesda, Maryland 20814

Thomas B. Clarkson, Department of Comparative Medicine and Arteriosclerosis Research Center, Bowman Gray School of Medicine, Wake Forest University, Winston-Salem, North Carolina 27103

Christopher L. Coe, Department of Psychiatry and Behavioral Sciences, Stanford University School of Medicine, Stanford, California 94305

W. Richard Dukelow, Endocrine Research Center, Michigan State University, East Lansing, Michigan 48824

Dorothy Munkenbeck Fragaszy, Department of Veterinary and Comparative Anatomy, Pharmacology, and Physiology, Washington State University, Pullman, Washington 99164

Daniel L. Gallina, Department of Nutrition, Harvard School of Public Health, Boston, Massachusetts 02115; *Present Address*: Division of Diabetes, Department of Medicine, University of Massachusetts Medical School, Worcester, Massachusetts 01605

Michael B. Hennessy, Department of Psychology, Wright State University, Dayton, Ohio 42435

Gerald H. Jacobs, Department of Psychology, University of California, Santa Barbara, Santa Barbara, California 93106

S. S. Kalter, Southwest Foundation for Research and Education, San Antonio, Texas 78284

Seymour Levine, Department of Psychiatry and Behavioral Sciences, Stanford University School of Medicine, Stanford, California 94305

John D. Newman, Laboratory of Comparative Ethology, National Institute of Child Health and Human Development, National Institutes of Health, Bethesda, Maryland 20205

Robert J. Nicolosi, Harvard Medical School, New England Regional Primate Research Center, Nutrition Division, Southborough, Massachusetts 01772

Leon T. Rosenberg, Department of Medical Microbiology, Stanford University School of Medicine, Stanford, California 94305

Leonard A. Rosenblum, Primate Behavior Laboratory and Department of Psychiatry, Downstate Medical Center, Brooklyn, New York 11203

Gary G. Schwartz, Primate Behavior Laboratory and Department of Psychiatry, Downstate Medical Center, Brooklyn, New York 11203

Erla R. Smith, Department of Physiology, Stanford University School of Medicine, Stanford, California 94305

Harriet L. Strickland, Laboratory of Behavioral Medicine, Southwest Foundation for Biomedical Research, San Antonio, Texas 78284

Richard W. Thorington, Jr., Department of Vertebrate Zoology, National Museum of Natural History, Smithsonian Institution, Washington, D.C. 20560

Sandra G. Wiener, Department of Psychiatry and Behavioral Sciences, Stanford University School of Medicine, Stanford, California 94305

Preface

As the editors of the first book on the squirrel monkey prophesied in 1968,* there has been an incredible expansion in primate research during the past 16 years. Their projection that the squirrel monkey would play an increasingly important role in this research effort has also come to be true during the ensuing years. One inadvertent result of the rapid growth, however, is that it has become more and more difficult for investigators to keep track of new information, both in their own disciplines and in related fields. For scientists who study and use the squirrel monkey in research, this problem is particularly pronounced, because articles are often published in specialized and disparate journals.

We felt that a new synthesis of the vast amount of information on *Saimiri* would resolve this problem and would provide an extremely valuable companion volume to the first book. The idea grew out of a small symposium held at the IX Congress of the International Primatological Society in Atlanta, Georgia, during August, 1982. Following the format of *The Squirrel Monkey*, additional authors were invited to discuss advances in areas which had experienced exceptional growth or to review basic information that would be of practical value to future researchers. Even with focused topics and synthetic reviews, the wealth of new data resulted in many long manuscripts. In response to the continuing problems with *Saimiri* nomenclature, Richard Thorington has provided us with a definitive statement on squirrel monkey taxonomy. To aid investigators who are unfamiliar with the squirrel monkey, we have compiled photos of the commonly used varieties and indicated the appropriate name as designated by Thorington in Chapter 1 (see Figure 1). We have also added the city of export in the parentheses, following the suggestion of Cooper (1968).† The natural behavior of the squirrel monkey is discussed in the second chapter by John Baldwin, who has provided us with the most thorough field studies on *Saimiri*. In part, due to his inspiration, there are two long-term field studies currently underway which will further our knowledge of squirrel monkeys in Costa Rica and southern Peru.

The focus of the book then shifts to laboratory studies on the squirrel monkey with Dorothy Fragaszy's review of learning and cognition. However,

* Rosenblum, L. A., and Cooper, R. W., 1968, *The Squirrel Monkey*, Academic, New York.

† Cooper, R. W., 1968, Squirrel monkey taxonomy and supply, in: *The Squirrel Monkey* (L. A. Rosenblum and R. W. Cooper, eds.), Academic, New York, pp. 1–30.

Fig. 1. (1) Guyanese type (*Saimiri sciureus sciureus*, Georgetown); (2) Brazilian or Colombian type (*Saimiri sciureus sciureus*, Leticia); (3) Peruvian type (*Saimiri sciureus boliviensis*, Iquitos); (4) Bolivian type (*Saimiri sciureus boliviensis*, Santa Cruz); (5) Central American type (*Saimiri sciureus orstedii*) (by F.D. Schmidt, San Diego Zoo);(6) Hybrid offspring of Guyanese x Bolivian cross (Stanford, CA).

we see an important shift in emphasis from early learning research to an orientation that pays homage to the organism's natural abilities and life style. Similarly, John Newman's review of vocal communication shows the insights one can gain by maintaining an awareness of an animal's natural behavior. The next chapter on the response of infant squirrel monkeys to maternal loss also takes advantage of a natural response, but in this case with the aim of assessing stress physiology. In 1968, the focus of infant studies was on general development; today we have detailed information available on many aspects of the squirrel monkey's psychological and physiological maturation. This point is amply illustrated in Michael Hennessy's chapter on the effect of surrogate-rearing. Not too long ago, our knowledge was virtually restricted to a single species, the rhesus macaque, but due to the efforts of J. Kaplan, S. Hopf, and M. Hennessy, we now must question many of the generalities about attachment processes and the impact of early experiences in primates.

Research on the reproductive physiology of the squirrel monkey has undergone the same renaissance. Richard Dukelow, who pioneered many of these investigations, has reviewed the new information on female physiology as it pertains to ovulation and breeding. General information about the endocrine system of the squirrel monkey and the relationship between hormones and behavioral processes are reviewed by Christopher Coe, Erla Smith, and Seymour Levine. A topic which may be less familiar to most readers is thermoregulation. Eleanor Adair's review delineates the important contribution of the squirrel monkey to the development of this scientific area. The significance of thermoregulatory processes for understanding other behavior systems is described by Gary Schwartz and Leonard Rosenblum, who also illustrate the utility of approaching laboratory studies with a solid grounding in evolutionary biology.

Another sensory system which obviously has a primary function in primates is vision. This topic is reviewed by Gerald Jacobs, who has studied the visual system of the squirrel monkey for many years. More recently, the squirrel monkey has figured prominently in cardiovascular research. The important findings and progress in this area are reviewed by Harriet Strickland and Thomas Clarkson. Next, James Barrett discusses the long-standing and surprisingly widespread use of squirrel monkeys in studies of behavioral pharmacology. Advances in our understanding of nutrition in the squirrel monkey are then described by Lynne Ausman. She shows how the unusual physiology of the squirrel monkey allows one to blend the acquisition of basic information with the pursuit of creative research questions. The immune system and diseases of the squirrel monkey are reviewed, in turn, by Sy Kalter. The encyclopedic breadth of this chapter speaks for itself. Finally, we are fortunate to have Christian Abee's detailed and comprehensive chapter on medical care and laboratory management of squirrel monkey colonies.

These papers have more than fulfilled our initial plans for the book, and we hope that the reader feels that our goal of providing an up-to-date resource volume has been met. Thanks obviously must go to the contributors and to the many individual researchers who made a second book on the squirrel monkey a necessity. Let us conclude with an apology to the many important researchers who could not be included in the *Handbook*. Perhaps the most significant comment one can make about the continuing role of the squirrel monkey in scientific research is that, in spite of the prodigious scope of the current volume, it was inevitable that many topics could still not be included.

C.L.C.
L.A.R.

Contents

Chapter 3

Cognition in Squirrel Monkeys: A Contemporary Perspective

Dorothy Munkenbeck Fragaszy

Chapter 4

Squirrel Monkey Communication

John D. Newman

Chapter 5

Physiological Consequences of Maternal Separation and Loss in the Squirrel Monkey

Christopher L. Coe, Sandra G. Wiener, Leon T. Rosenberg, and Seymour Levine

Chapter 6

Effects of Surrogate-Rearing on the Infant Squirrel Monkey

Michael B. Hennessy

Chapter 7

Reproductive Cyclicity and Breeding in the Squirrel Monkey

W. Richard Dukelow

Chapter 8

The Endocrine System of the Squirrel Monkey

Christopher L. Coe, Erla R. Smith, and Seymour Levine

Chapter 9

Thermoregulation in the Squirrel Monkey

Eleanor R. Adair

Chapter 10

Sneezing Behavior in the Squirrel Monkey and Its Biological Significance

Gary G. Schwartz and Leonard A. Rosenblum

Chapter 11

Visual System of the Squirrel Monkey

Gerald H. Jacobs

Chapter 12

Use of Squirrel Monkeys in Cardiovascular Research

Harriet L. Strickland and Thomas B. Clarkson

Chapter 14

Nutrition and Metabolism of the Squirrel Monkey

Lynne M. Ausman, Daniel L. Gallina, and Robert J. Nicolosi

Chapter 15

Immunology and Pathology of the Squirrel Monkey

S. S. Kalter

Chapter 16

Medical Care and Management of the Squirrel Monkey

Christian R. Abee

Handbook of Squirrel Monkey Research

1

The Taxonomy and Distribution of Squirrel Monkeys (Saimiri)

Richard W. Thorington, Jr.

1. INTRODUCTION

Squirrel monkeys of the genus *Saimiri* are a distinct group of New World primates. They are easily recognized visually or by their vocalizations. Anatomically they have a number of peculiarities, such as an interorbital fenestra, a double brachioradialis muscle in the arm, and a prominent articulation between the calcaneus and navicular bones of the foot. These specializations suggest that squirrel monkeys have a long evolutionary history independent of other primates. This hypothesis is strengthened by the dearth of derived features shared with other New World primates. The search for the closest living relative of *Saimiri* (its sister group) has not led to convincing conclusions. *Saimiri* is commonly placed in the same subfamily with *Cebus*, as in the classifications of Pocock (1925), Simpson (1945), von Pusch (1942), and Rosenberger (1981). But Hershkovitz strongly dissents, and this classification is not in agreement with Ford's analysis of the postcranial skeleton. Ford (1980) would link *Saimiri* more closely with *Callicebus* and *Aotus*. Hershkovitz (1977) separates it from all other cebids at the subfamily level. I presently adopt the same classification of *Saimiri*, as the sole member of the subfamily Saimirinae, but mine is a position of agnosticism. I am not convinced by any of the arguments to the contrary.

Squirrel monkeys are widespread in South America (Fig. 1). Their distribution is basically the Amazon basin, with extension into the Guianas.* But there is a disjunct population in western Panama and eastern Costa Rica (isolated by more than 600 miles from the South American populations), which clearly demonstrates that *Saimiri* once had a more extensive distribution.

* Guiana (or "the Guianas") is the region of northeastern South America made up of Guyana (formerly British Guiana), Surinam (formerly Dutch Guyana) and French Guiana.

Richard W. Thorington, Jr. • Department of Vertebrate Zoology, National Museum of Natural History, Smithsonian Institution, Washington, D. C. 20560.

Hershkovitz (1972) has postulated that the Central American squirrel monkeys were introduced by man, but this is a distinctive form of *Saimiri*, which clearly has been isolated for a much longer period of time than suggested by Hershkovitz' hypothesis.

Table I shows the way *Saimiri* has been classified by five reviewers during the 20th century. *Saimiri* was treated as a species of *Cebus* by von Pusch, but as a distinct genus by all other reviewers. One to seven species have been recognized and up to 16 different subspecies. A comparison of the classifications of Cabrera (1958) and Hill (1960) shows that there has been no approach to consensus, either with reference to how many species and subspecies exist or to the names that should be applied to them. My study was prompted by this unsatisfactory situation and consists of a taxonomic and systematic review of the genus based on an examination of museum specimens.

2. *METHODOLOGY*

This study is based on an examination of skins and skulls of squirrel monkeys in museums. The objective of the study is to determine how many species and subspecies of squirrel monkeys there are. Since species and subspecies are not directly observable in museum specimens, a brief consideration of research strategies and logic are presented here.

The species definition I use is that of Mayr's (1963) and many other authors, that species are "groups of populations that actually or potentially interbreed with each other." Our ability to use morphology to infer interbreeding is dependent on the logic that if there is interbreeding between two morphologically distinct populations, the distinctions will gradually be reduced, blurred, or eliminated, but if there is no interbreeding, the distinctions will be maintained. The clearest demonstration that two species exist is the occurrence of two kinds of animals over a significant geographic range, with no morphological intergradation between them. The commonest demonstration that two morphologically different populations belong to the same species is that geographically intermediate animals are morphologically intermediate. There are weaknesses in both demonstrations, some of which will be discussed with reference to particular examples, but here I wish to emphasize that it is the pattern of geographic variation that provides the clues.

The laboratory scientist who studies squirrel monkeys from Guyana, Peru, Brazil, and Bolivia can readily see the differences among animals from geographically remote areas, and may wonder why they are considered the same species. He or she may disparage the arcane acts of the museum taxonomist or blame that specialist for ignoring the sophisticated studies of chromosomes, biochemistry, behavior, etc., that demonstrate these differences. But these data are very difficult to interpret without evidence from geographically intermediate populations. As data become available from more populations of *Saimiri*, and from known proveniences, these sorts of information will become increasingly valuable. But at present, it is still the museum collections that provide

Table I
Saimiri Synonymy

Elliot (1913)	Lönnberg (1940)	Von Pusch (1942)	Cabrera (1958)	Hill (1960)
S. oerstedi	—	*Cebus sciureus oerstedi*	*S. oerstedi* or	*S. o. oerstedi*
(= *S. oerstedi*)	—	*C. sciureus citrinellus*	*S. s. oerstedi*	*S. o. citrinellus*
S. sciureus	*S. sciurea sciurea*	*C. sciureus sciureus*	*S. s. sciureus*	*S. s. sciurea*
—	—	(= *C. s. sciureus*)	*S. s. collinsi*	*S. s. collinsi*
S. cassiquiarensis	—	*C. s. cassiquiarensis*	(= *S. s. sciureus*)	*S. s. cassiquiarensis*
—	*S. s. codajazensis*	(= *C. s. sciureus*)	(= *S. s. ustus*)	*S. s. codajazensis*
—	—	(?= *C. s. albigena*)	(= *S. s. macrodon*)	*S. s. caquetensis*
—	—	*C. s. albigena*	(= *S. s. macrodon*)	(= *S. s. caquetensis*)
S. macrodon	—	(= *C. s. sciureus*)	*S. s. macrodon*	*S. s. macrodon*
—	—	—	(= *S. s. macrodon*)	*S. s. petrina*
S. ustus	—	(= *C. s. sciureus*)	*S. s. ustus*	*S. usta*
S. madeirae	*S. m. madeirae*	(= *C. s. sciureus*)	(= *S. s. ustus*)	*S. m. madeirae*
—	*S. m. juruana*	(= *C. s. sciureus*)	(= *S. s. nigriceps*)	*S. m. juruanus*
S. boliviensis	*S. b. boliviensis*	*C. sciureus boliviensis*	*S. s. boliviensis*	*S. b. boliviensis*
—	*S. b. pluvialis*	*C. sciureus pluvialis*	(= *S. s. nigriceps*)	*S. b. pluvialis*
S. b. nigriceps	—	—	*S. s. nigriceps*	*S. b. nigriceps*
—	*S. b. jaburuensis*	—	(= *S. s. ustus*)	*S. b. jaburuensis*
8 forms/7 species	7 forms/3 species	7 forms/1 species	7 forms/1–2 species	16 forms/5 species

the broadest and most detailed view of the patterns of geographic variation. Therefore it is on these that I have relied principally in this study.

The definition and concept of the subspecies are more slippery. Subspecies are normally thought of as recognizably distinct populations within a species. But with increasingly sophisticated morphometric, biochemical, and immunological techniques, it has become possible to recognize smaller and smaller demes, some of them with the degree of confidence that was once reserved for subspecies [e.g. the 75% rule, as described by Mayr *et al.* (1953)]. I see little reason to subdivide species into more and more subspecies, based just on our ability to recognize them. Another problem with subspecies has been the practice of using them to describe gradual clinal geographic variation. In such a case subspecies become completely arbitrary. A geographic boundary for the subspecies must be defined and identification of the animals must subsequently be based on their geographic origin. Different taxonomists might draw the border at different places, with equal validity. It should be obvious that recognition of subspecies is a poor way to describe clinal geographic variation. A further complication is that clines commonly differ for different characters: size may vary from east to west and coast color from north to south.

The subspecies category remains useful for the description of two evolutionary phenomena (Mayr, 1982). One is the recognition of geographically isolated populations that are distinctly different from each other. It is usually difficult to know whether these are valid species or not, because one must argue by analogy with the degree of morphological differences between or within demonstrable species. The other phenomenon is secondary intergradation. If two populations of animals have been isolated from one another in the past and have diverged during isolation, they may form a zone of intergradation where their ranges grow together again. Among museum specimens this may be recognized as an area of abrupt geographic variation, usually called a step cline. Geographic variations in a number of biological characteristics are likely to be concordant at the step, which thus serves as a good subspecific boundary between the previously isolated populations. These are the two types of subspecies I recognize in this paper. They represent a much broader concept of the subspecies than that used by previous authors who have studied *Saimiri*, or that used by many mammalogists today.

The strategy of this study was to examine the geographic pattern of coat color variation exhibited by museum skins. From these observations hypotheses were formulated about the location of species boundaries, areas of secondary intergradation, and areas of clinal variation. Then the skulls were measured and the craniometric data were analyzed for concordant and discordant patterns. Where the geographic patterns of variation in pelage characters and in craniometry differed, both were reexamined, and the hypothesis was modified or defended. The resulting taxonomic arrangement is therefore based on the evaluation of two anatomical systems. The weaknesses and strengths of each taxonomic decision are described, in the hope that this will encourage others to take them as hypotheses that should be challenged by other kinds of data.

3. METHODS

Coat color descriptions are based on notes taken on individual specimens. I attempted to quantify differences between individuals by counting the numbers of dark bands on agouti hairs and measuring the lengths of the bands, but I became discouraged by the magnitude of the task—the need to measure numerous hairs from several parts of the pelage of each animal. Instead, the notes are qualitative and based generally on comparisons of individuals within the different museum collections. Notes were made on most, but not all, animals studied.

Nineteen measurements were taken with dial calipers to an accuracy of 0.1 mm on skulls of adult animals. The measurements were:

1. Total length (TL), greatest length of skull, measured from the most anterior point on the premaxilla.
2. Greatest width (GW) of brain case, measured dorsal to the mastoid region.
3. Zygomatic width (ZW), the greatest outside width across the zygomatic arches.
4. Orbital width (OW), the greatest outside width of the orbits.
5. Postorbital width (POW), the least outside width across the postorbital constriction of the skull.
6. Palatal width (PW), measured between the outside edges of the alveoli of the anterior roots of the first upper molars, thus including the widths of the molars.
7. Length of tooth row (LTR), from the anterior edge of the canine at the alveolus to the posterior edge of the third molar at its alveolus.
8. Palatal length (LP), measured in the midline from between the incisors, thus close to the anterior edge of the premaxilla, to the posterior edge of the palatal spine.
9. Occipital length (OL), a midline length of the occipital bone from the occipital–parietal suture to the posterior edge of the foramen magnum.
10. Bicaninal width (CCW), the outside width across the upper canines at the alveoli.

Measurements 11–19 were taken between six points in the midsagittal plane of the skull: *A*, the most anterior point on the premaxilla; *B*, the most dorsal point of the nasal–frontal suture; *C*, the point on the frontal bone furthest from *F*, and thus on the bulge of the forehead at approximately the point of greatest convexity (or, if difficult to determine, it was defined as the midline point on the frontal at the level of the upper edge of the orbits); *D*, the point defined by the intersection of the frontal and parietal sutures, or by the most anterior point on the interparietal suture if bregmatic bones are present; *E*, the point defined by the interparietal and parietaloccipital sutures; *F*, a point in the foramen magnum between the dorsal edges of the occipital condyles, defined by the edge of the calipers when it is placed in the notch behind the condyles. The measurements were the following chords: *AE*, *AF*, *BE*, *BD*, *BF*, *CD*, *DE*, *CF*, *DF*.

A variety of univariate and multivariate analyses were conducted on these measurements. Because of sexual dimorphism, males were always compared with males and females with females. Significance levels for univariate comparisons are based on *T*-tests, *F*-tests having been checked for near equivalence of variances. Multivariate analyses are portrayed graphically with a cross for each sample. The mean for each lies at the intersection of the arms of the cross, and the lengths of the arms are equivalent to 1.5 standard errors. As a rule of thumb, the arms overlap in cases in which there is not a significant difference between the means ($p > 0.05$); if the two samples have similar standard errors along an axis, significance at the 5% level is shown by nonoverlapping arms. Ratio diagrams (Simpson *et al.*, 1960) are based on logarithmic transformation of the data, and hence illustrate the relative differences between samples for each measurement.

4. RESULTS

4.1. *Saimiri sciureus sciureus*

4.1.1. Taxonomy

Linnaeus (1758) described *Simia sciurea* on p. 29 of *Systema Naturae*, noting its prominent occipital region and the ferruginous color of the hair over the ulna and tibia. On pp. 25 and 26 he described *Simia apedia*. Until Fooden rediscovered the skull and positively identified it as a squirrel monkey (Fooden, 1966), its identity was unrecognized. It is thus treated as a "forgotten name" in nomenclature, even though *apedia* has page preference over *sciurea*.

Thomas (1911) noted that Guianese animals could be taken as typical of Linnaeus's *sciurea*, and Tate (1939) formalized this by restricting the type locality to Kartabo, British Guiana, because Linnaeus' locality, India, was obviously wrong. Thus, by definition, squirrel monkeys from Guyana are *Saimiri sciureus sciureus*. The masculine ending is used because *Saimiri* is considered to be a barbaric name without gender, and is therefore defined as masculine [as argued by Cabrera (1958)].

A number of similar squirrel monkeys have been described from the northern half of South America. They include *Pithesciurus saimiri* Lesson (1840) from French Guiana, *Saimiri sciureus collinsi* Osgood (1916) from Marajo Island in Brazil, *Cebus sciureus albigena* von Pusch (1942) from Medina, Colombia; *Saimiri caquetensis* Allen (1916) from Florencia, Colombia; *Saimiri macrodon* Elliot (1907) from the Copataza River, Ecuador; *Saimiri sciurea petrina* Thomas (1927a) from Yurac Yacu, Peru; and *Saimiri madeirae juruana* Lönnberg (1940), from the Rio Jurua, Brazil. As shown in Table I, they have received varying degrees of recognition by previous authors.

4.1.2. Coat Color

The squirrel monkeys from Guyana have gray crowns, yellowish to tawny backs, a broad patch of white over the eyes [the "gothic pattern" described by MacLean (1964)], and yellow on the forearms to the elbow. Hill (1960) provides a more detailed description. Cooper (1968; Fig. 3) provides photographs. Some older females have a patch of dark hairs in front of the ear and a dark edge to the crown. Squirrel monkeys having this general appearance occur throughout the Guianas and along the northern side of the Amazon as far west as the Rio Negro. They also occur south of the Amazon from the mouth of the Amazon to the Xingu River. Further west, along the Rio Tapajos and between there and the Rio Madeira, the situation is more complicated, and will be discussed with reference to *Saimiri madeirae*, but squirrel monkeys looking generally like the ones from Guyana occur as far west as the eastern bank of the Rio Madeira. The animals to the south of the Amazon River differ only in being slightly lighter yellowish on the back and having a bit more yellow suffusion in the gray of the crown. Squirrel monkeys from northern Colombia, 1000 miles to the east of Guyana, are also very similar in general appearance to the Guianese form. Animals from Villavicencio differ principally in having much less yellow on the forearm (on some individuals yellow hairs are restricted to the hands and wrists), and in having the yellow of the back more saturated, so that they appear to be more orange on the dorsum. Old females develop a pattern of black hairs in front of the ears and long the edge of the crown, as in the Guianese animals. Animals from southern Colombia and neighboring Peru and Brazil differ from the animals in northern Colombia in having more yellow on the forearms and in having a less saturated coloration of the yellow on the back. In these respects they are more similar to the Guianese *Saimiri* than are the animals from Villavicencio. Among these more southern animals are the squirrel monkeys that were formerly imported from Leticia, illustrated by Cooper (1968, Fig. 1), and probably captured in both sides of the Amazon near Tabatinga and Benjamin Constant. This phenotype is found to the south and east, probably to the left bank of the Rio Jurua. Lönnberg (1940) described a series of animals with this morphology as the subspecies *juruana*. Unfortunately, he allocated them to the species *Saimiri madeirae*, although he noted that they did not have the naked ears of *madeirae*. Along the Rio Ucayali in Peru, there is a transition in squirrel monkey morphology from the Colombian form in northern Peru, to the coat color and pattern well known from the many monkeys shipped out of Iquitos, Peru during the 1960s and early 1970s [illustrated in Cooper (1968 Fig. 2], to the black-capped *boliviensis* in southern Peru. This transition will be treated separately.

4.1.3. Hypotheses

The squirrel monkeys are very similar in coat color over a large portion of northern South America, including the Guianas, Brazil north of the Amazon and east of the Rios Branco and Negro, Brazil south of the Amazon and east of the Rio Xingu, Colombia, Ecuador and Peru east of the Andes and north of approximately the Rio Marañon, and Brazil west and north of the Rio Jurua. This coat color variation suggests that there is very little genetic variation among populations over this whole area, in spite of the fact that eight forms—*sciureus*, *albigena*, *Saimiri*, *caquetensis*, *macrodon*, *petrina*, *juruana*, and *col-*

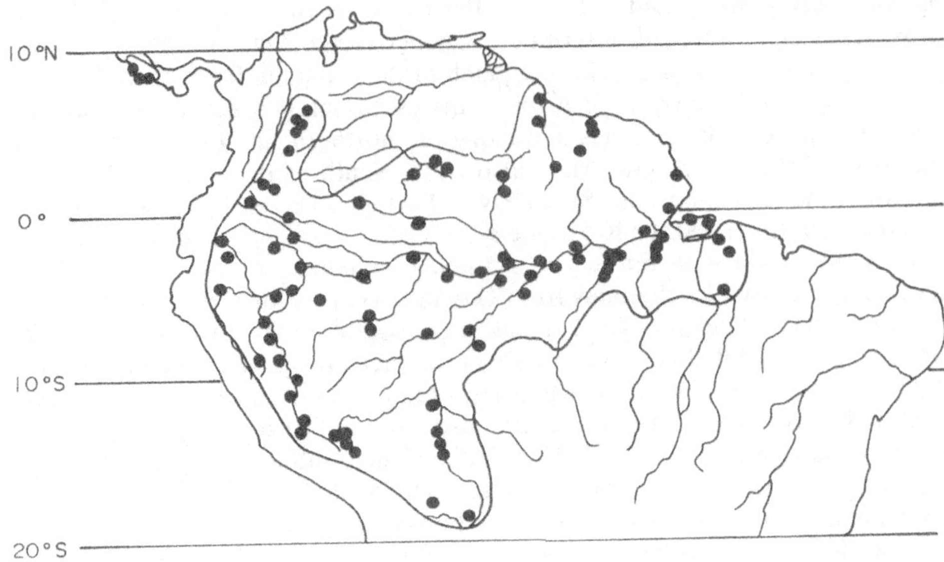

Fig. 1. *Saimiri* distribution. Localities represented in museum collections. Localities are listed in Appendix 2.

linsi—have been recognized from this region. My hypothesis is that very little variation will be found in other characteristics of squirrel monkeys over this vast geographic area and that they should be treated as a single subspecies, *Saimiri sciureus sciureus*. The most remarkable aspect of this hypothesis is that there is a 1000-mile gap in the distribution of this proposed subspecies, between Guyana and Colombia. (Squirrel monkeys from southern Venezuela are different in appearance from Guianese and Colombian *Saimiri* and are treated as a different subspecies.) This disjunct distribution implies that squirrel monkeys formerly had a continuous range across northern South America, presumably through northern Venezuela, where they do not presently occur, and that the isolation of the Guianese and Colombian animals from each other is a relatively recent event, perhaps late Pleistocene.

4.1.4. Craniometric Analyses

A univariate analysis of skull measurements of males from Colombia and from Guyana ($N = 30, 14$) demonstrated that the skulls are very similar. There were no significant differences in the 19 measurements. A comparison of the skulls of females ($N = 20, 6$) showed that the skulls of the Guianese animals are smaller than those of the Colombian animals in six measurements ($p < 0.05$), mostly measures of skull length, and notably smaller in the length of the tooth row ($p < 0.01$). Canonical variate plots (Figs. 2) were conducted on the same data from these two populations and from four other populations from Venezuela, northern Peru, and neighboring Brazil. In the analysis of the males the Colombian and Guianese samples were indistinguishable from each other and fit in the cluster of samples from northern Peru and neighboring Brazil, contrasting with the sample from southern Venezuela, which was separated from them on the first axis. In the analysis of the female skulls, the Guianese and Colombian samples were distinguished clearly, but both samples fit in the cluster of samples from northern Peru and neighboring Brazil. Again, the Venezuelan sample was separated from the others on the first axis. Thus the craniometric analyses exhibit the same geographic pattern as coat color, linking the Guianese and Colombian populations of *Saimiri* and distinguishing them sharply from the geographically intermediate Venezuelan animals.

Craniometric analyses of the animals from the Guianas and neighboring Brazil demonstrate a considerable amount of heterogeneity among these populations (Fig. 3) but few patterns. In both Guyana and Surinam there is a notable difference between animals from inland and animals from the coastal plain, demonstrated on axis 1. Values on this axis correlate highly with measures of skull length, and the differences between the samples are readily seen as major differences in size. Coastal plain animals are smaller than animals from inland. Broader geographic patterns are not obvious.

Using Fig. 3 as a measure of local geographic variation in *Saimiri sciureus sciureus*, I computed the positions of other populations on the first two axes of this graph. Samples from the north bank of the Amazon River, east of the Rio Negro, would fall on the graph between samples 5 and 6, close to sample 7. A sample from the island of Marajo overlaps extensively the position of sample 7. A sample from the mouth of the Xingu River, on the south side of the Amazon, does not differ significantly from samples 4 and 6. This suggests that there are no craniometric differences among these populations in excess of the local geographic variation seen in the Guianas.

4.1.5. Conclusions

I treat the squirrel monkeys from the Guianas, from Brazil north of the Amazon and east of the Rio Negro, and from Brazil south of the Amazon and east of the Xingu River as a single subspecies, *Saimiri sciureus sciureus*. Animals from Colombia, Ecuador, northern Peru, and adjoining Brazil are also included in this subspecies. There is geographic variation within this large area, both in coat color and skull morphology, but I do not detect any concordance between coat color and craniometric differences that makes the recognition of more than one subspecies a compelling argument. Between the Madeira and Xingu Rivers, south of the Amazon in Brazil, squirrel monkeys present a complicated biological situation, which is analyzed and discussed in Section 4.5 on *Saimiri madeirae*. The conclusion of that analysis is that *Saimiri sciureus sciureus* is found south of the Amazon as far west as the Rio Madeira. Another complicated area is found in Peru. As described below, I interpret this as a region of intergradation between *Saimiri sciureus sciureus* and *S. sciureus boliviensis*.

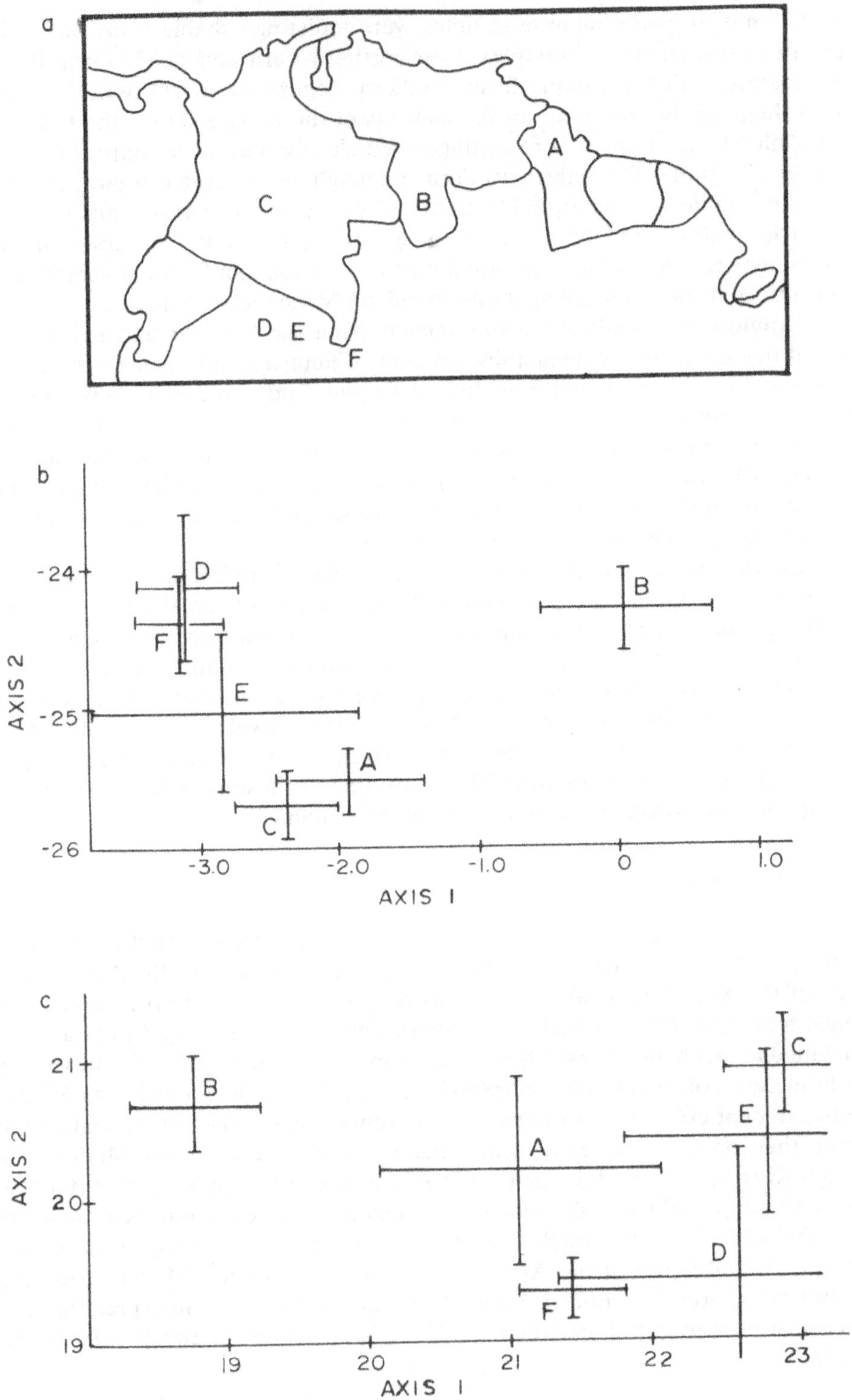
a
A
B
C
D
E
F
b
AXIS 2
AXIS 1
-24
-25
-26
-3.0
-2.0
-1.0
0
1.0
c
21
20
19
19
20
21
22
23

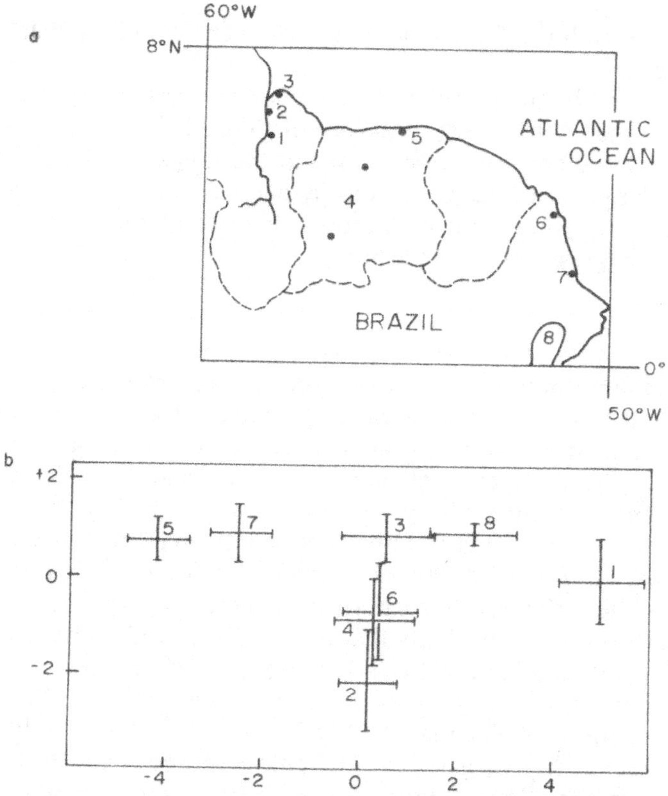

Fig. 3. (a) Map of localities plotted on graph. (b) Principal coordinates analysis based on correlation matrix for 19 cranial measurements of male *Saimiri* from the Guianas and neighboring Brazil ($N = 37$). For each sample, mean ± 1.5 S.E. is plotted for axis 1 and axis 2.

4.2. *Saimiri sciureus boliviensis*

4.2.1. *Taxonomy*

Several names have been applied to black-capped squirrel monkeys from Bolivia, Peru, and Brazil. These include *Callithrix boliviensis* D'Orbigny (1834) and *Callithrix entomophagus* D'Orbigny (1835) for animals from Bolivia; *Chrysothrix nigrivittata* Wagner (1848) for animals from Tefé, Brazil; *Saimiri boliviensis nigriceps* Thomas (1902) for animals from Cosnipata, Peru; and *Saimiri boliviensis pluvialis* Lönnberg (1940) for animals from the Rio Jurua in Brazil. Lönnberg (1940) also applied the name *Saimiri boliviensis jaburuensis* to gray-

Fig. 2. (a) Map of localities plotted on graphs. (b) Canonical variate analysis of 19 cranial measurements of male *Saimiri* from six localities. (c) Canonical variate analysis of 19 cranial measurements of female *Saimiri* from six localities. Localities: A, Guyana ($N = 13$ males, 6 females); B, southern Venezuela ($N = 16, 14$); C, Colombia, north of 0° ($N = 30, 20$); D, Rio Curaray, Peru ($N = 8, 6$); E, Puerto Indiana, Peru ($N = 7, 6$); F, laboratory animals derived from Leticia, Colombia ($N = 40, 25$).

capped animals with the Roman-arch pattern over the eye from Jaburu, on the Rio Purus, Brazil.

Various combinations of these have been recognized by various authors (see Table I), and clearly most forms are distinguishable from the others. Previous authors have not been able to delimit the ranges of the different forms, however, in large part because of the inadequacy of the material they were studying. It seems likely that with more adequate material, the delimitations will become progressively more arbitrary.

4.2.2. Coat Color

The combination of a black crown and lack of a white patch of hairs over the eyes (producing the Roman-arch superciliary pattern) distinguishes the *boliviensis* group from *sciureus* and *madeirae*. Not all Roman-arch animals have black crowns, but the range of all animals with Roman arches is geographically continuous and this superciliary pattern is unique to this group of animals. Therefore I suspect that it is an important genetic marker. In Central Peru and in the vicinity of Jaburu in Brazil, *Saimiri* are found with the Roman-arch pattern but without the black crown. In view of the geographic distance between these populations, it seems likely that this combination of characters has been derived independently in these two populations and should not be used to justify combining them into a single subspecies [as suggested by Hill (1960)]. Furthermore, it seems probable that the black crown is a derived feature in *Saimiri*, derived independently in *oerstedii* in Central America and in *boliviensis*. In both areas some populations tend to have gray crowns, and also the jet black crown is an adult character, not found in juveniles. If the black crown is the derived morphology, then the gray crowns of other animals are shared primitive characters, not coat color markers indicating genetic similarity.

Among the animals with the Roman-arch morphology, there is variation in the degree of saturation of the orangish or yellowish color of the back and limbs. As suggested by Lönnberg, this degree of saturation seems to correlate with the amount of rainfall, that being the reason he gave the name *pluvialis* to the dark animals from the Rio Jurua. The animals with more saturated body colors also have a more extensive distribution of black hairs, particularly down the midline of the back. The darkest animals I have seen are those from Tefé, on the south bank of the Amazon in Brazil. The degree of saturation and extent of black coloration on the back appear to vary clinally, and hence to be a poor basis for recognition of the different described subspecies. The populations called *boliviensis*, *pluvialis*, and *nigrivittatus* are distinguishable from one another, but they are probably linked clinally by populations of intermediate degrees of saturation.

Dramatically, at the Jurua River in Brazil, there are black-capped, Roman-arch squirrel monkeys of the *boliviensis* type on the right bank of the river and gray-capped Gothic-arch forms on the left bank. This gives one the impression that the two forms act as good species and do not intergrade, but other interpretations are possible.

In Peru, there appears to be a more gradual transition from the morphology of *boliviensis* to the morphology of *sciureus* as one progresses north along the Ucayali River. This is a critical area, so a more detailed review of the situation is warranted. A series of animals from localities 12–14° S latitude, including the type specimen of *nigriceps*, are very similar to the animals from Bolivia. The crowns of adult males and females are jet black, the superciliary pattern is Roman arch. Near 10° S, from localities Cumeria and Chicosa, all animals have the Roman-arch pattern, and the adult females have almost fully jet black crowns. However, the adult males do not have jet black crowns; there are black hairs along the edge of the crown and scattered through the top of the crown, but many hairs in the crown do not have the long black tips that give the jet black appearance of the Bolivian animals. These animals were described by Thomas (1928) and seem to have caused him some confusion. A number of the animals he described were immature and have predominantly gray crowns, so he was faced with a sample showing all grades of crown color from gray to black, which led him to broaden his definition of the subspecies *nigriceps*. At approximately 9° S, at Tingo Maria, the animals are quite similar. The adult females have jet black crowns, and the males have grayish crowns, darker at the edge than in the center. All animals have the Roman-arch superciliary pattern. From slightly further north, at Yarinacocha, a series of males exhibit gray crowns with varying degrees of darkness on the edge of the crown and scattered through the crown, caused by hairs with long black tips. At Contamana and Sarayacu, approximately 7.5° S, the males are similar in the appearance of the crown, and most of them appear to have the Roman-arch superciliary pattern, but one specimen (AMNH 76376) has the Gothic-arch pattern. The females differ from those further to the south in having much less black in the crown. Among seven adult females from Sarayacu, one had no black on the crown, two had black along the edge of the crown, with only a few black hairs in the middle of the crown, and three had much darker crowns with long, black-tipped hairs throughout the crown.

Unfortunately, I have not seen *Saimiri* specimens with known provenience from the Ucayali River between Sarayacu at 7° S and the mouth of the Maranon at almost 4° S. Animals from Yurac Yacu in the Huallaga–Maranon drainage at 6° S described by Thomas (1927a) under the name *petrina* have all-gray crowns and the Gothic-arch superciliary pattern. As he noted, there was no black in the crowns of these animals, but female squirrel monkeys from further north near Nauta, ~4.5° S, do have a significant amount of black in the crown (FMNH 86972 and 86973). Among specimens north of 4° S in Peru, I have not seen any females with completely black crown. Specimens from along the Rio Napo (Puerto Indiana and mouth of the Rio Curaray) include seven females with black hairs along the edge of the crown and in the tuft in front of the ear. In other populations that have been maintained in captivity, this pattern of black hairs is age-related, becoming prominent only in older adult females. Thus these animals from the vicinity of Iquitos and north do not differ significantly from the populations I have already labeled as *sciureus*.

4.2.3. Hypotheses

The geographic pattern of coat color variation in squirrel monkeys in Peru looks to me like the kind of clinal variation typical of secondary intergradation between previously isolated populations. As one moves from south to north, first the crown color of the males turns from black to gray, then the crown of the females becomes less black and some females remain gray-crowned as adults, the superciliary pattern changes from Roman arch to Gothic arch, and finally the black edges to the crown in females is found only as a character of old age. Thus, from south to north, we progress from the morphology of *boliviensis* to that of *sciureus*. On the basis of this evidence, I hypothesize that there is genetic intergradation between the *sciureus* and *boliviensis* and that the two should therefore be considered subspecies.

The border between *sciureus* and *boliviensis* appears to be very abrupt along the Jurua river. It is doubtful that this or any of the rivers is an absolute barrier to gene flow, because the rivers change their course so frequently, but they may restrict the amount of gene flow. If the ranges of *sciureus* and *boliviensis* were restricted to refugial forests during the Pleistocene, their ranges would have expanded with the subsequent expansion of the forests. When the range expansion of one of the subspecies reached the edge of a large river, it would temporarily be stopped, until a propagule of monkeys was transferred from one side to the other by a changing river meander. If there were no *Saimiri* on the other side, the transplanted propagule would be free to expand its population and the range of the subspecies. If there were already squirrel monkeys there, however, the propagule would interbreed with the residents and we would see only the evidence of gene flow. I hypothesize that this is what happened at the Rio Jurua, and that this is the area in which the expanding ranges of *boliviensis* and *sciureus* met. Any propagules of *boliviensis* transferred to the left bank are probably genetically swamped by interbreeding with the large *sciureus* population on that side of the river, and similarly any *sciureus* propagules transferred to the right bank would be swamped by interbreeding with the much larger *boliviensis* population there. If this is true, then a survey along the Jurua should find varying degrees of intergradation between the two subspecies on both banks of the river, the amount of intergradation depending on the recency of changes in the river meanders and the sizes of meanders, and thus of the populations of *Saimiri*, that have been moved from one bank to the other. If instead, *sciureus* and *boliviensis* are valid species, then there should be many places along the Jurua where the two forms can be found side by side without any evidence of interbreeding.

On the basis of very little evidence, I map the distribution of the subspecies *boliviensis* between the Jurua and Purus Rivers, south of the Amazon. Because *Saimiri* live in forest along the banks of rivers, they should be commonly switched from side to side by changes in the meanders, not only along the Jurua but also along the Amazonas and Purus Rivers. Along all three rivers there should be gray-capped Gothic-arch squirrel monkeys on the opposite bank. With such dramatic coat color markers, it should be possible to document clearly what happens to propagules of monkeys transferred to the opposite bank from their mother population. The gray-capped *boliviensis* found at Jaburu on the Purus River may be another example of intergradation between *boliviensis* and a gray-capped form, although it is not possible to tell which, because two forms of *Saimiri* are known from the right bank of the Purus.

4.2.4. Craniometric Analyses

For the craniometric studies, the Bolivian and Brazilian specimens of *boliviensis* were divided into five geographic samples. There is significant heterogeneity among the three Bolivian samples. Animals from the Rio Beni and from the Rio Mamoré, south of 14° S, are very similar, but both differ significantly in size from the squirrel monkeys found north of 14° S on the Rio Mamoré, which are smaller. Proportional differences were examined by dividing each measurement by the condylobasal length of the same skull. The northern Mamoré population does not differ from the other two samples in these proportions, suggesting that size alone differentiates it. The Brazilian specimens from Jaburu and the Rio Jurua differ little from the Bolivian animals. The Jurua animals are slightly larger than the Bolivian *Saimiri* but do not differ in proportions.

There are significant differences between the *Samiri* on opposite banks of the Rio Jurua, with regard to both size and proportions. (Adequate cranial samples, 14 and 15, are available for females only.) The gray-capped *sciureus* on the left bank may be characterized as having a longer skull (total length, condylobasal length, and length of tooth row), much longer occipital bones ($T = 6.2$, $p < 0.001$), and narrower zygomatic arches than the black-capped *boliviensis* on the right bank.

A principal objective of the craniometric analyses was to determine if they demonstrate intergradation between *boliviensis* and *sciureus*, or if they suggest that the two forms do not interbreed where their ranges meet in Peru. The first analysis was a univariate comparison of three samples, from Bolivia, Peru, and Colombia (Fig. 4). This demonstrates that there is a significant difference in many cranial measurements between the Bolivian and Colombian animals and that the skulls of the Colombian animals are bigger than those of the Bolivian animals in a number of measurements. Specimens from the Ucayali in Peru are intermediate. Of particular interest is measurement 9, the occipital length of the skull, which is disproportionately smaller in the Bolivian skulls than in the Colombian ones.

Two discriminant multivariate analysis were conducted on these data, one for males and one for females (Fig. 5). Geographically intermediate populations were plotted on these axes. These demonstrate that the geographically intermediate populations are morphologically intermediate between the Bolivian and Colombian animals. The discriminant axis for the males is defined differently than that for the females (Table II), but the geographic pattern is virtually identical for the two analyses. A more detailed analysis of the transitional area between 3° and 10 ° S is desirable, but it is not warranted at present, because of the lack of appropriate samples and the small sample sizes available for the represented populations.

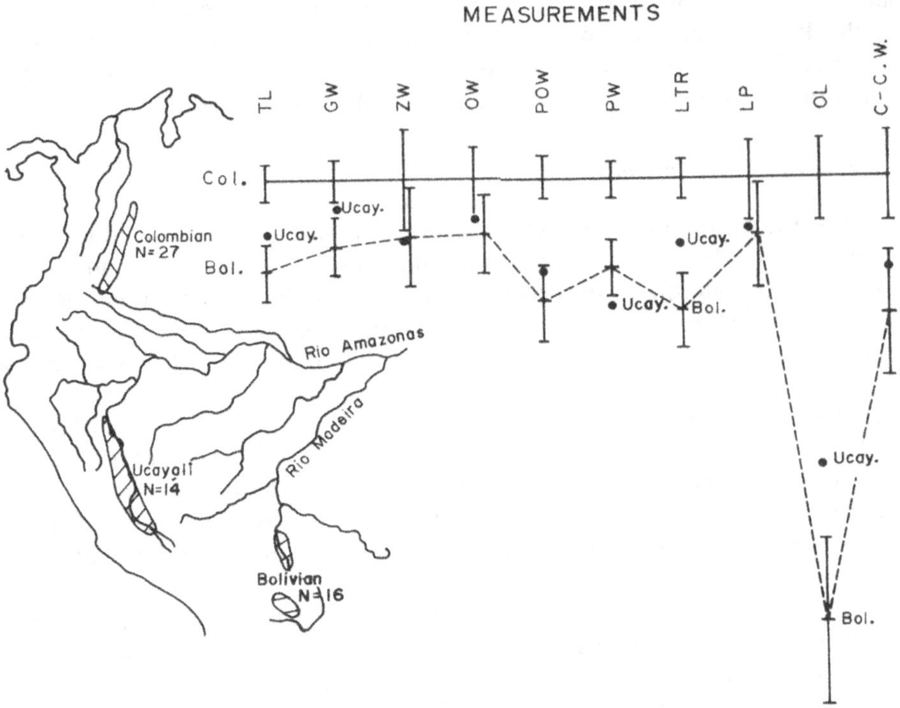

Fig. 4. Ratio diagram showing relative differences in ten cranial measurements between *Saimiri* from Colombia, Ucayali River in Peru, and Bolivia. Vertical axes have logarithmic scales and are adjusted so that all means of the Colombian sample lie on the same horizontal axis. Means and two standard deviations portrayed on vertical bars. Means only are shown for the Ucayali sample. Measurements are listed in Section 3.

4.2.5. Conclusions

Saimiri sciureus boliviensis is treated as a single subspecies, with *nigriceps*, *nigrivittatus*, *pluvialis*, *entomophagus*, and *jaburuensis* as synonyms. Because of the documented clinal variation in coat color and cranial measurements along the Ucayali River and the lack of sympatry with Gothic-arch *Saimiri*, it is considered to be a subspecies of the species *Saimiri sciureus*. The characteristic that best seems to distinguish *boliviensis* from other squirrel monkeys is the Roman-arch pattern over the eye. Using this as a key to the subspecies, we find that the form ranges through much of Bolivia, in Peru north to approximately 7° S, and in Brazil between the Jurua and Purus Rivers north to the Amazon.

4.3. *Saimiri sciureus cassiquiarensis*

4.3.1. Taxonomy

Humboldt (1811–1812) described the squirrel monkeys from the Cassiquiare River, Venezuela, as *Chrysothrix sciureus cassiquiarensis*. Geoffroy

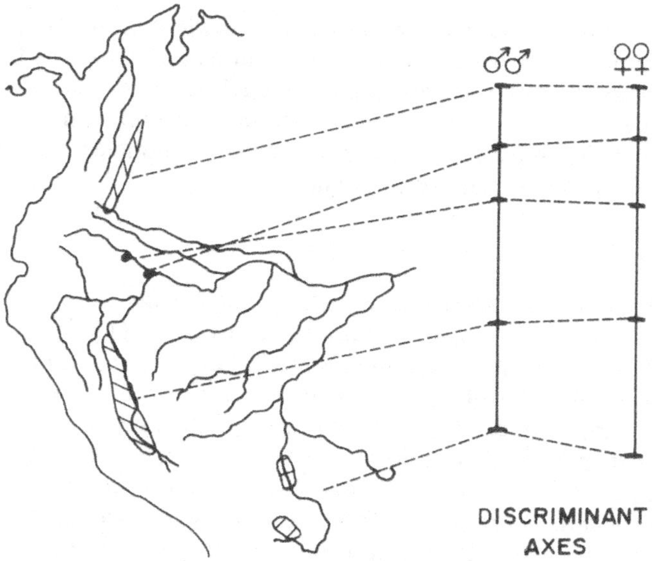

Fig. 5. Discriminant analyses of Colombian and Bolivian *Saimiri*, based on ten cranial measurements. Males and females were analyzed separately. Three geographically intermediate populations are also plotted on the discriminant axes and are shown to be morphologically intermediate. From north to south, samples are from Colombia, north of 0°; Rio Curaray, Peru; Puerto Indiana, Peru; Ucayali River, Peru; and Bolivia.

(1845) described squirrel monkeys from the banks of the Orinoco as *Saimiri lunulatus*. These are virtually the same populations. Lönnberg (1940) described a very similar squirrel monkey from Codajaz, Brazil, on the left bank of the Amazon above Manaus. He probably did not have specimens of *cassiquiarensis* for comparison, and he named it a new subspecies, *codajazensis*. He noted particularly the fact that there was a collar of yellowish white hairs separating

Table II
Scaled Vectors from Discriminant Analyses of Colombian and Bolivian Saimiri

Measurements[a]	Males	Females
TL	−121.9	−0.27
GW	−1.5	0.15
ZW	35.9	1.56
OW	2.5	0.13
POW	34.7	0.42
PW	17.1	0.18
LTR	73.0	2.12
LP	−58.3	−2.04
OL	108.9	1.38
CCW	67.6	0.85

[a] See Section 3.

the crown from the back. He overstated this, however. In his specimens the crown is continuous with the back: in five animals the width of the band of grizzled hairs connecting the crown and the back is 15, 20, 20, 12, and 20 mm. In the type specimen it is 20 mm. He noted that the single female did not have a black tuft of hairs in front of the ear (as is seen in females from Venezuela), but this specimen is a male, not a female.

4.3.2. Coat Color

This subspecies is described in detail under the names of *cassiquiarensis* and *codajazensis* by the original authors and by Elliot (1913) and Hill (1960). What strikes me most about its appearance is the tawny orange color of the back and the suggestion of tawniness in the crown. To my eye, this gives a very different appearance to the whole animal and makes it distinctive from *sciureus* and other forms. The coloration of the back is similar to that of *sciureus* from the llanos of Colombia, but the olive to tawny color of the crown in *cassiquiarensis* differs distinctly from the gray crown of the Colombian *sciureus*, and the extensive distribution of orange hairs on the forearm of *cassiquiarensis* contrasts with the gray coloration of the forearm of *sciureus* from the llanos. Like most other forms of *Saimiri*, *S. cassiquiarensis* has a patch of white hairs over the eye, giving it a Gothic-arch appearance.

The differences between *cassiquiarensis* on the right bank of the Rio Negro and Brazilian *sciureus* to the east of the Rio Negro are more subtle. The *sciureus* are more gray-crowned and the *cassiquiarensis* are more yellow or tawny in the crown.

4.3.3. Hypotheses

I have seen adequate samples of this subspecies from only three localities: Amazonas, Venezuela; Codajaz, Brazil, on the north bank of the Amazon west of the Rio Negro; and Cacão Pereira Igarapé, Brazil, west of the Rio Negro, near Manaus. Coat color similarities suggest that they are part of a genetically homogeneous population that ranges from the Amazon west of the Rio Negro north into Venezuela. They are distinctly different from the Colombian and Guianan populations but probably intergrade with the *sciureus* populations east of the Rio Negro in Brazil. Unfortunately, I have seen little material from the zone in which I would expect intergradation—the left bank of the Rio Negro and further to the east. Therefore I am unsure whether the intergradation is very gradually clinal or more abrupt. I propose to treat these two as separate subspecies, as if this were a zone of secondary intergradation east of the east bank of the Rio Negro, but this should be recognized as a weak hypothesis needing testing. I must be even more vague about the proposed intergradation between *cassiquiarensis* and the *sciureus* populations to the west in southern Colombia, because I have not seen specimens from the intervening areas.

4.3.4. Craniometric Analyses

The analyses of skull measurements concur with the similarities and differences observed in coat colors. In Fig. 2, one can see that the Venezuelan

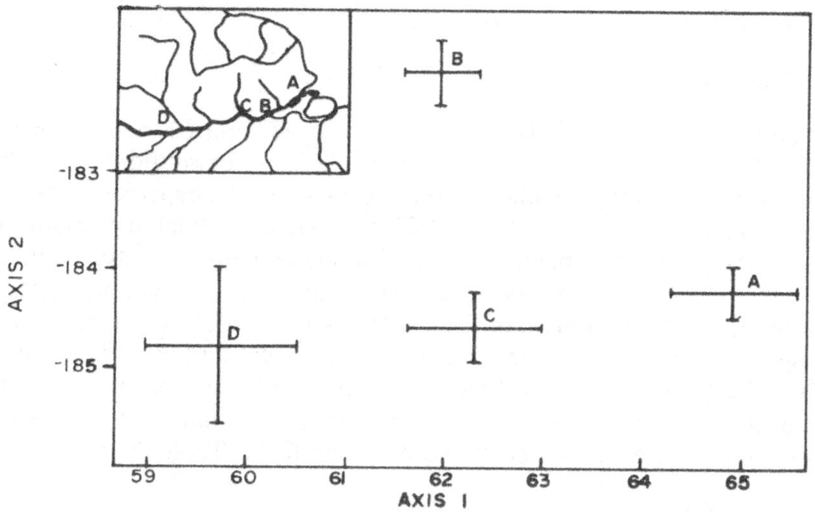

Fig. 6. Discriminant analysis of male *Saimiri* from eastern Brazil, north of the Amazon. A, Amapá (N = 13); B, Jacuara and Urumanduba (N = 12); C, other combined localities in Para (N = 9); D, Cacão Pereira Igarapé, west of Rio Negro (N = 9).

Saimiri are quite distinct from those of Colombia and Guiana. Figure 6 shows a discriminant analysis between the *Saimiri* west of the Rio Negro and three samples east of the Rio Negro. The analysis aligns these samples approximately in geographic order, placing the samples from Amapá and Cacão Pereira Igarapé at the extremes and the geographically intermediate populations between them.

4.3.5. Conclusions

The recognized range of *Saimiri sciureus cassiquiarensis* should be enlarged to include the population recognized by Lönnberg as *codajazensis*. The intergradation of this subspecies with *Saimiri sciureus sciureus* to the east and to the west needs to be examined carefully. If it can be documented to be gradual clinal variation without evidence of secondary intergradation, then it may be appropriate to treat *cassiquiarensis* as a distinct population of *Saimiri sciureus sciureus*, but not worthy of subspecies recognition. As described in the discussion of *Saimiri madeirae* (Section 4.5), *Saimiri sciureus cassiquiarensis* occurs south of the Amazon at Berury, sympatric with *madeirae* and apparently not intergrading with it. At some places north or south of the Amazon, between the mouths of the Jurua and Purus, it probably is sympatric with *Saimiri sciureus boliviensis*. I would expect a zone of hybridization and intergradation where their ranges meet. Possibly the *boliviensis* population from Jaburu, lacking the black caps of *boliviensis* populations from further south and west, occur in such an area of intergradation.

4.4. *Saimiri sciureus oerstedii*

4.4.1. *Taxonomy*

Reinhardt (1874) described squirrel monkeys from Chiriqui (Panama) as *Chrysothrix*, noting their black cap and the golden red coloration of their backs. Thomas (1904) named specimens from Costa Rica a subspecies, *Saimiri oerstedi citrinellus*, and noted that they differed principally in having gray or nearly gray crowns rather than black, and that the limbs were more gray rather than suffused with orange-yellow as in Reinhardt's animals. Since then, taxonomists have varied in their treatment of the Central American squirrel monkeys, some treating them as a distinct species, *Saimiri oerstedii*, and others treating them as a subspecies of *Saimiri sciureus*. They have also differed in whether they recognized two subspecies of Central American *Saimiri*, as proposed by Thomas, or whether they recognized only one form (Table I).

4.4.2. *Coat Color*

There is little to be added to the original descriptions of the Central American *Saimiri*. The golden-orange color of the back is caused by the guard hairs being orange over most of their length, with only a very short black tip. In *oerstedii* the orange band is approximately 2 cm in length, while in *Saimiri sciureus* from the Colombian llanos the band is only 5–7 mm long. The black band at the tip of the hair is 2–4 mm in *oerstedii*, and thus only 10–20% of the length of the orange band, whereas in *sciureus* the black band is commonly half the length of the yellow or orange band. This difference gives the Central American *Saimiri* a very different appearance from any of the South American populations. The black crown is much like that of animals from Bolivia.

In the series from Pazo Azul, Costa Rica, described by Thomas, adult females have black crowns, and males and juveniles have gray crowns. This is analogous to the situation in central Peru, where *Saimiri sciureus boliviensis* exhibits the same pattern.

4.4.3. *Hypotheses*

It is possible that the striking differences in coat color and pattern between *oerstedii* in Central America and *sciureus* in northern South America have a simple genetic basis, although the mechanism is quite unknown. The red color of the back could be obtained by intensifying the color and elongating the orange band of the dorsal hairs of Colombian squirrel monkeys. The black coloration of the crown is not universal among *oerstedii* populations, and seems to be a maturational melanism like that seen in *boliviensis*. Two hypotheses are possible, if we presume that the black crown is a derived feature: either that *oerstedii* and *boliviensis* are closely related or that the black crown has evolved twice in *Saimiri*. The latter seems more probable on geographic grounds and because *oerstedii* shares the high Gothic-arch superciliary pattern with *sciureus*, rather than the Roman-arch pattern (which seems a more basic characteristic of *boliviensis* than the black crown). Therefore it is not obvious that

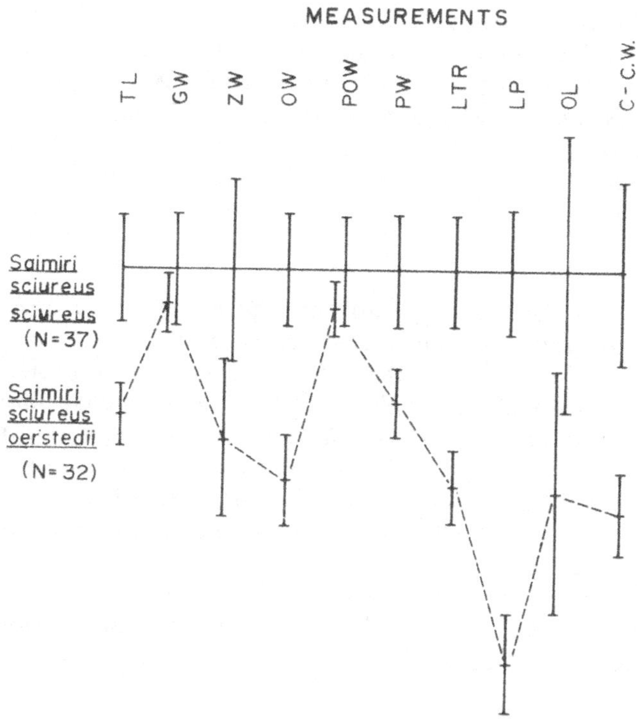

Fig. 7. Ratio diagram of ten skull measurements comparing male *Saimiri* from the Guianas (*Saimiri sciureus sciureus*) and from Panama (*S. sciureus oerstedii*). Measurements are listed in Section 3.

oerstedii is sufficiently different from the South American squirrel monkeys to be considered a distinct species. I postulate that it is no more distinct in other ways from *Saimiri sciureus sciureus* than the latter is from *Saimiri sciureus boliviensis*, and that by analogy it, too, should be considered a subspecies of *Saimiri sciureus*.

4.4.4. *Craniometric Analyses*

The skulls of squirrel monkeys from Panama are slightly smaller than those from Costa Rica, but are similarly proportioned. The Costa Rican skulls are approximately the same size as skulls of *Saimiri sciureus sciureus* from the Guianas. Proportions are shown in a ratio diagram (Fig. 7) that compares *oerstedii* from Panama and *Saimiri scrureus sciureus* from the Guianas. This suggests that *oerstedii* is disproportionately larger in two width measurements, greatest width of the brain case and postorbital width. However, in a comparison of *oerstedii* from Costa Rica with *sciureus* from the Guianas, these differences do not exist. The disproportionately short palate of *oerstedii* shown in Fig. 7 is seen in both samples of Central American *Saimiri*. This reflects the lack of development of a spine at the back of the palate in *oerstedii*. The occipital length in *oerstedii* is 31% of the total length of the skull, which is similar to its relative length in *sciureus* (32% in the Guianas, 31% in a Leticia sample) and much greater than in *boliviensis* (27–28%). Thus, the skulls of *Saimiri sciureus oerstedii* are similar to those of *Saimiri sciureus sciureus*, except for the relative length of the palate. They are dissimilar to *S. s. boliviensis* in the same ways that *sciureus* differs from *boliviensis*.

4.4.5. Conclusions

Allopatric populations frequently present unresolvable problems. I propose to treat *oerstedii* as a subspecies of *Saimiri sciureus*, but there is no satisfactory way to test this hypothesis or the opposite hypothesis that *oerstedii* is a valid species. Instead, I can only argue by analogy. In coat color and pattern and in craniometric analyses, *oerstedii* seems no more different from *Saimiri sciureus sciureus* in Colombia than does *boliviensis*. Because I am treating *boliviensis* as a subspecies of *Saimiri sciureus*, I should treat *oerstedii* as a subspecies of *sciureus* as well. A demonstration that *boliviensis* is a valid species would change the way I treat *oerstedii*. A demonstration that *oerstedii* is much more different from Colombian *sciureus* than is *boliviensis* in their biochemistry, immunology, or other characteristics would also seriously compromise my argument that *oerstedii* should be treated as a subspecies of *sciureus*.

4.5. Saimiri madeirae

4.5.1. Taxonomy and Coat Color Descriptions

Thomas (1980) named a species of squirrel monkey from Humayta, on the west bank of the Rio Madeira, accurately describing it as having a clear gray crown with a blue tone and "with the fulvous of the hands not extending above the wrists." Squirrel monkeys meeting this description occur along the Rio Madeira, west to the Rio Purus, and east toward the Rio Tapajos. In addition to the characters noted by Thomas, they are distinctive in having nearly naked ears.

Lönnberg (1940) described specimens from east of the Rio Purus as being hybrids between *madeirae* and "*codajazensis*" (= *Saimiri sciureus cassiquiarensis*), the latter being the form of squirrel monkey found north of the Rio Amazonas in this region. However, this is an inaccurate description. Lonnberg had two specimens from Lago Berury, one of which (no. 1136) is a *madeirae* and the other of which (no. 1137) is a "*codajazensis*," and a single specimen from Igarape do Castanha, which is a "*codajazensis*." Thus his specimens document that there are two kinds of squirrel monkeys in the woods on the east bank of the Rio Purus. On the east side of the Rio Madeira, there are again two kinds of squirrel monkeys. The Olalla brothers obtained specimens from Borba, all but one of which have olivaceous crowns, hairy ears, and fulvous forearms. Along a stream near Borba, at Auara Igarape, they obtained typical *madeirae*, with gray crowns, naked ears, and blue-gray forearms. One of the animals reported from Broba is also a specimen of *madeirae*. It would appear that there are two kinds of squirrel monkeys, probably living in different habitats, in the vicinity of Borba. Further east, on the south bank of the Amazon in the vicinity of Villa Bella do Imperatriz, the Olalla brothers obtained *Saimiri madeirae*, and nearby at Serra de Parintins they obtained the other form of *Saimiri*. The crown color of the animals from Parintins is more gray than those from Borba, however, and thus is more similar to the crown color of *madeirae*. But again the appearance is that there are two kinds of squirrel monkeys, inhabiting different habitats.

Saimiri from four sites on the east side of the Rio Tapajos (Fordlandia, Caxiricatuba, Piquiatuba, and Tauri) all show a mixture of the *madeirae* and *sciureus* characteristics: the naked ears of *madeirae* and the yellow forearms and yellowish gray crown color of *Saimiri sciureus*. In this area there appears to be only one kind of *Saimiri*. Further east, on the Rio Xingu, all the *Saimiri* have the morphology of *sciureus* without any of the characteristics of *madeirae*.

Toward the southern edge of the range of *Saimiri madeirae*, the animals have the same gray crowns and naked ears of animals further north, but the forearms are orange, like those of most other forms of squirrel monkeys (specimens from Calama and from Territorio Rondonia, across from the mouth of the Baures River). The Mamoré and Guaporé Rivers, which form the boundary between Bolivia and Brazil, also form the border between *Saimiri madeirae* and *boliviensis*. The contrast between the animals on the two sides of the rivers is dramatic, with the *boliviensis* being black-crowned, with no white patch over the eyes (Roman arch), and the *madeirae* being gray-crowned with a large white patch (Gothic arch) over the eyes. There is no indication that there is any intergradation between the two forms, but there are not enough specimens from the Brazilian–Bolivian border to demonstrate what happens where river cutoffs bring the two forms together, as they undoubtedly do. West of the Madeira River, near the Bolivian border, there is the possibility that the ranges of *madeirae* and *boliviensis* merge without there being a river separating them. No specimens are available from this area. The western boundary of *madeirae* is the vicinity of the Rio Purus, to judge from the fact that specimens from Jaburu, on the west side of the Purus, have hairy ears and the Roman-arch pattern over the eyes, lacking the broad, white superciliary patch of *madeirae*.

4.5.2. Hypotheses

Over much of the area between the Rio Purus and the Rio Tapajos, *Saimiri madeirae* gives the appearance of being a distinct species. There is no evidence that it intergrades to the south or west with *boliviensis*. There is evidence that it coexists with *cassiquiarensis* without intergrading with it, and there is evidence that it coexists with *sciureus* at two localities without intergradation. To the east, in the vicinity of the Rio Tapajos, there is only one kind of *Saimiri*, and it has a mixture of the characters of *madeirae* and *sciureus*. This suggests that there is a zone of intergradation between *madeirae* and *sciureus*, perhaps in the vicinity of the Rio Tapajos.

Table III
Skull Lengths of Saimiri at the Rio Madeira

Species	Sex	*N*	$\overline{X}$ ± SE
S. madeirae	Males	11	66.43 ± 0.32
S. sciureus	Males	3	61.9 ± 0.68
S. madeirae	Females	18	62.47 ± 0.40
S. sciureus	Females	3	58.97 ± 0.48

4.5.3. Craniometric Analyses

Examination of the skulls of *Saimiri* support some of the hypotheses based on pelage. At Berury, Borba, and Villa Bella do Imperatriz, the *madeirae* animals have large skulls and the others have smaller skulls. Sample sizes are sometimes small, but some of the differences are large. Combining the samples from Rosarinho, Borba, and Auara Igarapé, as appears to be justified, we find the skull lengths of *S. madeirae* to be on the average larger than those of *S. sciureus* by 7% in males and 6% in females (Table III), a significant difference. A Lago Berury, on the Rio Purus, there is only one skull of each for comparison, but there is also a 7% size difference. The difference between the skulls of the males from Villa Bella and Serra de Parintins is only 3.6%, which is marginally significant ($p < 0.05$).

The geographic pattern of skull size among males is shown in Fig. 8. There is not a significant difference in skull length among animals identified by pelage as *Saimiri sciureus* from Borba, Serra de Parintins, Rio Tapajos, and the island of Marajo. The males identified as *S. madeirae* have significantly larger skulls, but the *S. madeirae* from Villa Bella are intermediate in size between the *S. madeirae* from Rosarinho and *S. sciureus*, which could be interpreted as evidence of intergradation or as evidence that *S. madeirae* becomes smaller to the east. The *Saimiri* from the Rio Tapajos, which have pelage characteristics of both *sciureus* and *madeirae*, have skulls of the same size as other *sciureus*. The same pattern is exhibited by the skull size of female squirrel monkeys: *S. madeirae* is larger than *S. sciureus*; specimens of *S. madeirae* from Villa Bella are smaller than those from Rosarinho, and the *Saimiri* from the Rio Tapajos are not intermediate in size between the *S. madeirae* to the west and the *S. sciureus* to the east. In fact, females from the Rio Tapajos have significantly smaller skulls than those from the island of Marajo.

An examination of the univariate differences for 19 skull measurements between these populations is shown in Table IV. For the male skulls, it is evident that there are many more differences between the animals from Rosarinho and these from other localities than there are among the other three samples. For the female skulls, the Rosarinho and Villa Bella samples are more similar to each other than either is to the Tapajos samples; the Tapajos sample stands out as being the most distinctive; and the Villa Bella and Marajo samples are surprisingly similar. Again this supports the hypothesis that the Villa Bella *madeirae* are more similar to *sciureus* than are the *madeirae* from further west.

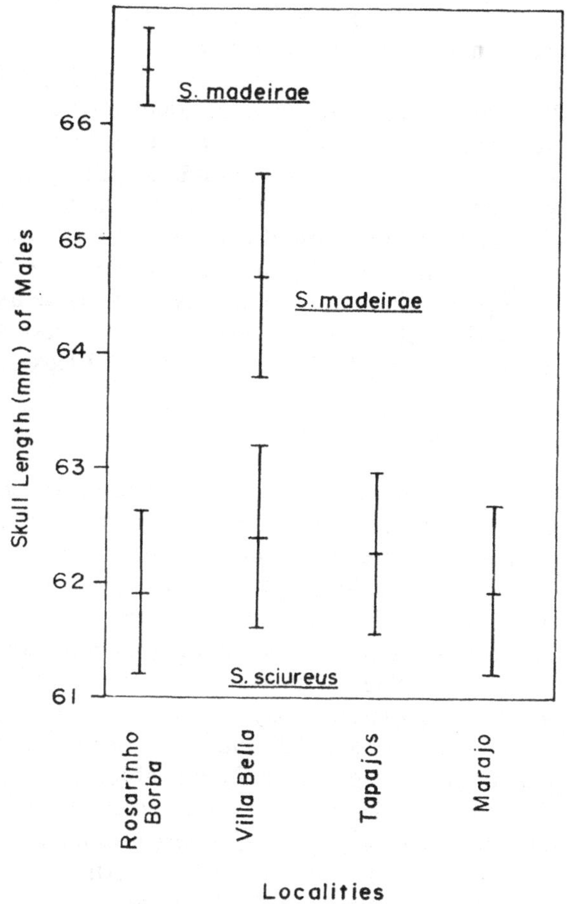

Fig. 8. Skull sizes of *Saimiri madeirae* and *Saimiri sciureus* from four localities south of the Amazon, arranged left to right, east to west.

Table IV
Number of Measurements (from a Total of 19) Differing Significantly between Populations of Saimiri, and the Level of Significance

	Villa Bella			Tapajos			Marajo		
p =	0.05	0.01	0.001	0.05	0.01	0.001	0.05	0.01	0.001
Males									
Rosarinho	4	5	0	1	1	10	1	5	8
Villa Bella	—	—	—	0	2	0	1	0	0
Tapajos	—	—	—	—	—	—	2	0	2
Females									
Rosarinho	6	1	1	1	2	13	0	4	10
Villa Bella	—	—	—	1	7	5	4	2	0
Tapajos	—	—	—	—	—	—	2	0	2

and it does not support the hypothesis that the Tapajos sample is an intergrade between *sciureus* and *madeirae*.

Several multiviariate analyses (including principal component analyses and discriminant analyses) were conducted on the craniometric data. Each analysis emphasized a different combination of length and width measurements, and therefore seemed to be examining a different set of shape variables, but all gave similar results. They aligned the samples on the first axis in approximately the same way they are aligned west to east on a map, and on the second axis they separated the Tapajos sample from the others. The Rosarinho and Villa Bella populations are readily distinguishable as *Saimiri madeirae* from the animals with pelage of *S. sciureus*, but the more eastern populations of *madeirae* are more similar to *sciureus* than is the Rosarinho population.

4.5.4. Conclusions

The populations of *madeirae* are treated here as a distinct species, *Saimiri madeirae*, because of the demonstration of sympatry in three localities and the equivocal nature of the demonstration of intergradation with *sciureus* to the east. Where the geographic variation of pelage suggests there is intergradation, the craniometric data seem to deny it; where the craniometric data suggest intergradation may occur, there is no convincing evidence from the geographic pattern of pelage. Obviously the status of *Saimiri medeirae* is deserving of much more intense study.

No subspecies of *S. madeirae* are recognized. Animals from the southern part of the range are distinguishable from the typical forms of Humaita and Rosarinho by the extension of the orange wash from the wrists up to the elbow. The subspecies *juruana* Lönnberg is not considered to be a form of *S. madeirae*. Contrary to Lönnberg's (1940) description, the specimens he studied have a great deal of orange on the forearm and the crown color is more greenish than in typical *madeirae*. As noted by Lönnberg, the specimens of *juruana* have hairy ears, unlike *madeirae*. I consider these animals to belong to the species *Saimiri sciureus*.

The animals from the east bank of the Rio Tapajos are not included in the species *S. madeirae*. They have the nearly naked ears of *madeirae*, but in every other characteristic they seem like *S. sciureus*. If by other means it can be demonstrated that this population is genetically intermediate between *madeirae* and *sciureus*, then there would be a good case for reducing *madeirae* to a subspecies of *sciureus*.

There is an unresolved taxonomic problem associated with this population from the Tapajos. Geoffroy (1845) gave the name *Saimiris ustus* to an animal obtained in Portugal. Because it was an animal with very little hair on its ears, Cabrera (1958) defined its type locality as the Rio Madeira, but it is not like any squirrel monkeys that occur on the Madeira, to judge from the illustration and description; therefore this designation was in bad judgment. Instead, it is more likely that Geoffroy's animal came from the Tapajos. If the Tapajos animals are considered to be *Saimiri sciureus*, then this name is a junior synonym of *sciureus* or is available for this population if the Tapajos animals are treated

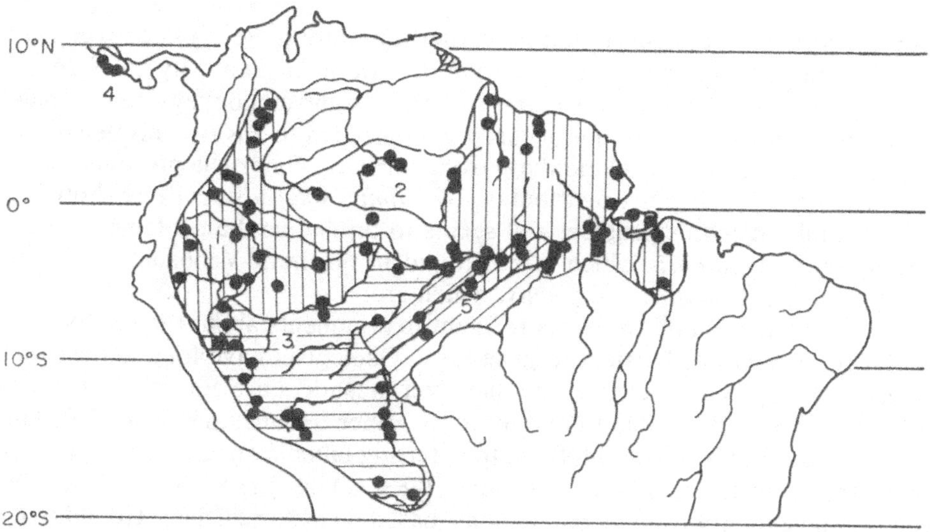

Fig. 9. Geographic ranges of species and subspecies of *Saimiri* recognized in this chapter. 1, *Saimiri sciureus sciureus*; 2, *S. s. cassiquiarensis*; 3, *S. s. boliviensis*; 4, *S. s. oerstedii*; 5, *Saimiri madeirae*.

as distinct from the Guianese animals. If the Tapajos animals were to be treated as belonging to the same taxon as the Madeira animals, then the correct name for the species would be *Saimiri ustus*, as indicated by Cabrera (1958). I have not seen the type of *ustus*, but because of its olive crown and reddish back, I exclude it from the synonymy of *madeirae*, which have gray backs and crowns, and list it as a synonym of *sciureus*. Geoffroy suggested that the animal may have been in captivity for several years, in which case the true origin of the type specimen might be indeterminable.

5. DISCUSSION

The distribution of the two species and five forms of *Saimiri* proposed in this chapter are shown in Fig. 9. These distributions raise a number of zoogeographic questions. The most conspicuous question is why *Saimiri* has a disjunct distribution in Central America. As noted before, Hershkovitz (1972) suggested that these animals were introduced by man. The morphological distinctiveness of the Central American squirrel monkey makes this hypothesis seem very improbable. In fact, it probably focuses attention on the wrong question, dispersal, rather than on the more interesting and testable questions of extinction. The known climatic fluctuations of the Pleistocene must have greatly affected the distributions of mammals in the past, and dispersal of *Saimiri* into Central America could easily have occurred during a pluvial period when the forests were more extensive. Questions about the presumed extinction of squirrel monkeys in eastern Panama and the Choco region of Colombia, where *Saimiri* surely occurred at an earlier time, are more provocative. Five other species of nonhuman primates occur in eastern Panama, why not *Saimiri*? Detailed ecological studies of Central American squirrel monkeys, and of habitats where they do and do not occur, will be needed to answer this question.

The absence of squirrel monkeys from northern Venezuela presents a similar problem. The disjunct distribution of *Saimiri sciureus sciureus* shown in Fig. 9 emphasizes this problem and seems to dramatize the ecological differences between *Saimiri*, which is not found in northern Venezuela, and the genera *Ateles, Alouatta*, and *Cebus*, which are.

The distribution of *Saimiri* is relevant to arguments about the relative importance of rivers as barriers to gene flow. Geologists have long known that rivers regularly change course and that river meanders are cut off, thus transferring land from one side of the river to the other (Holmes, 1964, p. 536). The significance of these river cutoffs in transferring fauna from one side of a river to another is unclear, although Clemens (1883, Chapter 1) explicitly described the phenomenon for primate populations, based on historical data. Hershkovitz (1977, p. 97) illustrates river meanders, remarking that they are means of passive transfer of animals from one side to the other of a river, but also claims that rivers are borders between different subspecies of *Saguinus fuscicollis*, implying that rivers are barriers for these tamarins. Hershkovitz (1972, 1983) and Thorington (1976) have pointed out other cases in which primate populations must have been transferred across rivers by cutoffs. Such cutoffs may affect different species differently, depending upon their ecology. Species that live in *terra firma* habitats in the Amazon forest may be less subject to passive transfers than those that are common river-bank inhabitants. Rivers do not appear to be barriers to *Saimiri*, as demonstrated by the occurrence of *Saimiri sciureus sciureus* on both sides of the lower Amazon, by the presence of *S. s. cassiquiarensis* on both sides of the Amazon near Codajas, and by the occurrence of *Saimiri madeirae* on both sides of the Rio Madeira. These distributions probably reflect the fact that *Saimiri* lives in riverside habitats and is frequently included in river cutoffs.

The more meanders there are in a river, the more frequently cutoffs should occur, and the greater is the likelihood that there will be passive transfer of fauna from one bank to the other. Therefore it is rather surprising that the Rio Jurua, which is full of meanders, appears to be an abrupt boundary between *Saimiri sciureus sciureus* and *S. s. boliviensis*. The few samples from along the river undoubtedly give us a simplified perspective. More samples should demonstrate a more complex genetic and taxonomic picture.

It is difficult to assess the role of Pleistocene refugial forests in determining the present distribution and geographic variation of *Saimiri*. Refugial forests seem to be the modern equivalent of landbridges, which were constructed in numbers by past zoogeographers to explain the transoceanic dispersal of terrestrial animals. Refugia, too, have been multiplied indiscriminantely, so that there are enough of them postulated for South America to support almost any hypothesis. This dilutes the usefulness of the concept, but is not cause for denying the reality of the refugia. *Saimiri* would seem to be an unlikely candidate for restriction to refugial forests during the Pleistocene, due to the fact that squirrel monkeys survive in gallery forests and in small forest patches in the Colombian Ilanos. However, the gaps in the range of *Saimiri* suggest that it does not survive well under some ecological conditions, and the disjunct population in western Panama and Costa Rica gives us cause to believe that other populations have been isolated in the past. Therefore, the concurrence of the ranges of some of the forms recognized in this chapter with the postulated Pleistocene refugia may be significant. The range of *S. madeirae* includes the Madeira refuge and the range of *S. sciureus cassiquiarensis* includes the Imeri refuge (Haffer, 1969; Vanzolini, 1973). Isolation in the refugia may have led to the differentiation of these forms. The disjunct range of *S. s. oerstedii* is a geographic pattern seen in other species as well (Müller, 1973), probably demonstrating some similar ecological requirements. The widespread but disjunct distribution of *S. s. sciureus* is difficult to interpret, but the zone of secondary integradation between it and *S. s. boliviensis* is intriguingly close to the border between two species of *Aotus* recognized by Hershkovitz (1983) and between subspecies of *Saguinus fuscicollis* as mapped by Hershkovitz (1977). Such repetitious geographic patterns are most likely the results of past geographic distributions of forests. However, it is difficult to frame testable and falsifiable hypotheses about the effects of refugia on present distributions of animals like squirrel monkeys.

APPENDIX 1. TAXONOMIC BORBORYGMI

Lönnberg (1940) described a number of new taxa of *Saimiri*, but did not designate type specimens. Among the specimens he examined, types are designated on the labels. I presume these to be Lönnberg's designations, which should be given in the literature. The type of *Saimiri sciureus codajazensis* Lönnberg (1940) is NHMS 991. The type of *Saimiri madeirae juruana* Lönnberg (1940) is NHMS 2196. The type of *Saimiri boliviensis jaburuensis* Lönnberg (1940) is NHMS 1247. The type of *Saimiri boliviensis pluvialis* Lönnberg (1940) is NHMS 2446.

APPENDIX 2. SPECIMENS EXAMINED

Specimens from the folowing museums were examined: Museum of Comparative Zoology, Harvard; American Museum of Natural History; U.S. National Museum of Natural History; Field Museum of Natural History; British Museum (Natural History); Naturhistorische Riksmuseet, Stockholm; Staatsammlung, München; Museu Nacional, Rio de Janeiro; Museu Goeldi, Belém; Museu de Zoologia, Universidade de São Paulo. In the following lists, the numbers in parentheses indicate the number of specimens examined.

Saimiri sciureus sciureus (453)

Colombia. Arauca: Arauca (1). *Vichada*: ? (1). *Meta*: Restrepo (1); Villavicencio (15); Guaycaramo (2); Barrigou (3); San Martin (2); Los Micos, San

Leraìn de Arama (11); La Macarena (4). *Vaupes*: Cerro de los Pinturas, lower Rio Guayabero (2); Tahuapunta, Rio Vaupes (1). *Huila*: Acevedo, Rio Suaza (1); Acevedo, San Adolfo (10); Pitalito (3). *Caqueta*: Florencia (1); Rio Mecaya (4); Rio Caqueta, La Tagua, Tres Troncos (10). Locality not designated: laboratory animals acquired from Leticia (40); other (2).

Ecuador. Upper Rio Payamino (5); Copataza, upper Rio Pastaza (4); Rio Pastaza (2); Santa Rosa de Sucumbios (1); Rio Tigre (3).

Peru Loreto. Boca Rio Curaray (20); Rio Napo (3); Puerto Indiana (18); Iquitos (3); Iquitos, Santa Rita (2); Rio Morona, Quebrada Pushaga (1); Nauta, Rio Samiria, Santa Elena (5); Rio Samiria (12); Rio Marañon (mouth) (1); Rio Yavari Mirim, Quebrada Esperanza (1); Rio Yavari, San Fernando (1); Rio Yavari, Boca Rio Yaquerana (1). *San Martin*: Yurac Yacu (6).

Guyana. Georgetown (2); Essiquibo River (5) Buck Hall, Essiquibo River (4); Rupainuni, 16 miles E of Dadanawa Forest (1); Kuitaro, 20 miles E of Dadanawa Forest (1); Kalacoon (1); Coast Region, Demarara (7); Moon Mountains (2).

Surinam. Kaiserberg airstrip, E of Zuid River (4); Saramaca, La Poule (10); Brokopondo, Loksie Hattie (Saramacca River) (1); Brokopondo, Finisanti, Saramacca (2); Paramaribo (1).

Brazil. Amapá. Rio Vila Nova Mazagão (2); Amapá (5); Rio Iratapuru Municitio Mazagão (1); Igarapé Novo, Mazagão (1); Igarapé Rio Branco, Rio Maracá (1); Vila Velha do Cacipore, Municitio Oiapoque (11); Fazenda Nova California, Rio Araguari (1); Rio Amapari de Macapa (1); Iracema, Rio Aporema (2); Rio Tracajatuba (7). *Marajo*: Souré (2); Igarapé Tapereba, Municitio Chaves (3); Isla Mexiana (11); Caldeirao (8). *Para*: Capim (1); Bom Jardin (4); Belem (1); Civiuatinga, Curua-urra de Prainha (1); Utinga, near Utinga Bei (3); Nova Timboteua (4); Villarinha do Monte (4); Tapara (2); Porto de Moz (2); Recreio (2); Tavio (1); Tapaiuna (1); Santarem (36); Fordlandia (4); Bravo (3); Itapoama (1); Fazenda Monte Christo (2); Taperinha (2); Faro, Rio Jamunda (4); Rio Jamunda (19); Jacuara (20); Bravo (2); Urumanduba (12); Boiussu (6); Obidos (2); Santa Rita (3); Lago Cuiteua (3); Prata (1); Paru, Este Uchizal Almeirim (3); Cametá (1); Caxiricatuba (6); Tauary (3); Piquiatuba (1); Aveiros (3); Rio Arapiuns (2); Rio Trombeta (1); Itaituba (1); Cachoeria du Tronco, Rio Erepecuru (1); Rio Traquateua, Municitio Ananindeua (2); Providencia, Braganca RR (1). *Rio Branco*: Conceicão (2); *Maranhão*: Imperatriz (2). *Amazonas*: near Tabatinga (2); Estirão do Ecuador, Rio Javari (5); Ilha da Uricurituba (1); Itacoatiara (3); Igarapé Aniba (5); João Pessôa, Rio Jurua (28); Igarapé do Gordão, Rio Jurua (1); Fonte Boa (1); São Gabriel, upper Rio Negro (1); Benjamin Constant (3); São Paulo de Olivenca (2).

Saimiri sciureus boliviensis (265)

Peru. *Loreto*: Rio Ucayali, Sarayacu (15); Contamana (1); Rio Pisqui (1); Yarinacocha (7); Cumeria (9); Chicosa (2); Lagarto (8); Alto Ucayali (1); Rio

Urubamba (boca) (6) (reported to be a fraudulent locality); Rio Inuya (boca) (2); Rio Maniti, Santa Cecilia (1); Marupa (2); locality not designated: laboratory animals acquired from Iquitos (20). *Huanuco*: Tingo Maria (8); Rio Zungaro, Rio Pachitea (1). *Cusco*: Rio Comberciato (4); Cosnapata (1); Quince Mil (6); Marcapata (2). *Madre de Dios*: Tahuamanu River (1); Rio Manu, Alta Mira (2); Itahuanaia (1); Zona Boca Colorado (3).

Bolivia Beni. Rio Mamoré: opposite Coimbra (1); 5–6 km S of Buena Hora (7); 5 km NW of Alejandria (1); opposite Cascajal (3); 12° 26′ S (7); Puerto Siles (5); 10 km N of Santa Rosa (1); Rio Ibaré, 26–27 km from mouth. *Rio Mamoré*: 17 km SE of Puerto Julio (5); 5 km NE of mouth of Rio Grande (3); left bank Vienna (2). *Santa Cruz*: Rio Ichilo, 51 km S of mouth of Rio Chaparé (2); 70 km S of mouth of Rio Chaparé (7); Buena Vista (14); Matas arredores (1). *Cochabamba*: Todos Santos (13); Rio Chipiri (2); Rio Beni, Puerto Salinas (14). *Provincia Guarayas* (2). *Provincia Sara* (5); 7 km S of 7 Island (2); Santa Cruz de la Sierra (1).

Brazil. Jaburu, W. of Rio Purus (7); Lago Grande, Rio Jurua (16); Santa Cruz, Rio Eiru, Rio Jurua (26); Santo Antonio, Rio Eiru, Rio Jurua (11); Rio Jurua, opposite João Pessôa (1); Tefé (2).

Saimiri sciureus cassiquiarensis (111)

Venezuela. Belen, Rio Cunucunuma, 56 km NNW of Esmeralda (7); Boca Mavaca, 84 km SSE of Esmeralda (1); Capibara, Brazo Cassiquiare, 106 km SW of Esmeralda (4); Rio Mavaca 108, km SW of Esmeralda (9); San Juan, Rio Manapiare, 163 km ESE of Pto Ayacucho (24); Tamatama, Rio Orinoco (12); Rio Cassiquiare (4); Mt. Duida (4); Rio Orinoco, Boca del Rio Icamo (4).

Brazil. Tabocal (1); Cacão Pereira Igarapé (21); Codajaz (14); Lago Berury (1); Igarapé do Castanha (1); Rio Alegria, at Rio Negro (2); Manacaparu (2).

Saimiri sciureus oerstedii (137)

Costa Rica. El Pozo de Terraba (1); Pozo Azul (11); Palmar (15).

Panama. Boqueron (98); Burica Peninsula (4); Chiriqui, Progresso (4); Sevilla Island, opposite David (2); Almijas Island (2).

Saimiri madeirae (89)

Brazil. Lago Berury (1); Lago Tapayuma (6); Lago do Baptista (25); Marai (2); Borba (1); Auara Igarapé (8); Rosarinho (26); Villa Bella do Imperatriz (15); Humaytha (3); Calama (1); Guaporé River, opposite mouth of Baures River (1).

REFERENCES

Allen, J. A., 1916, New South American mammals, *Bull. Am. Mus. Nat. Hist.* **35**:83–87.

Cabrera, A., 1958, Catalogo de los mamiferos de America del Sur, *Rev. Mus. Argent. Cienc. Nat. Bernardino Rivadavia Cienc. Zool.* **4**(1):i–iv, 1–307.

Clemens, S., 1883, *Life on the Mississippi* [reprinted in: *The Family Mark Twain* (Harper, New York, 1935)].

Cooper, R. W., 1968, Squirrel monkey taxonomy and supply, in: *The Squirrel Monkey* (L. A. Rosenblum and R. W. Cooper, eds.), Academic Press, New York, pp. 1–29.

D'Orbigny, A., 1834, *Nouv. Ann. Mus. Paris* **3**:89. [Not seen. As cited by Cabrera (1958)].

D'Orbigny, A., 1835, *Voyage dans l'Amerique meridionale . . . Atlas Zoologigue* [Not seen. As cited by Cabrera (1958).]

Elliot, D. G., 1907, Descriptions of apparently new species and subspecies of mammals belonging to the families Lemuridae, Cebidae, Callitrichidae, and Cercopithecidae in the collection of the Natural History Museum, *Ann. Mag. Nat. Hist. Ser. 7*, **20**:185–196.

Elliot, D. G., 1913, *A Review of the Primates*, Volume 1, American Museum of Natural History, New York.

Fooden, J., 1966, Identification of the type specimen of *Simia apedia* Linnaeus, 1758, *Mammalia* **30**:507–508.

Ford, S. M., 1980, A systematic revision of the Platyrrhini based on features of the posteranium, Unpublished Ph.D. dissertation, University of Pittsburgh, Pittsburgh.

Geoffroy Saint-Hilaire, I., 1845, Description des mammifères nouveaux ou imparfaitement connus de la collection du Muséum d'Histoire Naturelles, *Arch. Mus. Paris* **4**:5–42.

Haffer, J., 1969, Speciation in Amazonian forest birds, *Science* **165**:131–137.

Hershkovitz, P., 1972, The Recent mammals of the Neotropical region: A zoogeographic and ecological review, in: *Evolution, Mammals, and Southern Continents* (A. Keast, F. C. Erk, and B. Glass, eds.), State University of New York, Albany, pp. 311–431.

Hershkovitz, P., 1977, *Living New World Monkeys (Platyrrhini)*, Volume 1, Chicago University Press, Chicago.

Hershkovitz, P., 1983, Two new species of night monkeys, genus *Aotus* (Cebidae, Platyrrhini): A preliminary report on *Aotus* taxonomy, *Am. J. Primatol.* **4**:209–243.

Hill, W. C. O., 1960, *Primates, Comparative Anatomy and Taxonomy. IV. Cebidae. Part A*, Interscience, New York.

Holmes, A., 1964, *Principles of Physical Geography*, Ronald Press, New York.

Humboldt, A. de 1811–1812, in: *Recueil d'observations de zoologie et d'anatomie comparée, faites dans l'ocean atlantique, dans l'interieur du nouveau continent at dans la mer du sud pendant les années 1799, 1800, 1801, 1802 et 1803*, Volume 1 (A. de Humboldt and A. Bonpland, eds.), Paris.

Lesson, R.-P., 1840, *Species des mammifères: Bimanes et quadrimanes; suivi d'un mémoire sur les Oryctéropes*, Paris.

Linnaeus, C., 1758, *Systema Naturae . . . Tomus I. Regnum Animalium*, 10th ed., reformed, Holm.

Lönnberg, E., 1940, Notes on some members of the genus *Saimiri*, *Ark. Zool.* **32A**(21):1–18.

MacLean, P. D., 1964, Mirror display in the squirrel monkey, *Saimiri sciureus*, *Science* **146**:950–952.

Mayr, E., 1963, *Animal Species and Evolution*, Harvard University Press, Cambridge.

Mayr, E., 1982, Of what use are subspecies?, *Auk* **99**:593–595.

Mayr, E., Linsley, E. G., and Usinger, R. L., 1953, *Methods and Principles of Systematic Zoology*, McGraw-Hill, New York.

Müller, P., 1973, *The Dispersal Centres of Terrestrial Vertebrates in the Neotropical Realm*, W. Junk, The Hague.

Osgood, 1916, Mammals of the Collins–Day South American expedition, *Field Mus. Nat. Hist. Publ. Zool. Ser.* **10**:199–216.

Pocock, R. I., 1925, Additional notes on the external characters of some platyrrhine monkeys, *Proc. Zool. Soc. Lond.* **1925**:27–47.

Reinhardt, J., 1874, Et Bidrag til Kundskab om Aberne i Mexiko og Centralamerika, *Vidensk. Medd. Dansk. Natruhist. Forens.* **1872**:150–158.

Rosenberger, A., 1981, Systematics: The higher taxa, in: *Ecology and Behavior of Neotropical Primates* (A. F. Coimbra-Filho and R. A. Mittermeier, eds.), Academia Brasileira de Ciencias, Rio de Janeiro.

Simpson, G. G., 1945, The principles of classification and a classification of mammals, *Bull. Am. Mus. Nat. Hist.* **85**:i–xvi, 1–350.

Simpson, G. G., Roe, A., and Lewontin, R. C., 1960, *Quantitative Zoology*, Harcourt, Brace and Company, New York.

Tate, G. H. H., 1939, The mammals of the Guiana Region, *Bull. Am. Mus. Nat. Hist.* **76**:151–229.

Thomas, O., 1902, New forms of *Saimiri, Oryzomys, Phyllotis, Coendou*, and *Cyclopes, Ann. Mag. Nat. Hist. Ser. 7* **10**:246–250.

Thomas, O., 1904, New forms of *Saimiri, Saccopteryx, Balantiopteryx*, and *Thrichomys* from the Neotropical Region, *Ann. Mag. Nat. Hist. Ser. 7* **13**:250–255.

Thomas, O., 1908, Four new Amazonian monkeys, *Ann. Mag. Nat. Hist. Ser. 8* **2**:88–91.

Thomas, O., 1911, The mammals of the tenth edition of Linnaeus; An attempt to fix the types of the genera and the exact bases and localities of the species, *Proc. Zool. Soc. Lond.* **1911**:120–158.

Thomas, O., 1927a, The Godman–Thomas expeditio to Peru. V. On mammals collected by Mr. R. W. Hendee in the province of San Martin, N. Peru, mostly at Yurac Yacu, *Ann. Mag. Nat. Hist. Ser. 9* **19**:361–375.

Thomas, O., 1927b, The Godman–Thomas expedition to Peru. VI. On mammals from the upper Huallaga and neighboring highlands, *Ann. Mag. Nat. Hist. Ser. 9* **20**:594–608.

Thomas, O., 1928, The Godman–Thomas expedition to Peru. VII. The mammals of the Rio Ucayali, *Ann. Mag. Nat. Hist. Ser. 10* **2**:249–265.

Thorington, R. W., Jr., 1976, The systematics of New World monkeys, in: *First Inter-American Conference on Conservation and Utilization of American Non-human Primates in Biomedical Research*, pp. 8–18, Pan American Health Organization, Washington, D. C.

Vanzolini, P. E., 1973, Paleoclimates, relief, and species multiplication in equatorial forests, in: *Tropical Forest Ecosystems in Africa and South America: A Comparative Review* (B. J. Meggers, E. S. Ayensu, and W. D. Duckworth, eds.) Smithsonian Institution Press, Washington, D. C., pp. 255–258.

von Pusch, A., 1942, Die Arten der Gattung *Cebus, Z. Saugetierkhd.* **16**(volume for 1942): 183–237.

Wagner, J. A., 1848, Beiträge zur Kenntniss des Säugethiere Amerikas. Dritte Abtielung. Vierte Ordnung. *Affen. Abh. Bayer. Akad. Wiss. München* **5**:405–480.

2

The Behavior of Squirrel Monkeys (*Saimiri*) in Natural Environments

John D. Baldwin

1. INTRODUCTION

This chapter presents a brief summary of the available information on *Saimiri* in their natural environments, including data on ecological specializations, daily behavior patterns, group structure, social behavior and organization, developmental patterns, and relations with other species.

The field studies on *Saimiri* are listed in Table I, along with basic data from each study. Most reports have only included general observations, censuses, or survey data. Less than 1000 hr of systematic observation on behavior has been made in natural environments in Central and South America. In evaluating and interpreting the limited behavioral data from natural settings, it has been useful to draw upon studies in two seminatural environments: a 1.6-ha (hectare) zoological park in southern Florida, and Santa Sophia Island (a stocked island in the Amazon River, near Leticia, Colombia). In the seminatural environments, detailed observations have been possible at close range for prolonged periods. Much of the behavior and social organization observed in these seminatural environments appears to be quite similar to that observed under natural conditions. However, data from seminatural environments must be used with caution, since differences in food abundance, population density, climate, and other factors clearly affect certain features of *Saimiri* behavior and social organization (e.g., allowing greater leisure for social interactions such as play, inducing greater physical proximity, and reducing the probability of group fission). Nevertheless, the observations from seminatural environments have been useful in piecing together the limited data from the wild.

2. SPECIALIZATIONS

Saimiri are small, agile, inquisitive, arboreal primates that spend many hours of the day moving actively through the trees in search of various types

---

John D. Baldwin • Department of Sociology, University of California at Santa Barbara, Santa Barbara, California 93106.

Chapter 2

Table I
Field Studies on Saimiri

Study site	General time	Study hours	Group size	Home range (ha)	Day path (km/day)	Population Density (number/ km^2)	Type of study[a]	Reference
Natural environments								
Surinam, etc.	1938	—	? to 550	—	—	—	GO: ms	Sanderson (1949, 1957)
Colombia	1955?	—	50 maximum	—	—	—	S: ms	Hernández-Camacho and Cooper (1976)
Panama; Costa Rica; Colombia	1960s	—	20–50	—	—	—	GO: ms	Moynihan (1976)
Rio Putumayo, Colombia	1960–1962	—	18 average	—	—	—	S: ms	Tokuda (1968)
Monte Seco, Colombia	1965	500	22	14.7	—	149	B&E: *Saimiri*	Thorington (1967, 1968)
La Macarena, Colombia	1967–1968	55	20–40	65–130	—	19–31	B&E: ms	Klein and Klein (1973), Klein (1974)
Southwestern Panama	Late 1960s	—	33	—	—	—	GO: *Saimiri*	Hladik and Hladik (1969)
Southwestern Panama	1968	—	10–35	—	—	—	S, GO: *Saimiri*	Baldwin and Baldwin (1971, 1976)
Southern Peru	1969	—	8–23	—	—	—	S: ms	Durham (1975)
Peru; Brazil; Colombia	1969	—	20–300	—	—	—	S, GO: *Saimiri*	Baldwin and Baldwin (1971)

Barqueta, Panama	1970–1971	261	23–27	17.5–24+	2.5–4.2	130	B&E: *Saimiri*	Baldwin and Baldwin (1972a)
Rio Peneya, Colombia	1971–1974	—	20–200	—	0.6–1.1	—	S, GO: ms	Izawa (1975, 1976)
Osa Peninsula, Costa Rica	1973	—	15–80	—	—	—	GO: *Saimiri*	Jones (1973)
Loreto, Peru	1973–1974	—	20–50	—	—	151–528	C: ms	Neville *et al.* (1976)
Peru and Colombia	1974	—	30–50	—	—	—	C: ms	Freese (1975)
Peru and Bolivia	1974–1975	—	35–50	—	—	8–100	C: ms	Freese *et al.* (1982)
Guyana	1975	—	15–60	—	—	—	C: ms	Vessey *et al.* (1978)
Surinam	1976–1978	—	? to 50+	—	—	—	B&E: ms	Fleagle and Mittermeier (1980), Fleagle *et al.* (1981), Mittermeier and van Roosmalen (1981)
Pando, Bolivia	1978–1980	—	14–38+	—	—	—	S, GO: ms	Izawa and Bejarano (1981), Izawa and Yoneda (1981)
Seminatural environments								
Southern Florida	1960–1967	270+	110	1.6	—	6875	B&E: *Saimiri*	DuMond (1968)
Southern Florida	1966–1967	935	110	1.6	—	6875	B&E: *Saimiri*	Baldwin (1968, 1969, 1971)
Puerto Rico	1969	33	5	—	—	—	GO: *Saimiri*	Winter (1972)
Santa Sophia, Colombia	1972–1973	—	41–50 average	—	—	320–360	C: *Saimiri*	Bailey *et al.* (1974), Sponsel *et al.* (1974)
Southern Florida	1975–1978	—	45–200	1.6	—	8000–12000	B&E: *Saimiri*	Scollay and Judge (1981)

[a] B&E, Behavior and ecology; GO, general observations; C, census; S, survey; *Saimiri*, on squirrel monkeys; ms, multiple species studied.

of food, At birth, infants usually weigh 80–140 g; by adulthood, the females typically reach 500–750 g, while the males attain weights of 700–1100 g (Long and Cooper, 1968; Cooper, 1968; Middleton and Rosal, 1972).

2.1. Diet

Saimiri are omnivores that feed mostly on fruit and insects (Thorington, 1968; Izawa, 1975; Mittermeier and van Roosmalen, 1981). They prefer insects to vegetable matter, but will forage predominantly on fruit and other plant material when insects are scarce (Baldwin and Baldwin, 1972a; Jones, 1973). *Saimiri* have a relatively short digestive tract that is better adapted to digesting insect food than plant material (Hill, 1960; Fooden, 1964). *Saimiri* appear to value fruit less than insects, since they are more likely to drop fruit than insects. Protein intake is mostly in the form of small arthropods: caterpillars, butterflies, flies, larvae, grasshoppers, cicadas, ants, and spiders (Thorington, 1968; Izawa, 1975). *Saimiri* have also been observed to eat tree frogs, snails, and crabs (Sanderson, 1957). Juveniles and adults are relatively successful in catching a variety of insects, including flying insects in the air. *Saimiri* appear to obtain much of their water from the food they eat; but they also lick droplets from leaves, dip their hands into tree holes containing water, and occasionally come to the ground to drink.

2.2. Locomotion

Saimiri spend most of their time in the trees. The animals are very agile and can move with great dexterity and speed through all but the smallest branches. A quadrupedal walk or run is the typical mode of locomotion (Fleagle and Mittermeier, 1980); but the animals make frequent leaps of 1 or 2 m. After infancy, the animals become quite skillful at jumping, and adults are not reluctant to make leaps of up to 5 m. After some hesitation—with repeated intention movements—adults have been observed to jump as far as 5–7 m forward during an 8–13 m fall (Baldwin, 1967). Fleagle *et al.* (1981) estimated that during travel, over 40% of locomotor activities involved leaping (though leaping is less common during foraging and feeding). The tail is used for balance, usually trailing behind the animal and being moved from side to side when the animal is moving along small or unstable arboreal supports. The tail is sometimes used as a brace or support while foraging; and the infant wraps its tail around the mother's abdomen or tail to gain extra support. The monkeys sometimes stand bipedally when on the ground in tall grass or foraging for food above them.

2.3. Habitat

Saimiri can be found in many types of tropical wet and dry forests in their range, including virgin forests, the continuous rainforests of the Amazon drainage, secondary forests, mangrove swamps, and riparian and remnant forests. In a study in Bolivia, *Saimiri* were found more commonly in low forests (with thickets, vines, shrubs, etc.) than in the taller inland forests (Izawa and Benjarano, 1981).

The monkeys utilize all levels of thickets and mangrove swamps. In taller forests they usually utilize the understory and lower portions of the main canopy (Tokuda, 1968; Fleagle *et al.*, 1981; Mittermeier and van Roosmalen, 1981). However, at times they come to the ground or climb to the highest levels of the canopy. Since *Saimiri* leap gaps of several meters with ease, they are well suited to life in the understody with its discontinuous arboreal supports. While foraging, *Saimiri* tend to move frequently from spot to spot, searching the foliage for available food, rather than spending a long time in one location. As they move, they rummage in the twigs and leaves, looking for insects, fruit, and other foodstuffs. They unfurl leaves in search of insects and spiders; and they stalk more mobile prey, then pounce and grab them. They are among the smallest and least powerful of the cebids, and usually do not strip bark, break branches, or open hard nuts in search of food, as *Cebus* do (Fleagle *et al.*, 1981). Instead of using strength, *Saimiri* couple alertness with speed and agility.

3. ACTIVITY PATTERN

Adult *Saimiri* spend many hours each day actively foraging, relying on their inquisitiveness and speed to obtain the insects that are their preferred foods. The daily activity pattern is somewhat variable (Thorington, 1968; Baldwin and Baldwin, 1972a). The monkeys tend to become active at first light (about 40–60 min before sunrise). During the first hour or two of the morning, they often forage primarily on fruit (without neglecting any insects they chance to discover). Where food is abundant, the animals often switch to a more leisurely mode of foraging after satiating on fruits; and they spend more time looking for insects and resting. In forests where insects are scarce, the monkeys continue to rely on plant food throughout the day. At Monte Seco, where fruits and insects were not scarce, the monkeys descended to lower parts of the forest and became inactive and somnolent for 1–2 hr during the heat of the day, when the temperature rose to 29–32°C in the afternoon (Thorington, 1968). Foraging and travel resumed after the midday rest. At Barqueta, where food was scarce, rest periods were brief and infrequent (Baldwin and Baldwin, 1972a). The pace and intensity of afternoon and evening foraging probably also depend on the availability of food. At Barqueta, active foraging and travel continued until 1.5–2 hr after sunset.

The animals show little change in behavior during light rains; but heavy rain depresses general activity and increases the likelihood of huddling, shaking, back-rolling, and scratching.

Groups tend to travel and rest as a single unit in the early and late parts of the day, but to fragment during the more leisurely periods of daytime foraging (Thorington, 1968; Izawa, 1976). Because much of the monkeys' food—i.e., most insects and some fruit—is dispersed, groups frequently split into small foraging parties of 2–8 individuals, moving through the forest as scattered subgroups. However, groups do not always fragment during the day, especially when the monkeys converge at clustered sources of fruit or congregate to rest together. At Barqueta, where insect food was scarce and the animals' diet consisted mostly of fruit, the animals often came into close proximity when the group located a cluster of fruit, and sometimes they even walked over each other's bodies while attempting to pick the food (Baldwin and Baldwin, 1972a). The frequency of overt social interactions and resting appears to be higher in groups that spend less time in intense foraging. Food stealing is uncommon, but occurs when animals are foraging in close proximity. Food stealing rarely leads to fights, even in situations where food is scarce.

At night, *Saimiri* sleep on top of branches, hunched over in a huddled position with the tail curled over the shoulder. It is not clear to what degree they have regular sleeping sites. In a seminatural environment, DuMond (1968) obtained some evidence of stable sleep groupings. In the laboratory, huddle groups are stable for long periods and reflect social preferences seen in other interactions (Ploog *et al.*, 1963; Fairbanks, 1974; Hopf, 1978). At Barqueta, the sleep area used by the study group throughout the 10-week study was no larger than 0.4 ha (or 2.3% of the total home range), and it was perhaps only half that size (Baldwin and Baldwin, 1972a). A second group slept in the study group's home range on one evening, and spent that night near the study group's sleep area. Thus it appears that social groups and sleep areas are relatively stable, but not inflexible.

4. GROUP SIZE

The data on group size are summarized in Table I. Group size seems to correlate to some degree with habitat conditions (Baldwin and Baldwin, 1971), but it is also influenced by the social dynamics of group growth and fission (Izawa, 1976; Scollay and Judge, 1981). Group sizes of 10–35 are common in smaller forests. In large expanses of forest, group size appears to be much more variable, ranging from 15–20 to several hundred animals. Nevertheless, group counts are often small (e.g., 2–8) in the middle of the day because of the fragmentation of large groups into smaller foraging parties. In addition, subadult males and adult males sometimes leave their groups and form small all-male subgroups.

In natural environments, population density has been reported to range from 8 to 528 monkeys per km^2 (Table I). Group composition varies, but a typical group contains 20% infants, 32% juveniles, 29% adult females, 13% subadult males, and 6% adult males.

The formation of new groups has not been observed in the wild. Observations in the seminatural environment in Florida suggest that large groups may divide when the core group of adult females with their young begins to split and regularly follow different foraging routines (F. V. DuMond, personal communication; Scollay and Judge, 1981). Animals in the other age–sex classes follow the female core group for which they have the greater affinity, and avoid animals with whom they have had aversive contacts. A year or more after group fission, the separate groups can rejoin and travel as an integrated large group. This type of group fusion has not been observed in the wild. However, observations on groups of *Saguinus* (tamarins), *Pithecia* (sakis), and *Lagothrix* (woolly monkeys) traveling together have led Izawa (1978, p. 273) to conclude that in "the New World monkeys inhabiting the neotropical forest left untouched by man . . . friendly relations rather than antagonistic ones can be seen among adjoining unit groups." The fact that groups of *Alouatta* (howler monkeys) have also been observed to merge and travel together (Baldwin and Baldwin, 1972b) lends support to this hypothesis. If *Saimiri* follow this pattern, it is possible that two or more groups could temproarily join and travel together. This may help explain the observations of large groups of 200–550 animals in some parts of South America. These large groups could arise in at least three different ways. They might be unit groups that have not been affected by the causes of group fission. They might consist of two or more fragments of a group that had divided earlier, yet at some times join and travel together. They might represent the temporary merger of two or more friendly adjacent groups.

5. *HOME RANGE*

Home range data for *Saimiri* are available from three natural environments. The group of 22 animals observed at Monte Seco utilized all of its 14.7-ha forest (Thorington, 1968). At Barqueta, the main study group (consisting of 23 animals) had a home range of 17.5 ha, but spent 80% of its daylight hours in a core area of 9.2 ha (Baldwin and Baldwin, 1972a). A second group (consisting of 27 animals) could not be followed throughout its entire range, but its range appeared to be as large as 24–40 ha or more. No major shifts in range were noted over the 10-week study period. The *Saimiri* at Barqueta utilized 6–8 ha on days when food was found primarily near the sleeping area, and 13–16 ha on days when they foraged farthest from the sleep area. Klein and Klein (1976) estimate that the groups at La Macarena had home ranges of 65–130 ha.

The path length (or day range) of *Saimiri* groups has been reported at 0.6–1.1 km/day at Rio Peneya, Colombia (Izawa, 1976). At Barqueta, Panama, it was 2.5–4.2 km/day (Baldwin and Baldwin, 1972a). It is likely that the day range in most environments is less than that at Barqueta, since the limited food supply at Barqueta forced the animals to spend an unusually large amount of time foraging and moving (cf., Davidge, 1978).

Territoriality has not been reported in any of the studies. At Monte Seco, there was some degree of home range overlap because a second group occasionally entered the forest during the study (Thorington, 1968). At Barqueta, the larger group entered the home range of the main study group on six of the 69 study days, and ranged through 90% of its area (Baldwin and Baldwin, 1972a). At Barqueta, there were no signs of territoriality. There was no evidence that either group defended any part of its range. The two groups sometimes came as close as 70–100 m, but no closer. There was no indication that the animals of either group emitted special warnings, threats, or distance-maintenance signals. Instead, it appears that they detected the normal vocalizations of the other group and avoided coming closer than 70 m to the other group.

Extensive home range overlap without fighting has also been observed in two seminatural environments (R. C. Bailey, personal communication; Scollay and Judge, 1981).

Most of the data indicate that female and immature *Saimiri* form relatively closed groups, except when previously split groups rejoin. Genetic exchange occurs when males move from one group to another. It appears likely that a significant number of maturing subadult males leave their natal groups and join new groups, travel alone, or join all-male groups. Peripheral males and all-male subgroups have been observed in enough locations to suggest that males do not always find it easy to join new groups.

6. SOCIAL BEHAVIOR AND ORGANIZATION

The social organization of *Saimiri* groups is influenced by several factors, such as the affinity that individuals have for others of the same age and sex, the affinity that all individuals have for adult females, and the phase of the yearly reproductive cycle.

First, *Saimiri* tend to travel near and interact with individuals of same age and sex, with the important exceptions of mothers and infants, and opposite-sex adults in the mating season (Thorington, 1968; Baldwin, 1971; Izawa, 1976). Thus, most social interactions are within each age–sex class, as shown in Fig. 1. There is probably a more subtle level of organization based on each individual's preferences for favorite interaction partners (Scollay and Judge, 1981), but most field observers have not been able to identify enough individuals to determine the nature of such relationships.

Second, the different age–sex subgroups are organized into the unified group by the attractions that all individuals have for adult females. Of all the possible interactions *between* the different age–sex classes, the ones directed to the adult females are the most frequent and the least aggressive (Fig. 1). The infants are strongly attached to adult females and juveniles are moderately attached to them. Subadult and adult males are especially attracted to adult females in mating season. Adult females are attracted to each other more than to other classes of individuals, except for the attraction of mothers to their own young infants. The adult females regulate their activities and movements primarily in reference to other adult females and their own infants; and it is their activities that determine much of the group's movements, though no single adult female functions conspicuously as leader. The adult females unify the group not because they enforce any social order (such as in hierarchies), but because the other age–sex classes are more strongly attracted to them than *vice versa*.

Third, the basic social organization portrayed in Fig. 1 varies throughout the year, depending on the phase of the annual reproductive cycle (Baldwin, 1968, 1971). The yearly cycle typically consists of a mating season of 1–3 months, followed about 5.5 months later by a birth season. The responses of adult males and infants to adult females vary according to the phase of the yearly cycle. Since adult females serve as the core of *Saimiri* groups, their

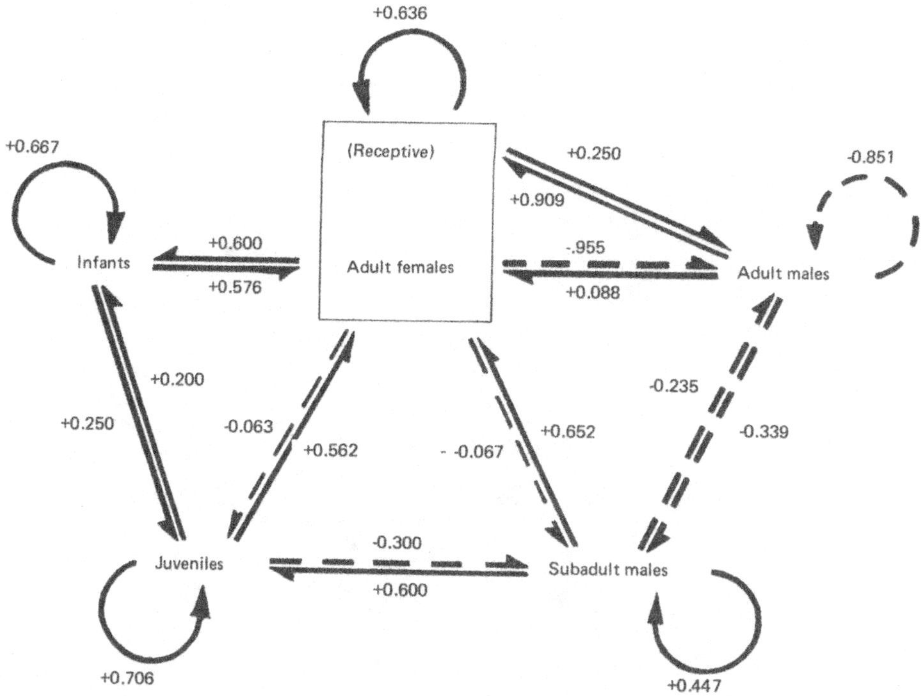

Fig. 1. The degree of friendliness of interactions between and within the major age–sex classes. The numbers represent the degree to which the interactions were purely friendly (+ 1.0) or purely agonistic (– 1.0). Friendly activities are represented with solid arrows; agonistic ones with broken arrows.

phase in the annual reproductive cycle has considerable influence on group activities and organization (Fig. 2). In birth season and the following several months, the adult females direct attention to each other and the infants. Juvenile females are somewhat attracted to young infants. Subadult males show a general attraction to the adult females, but there are few overt social interactions. The adult males are minimally attracted to the group, and sometimes travel in a peripheral position. With the onset of mating season, the adult males become increasingly attracted to the adult females, but most adult females—except receptive ones—respond hostilely to the males' approaches. After mating season, the pregnant females gradually show decreasing interest in their maturing infants, though the infants, juveniles, and subadult males continue to show a general attraction to them.

6.1. Mating Activities

There is some evidence that the yearly reproductive cycle is synchronized by cycles of rainfall (DuMond and Hutchinson, 1967; Baldwin and Baldwin,

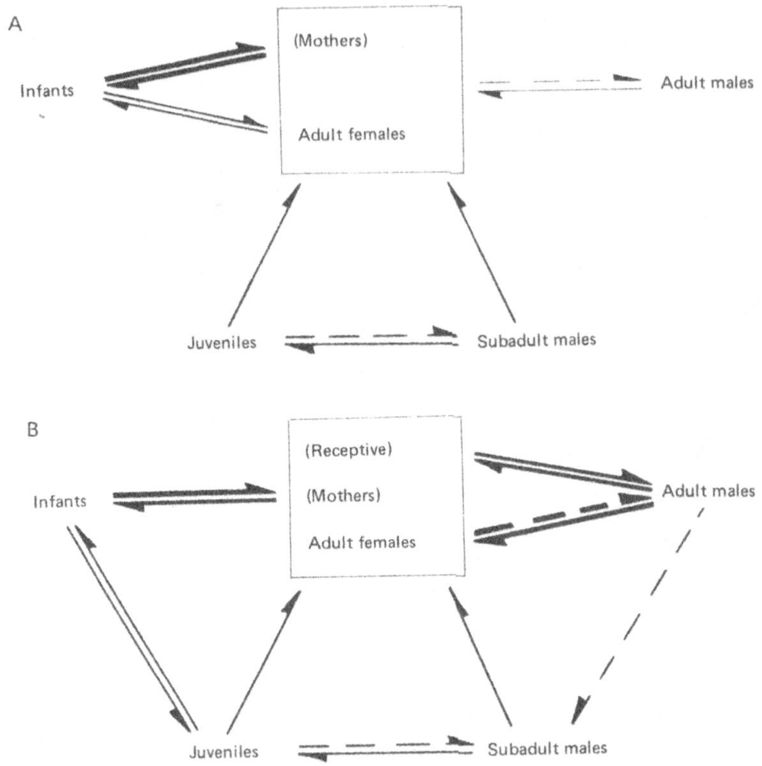

Fig. 2. The interactions between the various age–sex classes in (A) birth season and (B) mating season. Friendly activities are represented with solid arrows; agonistic ones with broken arrows. The width of the arrows reflects the frequency of the friendly or agonistic activities. Minor interactions are omitted.

1981b). However, other environmental cues, such as changes in illumination, may also be involved (Follett and Follett, 1981). Available evidence suggests that adult females are more responsive to the environmental timing cues than are males and that the reproductive activity of the adult males is at least in part stimulated and synchronized by olfactory and other cues from the adult females (Baldwin, 1970). In general, birth season falls in wet season, which may be adaptive in synchronizing the mother's and/or infant's nutritive needs with annual cycles in food abundance (Lancaster and Lee, 1965).

At the beginning of mating season, adult males become interested in the odors of the adult females and begin to approach adult females with increasing frequency (Baldwin, 1968). It appears that the female odors excite the males behaviorally, and they may be partially responsible for the stimulation of male reproductive activity. Gradually, the males become spermatogenic and develop a "fatted" condition typical of mating season (DuMond and Hutchinson, 1967). There is a significant gain in body weight (e.g., 85–222 g)—especially subcutaneously in the upper torso, shoulders, and arms—giving males the fatted appearance. The adult males also become more active, reactive, and irascible. There are increasing numbers of genital displays and fights among the males, and dominance relations become conspicuous. The genital display—in which a more dominant male displays an erect penis to a less dominant one—is the most frequent indicator of dominance relations. In both the laboratory and the field, hierarchies are usually linear, though the rank order of the males may change as often as three times per month (Baldwin, 1968). Changes in rank orders often involve serious fights. Older subadult males and young adult males are usually excluded from the adult male heirarchy by the threats, chases, and fights. It is not uncommon for young adult males to become peripheralized or to leave the group entirely at this time. Some may even die from wounds.

As the males become increasing fatted and socially active, they approach the females more often. One or more adult males may approach a given female to smell her odors. Since the adult females travel together and most are nonreceptive on any given day, male approaches to a receptive female may elicit hostile responses from nearby females. Nonreceptive females sometimes form coalitions to threaten and chase the males away. After the males have retreated and the females have returned to foraging or other activities, the males typically return to make new approaches.

The female odors doubtless help the males to discriminate the reproductive condition of each female. In addition, there are behavioral cues that the males could use from a distance to discriminate receptive females from nonreceptive ones. For example, nonreceptive females are not very tolerant of the adult males, and several females may band together to threaten or chase a male that comes too close. In contrast, a receptive female is more tolerant when a male approaches, and often drops back in group progressions, allowing herself to become separated from other females to associate with the male.

When a male approaches a receptive female, the interaction can proceed in a variety of different ways. If nonreceptive females discover the male, they may threaten and chase the male away. This is most likely when the receptive female is traveling close to nonreceptive females and/or the approaching male makes himself too conspicuous. However, a relatively quiet approach by one male may induce a receptive female to separate from the other females and join the male for a period of minutes—and sometimes over 1 hr—for consorting and one or more copulations. Consorting can be very brief. Short sexual interactions of less than 1 min (including copulation) were observed at Barqueta (Baldwin and Baldwin, 1972a). Since it is possible for a male and a receptive female to find each other and (if no other animals are nearby) to copulate almost immediately, the length of consorting is determined, in part, by the time spent waiting for other animals to leave the pair in privacy.

It is not clear whether high rank in the adult male dominance hierarchy is an important determinant of male reproductive success. In the seminatural environment in Florida (the only location from which relevant data are available), none of the 13 observed copulations involved the two males who occupied the alpha position in the dominance hierarchy (Baldwin, 1968). The two males responsible for all 13 observed copulations tended to be quiet, nonaggressive animals, and these behavioral qualities seemed to facilitate their successful interactions with adult females.

After mating season, the males lose their fatted condition and become more asocial. They lose interest in the females, cease approaching them, and often drift to a more peripheral position in the group. The adult females no longer have to threaten and chase the males. As their infants from the previous birth season become increasingly independent, mothers interact less frequently with them. During pregnancy, the adult females often engage in the minimal number of overt social interactions, although they still tend to travel, forage, and spend time together. Thorington (1968) found pregnant females and nursing females to be more shy and less active than other animals: They rested longer and traveled shorter distances. During this period of the year, play among the young is typically the most common form of overt social interaction.

6.2. Birth Season

Gestation has been estimated at 167, 165, 170, and 162.6 days (Beischer and Furry, 1964; Wiswell, 1965; Goss *et al.*, 1968; and Lorenz *et al.*, 1973). The onset of birth season brings several changes in social behavior and organization. The changes typical of birth season and the following months can be understood by tracing early infant development and female–infant social relations (Baldwin, 1969).

Although birth has not been observed in natural environments, it has been studied and filmed in the laboratory (Takeshita, 1962; Bowden *et al.*, 1967; Hopf, 1967). The newborn infant typically assumes a dorsal-ventral riding position with its tail wrapped around the mother's lower abdomen or tail for support. The neonate appears to find the nipples by itself by crawling part way around the mother's side. The mother directs little overt attention to a healthy, normal newborn infant; however, laboratory studies have shown that mothers do attend to their infants when the infants are incapacitated (Rumbaugh, 1965; Kaplan, 1973).

During the first week of life, the infant is relatively inactive, either sleeping on the mother's body or nursing. In the wild, there has been no indication that other group members show much attention to either the neonate or the mother. The infant does not have special coloration that might draw attention to it. Gradually in the second week, it increases the amount of time it spends exploring the mother's body and gazing at the environment.

As the infants become increasingly active and socially responsive during the third and fourth weeks, juvenile and adult females show increasing interest in the infants. Adult females without infants of their own sometimes form close and stable relations with certain mothers and serve as "aunts" to their infants (Baldwin, 1969). An adult aunt often shows maternal responses to the infant, inducing the infant to ride on her back and carrying it. Some mothers are more accepting of aunt interactions than are others. Juvenile females also become interested in infants once the infants become more active and interactive; however, compared with the aunts, the juveniles are more playful and less maternal with the infants. Although they sometimes handle the infants gently, their more playful actions can disturb the infants. When an infant is disturbed, it gives peeps or shrieks, which usually cause the mother to approach and retrieve it. Mothers often follow juveniles closely, watching the infants. These cases of mothers' attending to and retrieving their infants from juveniles (and less often from adult aunts) are the first regular signs of overt maternal concern shown by mothers.

During weeks 5–7, the infant begins to venture off its mother's back and explore the areas near her. The mother becomes increasingly attentive to and protective of the infant when it begins leaving her. During this period, most mothers retrieve their infants from dangerous situations or from circumstances that cause the infant to vocalize: for example, when the infant reaches an arboreal gap that it cannot jump or when another group member disturbs the infant. The juvenile females and adult aunts continue to show interest in 5- to 7-week-old infants, but the increasingly mobile and independent infants are less likely to ride passively and the females gradually lose interest. Immature males sometimes show interest in infants, but adult males have not been seen to care for infants.

6.3. *After Birth Season*

From the eighth week through the fourth month of the infants' lives, the mothers gradually become less protective. Aunts and immature females stop interacting with infants early in this period. The infants are maturing and gaining many skills for independence as they explore further away from their mothers. Peer play among infants begins to become common, and infants may spend several hours of each day engaged in exploration or social play. However, the infants still return to their mothers when disturbed or aroused. Social grooming is uncommon in *Saimiri*; however, mothers sometimes groom infants during this developmental period. The grooming may be an affiliative behavior or a method of pacifying agitated infants.

Social play is not equally common in all groups (Baldwin and Baldwin, 1974). It is most likely to appear and to be relatively frequent if there is abundant food, little danger from predators, and a large enough group of same-age peers to provide ample play partners. During the study at Barqueta, where there was a limited food supply, not a single bout of social play was observed during the 10-week study (Baldwin and Baldwin, 1972a). As Müller-Schwarze *et al.* (1982) suggest, play appears to be "behavioral fat" that is reduced when food is scarce.

Play is most common among individuals of the same age, size, strength, and skills. Hence, infants tend to play mostly with other infants, and juveniles with juveniles. However, in small groups with few individuals in each age cohort, there is more play between individuals of different ages.

Early social play consists of tentative explorative movements, as the infants gently touch, pull, and push each other. Gradually the infants progress to more active forms of wrestling play: e.g., holding and pulling each other with hands and feet while wrestling, sham biting, and rolling. Wrestling play often occurs in mats of vines and leaves or while the players hang by the hind feet. As the infants gain strength and skill, they develop more active forms of play, including tackling, tumbling, and chasing. They jump through the branches, maintaining larger distances and spending less time in direct contact.

When the infants are about 4–6 months of age, the adult males of the group begin to approach the adult females for mating season activities. Because infants of this age still travel near their mothers, they often come in close contact with the adult males. Typically, the infants become quite excited by the presence of the adult males, and will vocalize and direct genital displays toward individual males. In most cases, the adult males avoid the infants; but in some cases, they threaten the infants with vocalizations and branch shaking before leaving.

During the mating season, adult females direct increasing attention to adult males (either chasing them or consorting with them). Mothers pay progressively less attention to their infants: They become less protective, and less likely to accept their infants' approaches to ride or nurse. By the time infants are 6 months of age, nursing interactions become relatively uncommon, although bouts of nursing occur infrequently during the next several months. Mothers become increasingly likely to threaten their infants and rebuff their approaches, especially when the infants are very active. The mother typically rebuffs and weans her infant by giving threat vocalizations, putting one or both hands on the infant firmly, or grabbing at the infant. More intense forms of rejection are uncommon.

During mating season and the months immediately thereafter, the infant makes rapid strides toward independence. The frequency of daytime contact with its mother drops to a low level, although the infant occasionally comes to its mother for nursing or sleeping at night. It spends increasing numbers of hours with age-mates. In small groups with few infants, the infant may be slower to break contact with its mother.

6.4. *Juveniles*

The juvenile age period lasts from approximately 11–30 months. During the first months of the juvenile period, the monkeys gain complete independence (Baldwin, 1969). They travel mostly with other juveniles, although they continue to show an attraction to the adult females of the group. Play tends to become rougher and more active during this period. Although some sexual elements appear in play in the first year, they become increasingly prevalent during the juvenile period. Rowdy and even aggressive play begins to appear, especially among the males. As the animals become stronger and play escalates to ever more active levels, juvenile females begin to withdraw from play and spend increasing amounts of time exploring alone or traveling near the adult females. By the age of 2.5 years, females have usually withdrawn completely from active play. As the females spend increasing amounts of time with the adult females, they become assimilated into the adult female subgroup and learn to behave more like the adults. Males continue their active forms of exploration and play through the juvenile period.

6.5. Sexual Maturation

The females become sexually mature at approximately 2.5 years. Their first pregnancy and birth facilitate their assumption of the behavior repertoire typical of adult females (Baldwin and Baldwin, 1981a, pp. 140ff).

Males do not become fully adult at 2.5 years. Rather, they enter a 2- to 3-year period of subadult status. There is no sharp boundary demarcating the end of the subadult phase, and a given male's age of transition to full adulthood may depend on a number of factors, such as the age, personality, and the aggressiveness of the subadult and adult males in his group. Males continue to play all through the subadult stage, though with decreasing frequency. (Even adult males have been seen to participate in short bouts of play with younger individuals.) Younger subadult males sometimes play gently with juveniles, engaging in self-handicapping that helps them avoid losing their play partners. The play of older subadults is oriented more to sex play. As they approach adulthood, some subadult males attempt to enter the adult male subgroup. The adult males often threaten, chase, and fight with these subadults. As a result, some leave the group, although others succeed in becoming integrated into the adult male subgroup (Baldwin and Baldwin, 1981b).

6.6. Communication

A detailed overview of the *Saimiri* repertoire of communicative behavior is presented elsewhere (Baldwin and Baldwin, 1981b, Tables IV–VII; and Newman, this volume, Chapter 4).

7. RELATIONS WITH OTHER SPECIES

The relation of *Saimiri* with nonprimate species has not been well established. *Saimiri* have been observed to give alarm peeps, twitters, and trills ("chitter" calls) at the sight of hawks, vultures, toucans, and parrots flying nearby (Thorington, 1968; Baldwin and Baldwin, 1972a). These birds evoked little or no response from the *Saimiri* when the birds were perched. Small hawks (such as *Harpagus bidentatus*) have been seen to travel with *Saimiri* groups and catch insects stirred up by the foraging activities of the monkeys (Thorington, 1968; Klein and Klein, 1973). In Central America, squirrels (*Sciurus variegatoides* and *S. granatensis*) sometimes join *Saimiri orstedii* groups and the monkeys usually ignore them (Moynihan, 1976). *Saimiri* have been observed to mob and give yap ("chaun" or "chunk") calls at dogs, snakes, tayras, cats, and humans (Thorington, 1968; DuMond, 1968; Baldwin and Baldwin, 1972a), although there is local variation in the responses the monkeys give to different species.

In South America, *Cebus apella* and *C. albifrons* (capuchins) sometimes form mixed groups with *Saimiri sciureus* (Izawa, 1975, 1976, 1980; Moynihan,

1976; Neville *et al.*, 1976; Fleagle *et al.*, 1981). Thorington (1967, 1968) observed that after *C. apella* and *S. sciureus* had congregated in a fruiting tree, they tended to move out of it in the same direction. Pregnant *S. sciureus* and mothers with infants often traveled more slowly than their group mates, which left them moving with the slower traveling *C. apella*. When an animal of one species was temporarily isolated from conspecifics, it would tend to seek out and remain with a group of the other species. Klein and Klein (1973) found that among the four primate species they observed, *C. apella* and *S. sciureus* had the most frequent intertaxa associations. No "friendly" interactions were observed between *C. apella* and *S. sciureus*, but agonistic interactions were observed. Based on observations in which small hawks caught and ate insects flushed by foraging primates, Klein and Klein speculated that the *C. apella* also may have benefitted by catching insects flushed by *S. sciureus*, or by being able to supplant individual *S. sciureus* from sources of insect prey. However, observations in Surinam led Mittermeier and van Roosmalen (1981) to the opposite conclusion. *Saimiri sciureus* appeared to benefit more from traveling in mixed groups than did *C. apella*. It was the *Saimiri* that benefitted from obtaining insects flushed by *C. apella*, and they may also have benefitted from "the larger and more frugivorous species' knowledge of the location of food sources." It is likely that each speices benefits from alarm calls given by the other.

In Central America, *S. oerstedii* shows a different relationship with *C. capucinus: S. oerstedii* frequently avoids *C. capucinus* (Baldwin and Baldwin, 1972a; Moynihan, 1976). The differences in *Cebus* and *Saimiri* relations in Central and South America could result from a variety of factors, such as genetic variables, ecological differences, or relative group sizes (since the *Cebus* groups are smaller than *Saimiri* groups in South America, but the opposite is true in Central America).

In South America, neither Thorington (1968) nor Klein and Klein (1973) reported an association between *S. sciureus* and *Alouatta seniculus* (howler monkeys). In central America the *S. orstedii* at Barqueta did show a slight tendency to travel near the *A. palliata* in their home range if the latter were vocal or otherwise conspicuous. Although animals of the two species often came within 3 m of each other, no direct physical contact or overt interactions were observed. The *S. sciureus* at Monte Seco did not associate with the *Callicebus moloch* (titis) in the forest; however, a lone male *S. sciureus* in a nearby forest did associate with the *C. moloch* in that forest. In the Florida seminatural environment, *S. sciureus* and *Cacajao calvus rubicundus* (uakaris) have been observed to groom and play together, and, on occasion, to engage in agonistic interactions (Baldwin, 1967; DuMond, 1968; Abordo *et al.*, 1975).

REFERENCES

Abordo, E. J., Mittermeier, R. A., Lee, J., and Mason, P., 1975, Social grooming between squirrel monkeys and uakaris in a seminatural environment, *Primates* **16**:217–221.

Bailey, R. C., Baker, R. S., Brown, D. S., von Hildebrand, P., Mittermeier, R. A., Sponsel, L. E., and Wolf, K. E., 1974, Progress of a breeding project for non-human primates in Colombia, *Nature* **248**:453–455.

Baldwin, J. D., 1967, A study of the social behavior of a semifree-ranging colony of squirrel monkeys (*Saimiri sciureus*), Ph.D. Dissertation, Johns Hopkins University.

Baldwin, J. D., 1968, The social behavior of adult male squirrel monkeys (*Saimiri sciureus*) in a seminatural environment, *Folia Primatol.* **9**:281–314.

Baldwin, J. D., 1969, The ontogeny of social behavior of squirrel monkeys (*Saimiri sciureus*) in a seminatural environment, *Folia Primatol.* **11**:35–79.

Baldwin, J. D., 1970, Reproductive synchronization in squirrel monkeys (*Saimiri*), *Primates* **11**:317–326.

Baldwin, J. D., 1971, The social organization of semifree-ranging troop of squirrel monkeys (*Saimiri sciureus*), *Folia Primatol.* **14**:23–50.

Baldwin, J. D., and Baldwin, J. I., 1971, Squirrel monkeys (*Saimiri*) in natural habitats in Panama, Colombia, Brazil and Peru, *Primates* **12**:45–61.

Baldwin, J. D., and Baldwin, J. I., 1972a, The ecology and behavior of squirrel monkeys (*Saimiri oerstedii*) in a natural forest in western Panama, *Folia Primatol.* **18**:161–184.

Baldwin, J. D., and Baldwin, J. I., 1972b, Population density and use of space in howling monkeys (*Alouatta villosa*) in southwestern Panama, *Primates* **13**:371–379.

Baldwin, J. D., and Baldwin, J. I., 1974, Exploration and social play in squirrel monkeys, *Am. Zool.* **14**:303–315.

Baldwin, J. D., and Baldwin, J. I., 1976, Primate populations in Chiriquí Panama, in: *Neotropical Primates: Field Studies and Conservation* (R. W. Thorington, Jr., and P. G. Heltne, eds.), National Academy of Sciences, Washington, D. C., pp. 20–31.

Baldwin, J. D., and Baldwin, J. I., 1981a, *Beyond Sociobiology*, Elsevier, New York.

Baldwin, J. D., and Baldwin, J. I., 1981b, The squirrel monkeys, genus *Saimiri*, in: *Ecology and Behavior of Neotropical Primates*, Volume 1 (A. F. Coimbra-Filho and R. A. Mittermeier, eds.), Academia Brasileira de Ciências, Rio de Janiero, pp. 277–330.

Beischer, D. E., and Furry, D. E., 1964, *Saimiri sciureus* as an experimental animal, *Anat. Rec.* **148**:615–624.

Bowden, D., Winter, P., and Ploog, D., 1967, Pregnancy and delivery behavior in the squirrel monkey and other species, *Folia Primatol.* **5**:1–42.

Cooper, R. W., 1968, Squirrel monkey taxonomy and supply, in: *The Squirrel Monkey* (L. A. Rosenblum and R. W. Cooper, eds.), Academic Press, New York, pp. 1–29.

Davidge, C., 1978, Ecology of baboons (*Papio ursinus*) at Cape Point, *Zool. Afr.* **13**:329–350.

DuMond, F. V., 1968, The squirrel monkey in a seminatural environment, in: *The Squirrel Monkey* (L. A. Rosenblum and R. W. Cooper, eds.), Academic Press, New York, pp. 87–145.

DuMond, F. V., and Hutchinson, T. C., 1967, Squirrel monkey reproduction: The "fatted" male phenomenon and seasonal spermatogenesis, *Science* **158**:1067–1070.

Durham, N. M., 1975, Some ecological, distributional, and group behavioral features of Atelinae in Southern Peru: With comments on interspecific relations, in: *Socioecology and Psychology of Primates* (R. H. Tuttle, ed.), Mouton, The Hague, pp. 87–101.

Fairbanks, L., 1974, An analysis of subgroup structure and process in a captive squirrel monkey (*Saimiri sciureus*) colony, *Folia Primatol.* **21**:209–224.

Fleagle, J. G., and Mittermeier, R. A., 1980, Locomotor behavior, body size, and comparative ecology of seven Surinam Monkeys, *Am. J. Phys. Anthropol.* **52**:301–314.

Fleagle, J. G., Mittermeier, R. A., and Skopec, A. L., 1981, Differential habitat use by *Cebus apella* and *Saimiri sciureus* in central Surinam, *Primates* 22:361–367.

Follett, B. K., and Follett, D. E., 1981, *Biological Clocks in Seasonal Reproductive cycles*, Halsted (Wiley), New York.

Fooden, J., 1964, Stomach contents and gastrointestinal proportions in wild-shot Guianan monkeys, *Am. J. Phys. Anthropol.* **22**:227–231.

Freese, C., 1975, A census of non-human primates in Peru, in: *Primate Censusing Studies in Peru and Colombia*, Pan American Health Organization, Washington, D. C., pp. 17–41.

Freese, C. H., Heltne, P. G., Castro, N., and Whitesides, G., 1982, Patterns and determinants of monkey densities in Peru and Bolivia, with notes on distributions, *Int. J. Primatol.* **3**:53–90.

Goss, C. M., Popejoy, L. T., Fusiler, J. L., and Smith, T. M., 1968, Observations on the relationship between embryological devemopment, time of conception, and gestation, in: *The Squirrel Monkey* (L. A. Rosenblum and R. W. Cooper, eds.), Academic Press, New York, pp. 171–191.

Hernández-Camacho, J., and Cooper, R. W., 1976, The nonhuman primates of Colombia, in: *Neotropical Primates: Field Studies and Conservation* (R. W. Thorington, Jr., and P. G. Heltne, eds.), National Academy of Sciences, Washington, D. C., pp. 35–69.

Hill, W. C. O., 1960, *Primates*, Volume IV, Part A, Edinburgh University Press, Edinburgh.

Hladik, A., and Hladik, C. M., 1969, Rapports tropique entre vegetation et primates dans la forêt de Barro Colorado (Panama), *Terre Vie* **1**:25–117.

Hopf, S., 1967, Notes on pregnancy, delivery, and infant survival in captive squirrel monkeys, *Primates* **8**:323–332.

Hopf, S., 1978, Huddling subgroups in captive squirrel monkeys and their changes in relation to ontogeny, *Biol. Behav.* **3**:147–162.

Izawa, K., 1975, Foods and feeding behavior of monkeys in the upper Amazon basin, *Primates* **16**:295–316.

Izawa, K., 1976, Group sizes and compositions of monkeys in the upper Amazon basin, *Primates* **17**:367–399.

Izawa, K., 1978, A field study of the ecology and behavior of the black-mantle tamarin (*Saguinus nigricollis*), *Primates* **19**:241–274.

Izawa, K., 1980, Social behavior of the wild black-capped capuchin (*Cebus apella*), *Primates* **21**:443–467.

Izawa, K., and Bejarano, G., 1981, Distribution ranges and patterns of nonhuman primates in western Pando, Bolivia, *Reports of New World Monkeys* (Kyoto University Primate Research Institute, Kyoto) **1981**:1–12.

Izawa, K., and Yoneda, M., 1981, Habitat utilization of nonhuman primates in a forest of the Western Pando, Bolivia, *Reports of New World Monkeys (Kyoto University Primate Research Institute, Kyoto)* **1981**:13–22.

Jones, C., 1973, Observations of *Saimiri orstedii*, in: *Course Book 73-2*, Organization of Tropical Studies, San José, Costa Rica, pp. 275–284.

Kaplan, J., 1973, Responses of mother squirrel monkeys to dead infants, *Primates* **14**:89–91.

Klein, L. L., 1974, Agonistic behavior in Neotropical primates, in: *Primate Aggression, Territoriality, and Xenophobia* (R. L. Holloway, ed.), Academic Press,New York, pp. 77–122.

Klein, L. L., and Klein, D., 1973, Observations on two types of Neotropical primate intertaxa associations, *Am. J. Phys. Anthropol.* **3**:649–654.

Klein, L. L., and Klein, D., 1976, Neotropical primates: Aspects of habitat usage, population density, and regional distribution in La Macarena, Colombia, in: *Neotropical Primates: Field Studies and Conservation* (R. W. Thorington, Jr., and P. G. Heltne, eds.), National Academy of Sciences, Washington, D. C., pp. 70–78.

Lancaster, J. B., and Lee, R. B., 1965, The annual reproductive cycle in monkeys and apes, in: *Primate Behavior: Field Studies of Monkeys and Apes* (I. DeVore, ed.), Holt, Rinehart and Winston, New York, pp. 486–513.

Long, J. O., and Cooper, R. W., 1968, Physical growth and dental eruption in captive-bred squirrel monkeys, *Saimiri sciureus* (Letecia, Colombia), in: *The Squirrel Monkey* (L. A. Rosenblum, and R. W. Cooper, eds.), Academic Press, New York, pp. 193–205.

Lorenz, R., Anderson C. O., and Mason, W. A., 1973, Notes on reproduction in captive squirrel monkeys, *Folia Primatol.* **19**:286–292.

Middleton, C. C., and Rosal, J., 1972, Weights and measurements of normal squirrel monkeys (*Saimiri sciureus*), *Lab. Anim. Sci.* **22**:583–586.

Mittermeier, R. A., and van Roosmalen, G. M., 1981, Preliminary observations of habitat utilization and diet in eight Surinam monkeys, *Folia Primatol.* **36**:1–39.

Moynihan, M., 1976, *The New World Primates*, Princeton University Press, Princeton.

Müller-Schwartze, D., Stagge, B., and Müller-Schwarze, C., 1982, Play behavior: Persistence, decrease, and energetic compensation during food shortage in deer fawns, *Science* **215**:85–87.

Neville, M., Castro, N., Mármol, A., and Revilla, J., 1976, Censusing primate populations in the reserved area of the Pacaya and Samiria Rivers, Department Loreto, Peru, *Primates* **17:**151–181.

Ploog, D. W., Blitz, J., and Ploog, F., 1963, Studies on social and sexual behavior of the squirrel monkey (*Saimiri sciureus*), *Folia Primatol.* **1:**29–66.

Rumbaugh, D. M., 1965, Maternal care in relation to infant behavior in the squirrel monkey, *Psychol. Rep.* **16:**171–176.

Sanderson, I. T., 1949, A brief review of the mammals of Suriname (Dutch Guiana), based upon a collection made in 1938, *Proc. Zool. Soc. Lond.* **119:**755–789.

Sanderson, I. T., 1957, *The Monkey Kingdom*, Chilton, Philadelphia.

Scollay, P. A., and Judge, P., 1981, The dynamics of social organization in a population of squirrel monkeys (*Saimiri sciureus*) in a seminatural environment, *Primates* **22:**60–69.

Sponsel, L. E., Brown, D. S., Bailey, R. C., and Mittermeier, R. A., 1974, Evaluation of squirrel monkey ranching on Santa Sophia Island, Amazonas, Colombia, *Int. Zoo Yearb.* **14:**233–240.

Takeshita, H., 1962, On the delivery behavior of squirrel monkeys (*Saimiri sciurea*) and a mona monkey (*Cercopithecus mona*), *Primates* **3:**59–72.

Thorington, R. W., Jr., 1967, Feeding and activity of *Cebus* and *Saimiri* in a Colombian forest, in: *Progress in Primatology* (D. Stark, R. Schneider, and H.-J. Kuhn, eds.), Gustav Fischer, Stuttgart, pp. 180–184.

Thorington, R. W., Jr., 1968, Observations of squirrel monkeys in a Colombian forest, in: *The Squirrel Monkey* (L. A. Rosenblum and R. W. Cooper, eds.), Academic Press, New York, pp. 69–85.

Tokuda, K., 1968, Group size and vertical distribution of new world monkeys in the basin of the Rio Putumayo, the upper Amazon, in: *VIIIth Congress of Anthropological and Ethnological Sciences: A-8, Primate Studies*, pp. 260–261.

Vessey, S. H., Mortenson, B. K., and Muckenhirn, N. A., 1978, Size and characteristics of primate groups in Guyana, in: *Recent Advances in Primatology*, Volume 1 (D. J. Chivers and J. Herbert, eds.), Academic Press, New York, pp. 187–188.

Winter, P., 1972, Observations on the vocal behavior of free-ranging squirrel monkeys (*Saimiri sciureus*), *Z. Tierpsychol.* **31:**1–7.

Wiswell, O. B., 1965, Gestation, parturition and maturation in the *Saimiri sciureus* (squirrel monkey), *Excerpta Med. Int. Congr. Ser.* **99:**E52.

3

Cognition in Squirrel Monkeys
A Contemporary Perspective

Dorothy Munkenbeck Fragaszy

1. INTRODUCTION

The aims of this chapter are to review the recent literature on cognition in squirrel monkeys and to evaluate the findings with reference to normal, directed activity in natural environments; that is, in biological perspective. A biological perspective involves recognition that the functions of the nervous system (including learning, attention, perception, memory, and other aspects of cognition) have an impact on the organism's life history, and that they must be evaluated in terms of that impact. We assume at the outset that the process of cognition provides the individual with potentially adaptable (and adaptive) ways to pick up information about the environment, to process this information, and to compare it against expectations (Neisser, 1976). This process guides behavior in a fundamental way. As Neisser (1976) puts it, cognition "is surely a matter of discovering what the environment is really like and adapting to it" (p. 9).

In accord with a biological perspective on cognition, the view adopted here is that "intelligence," or, more generally, "learning ability," is not a unitary process, and it cannot be considered in isolation from other knowledge-gaining processes possessed by living systems (Plotkin and Odling-Smee, 1979, 1982). Different animal species face different problems associated with ecological differences, and they bring to the task of solving these problems different sets of phylogenetic traits and developmental processes (Johnston, 1981, 1982). Thus, as in other realms of biology, there is unity in the nature of learning, and diversity in its expression, in accord with the evolutionary process. Nonhuman intelligence should be assumed to be multidimensional, as is human intelligence, and for the same reasons (Hunt, 1983; Sternberg and Salter, 1982). This premise leads directly to the view that a framework for specifying qualitatively different capacities must be adopted in order to make sense of species

Dorothy Munkenbeck Fragaszy • Department of Veterinary and Comparative Anatomy, Pharmacology, and Physiology, Washington State University, Pullman, Washington 99164.

differences in learning abilities (Warren, 1965, 1974). Adopting this perspective does not preclude the possibility of ordering species along a single dimension such as ability to deal with multiple reversals in discrimination learning tasks or ability to develop learning sets, as Rumbaugh (1968) has discussed for squirrel monkeys [see also Kintz *et al.* (1969)]. It simply precludes viewing any particular dimension as the only relevant one for all species.

One approach to the issue of species differences in learning abilities is to develop a hierarchical ordering of qualitative categories of intellectual function that permits meaningful interspecific comparisons (Thomas, 1980; Parker and Gibson, 1979). Another approach, which to my knowledge has not been explored in comparisons of nonhuman primates, is multidimensional representation of intellectual factors (Hunt, 1983). In the spirit of stimulating discussion, multidimensional representations of cognitive and sensorimotor characteristics underlying learning are presented here as an alternative heuristic approach to hierarchical and linear models of species differences in learning abilities. A multidimensional approach appears more likely than linear or hierarchical approaches to provide insight into both phylogenetic and ecological factors relevant to behavioral evolution.

The review begins with an overview of general sensory, motor, and cognitive characteristics that affect the way squirrel monkeys seek out and use information. The organization of this section reflects convenient categories, not independent capacities. Our knowledge of what squirrel monkeys can learn and what they know how to do will then be considered. Finally, the relationship between cognitive characteristics and behavior in natural environments will be assessed.

2. *SENSORY, MOTOR, AND COGNITIVE ATTRIBUTES UNDERLYING BEHAVIOR*

2.1. *Vision and Looking Behavior*

Vision is a primary sensory mode in squirrel monkeys. One indication of the importance of vision to squirrel monkeys is the finding that they engage in more visual scanning than several other primate species (Davis *et al.*, 1968). Visual scanning is particularly prominent in the presence of an external threat, e.g., a predator (Murray and King, 1973).

Visual acuity and sensitivity in squirrel monkeys are known to be reasonably good. They possess a visual angle threshold between 0.74 and 0.84 minutes of visual arc, a range between values for *Cebus* and *Macaca mulatta* [see Rumbaugh (1968) for review]. Photopic sensitivity for rod- and cone-based vision are similar in squirrel monkeys and humans (Jacobs, 1972, 1973), although female squirrel monkeys are significantly more sensitive to long-wavelength light (640 nm) than are males (Jacobs, 1977). The squirrel monkey's reliance on vision is emphasized by the findings that squirrel monkeys essentially stop responding in some operant tasks under darkness (Isaac, 1969) and respond more slowly than in the light in delayed-response tasks (Hornbuckle,

1972); that their levels of home-cage activity are highest under conditions of moderate illumination (Worthington, 1972); that squirrel monkeys will engage in protracted operant behavior with various light stimuli as the sole reward (Campbell, 1972); and that darkness is an effective aversive stimulus for squirrel monkeys (Parker, 1966). Darkness has a generally depressing effect upon vigilance in other sensory modes, such as decreasing detection of auditory stimuli (Delay and Isaac, 1980). Visual stimuli are more effective in establishing autoshaped responding in squirrel monkeys than are auditory stimuli (Schwam and Gamzu, 1975).

Visual acuity has been implicated by Harris and Meyer (1971) as a factor in squirrel monkeys' relatively poor performance on patterned-string problems, a traditional type of problem-solving task that has been used in comparative studies of animal learning (Davis, 1974). According to Harris and Meyer (1971), tracing strings as they cross and recross is a gross measure of visual acuity, and squirrel monkeys' last-place position on these tasks after chimpanzees, rhesus, and capuchins is a reflection of this sensory limitation. On the other hand, an identical ranking could be obtained on the basis of other characteristics. As will be discussed, attentional processes probably have as much to do with squirrel monkeys' performance on this task as does visual acuity.

Visual perception in squirrel monkeys has been studied at length by Marriott (1976). Squirrel monkeys spontaneously viewed pictorial stimuli for no reward other than the intrinsic value of seeing the picture, and like rhesus and chimpanzees, exhibited strong behavioral reactions to "natural" stimuli, indicating recognition of the representations (Humphrey, 1972; Humphrey and Keeble, 1974). As in rhesus monkeys, "interest" governed looking behavior more than "pleasure" (Humphrey, 1972). Pictures of insects and arthropod silhouettes were particularly attractive to squirrel monkeys. Pictures of feline silhouettes and snakelike forms and patterns evoked fear responses as well as heightened frequencies of looking (Murray and King, 1973). Pictures of other squirrel monkeys (full-body pictures especially, as opposed to pictures of faces only) elicited social responses correlated with the viewer's social status. Eyelike schemata inhibited looking and produced gaze aversion [see Coss (1970, 1972) for a review of this phenomenon]. Marriott concluded that squirrel monkeys' visual perceptual skills are fundamentally similar to those of Old World monkeys, apes, and humans.

Other studies concerning visual perception in squirrel monkeys have investigated social looking behavior. Squirrel monkeys look at an opposite-sex pairmate about half as much as do monogamous titis (*Callicebus moloch*), and rarely engage in mutual looking behavior (Phillips and Mason, 1976). The direction of species differences is not surprising, given that sustained visual regard is a basic feature of social attraction in primates (Mitchell, 1972), and that titi pairs are more socially interactive than are pairs of squirrel monkeys (Mason, 1971, 1974, 1975). On the other hand, Haude and Ray (1974) report that squirrel monkeys devote as much time to viewing pictures of conspecifics as do rhesus macaques under analogous viewing conditions, although the pattern of response frequencies as a function of stimulus duration is different. Specifically, squirrel monkeys look at slides most frequently at intermediate rates of slide change (125 sec), whereas rhesus monkeys look most frequently at stimuli of brief duration (5 sec). The authors suggest that the two measures, viewing time and viewing frequency, may reflect different aspects of visual performance. As we shall see, in squirrel monkeys, erratic viewing frequency (and consequently, duration per viewing bout) most likely reflects short-term attentional processes, whereas cumulative viewing time is probably a "purer" measure of interest in or habituation to a particular visual stimulus (Marriott, 1976; Humphrey, 1972).

Unfortunately, sensation and perception in sensory modalities other than vision have rarely been studied in squirrel monkeys. Investigations of chemical senses, especially their function in communication, have just begun (Candland *et al.*, 1980; Hennessey *et al.*, 1978; Epple, 1974). Audition has received some attention, especially in relation to vocal communication (Newman, this volume, Chapter 4). Green (1975) provides the only normative data on auditory function in squirrel monkeys to appear since Rumbaugh's (1968) review of sensory capacities. Equal-loudness contours in Green's study were similar in shape to those in macaques and humans. The audible frequency range extended from below 0.125 kHz to 46 kHz, with maximum sensitivity at 8 kHz. It is clear that the auditory capacities of squirrel monkeys must be well-developed, given the complexity of the vocal repertoire (Schott, 1975).

2.2. Attention

Attention is treated here as "a label for some of the internal mechanisms that determine the significance of stimuli" (Kahneman, 1973, p. 2), and is regarded as synonymous with cognitive "effort" (Kahneman, 1973). In Kahneman's model, an individual possesses a limited capacity for attention, which can be allocated to different tasks. Attention in squirrel monkeys has rarely been studied directly, but there are many findings that pertain to this issue. In general, squirrel monkeys appear to have shorter attention spans and/or are more distractable than monkeys of many other species (for example, macaques), which makes using them as subjects in repetitive laboratory tasks more difficult than using these other species. But this statement must not be construed as indicating that squirrel monkeys are somehow "deficient." As will be discussed below, squirrel monkeys are very good at tasks requiring certain types of attentional processes.

The first systematic work on "distractability" in squirrel monkeys was reported by Rumbaugh and McCormack (1967, 1969). These investigators found that the performance of squirrel monkeys and gibbons on visual discrimination tasks was more disrupted by the introduction of "irrelevant" foreground cues (produced by encasing discriminanda in plexiglass bins) than was the performance of great apes or macaques. Similarly, Abordo and Lee (1977) report that intrusion of irrelevant patterns or changes in size of a color cue disrupt performance of an established visual discrimination habit [see also Fragaszy (1981)]. More complex patterns produced greater disruption in Abordo and Lee's (1977) study, although qualitative attributes of specific stimulus patterns were also implicated in changes in performance.

Performance decrements when attention is disrupted are not restricted to the visual mode. Irrelevant auditory stimuli also disrupt performance: Hornbuckle (1972) found that the addition of white noise to the test area during the delay period decreased accuracy on a delayed-response task (location of a tone) by about the same proportions in squirrel monkeys and owl monkeys (*Aotus trivirgatus*).

The influence of distractive stimuli on attention is a general problem in cognition (Neisser, 1976), but it has received little attention from comparative psychologists. Rumbaugh and McCormack's (1967, 1969) suggestion that increased distraction by visual foreground cues in squirrel monkeys and gibbons versus rhesus or great apes is a correlate of arboreality is, to my knowledge, the only directly comparative hypothesis yet proposed. These authors suggest that arboreal travel requires greater attention to nearby cues (e.g., branches). than terrestrial travel. To develop this hypothesis, comparative data on distractibility in visual *and* other sensory modalities are needed from species designated as more and less arboreal. The magnitude of distractibility in the visual mode in arboreal species may be equal to that in other modalities, and greater overall than in terrestrial species. This would indicate that a sensory-nonspecific aspect of attention in arboreal species is responsible for a decrement in performance in all modalities in the presence of distracting stimuli, and support a unitary model of attentional processing. On the other hand, distractability in various modalities may vary according to specific environmental demands, as suggested by Rumbaugh and McCormack, indicating sensory-specific attentional processing. Comparative findings in this kind of framework are necessary to develop an ecological approach to the study of perception, an approach already evident in the study of learning (Johnston, 1981).

Just as attention to a task is disrupted by the presence of irrelevant cues, attention can be enhanced by emphasis on salient cues or by reduction of irrelevant cues. For example, Smith *et al.* (1976) found that squirrel monkeys acquired discrimination learning-set skills more rapidly when the discriminanda were the reinforcers (i.e., colored food) than when the discriminanda were colored blocks of wood. The authors attribute the superior performance of the experimental group to increased attention to the stimuli that were to be eaten, a biologically more meaningful situation for the monkeys than arbitrary stimuli. This point is emphasized by Menzel and Juno (1982) with another New World species, the brown-headed tamarin (*Saguinus fuscicollis*). These authors observed acquisition of learned visual discriminations in one trial in a naturalistic food-acquisition task.

A common finding with squirrel monkeys is that they are very restless, a characteristic that frequently interferes with (or at least indicates a lack of) concentrated attention to the task at hand. For example, King *et al.* (1968) report that fox squirrels outperformed squirrel monkeys in delayed visual discrimination tasks, apparently because the fox squirrels made efficient use of gross bodily orientation during the delays (up to 25 sec), whereas squirrel monkeys moved around the test box energetically during the delays. It is interesting that darkness during the delay interval interferred with squirrel monkeys' performance on King *et al.*'s (1968) visual discrimination task, even when dim cue

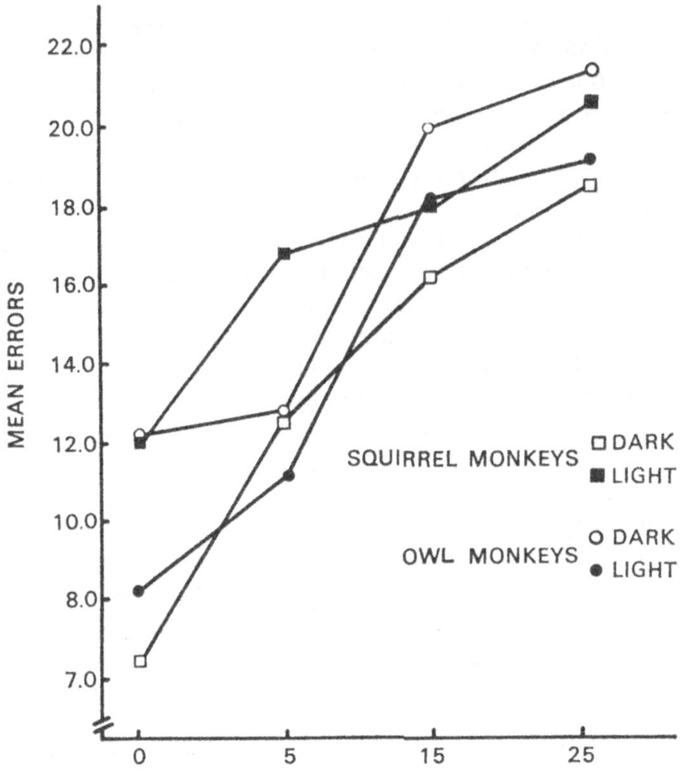

Fig. 1. Mean errors made by squirrel and owl monkeys under dark or light conditions in a delayed response task. Twenty errors equal 50% correct performance. Note that squirrel monkeys' scores are higher in the light; owl monkeys' scores are higher in the dark. [Reprinted from Hornbuckle (1972), with permission.]

lights remained on by the food wells, but that darkness enhanced performance on an auditory discrimination task using similar delay periods (Hornbuckle, 1972) (see Fig. 1). Perhaps the finding that attention to, or memory for, the location of a tone is less dependent on visual conditions than is attention to the location of an object is related to the use of vocal cues by squirrel monkeys in nature to orient toward unseen group members concealed by foliage (Winter, 1968). In line with this notion, Fragaszy (1980) suggested that squirrel monkeys were able to rely on auditory cues to solve a location problem where the subject had to move to a hidden distant goal area containing a familiar (vocal) conspecific. The correlation between naturalistic use of sensory information and attentional/perceptual capacities in different modalities, suggested in one form by Rumbaugh and McCormack (1967, 1969) (see above), crops up again here.

Squirrel monkeys tend to adopt patterned ways of responding when they are not attending to a given task. For example, Fife and Kamback (1970) note that squirrel monkeys often pull the string directly in front of the food reward

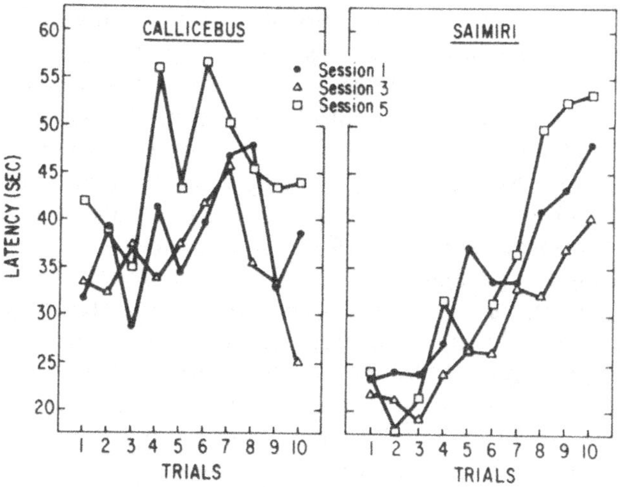

Fig. 2. Mean latency per trial within sessions by squirrel and titi monkeys to move 13.75 m along an above-ground pathway to a goal box. Data are presented for sessions 1, 3, and 5. Note the steep increase in squirrel monkeys' scores versus the erratic pattern of titis' scores. [Reprinted from Fragaszy (1980) with permission of the publisher.]

in patterned string tasks, and adopt position habits rather than using observational techniques, which require longer attention [and also involve tracing strings as they cross (Harris and Meyer, 1971)]. Similarly, Smith *et al.* (1976) note that random responding was associated with rapid acquisition of a discrimination learning set by squirrel monkeys, whereas slower acquisition was associated with greater position preferences expressed in doublets or triplets (i.e., making three responses to left, then three to right, etc.). The positive relationships between random responding and attention to the task has been observed in other species as well. Smith *et al.* (1976) propose that the deviation of responses from a random sequence is one potential behavioral measure of attention in repetitive choice tasks.

Subjects in repetitive choice tasks, however, have other means of displaying disinterest than responding in a position-patterned manner. They can refuse to respond at all. Squirrel monkeys seem prone to this syndrome. Investigators frequently comment (although not often in print) that squirrel monkeys seem difficult to "motivate" (i.e., do not work steadily) in comparison to most Old World species commonly used in laboratory experiments. Fragaszy (1981) found that squirrel monkeys balked (refused to respond within a generous time limit) significantly more frequently on a difficult discrimination task than on an easier task. Titi monkeys (*Callicebus moloch*) tested on the same problems, and experiencing the same dismal levels of success on the more difficult problems, did not exhibit increased balking. Similarly, squirrel monkeys apparently "lost interest" in repetitive travel tasks more rapidly than titis. This is indicated by steep increases in latencies to finish each trial as test sessions progressed for squirrel monkeys, but no systematic increase in latencies for titis (see Fig. 2) (Fragaszy, 1980).

A suggestive study relating attentional attributes and social status has been reported by Clark and Gay (1978). Socially dominant monkeys (from a group of eight young adult males) exhibited different working patterns than submissive animals. Specifically, socially dominant monkeys, although not larger or different in physical appearance, exhibited higher response rates in operant tasks, a steadier capacity to work under varying reinforcement schedules, shorter post-reward procrastination, and greater tolerance for the frustrations inherent in high work-to-reward schedules of reinforcement.

It is inviting to speculate on the relationship among attentional variables implicated in operant behavior and social status suggested by this study. It is reasonable to suppose that the capacity for steady attention might be positively correlated with the ability to follow another monkey's activities closely enough to be able to predict its behavior on the basis of subtle cues. This ability may allow the more dominant monkey to anticipate another's response to its own actions, and thus facilitate its attempts to gain control in social situations. Alternatively (and not exclusively), it is possible that high social status confers upon its possessor a physiological status conducive to sustained attention. Certainly the squirrel monkey's ready endocrine response to external stressors must be considered a physiological correlate of behavior in various test situations (Coe *et al.*, 1982). Perhaps subordinate monkeys are physiologically more labile than dominant monkeys, so that "normal" testing procedures elicit greater stress-related endocrine responses that interfere with optimal performance in them as compared to more dominant animals. Studies of the type reported by Williams *et al.* (1982) on physiological and neuroendocrine responsiveness to cognitive tasks in humans, appropriately adapted for nonhumans, could significantly improve our appreciation of the endocrine correlates of cognitive functioning in squirrel monkeys, and other species as well.

The influence of social setting on squirrel monkeys' attention to experimental tasks has been assessed in a few studies. Social contact with the home group facilitates task-directed behavior and acquisition of an instrumental response in comparison to testing in isolation, presumably because of the distracting effects of isolation (Jarosch, 1969). Isolation from the home group serves as a powerful stressor for squirrel monkeys, as indicated by vocal and behavioral signs of agitation, and by endocrine response (see Coe *et al.*, this volume, Chapter 5). Social contact serves as a powerful reward for animals tested away from the home group (Hupfer and Maurus, 1975).

2.3. Habituation and Sensitization

Habituation is "the elimination of a response as a result of continuous exposure to the stimulus which evoked the response" (Wolman, 1973). It is actually two distinct processes, occurring in short-term (minutes to hours) and long-term (hours to days) temporal modes (Groves and Thompson, 1970). Sensitization is the mirror-image counterpart of habituation, "the process of a receptor becoming more susceptible to a given stimulus" (Wolman, 1973). Both processes can occur simultaneously, providing for precise behavioral adjustment to environmental changes (Petrinovich and Patterson, 1981). The cellular basis for these simple forms of learning has been demonstrated in primitive invertebrates, and it seems likely that similar processes occur in all animals possessing complex nervous systems (Kandel, 1976).

In practical terms, habituation concerns most laboratory investigators as an aim to be achieved rather than as a process to study: Habituation of subjects to testing procedures must occur before experimental testing can begin. This is operationally determined by the return of selected behavioral indices to baseline levels, or by the ready elicitation of normal responses (to food, for example).

A review of the behavioral literature on squirrel monkeys from 1965 to the present turned up only three studies bearing directly on the issue of habituation (Marriott, 1976; Fragaszy and Mason, 1983; Murray and King, 1973), and none with sensitization. These are clearly topics deserving further investigation. The findings at hand allow few interspecific comparisons.

Marriott's (1976) work in visual perception is the most extensive study investigating habituation in squirrel monkeys. In initial studies, squirrel monkeys' cumulative viewing times declined from about 120–130 sec on the first session to 20–50 sec on the tenth session over successive 7.5-min presentations of two meaningful pictorial stimuli (spider, moth; see Figs. 3b and 3c). Terminal viewing times for these stimuli were similar to those for thoroughly familiar "base stimuli" (Fig. 3a). Viewing patterns for a snake stimulus pattern were initially low, then higher, reflecting flight responses followed by alternating approach and avoidance (Fig. 3D). Even to this fear-evoking stimulus, looking times declined by a factor of three (from 150 to 50 sec) over ten presentations. Habituation to a novel, but less complex and meaningless stimulus (a solid blue oblong) occurred more quickly (see Fig. 3E). Interestingly, very familiar "base stimuli" were looked at longer during the first several test sessions in which one novel picture was included in the series than during baseline sessions (compare Fig. 3A with Figs. 3B–3D). Overall, Marriott interprets her data as supporting Nunnally and Lemond's (1973) model of visual investigation, in which duration of looking is a monotonically increasing function of the degree of novelty, and of the degree of stimulus complexity of the visual stimulus. As in the rhesus monkeys studied by Humphrey (1972) and Humphrey and Keeble (1974), the decrease over time in amount and rate of looking depended on the specific content of the picture. In both species, individual looks were of about the same average length (about 6 sec).

Habituation to food stimuli was assessed in a comparative study by Fragaszy (1978; Fragaszy and Mason, 1983). When presented with familiar foods on a repetitive schedule, squirrel monkeys satiated more rapidly than titi monkeys. This was indicated by the number of selections made and the proportion of selected samples consumed by each species (see Figs. 4A and 4B). Differential responsiveness in this study was not due to variation in food preferences in the two species, as can be seen in the species' selection of the same foods in a paired-comparison situation, where different pairs of the same eight food items were presented on every trial (see Fig. 4C). Rapid habituation was also evident in Murray and King's (1973) study of snake avoidance in squirrel monkeys. Latency to retrieve food placed near a live snake (boa constrictor) decreased from >55 sec on day 1 to 15 sec on day 14 in wild-caught adults.

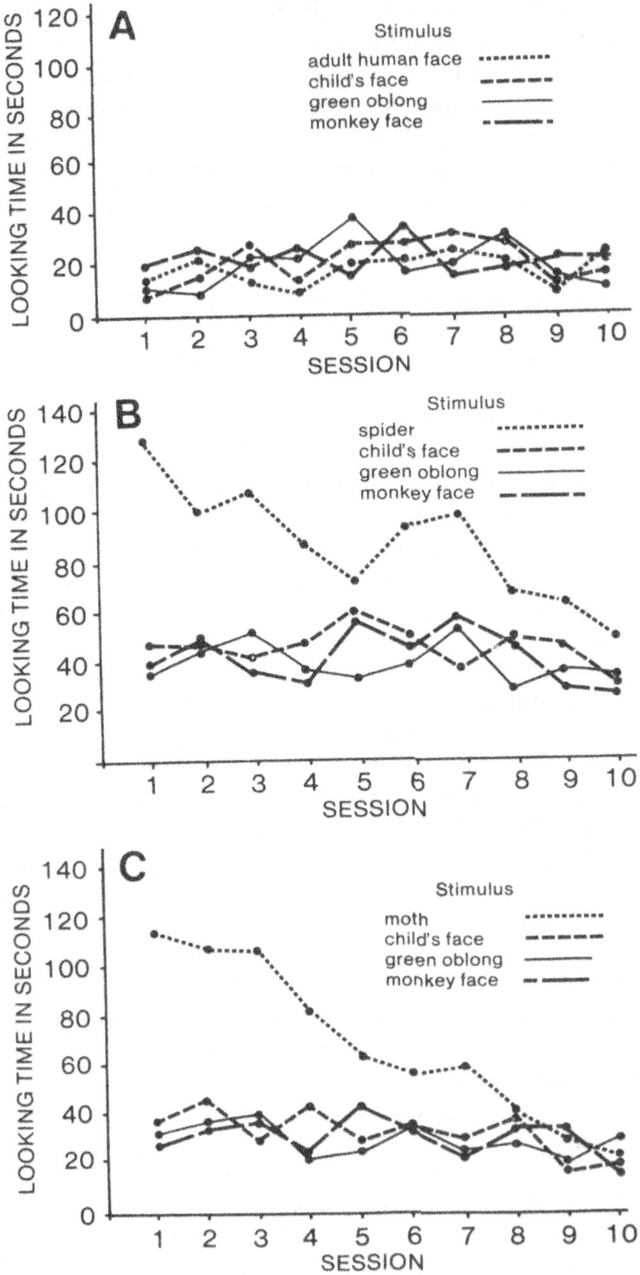

Fig. 3. Looking times (mean of eight subjects) at pictorial stimuli presented in the home cage. (A) Four base (familiar) stimuli across ten sessions, at the end of the training phase. (B–E) Experimental sessions in which (B) a snake, (C) a spider, (D) a moth, or (E) a novel, blue oblong were included in the pictorial stimuli. Each stimulus was presented for 7.5 min per session, in two sessions per day. See text for details. [Reprinted from Marriott (1976), with permission.]

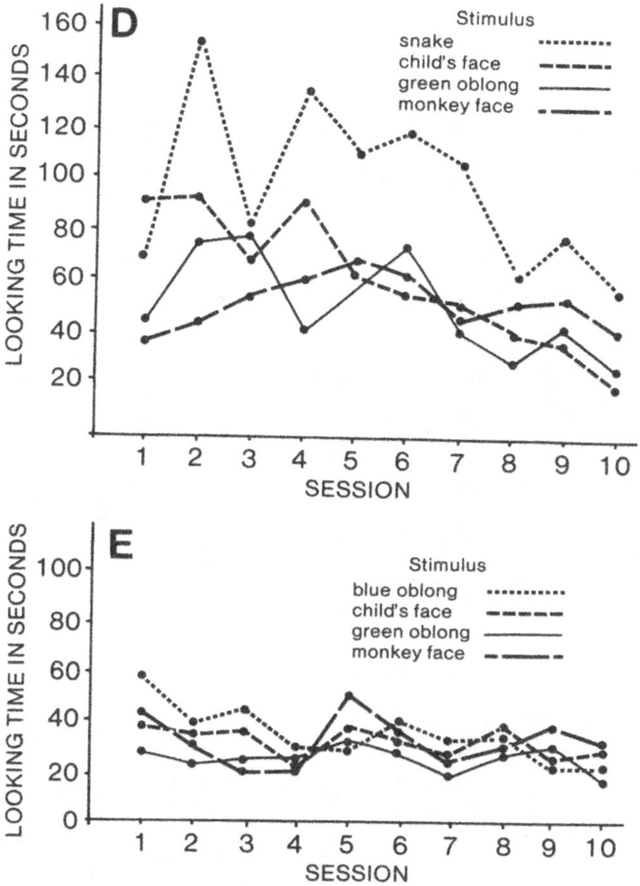

Fig. 3. *(continued)*

2.4. Memory

Like habituation, memory processes in squirrel monkeys have been addressed directly in only a few experiments. King (1971) studied memory via the effects of various testing procedures on learned performance in serial visual discrimination tasks. Increasing the interval between trials from 30 sec to 4 min had no effect on performance of experienced monkeys, indicating working memory skills equivalent to those of rhesus monkeys. Increasing the number of problems presented serially from one to eight produced only a small decrement (8%) in performance. In other words, squirrel monkeys are able to learn eight problems simultaneously almost as efficiently as they learn them successively. They share resistance to intralist interference (that learning one task interfers with learning another concurrently) with rock squirrels (King and Goodman, 1966).

Delayed response tasks are a popular method of assessing short-term memory (D'Amato, 1973). In general, increased delays (up to 2 or 3 min) between

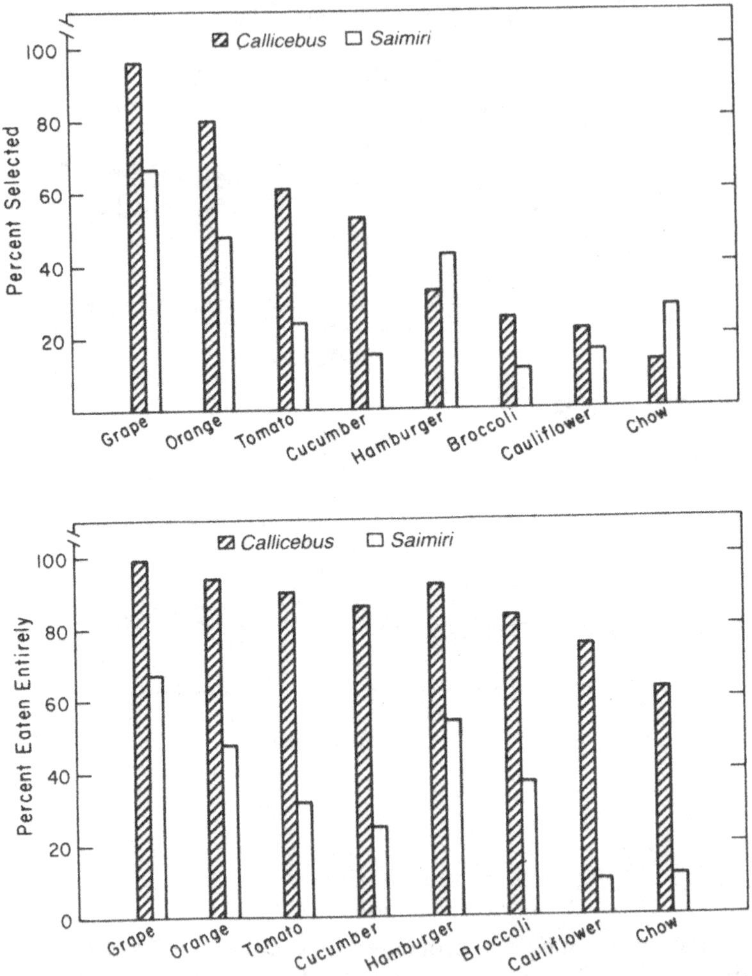

Fig. 4. Results of studies in which (A, B) a single piece of food was presented repetitively or (C, D) two different pieces of food were presented on each trial, to squirrel or titi monkeys. Preferences, as indicated by percentage of selections made of those possible, were quite similar in the two species in the paired comparison study (C). More rapid satiation by squirrel monkeys than titis in the repeated-item setting is evident in lower percentages of selections in comparison to the paired-comparison study (A to C) and in much lower consumption of selected items in squirrel monkeys than titi monkeys (B). [Reprinted from Fragaszy (1978), with permission of the publisher.]

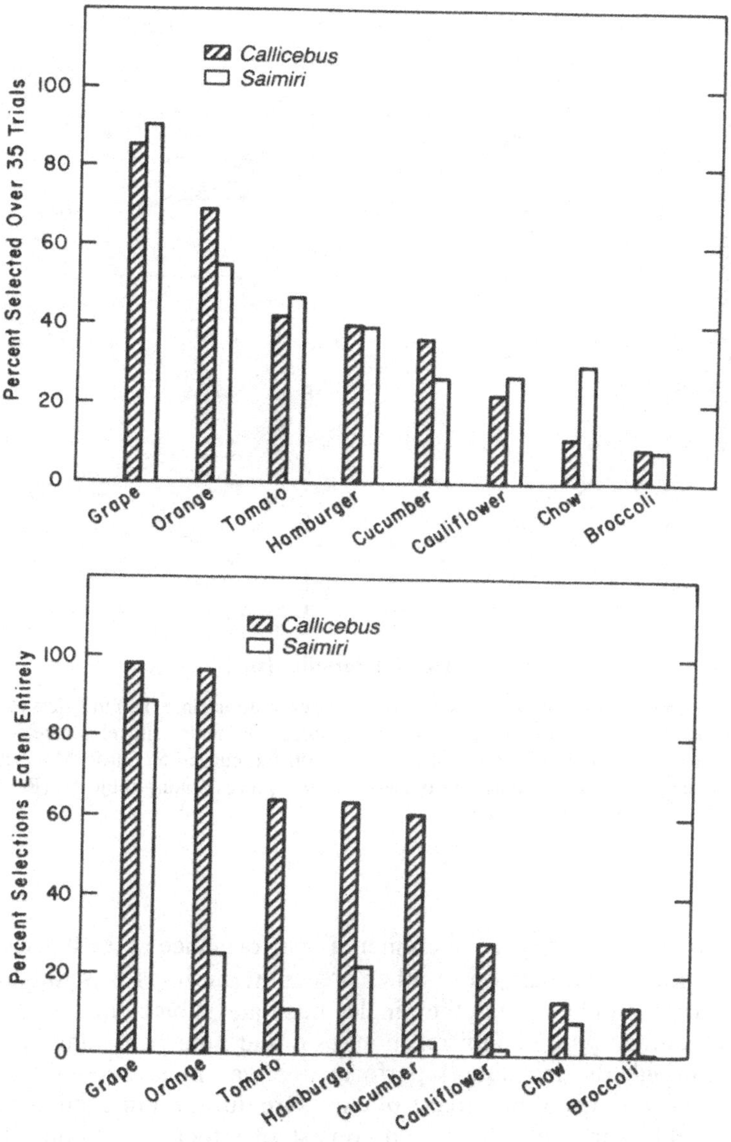

Fig. 4. *(continued)*

stimulus presentation and response opportunities produce decrements in performance (e.g., Hornbuckle, 1972; King *et al.*, 1968; Herzog *et al.*, 1977). However, the duration of sample presentation has been implicated in performance levels. Herzog *et al.* (1977) report that three experimentally naive adult male squirrel monkeys performed less accurately on a delayed matching-to-sample (DMTS) task when the stimulus was presented very briefly. A sample duration of 0.5 sec impaired performance in comparison to longer presentation durations, especially when there was a delay between stimulus presentation

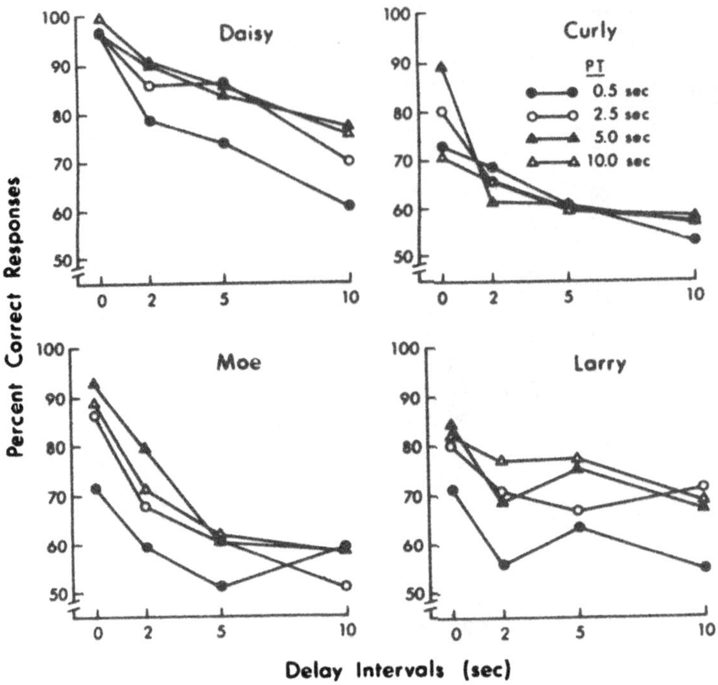

Fig. 5. Retention curves plotted for each length of presentation time (PT) in a delayed matching-to-sample task. Daisy is a macaque; the other three subjects are male squirrel monkeys. Each point on Daisy's curves is based on 128 trials; each point on the curves of Curly, Moe, and Larry is based on 96 trials. Note the similarity in the slope of the curves among subjects. [Reprinted from Herzog *et al.* (1977), with permission.]

and response (see Fig. 5). The decrement in performance at the 0.5 sec duration of presentation in comparison to other presentation times was roughly similar in the squirrel monkeys and the single macaque subject in this experiment (seven percentage points, squirrel monkeys, and nine percentage points, macaque), although the macaque's performance was more accurate overall. In view of other findings of no effects of exposure duration in capuchin monkeys (D'Amato and Worsham, 1972), and consistent effects in pigeons, Herzog *et al.* suggest that experience with brief presentation times in a matching task leads to the development of new memory techniques in monkeys. Pigeons apparently do not have the ability to develop new memory techniques. [See Roitblat (1982) for an extensive theoretical review of representational processes in animal memory.]

Unfortunately, as has been pointed out earlier, performance in delayed response tasks requiring spatial memory is negatively influenced by high levels of motor activity in squirrel monkeys (King *et al.*, 1968). What is needed, it seems, is more work with squirrel monkeys employing a delayed matching-to-sample (DMTS) procedure, where the subject must remember a *stimulus*, not a *place*, so that bodily orientation during the delay period is not relevant to performance (D'Amato, 1973). However, Herzog *et al.*'s (1977) data using the DMTS procedure suggest that memory in squirrel monkeys is on a par with Old World primate species, as represented by a single macaque.

2.5. *Response to Novelty (Curiosity)*

Squirrel monkeys respond to (nonfrightening) changes in their surroundings by approach and/or investigation. Visual responsiveness is particularly prominent (Glickman and Sroges, 1966). However, in a study comparing response to novel objects in the home cage by squirrel and titi monkeys, the species differed in the tempo and mode in which they investigated objects, but not in the overall frequency of exploratory acts (Fragaszy and Mason, 1978). Squirrel monkeys responded very quickly, orienting toward objects as they were being placed in the cage, and often approaching immediately. They also pulled, grasped, and rubbed objects vigorously. In comparison, titis' responses to the same objects were gentler and more frequently involved only chewing and close visual investigation. These differences in style of response to novel objects were mirrored in Visalberghi and Mason's (1983) study of problem-solving in the same species (see below) and in behavior in a novel environment (Fragaszy, 1979).

Response to novelty was assessed in a different way by Fragaszy (1980). Squirrel and titi monkeys were observed in a travel task where the monkeys moved along above-ground runways to a distant goal point. The squirrel monkey's general disinterest in repetitive tasks had been evident in the first two experiments in this setting, and the third experiment was an attempt to capitalize on their apparent preference for variety in travel routes. The monkeys were first trained on a spatially fixed runway system until position habits were developed (Fig. 6A). In subsequent test sessions, novel "shortcuts" were placed, one per test trial, along the fixed runway system. Squirrel monkeys responded to these shortcuts in ways clearly indicating their interest in novel pathways. Titis, although aware of the shortcuts (as indicated by latency data), were more likely than squirrel monkeys to continue to use the familiar "training" path (Fig. 6B). These differences cannot be explained simply in terms of greater effort on the part of squirrel monkeys to minimize travel distance, as results from the previous experiments in the series showed. When no novel paths were available, squirrel monkeys traveled along incomplete runways and then backtracked to reach the goal on about 30% of trials. Titis, on the other hand, became more proficient than squirrel monkeys at minimizing travel distance in this setting, where both paths were familiar to them (Fig. 7).

2.6. *Tempo of Motor Activity*

The squirrel monkey's rapid tempo of motor activity appears to be an integral part of its behavior in many laboratory settings. In studies of manual prehension in squirrel monkeys (where no choice was involved), the average latency from initial visual orientation to subsequent manual contact with an object placed about 15 cm from the subject was 0.47 sec (Fragaszy, 1983). This

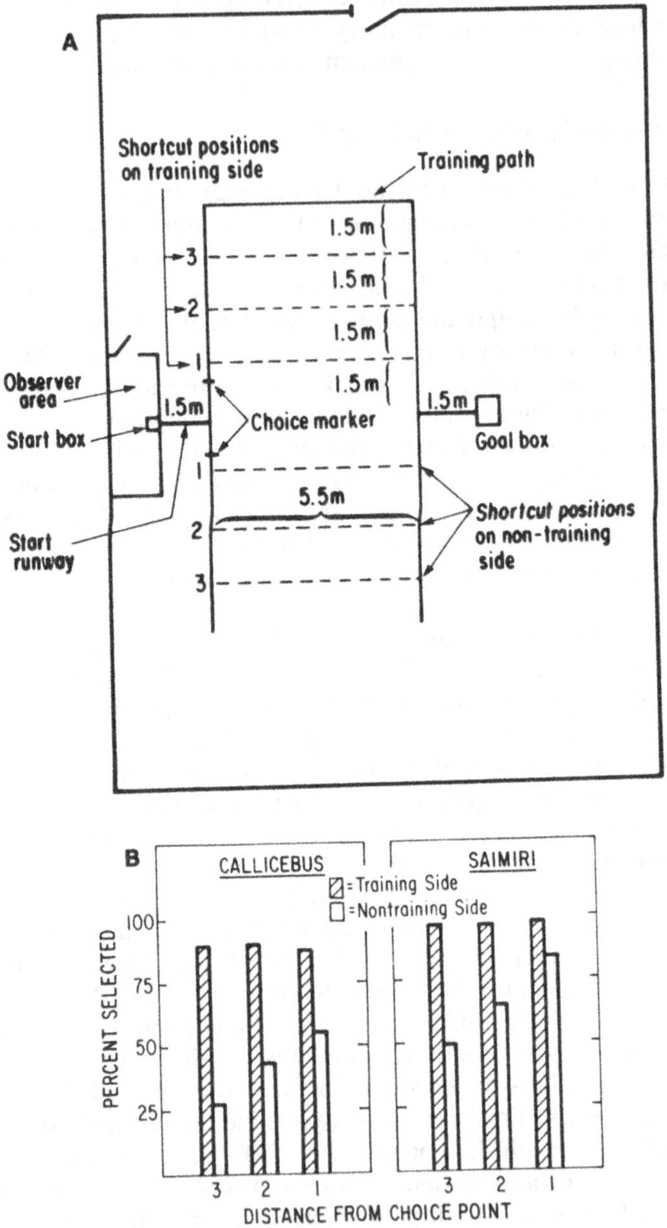

Fig. 6. (A) A sketch of the test apparatus in a path-choice experiment with squirrel and titi monkeys. The training path and its incomplete mirror image were placed in a fixed position and were present on every trial. Only the training path provided a complete path to the goal. A single shortcut at one of three distances from the choice point along either the training side or the non-training side was present on test trials. (B) Percentage of test trials in which the shortcut was selected, plotted for each distance and side of placement. Note that squirrel monkeys used the shortcuts more frequently at all distances on both sides. [Reprinted from Fragaszy (1980), with permission of the publisher.]

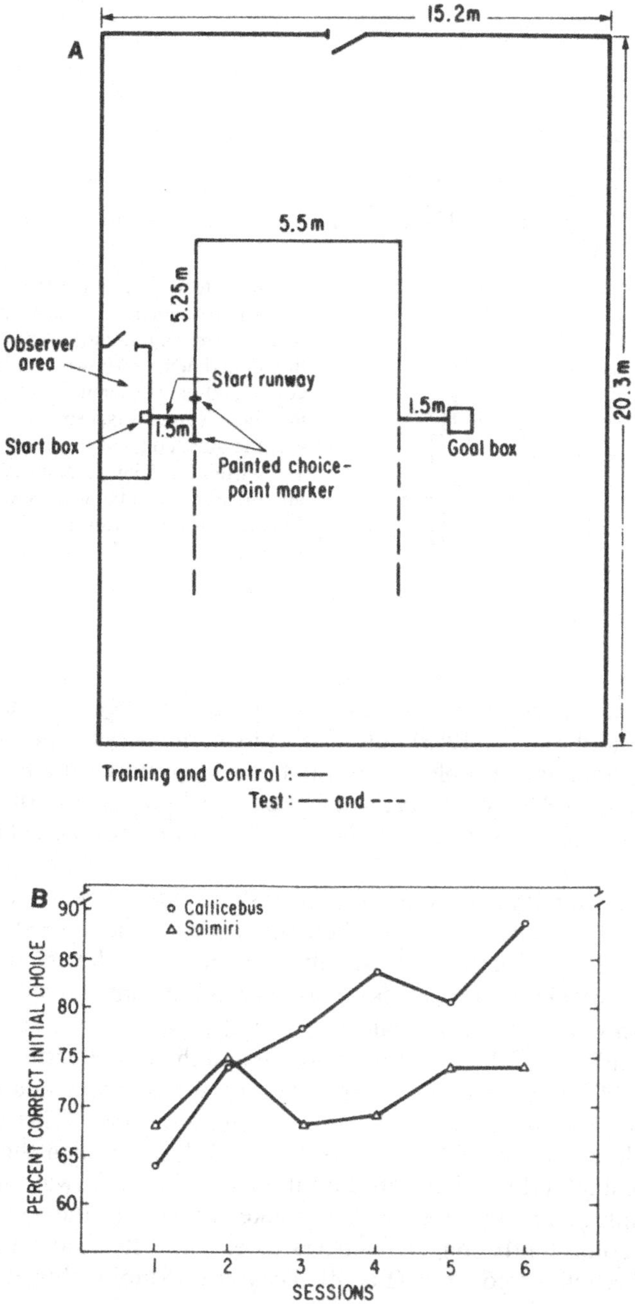

Fig. 7. (A) Sketch of the test apparatus in a path-choice experiment with squirrel and titi monkeys. A single complete path was available during control trials; during test trials an incomplete mirror-image path was also present. Left and right placement of the complete path varied quasirandomly over trials within sessions. (B) Percentage of trials in which the complete path was the initial choice, plotted by sessions. Note the improvement over sessions in titis' scores, but not squirrel monkeys'. [Reprinted from Fragaszy (1980) with permission of the publisher.]

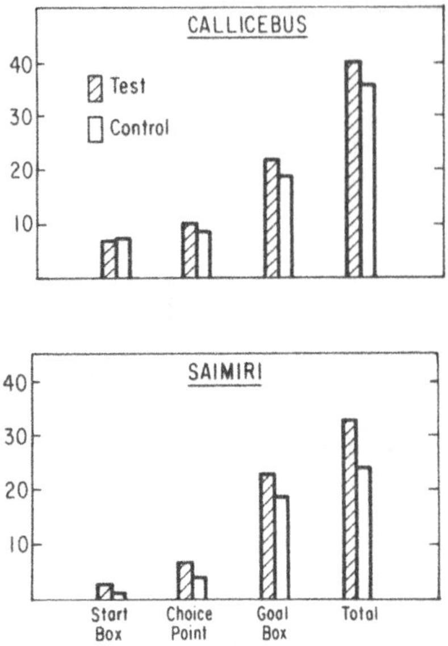

Fig. 8. Mean latency per trial for each of three component latencies in a path-choice experiment with squirrel and titi monkeys. The apparatus for this experiment is shown in Fig. 7A. Note the consistently lower latencies of squirrel monkeys in comparison to titis, especially in the latency to exit the start box and to pass the choice marker 1.5 m from the T-intersection of the start runway and the runway to the goal box. [Reprinted from Fragaszy (1980), with permission of the publisher.]

provides a new perspective on squirrel monkeys' ability to "snatch" objects. Visalberghi and Mason (1983) note that squirrel monkeys approached and contacted experimental "problems" in the home cage very rapidly (within 15 sec on 99.6% and 93.8%, respectively, of trials on which any proximity or contact occurred). In comparison, titis in the same situation approached and contacted problems more slowly.

Squirrel monkeys respond very quickly in most choice situations (if they respond at all), and in some cases their subsequent behavior indicates that they act before fully evaluating the available information relevant to the choice at hand. For example, in travel tasks where the subject moved past a choice point onto a pathway to reach a goal box several meters away, squirrel monkeys often sprinted out of the restraining box, down the initial pathway (1.5 m), and through a "T" intersection, before assessing the runways and making a final determination as to their travel route (Fragaszy, 1980). This occurred even though a delay period of 15 sec was enforced during which the subject could view the test area through a transparent window before it was allowed to leave the restraining box. In contrast, titi monkeys tested in the same experiments exhibited consistently longer latencies to exit the start box, pass the choice point, and reach the goal box (Fig. 8). They also exhibited fewer reversals after a choice had been made.

Squirrel monkeys' rapid choice behavior is also evident in situations in which the choice immediately determined outcome—i.e., in selection of food. In an experiment where subjects chose one of two pieces of food, squirrel monkeys responded more rapidly than titi monkeys (Fragaszy and Mason, 1983; Fragaszy, 1978). However, they also frequently failed to "follow through" on their choices, dropping much of the food they selected. Titis exhibited a very different pattern of making choices more slowly and consuming foods more thoroughly in the same setting (Fig. 4D).

Comparative data on latency to respond in a delayed response task has been reported for squirrel and owl monkeys by Hornbuckle (1972). Under lighting conditions appropriate to each species (light for squirrel monkeys, dark for owl monkeys), squirrel monkeys responded more rapidly than owl monkeys (4.5 versus 5.5 sec at 0-sec delay). Overall, it seems likely that the tempo of activity in squirrel monkeys is more rapid than that of other species in the family Cebidea (capuchins, howlers, spider monkeys, etc.). Casual observation suggests that Callitrichids (marmosets and tamarins) maintain a pace of behavior similar to that of squirrel monkeys, although quantitative data are lacking.

2.7. *Vigor and Frequency of Motor Activity*

In addition to a rapid tempo of activity, squirrel monkeys exhibit high levels of motor activity. This is reflected in a gross way in the activity profiles of several primate species presented by Davis (1974), and is clearly evident in studies where squirrel monkeys have been compared to more sedentary titi monkeys (Mason, 1974; Visalberghi and Mason, in press; Fragaszy and Mason, 1978, 1983; Fragaszy, 1979). For example, in a large enclosure, squirrel monkeys moved more often, and for longer durations per bout, than did titis over the course of 1-hr periods (Fig. 9). They covered much more distance in their travel (Fig. 10).

In comparison to another decidedly active species, tufted capuchins (*Cebus apella*), squirrel monkeys are not as forceful in their manipulative actions. However, they are almost incessantly active and manipulative in a superficial manner in natural environments (Thorington, 1967; Baldwin, this volume, Chapter 2). They are more forceful or vigorous than titi monkeys in their investigations of novel objects or problem situations in the laboratory (Visalberghi and Mason, 1983; Fragaszy and Mason, 1978).

2.8. *Variability of Motor Activity*

Comparative data on variability in gross behavioral patterns are generally lacking. Several studies have indicated greater variability in motor patterns in squirrel monkeys than in titis, particularly when an object is manipulated (Visalberghi and Mason, 1983; Fragaszy, 1979, 1981; Fragaszy and Mason, 1978). Squirrel monkeys typically exhibit relatively great variability in performance on laboratory tasks where repetitive, simple response patterns must be performed (Fragaszy, 1981; Gossette and Kraus, 1968; Nitsch and Jarosch, 1972). Fragaszy (1981) reported that performance by individual squirrel monkeys varied sufficiently across three visual discrimination experiments that interexperiment correlations of performance were close to zero (-0.08 to $+0.09$). Titis, on the other hand, were fairly consistent, as indicated by correlations ranging from $+0.34$ to $+0.60$ for the same comparisons.

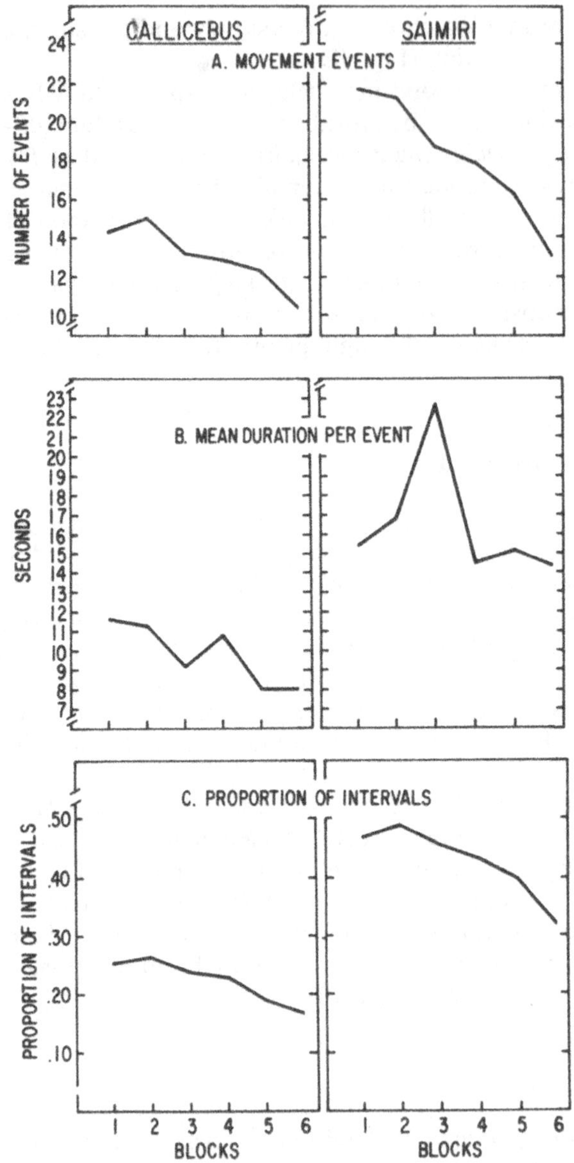

Fig. 9. Average locomotor activity per individual squirrel and titi monkey per 10-min blocks over four 1-hr sessions in a large, novel outdoor enclosure. The enclosure is shown in Fig. 19. (A) Frequency of movement events taken from event-recorder data. Each event was ≥1 sec and occurred after ≥3 sec of locomotor inactivity. (B) Mean duration per movement event. (C) Proportion of instananeous samples at 15-sec intervals in which movement was observed. Note the higher frequency and longer durations of squirrel monkeys' scores in comparison to titis'. [Reprinted from Fragaszy (1979), with permission of the publisher.]

In studies where finer aspects of movement were addressed, squirrel monkeys appear less variable. Three studies have investigated manual dexterity and fine motor coordination during manual prehension or manipulation (Welles, 1972, 1976; Bishop, 1964; Fragaszy, 1983). Welles (1972, 1976) found that squirrel monkeys' performances were better than or equivalent to titis' and talapoins', but less efficient than those of capuchins, rhesus, patas, chimpanzees, and humans on most prehensive tasks, especially those requiring inde-

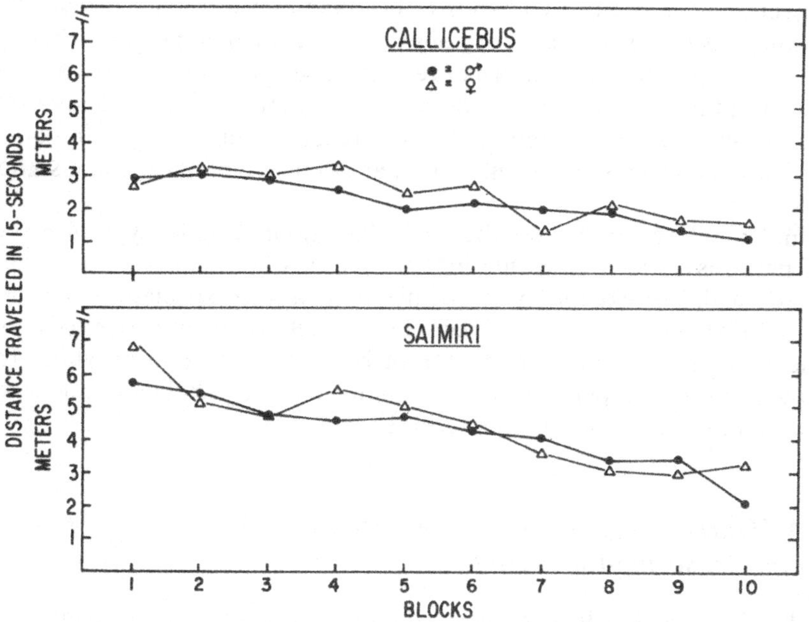

Fig. 10. Average distance traveled by individual squirrel and titi monkeys in each 15-sec interval per 6-min blocks based on point location data taken over four 1-hr sessions in a large, novel outdoor enclosure. The enclosure is shown in Fig. 19. Note the greater distances traveled by squirrel monkeys. [Reprinted from Fragaszy (1979), with permission of the publisher.]

pendent control of the digits or prehension without vision. They were intermediate at tasks requiring rotational guidance of the hand, whole-arm guidance through a point of access, and tactile discrimination. On the other hand, squirrel monkeys were more successful than all other nonhuman primate species except capuchins at prehending a rapidly moving object (a piece of food moving at 40 rpm on the rim of a 10-inch turntable).

Bishop (1964) observed only whole-hand and whole-arm control of objects in squirrel monkeys, although she noted that squirrel monkeys do vary hand positions to part fur while grooming, and that they feel along edges of objects with their digital touchpads. Subsequently, Fragaszy (1983) confirmed the predominance of whole-hand control during prehension in squirrel monkeys, even with moving or small objects. She also demonstrated that interindividual variability on most aspects of direct prehension was slight, in contrast to apparently high levels of variability in gross motor patterns. Variability of hand and arm form during the prehensive sequence was evident between moving and nonmoving objects, and for males, between two small objects (live cricket and a butterscotch chip) and three larger objects (wooden beads of two sizes and crumpled paper). Surprisingly, squirrel monkeys experienced the greatest difficulty in prehending a rhythmically swinging object. Blinking at contact with the object and the use of two hands rather than one were more frequent with moving objects than with stationary objects, whereas looking away while reaching was less frequent. The pattern of results suggests that small moving objects (which in natural circumstances would most likely be potential invertebrate prey capable of escape and/or defensive reaction) elicited greater effort or concentration during prehension, as well as defensive blinking, then stationary objects.

Unfortunately, except for the study just cited, variability *within* specific motor patterns, across or within individuals, has not been systematically determined, with the exception of vocalization, a very specialized motor skill (Schott, 1975). This is clearly a basic, biologically relevant aspect of behavior that needs more attention. Ignorance of behavioral variability within the individual or group hampers analysis of relations between behavior and resource exploitation, for example (Richard, 1981).

2.9. A Heuristic Approach to Interspecific Comparisons of Cognitive and Sensorimotor Attributes

The above review has emphasized a few cognitive and sensorimotor characteristics that are instrumental in the organization of the squirrel monkey's behavior in many settings [see also Fragaszy and Mason (1983)]. These same characteristics are relevant to behavior in other species, too, of course, although the particulars are different. How are we to compare species on these characteristics? Linear comparisons can be made on any particular dimension, but they do not convey the "whole cloth" of behavior. Hierarchical comparisons do not seem particularly appropriate, as the issue here is not whether or not a trait is present, but rather to what extent a trait is present, and how the trait meshes with others to produce a species-typical mode of interacting with the environment. A multidimensional approach is called for. In a multidimensional representation, relationships among dimensions can be discerned, as in a "landscape" of three dimensions. Sternberg and Salter (1982) suggest using the concept of niche space, drawn from ecology, as a heuristic metaphor for the constituents of intelligence in humans. This suggestion has obvious application in interspecific comparisons.

Two such multidimensional representations of behavioral characteristics are shown in Fig. 11. For reasons of pictorial clarity in a two-dimensional medium, Fig. 11 shows two three-dimensional "landscapes," labeled "attentional" and "sensorimotor." These are intentionally *not* task-specific labels. Although the units in each dimension are at present completely hypothetical, it is an empirical matter to develop a standardized unit that could be applied, as has been largely achieved in human psychometrics. For example, proportional performance decrement as a function of delay in a set of tasks could be transformed to a standardized "memory" score; proportional response decrement to a repeated stimulus in a set of situations could provide a standardized "habituation" score, and so forth. The choice of axes and scaling are arbitrary, but the axes should represent theoretically orthogonal characteristics. The scal-

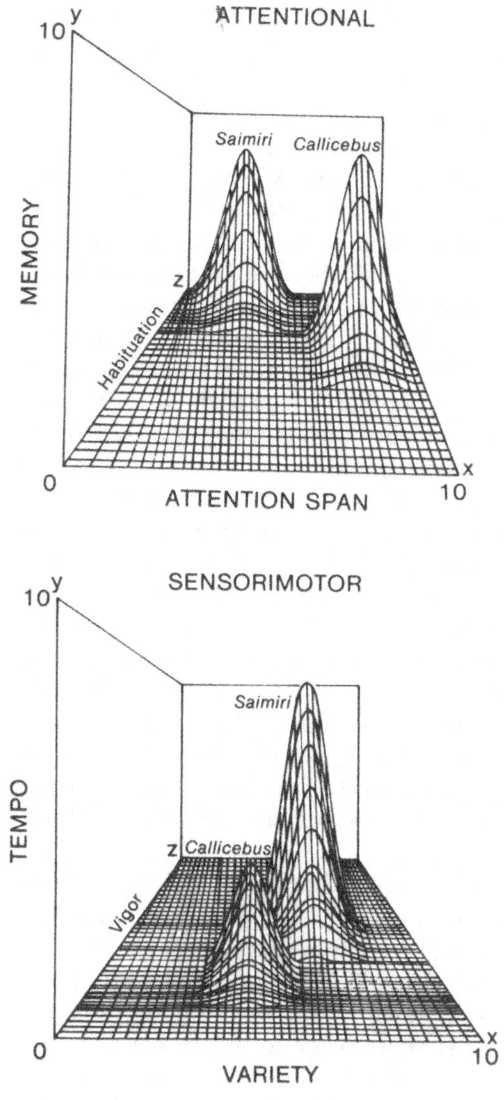

Fig. 11. Representations of hypothetical three-dimensional "landscapes" of cognitive and sensorimotor characteristics in squirrel and titi monkeys. A value of five in each dimension is assumed to represent the median for nonhuman primates.

ing in Fig. 11 assumes a value of five in any dimension to be the norm for nonhuman primates.

Speculative topographies in the landscapes are provided for populations of squirrel and titi monkeys because titi monkeys have been studied in comparable situations with squirrel monkeys more than any other species, and because the author is personally familiar with both species, so that she is comfortable with the occasionally intuitive basis for each species' placement. The form of each species' topography in these landscapes is arbitrarily presented as symmetric about the X and Z axes, a form analogous to actual landscapes. This, too, is a characteristic that can be empirically determined. For example, one might expect bimodal peaks on dimensions where sex differences are evident. Variability may be greater in one dimension than another, producing a "ridge." And the dimension shown on the *Y* axis may also vary in a similar way, rather than semisymmetrically, as shown.

The positions in space occupied by squirrel and titi monkeys in Fig. 11 reflect the findings of many comparative studies with these species conducted by W. A. Mason and colleagues [see Fragaszy and Mason (1983) for review]. These studies have illustrated consistent differences in the two species across many settings, reinforcing the notion that underlying factors are responsible for the distinctive features of each species' behavioral characteristics seen in any particular setting. Overall, titis move more slowly, interact with objects less vigorously, and exhibit less variability in locomotor and investigatory patterns than squirrel monkeys. Titis habituate to novel places or things more slowly, and sustain greater interest in familiar things than squirrel monkeys. They seem to attend to stimuli for longer consecutive periods. On the other hand, the two species appear to have roughly equivalent memory capacities, as indicated in performance on learning tasks (see Section 3).

The reader may recognize this as an outcome-oriented psychometric approach. One limitation of the psychometric approach is that, because it is outcome-oriented, its relationship to current theories of cognition is less direct than are process-oriented approaches (Hunt, 1983). Psychometric assessment of individual differences in human intellectual function also suffers from limitations inherent in the factor-analytic methods that are indispensable to it (Sternberg, 1977). However, given the current state of sophistication in comparative views of cognition, an outcome-oriented, descriptive categorization scheme is an appropriate place to begin. The level of analysis here is not the individual, but the species, vitiating the major criticisms of factor-analytic methods in human psychometrics.

What is the value of this heuristic approach? What does it offer that is not provided by linear or hierarchical schemes? In the author's view, it forces consideration of fundamental species-typical cognitive characteristics that contribute to performances assessed in particular tasks, as, for example, in learning-set experiments. It represents a lower level of analysis. The investigator must judge whether this level or another is appropriate for his or her aims; but it is useful to have the lower level expressly acknowledged. It also provides a means of comparing the range of intraspecific variability, an important factor rarely considered in comparative discussions.

Perhaps the strongest argument for the value of this approach is precisely its task independence. Variations in general behavioral functioning considered in this multidimensional framework are more amenable to evolutionary interpretations than task-specific or capacity-specific frameworks, for two reasons. First, evolution of cognitive capacities has surely not proceeded on a task-specific basis. Changes in the nervous system affect cognitive function only in very general ways, via changed sensory responsiveness, attentional mechanisms, and so forth. Second, we should not expect multiple dimensions necessarily to evolve in matched steps. The concept of dissociation in evolution and development (Gould, 1977) applies to the intricacies of the organization and function of the nervous system as well as to more obvious features of the organism. Task- or skill-specific frameworks are likely to be relatively insensitive to shifts in orthogonal dimensions of cognition unless they have very profound effects on behavior. Placing evolutionary shifts in one dimension in the context of others allows a more sensitive evaluation of the ways in which evolutionary changes in behavior are accomplished.

In the case of squirrel monkeys, the author's evaluation of these characteristics is that the tempo of behavior, interest in novelty, and attention span contribute the most to squirrel monkeys' distinctive patterns of behavior in laboratory tasks. Cooper and Regan (1982) suggest a similar interpretation of attentional and "strategic" factors as substantial determinants of ability differences in humans. In evolutionary terms, these characteristics would seem to require only "minor tuning" of "standard," ancestral cebid traits. The significance of these changes for behavioral ecology cannot be overestimated, however. In conjunction with small body size, these characteristics produce a distinctive active, mobile, surface-searching pattern of behavior that enables squirrel monkeys to fit comfortably somewhere among capuchins, titis, marmosets, and other sympatric primates in the use of their South American habitats. The connections between individual propensities and behavioral ecology are considered in more depth in Section 4.

3. *PERFORMANCE ON SELECTED LABORATORY TASKS*

3.1. *Discrimination and Discrimination-Reversal*

Sensory discrimination is the ability to recognize differences between stimuli; in discriminative learning, the subject learns that responses to one stimulus are reinforced but that responses to another stimulus are not reinforced or are punished (Wolman, 1973). This is a fundamental type of learning (Mackintosh, 1974). It represents one form of "stimulus control," where stimuli and responses covary. Discriminative learning is a "workhorse" task used in many operant conditioning studies, where the analytical focus is actually on the influence of various reinforcement or punishment regimes rather than on discriminative learning *per se*. Studies with squirrel monkeys focusing on reinforcement schedules and the effects of presentation of electrical shock are numerous; the reader is referred to the *Journal of Experimental Analysis of Behavior* for most of these. These studies usually suffer from a lack of ecological validity and at this point do not contribute to our understanding of the role of cognition in natural circumstances.

Many studies of discrimination learning in primates have a different focus than the operant emphasis described above. In the hands of Harlow (1949) and Rumbaugh (1974) in particular, discrimination learning studies contributed significantly to our understanding of comparative learning skills in nonhuman primates [see also Warren (1965, 1974)]. Harlow (1949) demonstrated that rhesus monkeys could "learn how to learn" with repeated experience solving similar discrimination problems, and subsequent work has shown that many primate and nonprimate species can acquire "learning sets" [see Hodos (1970),

Passingham (1981), and Warren (1974) for reviews]. Learning set is discussed in more detail below (Section 3.2).

Two recent studies have compared squirrel monkeys to other New World monkeys in discrimination tasks. Hornbuckle (1972) reported that squirrel monkeys and owl monkeys (*Aotus trivirgatus*) did not differ in the number of trials (with corrections allowed) required to reach criterion levels of performance on an auditory discrimination task (a 2800-cps tone versus a 4800-cps tone, both at 95 dB). Subjects of both species required about 400 trials to achieve 16 out of 20 successive correct choices, indicating relatively great difficulty in comparison to object discrimination (see below). The decrement induced by the addition of white noise at about 75–80 dB to the test chamber during the delay between stimulus presentation and response opportunity was equivalent for the two species (e.g., from about 80% correct to 70% in the 0-sec delay condition, and from about 52% correct to 48% in the 25-sec delay condition).

The performance of squirrel and titi monkeys on object, planometric (two-dimensional), and color discrimination problems was evaluated by Fragaszy (1981). The relative difficulty of these tasks for the two species was in line with findings from studies with other primate species (Warren, 1974), and much lower than that of the auditory discrimination task used by Hornbuckle (1972). Object discriminations were easier than planometric discriminations for both species. Variation in the size of the colored area affected performance in a learned red–green discrimination in similar ways in the two species: Reducing the area of the color cue on a 6.4 × 7.6 cm surface from 100% to 20% was associated with a decline of about ten percentage points in correct choice in squirrel monkeys, and about six percentage points in titis. Both species showed similar increases in correct performance as a function of experience with the problems, although titis' scores improved more slowly than squirrel monkeys' (see Fig. 12). The differential rate of improvement was associated with significantly higher cumulative levels of performance on object and color discriminations by squirrel monkeys than titis. However, the fact that titis' scores were still improving when testing terminated led Fragaszy to suggest that the two species probably shared similar discriminative abilities, but that the squirrel monkeys' more rapid adjustment to testing regimens gave them an advantage in this situation.

Rumbaugh (1968) proposed the transfer index as a means of comparing learning ability ("intelligence") based upon performance in discrimination learning tasks. In essence, the transfer index allows comparisons of the amount and direction of transfer from a discrimination task to its reverse (e.g., in a color discrimination task, the animal first learns red correct, green incorrect; subsequently it must learn red incorrect, green correct). The index avoids comparing absolute levels of performance, thus gaining generality in interspecific comparisons [see also Warren (1974)]. It provides an assessment of an individual's ability to form response "strategies" that apply to similar tasks. On this measure, squirrel monkeys exhibit very strong "negative transfer"; that is, they perform less well on the reversal task than on the original discrimination problem (Rumbaugh, 1968; but see Gossette and Kraus, 1968). Talapoins (*Miopithecus talapoin*, a small-bodied African monkey) are similar to squirrel

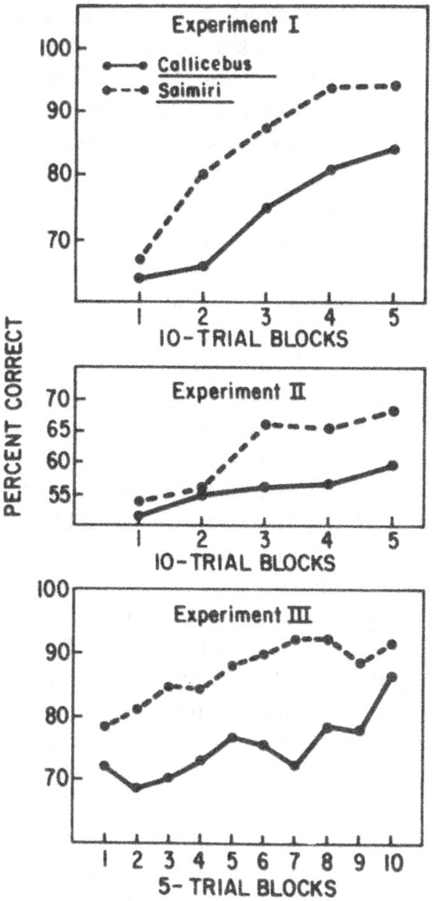

Fig. 12. Percentage of correct response over blocks in three visual discrimination experiments with squirrel and titi monkeys. The plotted points for experiments I and II represent average performance over five different problems per successive ten-trial blocks. Discriminanda were objects in experiment I and planometric patterns in experiment II. The points for experiment III represent average performance per five-trial blocks on the *same* problem (red–green discrimination) with the color cue covering varying areas of the stimulus surface (100% to 20% at 20% increments). [Reprinted from Fragaszy (1981), with permission of the publisher.]

monkeys on this measure [compare Rumbaugh (1968) and Rumbaugh (1974)]. Apes and macaques, on the other hand, show positive transfer, earning higher success scores for the reversal problem than for the original problem (Rumbaugh, 1974). This characteristic translates into greater behavioral flexibility in response to changing circumstances in apes and macaques than in squirrel monkeys and talapoins, a conclusion in line with our impressions of intellectual capacity in these primates.

Studies comparing species on successive reversals of discrimination tasks (SDR, successive discrimination reversal) have indicated that squirrel monkeys perform at intermediate levels on these tasks. For example, Gossette *et al.* (1968) found that squirrel monkeys were not significantly different from racoons, skunks, coatis, and kinkajous on these tasks, but were inferior to capuchins (*C. apella*) and cacomistles. Gossette and Slonim (1969) proposed that performance similarities between owl and squirrel monkeys and performance differences between squirrel and capuchin monkeys on SDR tasks provide support for the notion that *Saimiri* should be placed taxonomically with the

Aotinae rather than the Cebidae, as Hill (1960) suggested. No other investigators, to my knowledge, have taken a revisionist approach to taxonomy on the basis of performance on a learning task. Indeed, the approach is fraught with potential absurdities. Consider the implication of the following study: Treichler *et al.* (1970) report that squirrel monkeys performed more like cats than rhesus monkeys on SDR tasks involving varied reinforcement probabilities for the two choices (that is, rather than choice of an object being consistently reinforced or not reinforced, a 70:30 probability ratio, for example, was used). In this kind of task, squirrel monkeys, like macaques, eventually learn to choose consistently the more frequently reinforced alternative. However, they require many more training trials than macaques on the less disparate ratios. In this sense, their performance was more similar to that of cats than that of rhesus monkeys. One cannot entertain the notion that degree of phylogenetic relatedness is expressed in extent of similarity of performance when interpreting these findings. Indeed, Fragaszy (1981), after observing marginal differences in success on discrimination tasks in squirrel and titi monkeys, but great differences in behavior during testing, suggested that interspecific differences in discrimination tasks are more profitably viewed in some instances as the consequence of different underlying behavioral propensities than in terms of phylogenetic relations. As has been pointed out, general behavioral propensities are associated with broad features of behavioral ecology, allowing a different framework for interspecific comparisons. Rumbaugh's (1968, 1974) findings on transfer index scores for squirrel monkeys and talapoins, which are similar in many features of behavioral ecology and gross behavior, provide a case in point.

3.2. Concept Learning

Concept learning is demonstrated when an individual's performance can be shown to be guided by an internal "strategy" that transfers from one stimulus array to another. This is a qualitatively different kind of learning than stimulus-specific learning (Thomas, 1980). Discriminative learning set is the type of concept learning that has received the most attention in a variety of species, partly because of the procedural similarities to familiar discrimination testing, and partly because of the simplicity of the concept. To develop a discriminative learning set, the subject receives at least two trials in each of successive two-choice discrimination problems. As the "learning set" develops, the subject becomes increasingly more accurate as a consequence of adopting the response rule, "If A, not B," or "If B, not A." Primates in general do particularly well on these kinds of learning tasks (Warren, 1974).

Early work with squirrel monkeys demonstrated that they could acquire a discriminative two-choice learning set more readily than marmosets but less readily than cebus or spider monkeys, and much less readily than Old World monkeys or apes (Fig. 13) [reviewed in Rumbaugh (1968)]. One must be careful to make direct comparisons only among studies using the same procedures, as it is known that different numbers of trials or training procedures (i.e., to criterion versus a set number of trials; see below) affect the rate of acquisition of learning sets (Passingham, 1981). Figure 13 presents data on several mam-

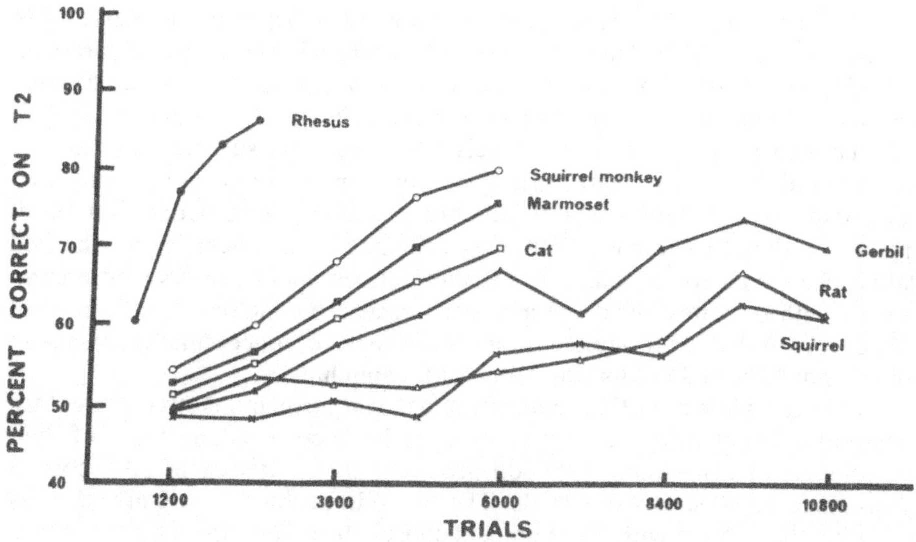

Fig. 13. Percentage correct on trial 2 (T2) as a function of the number of trials in discrimination learning-set tasks in different mammals. Each problem was presented for six trials; thus, 1200 trials represent 200 problems, etc. [Reprinted from Passingham (1981), with permission.]

malian species obtained in studies where six trials were presented on each of numerous problems. It should be noted that the six-trial procedure is not the most effective way to produce learning-set acquisition (see below), so that the data presented in Fig. 12 are not necessarily the "best" performance of these species. The plotted points represent performance on the *second* trial as a function of the number of trials (or divided by six, the number of problems) the animal has received. The subject has a 50:50 chance of being correct on the first trial of a new problem, but if the learning set is perfectly acquired, it should always be correct on the second and subsequent trials. Squirrel monkeys' performance peaked at about 80% correct on trial two after 1000 problems. However, it appears from the curves that performance levels could be improved above that achieved after 1000 problems in squirrel monkeys, marmosets, and cats.

Once acquired, learning-set skills are stable, so that months after initial training the learning set is still evident. However, various factors influence their acquisition, such as stimulus variables or whether successive problems are presented for an arbitrary number of trials or until a criterion of level of performance is reached (Warren, 1974). Squirrel monkeys, like other species, apparently do better with the latter procedure, especially when the task is difficult (Rumbaugh, 1968; Devine, 1970).

Discrimination learning sets represent one type of concept learning. Other forms of concept learning (e.g., oddity, absolute class, relative class, and numerousness) have been ingeniously and systematically studied in squirrel monkeys by Thomas and colleagues (Czerny and Thomas, 1975; Noble and Thomas, 1970; Thomas and Boyd, 1973; Thomas and Crosby, 1977; Thomas and Chase,

1980; Thomas *et al.*, 1980; Thomas and Ingram, 1979; Thomas and Kerr, 1976; Thomas and Peay, 1976). Thomas's research program is based upon a paradigm of learning as consisting of a hierarchy of qualitatively different abilities ranging from habituation (level 1) to complex concepts (levels 6–8) (Thomas, 1980, 1982). Thomas proposes that three levels of concepts are sufficient to describe conceptual abilities in primates: class concepts (including absolute and relative); conjunctive, disjunctive, and conditional concepts; and biconditional concepts. According to Thomas (1980), only primates have conclusively exhibited relative class conceptual learning. Further refinements of interspecific comparisons with primates await the necessary research. Thomas *et al.*'s current research emphasis is on demonstrating the range of conceptual processes in squirrel monkeys and comparing them with capuchins.

Noble and Thomas (1970) demonstrated that squirrel monkeys readily learn a one-odd concept (triangle correct; two circles incorrect) and three of their four subjects also mastered two-odd problems to a criterion of 90% correct performance on 40 successive trials. The two-odd problems were arranged by presenting the subject with three objects chosen at random from a pool of four identical pairs of objects—in this case, black or white crosses or diamonds. In any trial the subject received two objects of one form and one of the other, all forms of equal brightness (white or black). Comparison of the monkeys' performance on two-odd problems and learning-set training in oddity problems showed incomplete transfer from the two-odd problems. That is, performance of learning-set oddity problems was initially lower than that on the two-odd problems. This suggests that the monkeys did not use oddity alone as the basis for their responses in the two-odd problems. Acquisition of an oddity learning set, which all of Noble and Thomas' subjects achieved, is a more stringent criterion of oddity learning. In a later study, Thomas and Boyd (1973) demonstrated that capuchins (*Cebus albifrons*) and squirrel monkeys did not differ in their performance on oddity learning set, oddity reversal, or one-trial oddity tasks.

In a recent paper on the topic of oddity learning, Thomas and Kerr (1976) expand on the methodological requirements necessary to infer concept learning ("conceptual conditional discrimination"), rather than specific configuration learning, in oddity tasks. Their three-part training sequence involved oddity learning-set problems, five trials per problem, to criterion (36 out of 40 trials correct), then one-trial oddity problems to criterion (18 out of 20 correct) using a gray stimulus tray (on which objects were placed and presented to the subject), and finally conditional discrimination problems where tray brightness was the conditional cue, again to a criterion (18 of 20 correct). A white tray signaled that choice of the odd object would be rewarded; a black tray signaled that choice of either of the non-odd objects would be rewarded. New oddity problems were used on each trial, as in the previous one-trial oddity training procedure. The results from three subjects are shown in Fig. 14. Thomas and Kerr suggest that extensive oddity-problem experience by these subjects 2 years earlier [the same subjects were used in Thomas and Boyd's (1973) study; see above] contributed to their success in this series of tasks. In any case, this methodology is an appropriate means to provide conclusive demonstration of

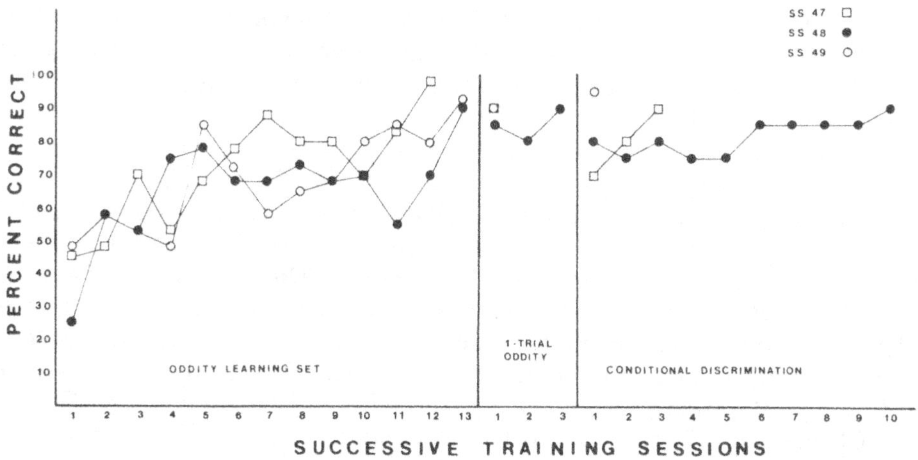

Fig. 14. Percent correct responses for each of three squirrel monkeys on three different conceptual tasks presented consecutively, plotted as a function of successive training sessions per task. Note that one subject passed criterion on the conditional discrimination task on the first session, and two subjects passed criterion on the first session of one-trial oddity problems. [Reprinted from Thomas and Kerr (1976), with permission].

conceptual conditional discrimination, as learning of specific stimuli is precluded.

Thomas and Crosby (1977) demonstrated that squirrel monkeys have the ability to learn both absolute and relative class concepts using the following tasks: An "absolute" discrimination was required, depending on the color of the stimulus tray. For example, when the tray was white, green was correct; when it was black, non-green was correct. A "relative" discrimination task was similar in design: When the tray was white, the odd object was correct, but when it was black, the even object was correct. All of the subjects could do these, but absolute judgments were easier, in the sense that fewer trials were required to reach criterion levels of performance on these as compared to relative judgments.

McGonigle and Jones (1978) summarized studies in which they evaluated the differences in perception of relative and absolute classes in squirrel monkeys using a different method than Thomas and Crosby's (1978) study. Their method involves varying the task requirements, and consequently the criteria necessary to evaluate stimuli, across subjects. Thus, some subjects must make relational judgments such as "larger than," while others must respond to specific cues such as a certain size or brightness, over a wide variety of stimulus conditions. Their results indicate that the relational code "larger or largest" was the most dififcult to disrupt once it had been learned, and thus they feel it represents the most "primitive" encoding form, which is used *unless* the task requirements demand more. However, different types of relational responses are not all equally easy. In fact, the concept of middleness was not grasped by McGonigle and Jones' squirrel monkeys. McGonigle and Jones

(1978) maintain that nonhuman perceivers, like human perceivers, "have some discretion as to the level of processing which they may carry out on a given stimulus" (p. 658). In other words, task requirements can change the way stimuli are encoded.

Space does not permit reviewing in depth other recent studies on conceptual learning in squirrel monkeys. The interested reader may refer to Thomas and Chase (1980) and Thomas *et al.* (1980) for studies of "numerousness" judgments [i.e., "one," "some," "all" (Thomas and Chase, 1980), or "less than" Thomas *et al.* (1980)]; Thomas and Peay (1976) for a study of conservation of length; Thomas and Ingram (1979) for a study of conceptual judgment of relative volume; and McGonigle and Chalmers (1977) for a study of transitivity of choice, that is, inference (i.e., train with A > B, B > C, C > D, D > E; test on B ? D). Squirrel monkeys succeed on all of these tasks. McGonigle and Chalmers (1977) point out that squirrel monkeys perform as well on transitivity tasks as do 4-year-old children.

Altogether, this body of literature leads to three conclusions. First, it illustrates the broad usefulness of discrimination learning as a procedure by which to probe other kinds of cognitive abilities. Second, it conveys forcefully that formal reasoning is not beyond the squirrel monkey's capacities, when the tasks are presented in designs carefully chosen for their clarity. Third, it supports the view that squirrel monkeys' intellectual capacities are in line with those of other monkeys, and in particular are apparently quite similar to capuchins'. Precise comparisons of squirrel monkeys and other species must await further comparative studies like those of Thomas and colleagues. Squirrel monkeys are now a "baseline" species in this kind of work against which other species' performances can be judged. This area of research with squirrel monkeys is thus likely to remain vigorous for some years to come.

3.3. Problem-Solving

Studies of problem-solving are inherently more open-ended than the discrimination learning or concept learning studies discussed above. In a problem-solving task, animals are left alone to arrive at an acceptable solution to a sensorimotor task, such as opening a container to retrieve food. In this type of study, the way a solution is developed is often more informative than what form it takes. Such studies have a long and venerable history in comparative psychology [see Riopelle (1967) for a historical anthology]. However, only two recent studies with squirrel monkeys have used this approach.

Davis and Leary (1968) compared squirrel monkeys with six other species on bent-wire problems, where the subject had to move a Lifesaver candy off a wire on which it had been threaded (Fig. 15). Squirrel monkeys are not particularly good at these tasks, performing less well than all other primate species used (Fig. 16). Davis and Leary (1968) note that squirrel monkeys were not responsive in this situation. Of seven subjects, one never responded, two never succeeded, another gave up early in training, and the other three balked ("failed to attempt a solution") on at least three-quarters of their trials. Nevertheless, Davis (1974) suggests that bent-wire detour problems tap an aspect of behav-

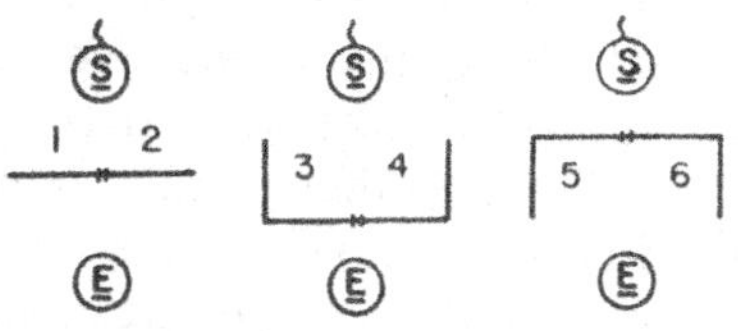

Fig. 15. Detour patterns used in bent-wire problems. A Lifesaver candy was threaded on the rod above or below the number that identifies the pattern. Odd-numbered patterns are to the left of center; even-numbered to the right. [Reprinted from Davis and Leary (1968), with permission].

ioral flexibility that is differentially present across species. In particular, he suggests that the learning that occurs in experience with bent-wire problems is motor learning, not perceptual learning. Inappropriate movements (such as pulling the candy) must be inhibited, and appropriate movements (including pushing the candy away) must be strenthened.

Squirrel monkeys are no doubt more capable of motor learning than suggested by Davis and Leary's (1968) study. For example, McGonigle and Flook (1978) found that squirrel monkeys could learn to use a specific hand to manipulate an object in a discrimination task (e.g., use left hand for color X, right hand for color Y). This learning represents the development of "limb-localized action schemes" (p. 97), an essential part of skilled manipulative motor sequences.

Squirrel monkeys are better able to deal with detours and barriers in a larger spatial scale than in bent-wire problems, as is evident in their performance in an experiment where the barrier (a missing section of runway) occurred several meters distant, but in full view, from the choice point (Fig. 8A) [Fra-

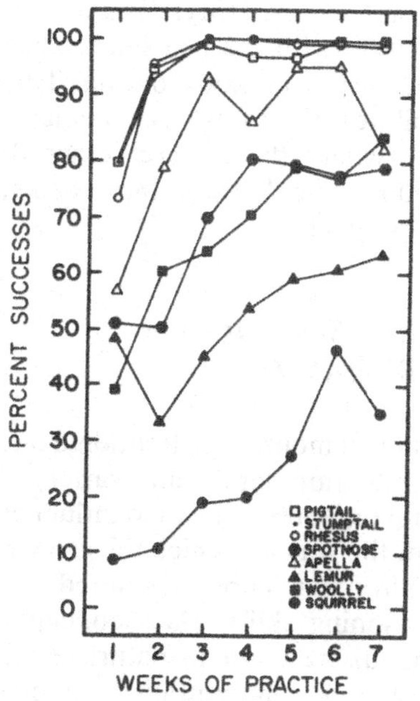

Fig. 16. Performance of eight species of nonhuman primates on bent-wire detour problems presented by Davis and Leary (1968). Note the squirrel monkeys' poor performance on these tasks. [Reprinted from Davis (1974), with permission.]

gaszy (1980), experiment II]. As noted above, in this setting, squirrel monkeys chose the complete runway first on about 70% of trials. Lack of improvement over test sessions in squirrel monkeys, compared to significant improvement by titis in the same task, was attributed to the squirrel monkeys' relative disregard for travel distances in comparison to titis. On the other hand, squirrel monkeys were less affected than titis by changes in runway configurations from one trial to the next (e.g., left path complete, right incomplete, to left incomplete, right complete). Overall the pattern of results suggested that titis make more effort to minimize travel distance than squirrel monkeys, but are more adversely affected by variable spatial configurations. The squirrel monkeys' ready use of novel shortcuts further reinforces this view (see p. 69).

Visalberghi and Mason (1983) have compared squirrel and titi monkeys in a series of sensorimotor problems. The problems required tipping over, lifting, or pulling an object, inserting a hand through a paper barrier, or extracting an object from behind a wire mesh in order to recover a food item. Squirrel monkeys were more successful at solving these problems than titis. Squirrel monkeys obtained the food in 86% of trials in which any contact occurred, versus 66% of trials for titis. They also solved them more rapidly (in an average of 20 sec versus 39 sec). The authors do not attribute this result to differing cognitive abilities. They suggest instead that the two species exhibited equivalent cognitive capacities in dealing with the problems. For example, both species occasionally reached directly for a food item that was fully visible behind a transparent barrier. These erroneous direct reaches occurred in the two species at equal percentages (about 37%) of all 15-sec interval samples in which the subject contacted the objects. Both species showed comparable reductions in the relative frequency of this response over sessions.

Why, then, were squirrel monkeys more successful? The authors suggest that problem-solving "style" is the most relevant factor. "Style" is a constellation of characteristics pertaining to interaction with the problems, such as speed of approach, force of manipulation, and variety of motor output. Titis were not handicapped by motor or cognitive abilities; they were as capable as squirrel monkeys of making precise responses where these were required. Nevertheless, titis were less successful than squirrel monkeys on these tasks, because they adopted a less vigorous style of approach.

4. *EXPRESSION OF COGNITIVE AND SENSORIMOTOR CHARACTERISTICS IN EVERYDAY BEHAVIOR*

The above review has suggested that for squirrel monkeys, attentional and sensorimotor characteristics (attention span, habituation, tempo and variety of motor activity, etc.) appear to underlie behavioral organization and to influence performance on laboratory tasks. In this section, the ways in which these characteristics contribute to behavior in natural environments are considered.

It is more difficult to translate formal reasoning skills (class concepts, inference, etc.) discussed earlier into spontaneous behavior in natural environments. Some steps in this direction have been made in naturalistic studies concerned with foraging behavior and location of food resources, although these studies do not deal with strictly conceptual skills (e.g., Kamil, 1978; Menzel, 1973). As Thomas (1982) notes, conceptual skills are not necessarily used in natural environments in ways that are tapped in laboratory tasks, and these skills may have been acquired in the absence of *direct* selection pressures. They are better viewed as selectively neutral characteristics than as "adaptations" until demonstrated otherwise (Gould and Lewontin, 1979). Although they are of evolutionary interest in their own right, it is still the case that developing "ecological validity" for conceptual learning skills would greatly broaden their interest to the behavioral and biological community. At the moment, the strongest statement to be made is that squirrel monkeys' species-typical behaviors do not reflect conceptual abilities that are unique to them.

4.1. Use of Space

Baldwin (this volume, Chapter 2) reports home range data for squirrel monkeys. Variability in home range size is extensive, from about 15 ha to more than 100 ha. Daily path length is also quite variable, from 0.6 km to more than 4 km (Baldwin, this volume, Chapter 2). This compares with an average daily path length of 0.6 km for titis (Mason, 1968) and 0.4 km for howler monkeys (Milton, 1980). In general, squirrel monkeys range more widely than sympatric *Cebus* and *Callicebus* groups (Thorington, 1968; Mason, 1968).

Wide use of space in squirrel monkeys was also evident in a laboratory study of squirrel and titi monkeys in a large outdoor enclosure containing an above-ground runway system (Fig. 17) (Fragaszy, 1979). Five heterosexual pairs of monkeys of each species were released, one pair at a time, into this enclosure for four 1-hr sessions over an 8-week period. As can be seen in Fig. 18, both species were highly selective in their use of space. A highly clustered distribution is evident for each species, but the species demonstrated very different preferences. Titis spent much more time than squirrel monkeys on the above-ground runway system; squirrel monkeys spent more time on the ground, especially along the western edge of the enclosure. The species also differed in the way their use of preferred areas changed over sessions, as they became more familiar with the enclosure (Fig. 19). Titis restricted their travels to a small area; squirrel monkeys (especially males) began to move into more areas of the cage. Squirrel monkeys traveled more than 1 km per hourly session, on average; titis traveled about half that distance. The data clearly suggest a trend toward wider ranging in squirrel monkeys and the restriction of ranging in titis, consistent with each species' pattern in natural environments. It appears that squirrel monkeys' high levels of motor activity, interest in novel places, and/or rapid habituation are primary contributors to their ranging patterns in nature.

4.2. Feeding and Activity Patterns

In natural environments, squirrel monkeys forage opportunistically for small invertebrate prey and fruit, and they appear to prefer insects to fruit.

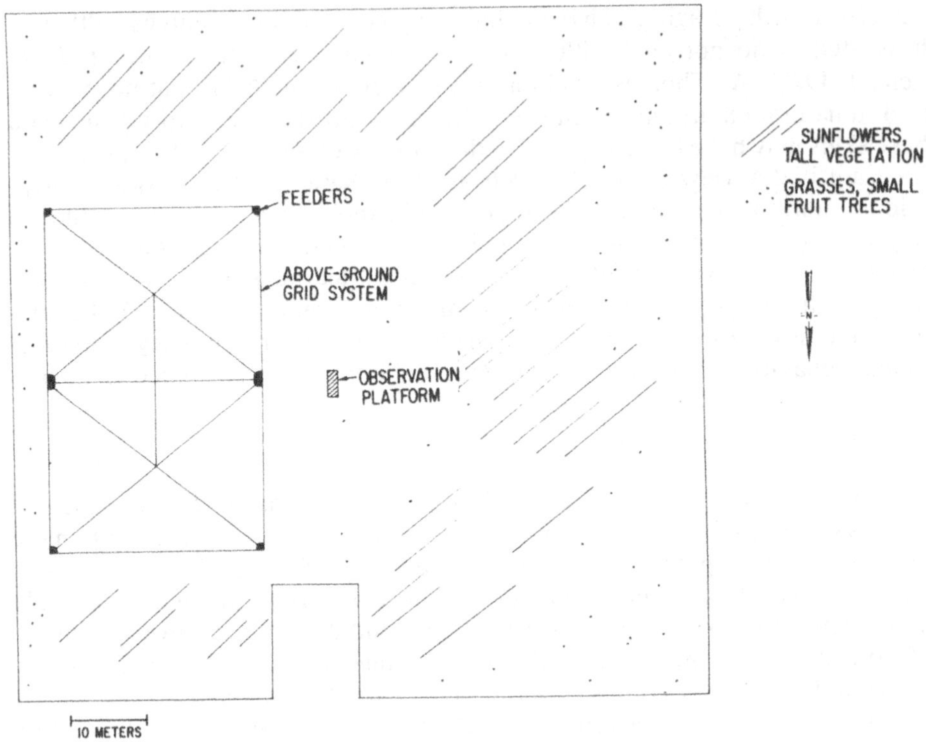

Fig. 17. A sketch of a large outdoor enclosure used as a release site in a study of squirrel and titi monkeys in a novel environment. The roof of the enclosure was 10 m high. The walls and roof were constructed of poultry wire supported by telephone poles. The enclosure was planted with grass, corn, sunflowers, and young fruit trees. A 49 m × 29 m above-ground grid of 2.5 cm × 10 cm lumber, with six feeders placed along the perimeter, was present on one side of the enclosure. No other artificial runways or paths were present. Prior to release sessions, the feeders were filled with desirable food items. [Reprinted from Fragaszy (1979), with permission of the publisher.]

Foraging is the predominant activity for much of the day [see Baldwin (this volume, Chapter 2) for review]. Yet squirrel monkeys remain only briefly in any particular tree (Thorington, 1967, 1968). In captive studies, these traits have obvious counterparts. For example, squirrel monkeys "forage" in many settings, even when that is not the task the experimenter has set for them (e.g., Fragaszy, 1980). "Curiosity," attraction to novel places and things, is evident here. High levels of motor activity are also important in the squirrel monkeys' foraging behavior. When in a novel outdoor enclosure (see Section 4.1) squirrel monkeys foraged actively, and obtained small items at a low, but steady rate over each 1-hr session. In the home cage, in the presence of abundant food, squirrel monkeys rapidly approach and take several pieces of food, but then seem to lose interest in their choices rapidly. They often drop most of their selections and return immediately to the food dish to make new ones (Fragaszy, 1978, Fragaszy and Mason, 1983; Leger *et al.*, 1981). Titis in this same setting are more thorough consumers. Interspecific differences in attention span

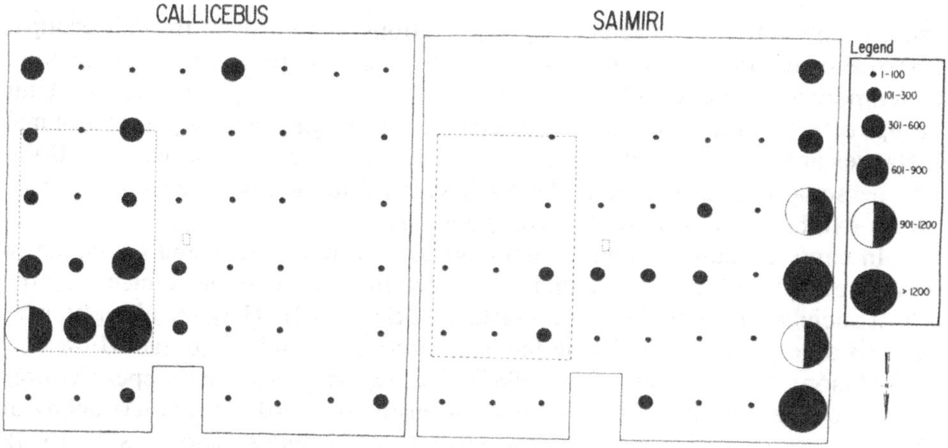

Fig. 18. Net occupancy of 169-m^2 quadrants per species for squirrel and titi monkeys over four 1-hr trials in a large enclosure, based on point locations of each subject at 15-sec intervals. Note the pronounced differences in preferred locations in the two species. [Reprinted from Fragaszy (1979), with permission of the publisher.]

and/or rate of satiation (probably both) are clearly implicated here, as in natural feeding patterns.

4.3. Group Cohesion and Dispersion

Attraction to conspecifics can be viewed as a special case of attraction to familiar objects. Baldwin (this volume, Chapter 2) notes that large *Saimiri*

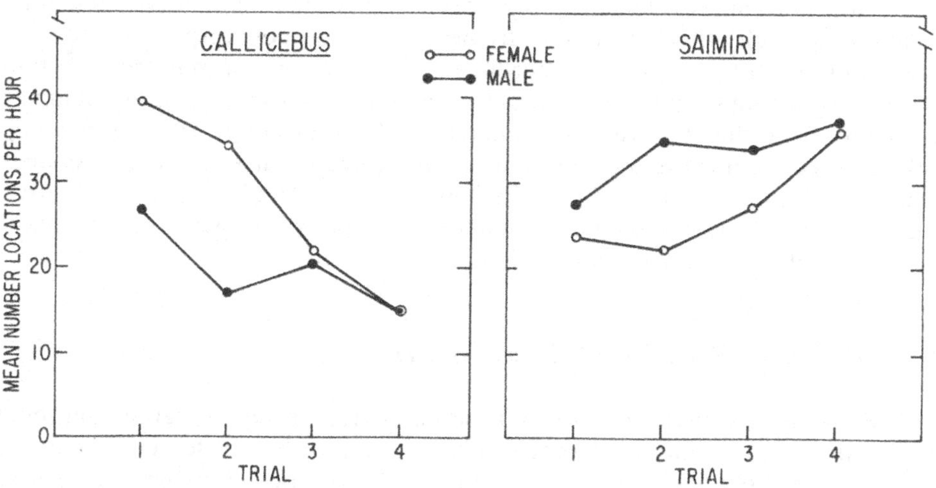

Fig. 19. Average number of different 3.75-m^2 locations entered per 1-hr trial by squirrel and titi monkeys in a large enclosure, derived from point location data taken at 15-sec intervals. Note the opposing trends in the two species. [Reprinted from Fragaszy (1979), with permission of the publisher.]

groups may actually be temporary aggregations of two or more unit groups. Intertroop relations range from simple avoidance where home ranges overlap to temporary formation of mixed groups or even mixed-species groups. Unit groups often split into smaller subgroups during foraging, and larger interanimal distances are maintained during foraging than while resting or moving (S. Boinski, personal communication). Overall, squirrel monkeys in natural environments exhibit relatively fluid social groupings.

In captive studies, squirrel monkeys are strongly motivated to remain in visual contact with familiar animals, as noted in several studies concerning the use of social reinforcers in operant tasks (see Section 3). They are also attentive to each other's activities during feeding, although avoiding contact (Fragaszy and Mason, 1983; Leger *et al.*, 1981). Yet squirrel monkeys appear almost oblivious to an opposite-sex pairmate in terms of coordinating their behavior when access to a resource is not at issue (e.g., Fragaszy and Mason, 1978), and when there is room, they travel independently of one another (Fragaszy, 1979). Squirrel monkeys are less tolerant toward a familiar cagemate than titis when access to a restricted resource is at stake. For example, in observations of feeding in squirrel and titi monkeys in the home cage when food was abundant, male squirrel monkeys routinely responded aggressively when their female pairmates approached the food dish. Positive social interactions were almost nonexistent. Titis, in contrast, although exhibiting occasional flares of aggressive behavior, also exhibited intimate positive interactions while feeding, including passive sharing of food. Providing a single piece of food rather than a full dish did not appreciably alter these contrasts (Fragaszy and Mason, 1983). Similar results have been obtained in studies where the restricted resource is a novel object or a food-retrieval problem (Fragaszy and Mason, 1978; Visalberghi and Mason, 1983).

Squirrel monkeys' tendency to avoid social contact, and at the same time to attend to each other's feeding activities, have their counterparts in the fragmentation of unit groups into subgroups and the spacing of individuals within subgroups during foraging activities. Presumably, such spacing allows animals to benefit from the discovery of abundant food sources (e.g., fruiting trees) by others while minimizing competition for more evenly distributed and smaller "packages" of invertebrate foods (Eisenberg *et al.*, 1972; Izawa, 1979). In fact, direct competition among squirrel monkeys for food is extremely rare in natural environments (S. Boinski, personal communication).

5. *TOPICS FOR FUTURE RESEARCH*

While reviewing the literature for this chapter, it became apparent that some topics in cognition have not received the attention they deserve. This is the case not only for squirrel monkeys, but for other species as well. I wish to close this review by briefly mentioning these topics. First, the relationship between social skills and cognition has been largely overlooked. We know very little about the cognitive requirement of various social tasks, such as monitoring current social status or evaluating the probability of friendly acceptance to social overtures. Yet there is no doubt that sociality requires a good deal of cognitive sophistication (Mason, 1982). Second, we know next to nothing about observational learning in primates, although it is surely a prominent type of learning in natural environments, especially among visually-oriented, social, manipulative primates. Restriction of attention to single-subject situations in studies of learning unduly limits our assessment of the potential for adaptive behavioral change. Furthermore, studies of observational learning in monkeys that have been conducted have frequently required perceptual learning [e.g., Mahan and Rumbaugh (1963) with squirrel monkeys] rather than sensorimotor learning, although observational learning in natural environments surely concerns the latter as often as the former. Third, fine motor skills and their relation to cognition have been generally ignored. Yet sensorimotor skills are critical elements in problem-solving situations, especially in the use of tools. They contribute to learned performance in the real world as importantly as do strictly conceptual acquisitions. Their importance to behavioral flexibility in humans is recognized by the common Piagetian phrase, "sensorimotor intelligence." Davis (1974) suggests, for example, that a species' ability to learn the motor adjustments necessary for solution of bent-wire problems is related to its ability to deal with perceptual transformations imposed by prism distortions of the visual field or by disrupted muscular sensations. These conditions, like bent-wire problems, require that the learner sort out information in complex visually guided interactions with objects.

6. CONCLUSION

The field of cognition, including studies of learning, is enjoying a period of unprecedented progress. Our knowledge of squirrel monkeys is richer because of this emphasis. We have moved well beyond attempts to place squirrel monkeys on a hypothetical phylogenetic scale. We recognize that underlying predispositions determine to a large extent how a squirrel monkey (or any species) will respond in a given experimental setting and we recognize that either interspecific comparisons must take into account these predispositions, or comparisons must be carefully arranged so that the effects of these predispositions do not bias the outcome.

When all qualifying statements have been made, we are presented with a diverse set of results indicating that squirrel monkeys are acutely aware of their surroundings, and very likely to act, and act rapidly, to investigate objects or places directly if they are attracted to them. On the other hand, squirrel monkeys are easily distracted and they have limited abilities to alter their manipulative patterns. Memory, perceptual, and conceptual abilities in squirrel monkeys are not particularly distinctive; we may expect these fundamental aspects of cognition to be evolutionarily conservative traits in the primates. However, the possibility of ecological correlates of interspecific differences in perceptual skills has yet to be addressed. More refined comparisons of cognitive skills across species require further research.

Primatologists are now in a position to try to bridge the gap from cognitive processes to two other topics of interest in the behavioral sciences: social behavior and sensorimotor skills. Studies of squirrel monkeys should contribute to this effort, as they present an array of typically primate physiological, motor, and social characteristics, and as we already have knowledge of their behavior in many laboratory settings. We should look forward to continued productive study of squirrel monkeys' cognitive attributes.

ACKNOWLEDGMENTS. The research by the author reported here was supported by an NSF predoctoral fellowship and a University of California Regents predoctoral fellowship, Grant RR-00169 from NIH, and Grant MH-33198-01 from the National Institute of Mental Health. Preparation of this report was supported in part by a Grant-in-Aid of Research from the San Diego State University Foundation. Judith Lawson provided very important bibliographic assistance.

REFERENCES

Abordo, E., and Lee, J., 1977, Effect of introducing novel configurational cues on a learned color discrimination in squirrel monkeys, *Percept. Mot. Skills* **44**:147–151.

Bishop, A., 1964, Use of the hand in lower primates, in: *Evolutionary and Genetic Biology of Primates* (J. Beuttner-Janusch, ed.), Academic Press, New York, pp. 133–225.

Campbell, H. J., 1972, Peripheral self-stimulation as a reward in fish, reptile, and mammal, *Physiol. Behav.* **8**:637–640.

Candland, D. K., Blumer, E. S., and Mumford, M. D., 1980, Urine as a communicator in a New World primate, *Saimiri sciureus*, *Anim. Learn. Behav.* **8**:465–480.

Clark, D. L., and Gay, P. E., 1978, Behavioral correlates of social dominance, *Biol. Psychiatry* **13**:445–454.

Coe, C. L., Franklin, D., Smith, E. R., and Levine, S., 1982, Hormonal responses accompanying fear and agitation in the squirrel monkey, *Physiol. Behav.* **29**(6):1051–1057.

Cooper, L. A., and Regan, D. T., 1982, Attention, perception, and intelligence, in: *Handbook of Human Intelligence* (R. J. Sternberg, ed.), Cambridge University Press, Cambridge, pp. 123–169.

Coss, R. G., 1970, The perceptual aspects of eye-spot patterns and their relevance to gaze behavior, in: *Behaviour Studies in Psychiatry* (S. J. Hutt and C. Hutt, eds.), Pergamon, Oxford, pp. 121–147.

Coss, R. G., 1972, Eye-like Schemata: Their Effect on Behaviour, Dissertation, University of Reading (unpublished).

Czerny, P., and Thomas, R., 1975, Sameness–difference judgments in *Saimiri sciureus* based on volumetric cues, *Anim. Learn. Behav.* **3**:375–379.

D'Amato, M. R., 1973, Delayed matching and short-term memory in monkeys, in: *The Psychology of Learning and Motivation: Advances in Research and Theory*, Volume 17 (G. H. Bower, ed.), Academic Press, New York, pp. 227–269.

D'Amato, M. R., and Worsham, R. W., 1972, Delayed matching in the capuchin monkey with brief sample durations, *Learn. Motiv.* **3**:304–312.

Davis, R. T. 1974, Monkeys as perceivers, in: *Primate Behavior*, Volume 3 (L. A. Rosenblum, ed.), Academic Press, New York, pp. 1–263.

Davis, R. T., and Leary, R. W., 1968, Learning of detour problems by lemurs and seven species of monkeys, *Percept. Mot. Skills* **27**:1031–1034.

Davis, R. T., Leary, R., Smith, M. D. C., and Thompson, R. J., 1968, Species differences in the gross behaviour of nonhuman primates, *Behaviour* **31**:326–338.

Delay, E. R., and Isaac, W., 1980, The effects of illumination, d-amphetamine, and methylphenidate upon vigilance performance of squirrel monkeys, *Bull. Psychon. Soc.* **15**:203–206.
Devine, J. V., 1970, Stimulus attributes and training procedures in learning set formation of rhesus and cebus monkeys, *J. Comp. Physiol. Psychol.* **73**(1):62–67.
Eisenburg, J. F., Muckenhirn, N. A., and Rudran, R., 1972, The relation between ecology and social structure in primates, *Science* **176**:863–874.
Epple, G., 1974, Olfactory communication in South American primates, *Ann. N.Y. Acad. Sci.* **237**:261–278.
Fife, D. D., and Kamback, M., 1970, The ability of squirrel monkeys to solve randomly presented string problems, *Psychon. Sci.* **18**(3):131–132.
Fragaszy, D. M., 1978, Contrasts in feeding behavior in squirrel and titi monkeys, in: *Recent Advances in Primatology*, Volume 1 (D. J. Chivers and J. Herbert, eds.), Academic Press, London, pp. 363–367.
Fragaszy, D. M., 1979, Squirrel and titi monkeys in a novel environment, in: *Captivity and Behavior* (J. Erwin, T. Maple, and G. Mitchell, eds.), Holt, Rinehart, and Winston, New York, pp. 172–216.
Fragaszy, D. M., 1980, Comparative studies of squirrel monkeys (*Saimiri*) and titi monkeys (*Callicebus*) in travel tasks, *Z. Tierpsychol.* **54**:1–36.
Fragaszy, D. M., 1981, Comparative performance in discrimination learning tasks in two New World primates (*Saimiri sciureus* and *Callicebus moloch*), *Anim. Learn. Behav.* **9**:127–134.
Fragaszy, D. M., 1983, Preliminary quantitative studies of prehension in squirrel monkeys (*Saimiri sciureus*), *Brain Behav. Evol.* **23**:81–92.
Fragaszy, D. M., and Mason, W., 1978, response to novelty in *Saimiri* and *Callicebus*: Influence of social context, *Primates* **19**:311–331.
Fragaszy, D. M., and Mason, W., 1983, Comparative studies of feeding in captive squirrel and titi monkeys, *J. Comp. Psychol.* **97**:310–326.
Glickman, S., and Sroges, R., 1966, Curiosity in zoo animals, *Behaviour* **26**:151–188.
Gossette, R. L., and Kraus, G., 1968, Successive discrimination performance of mammalia species on a brightness task, *Percept. Mot. Skills* **27**:675–678.
Gossette, R. L., and Slonim, P., 1968, Spatial SDR performance across three species of New World monkeys, *Psychon. Sci.* **14**:222–223.
Gossette, R. L., Kraus, G., and Speiss, J., 1968, Comparison of successive discrimination reversal (SDR) performances of seven mammalian species on a spatial task. *Psychon. Sci.* **12**:193–194.
Gould, S. J., 1977, *Ontogeny and Phylogeny*, Belknap Press of Harvard University Press, Cambridge, Massachusetts.
Gould, S. J., and Lewontin, R. C., 1979, The spandrels of San Marcos and the Panglossian paradigm: a Critique of the adaptationist programme, *Proc. Roy. Soc. Lond. B* **205**:581–598.
Green, S., 1975, Auditory sensitivity and equal loudness in the squirrel monkey (*Saimiri sciureus*), *J. Exp. Anal. Behav.* **23**:255–264.
Groves, P. M., and Thompson, R. F., 1970, Habituation: A dual process theory, *Psychol. Rev.* **77**:419–450.
Harris, D. G., and Meyer, M. E., 1971, The relationship between visual acuity and performance on patterned string problems by infrahuman primates, *Psychon. Sci.* **22**:160.
Harlow, H., 1949, The formation of learning sets, *Psych. Rev.* **56**:51–65.
Haude, R. H., and Ray, O. S., 1974, Visual observing behavior in the squirrel monkey, *Anim. Learn. Behav.* **2**:138–140.
Hennessey, M. B., Coe, C. L., Mendoza, S. P., Lowe, E. P., and Levine, S., 1978, Scent-marking and olfactory investigatory behavior in the squirrel monkey (*Saimiri sciureus*), *Behav. Biol.* **24**:57–67.
Herzog, H., Grant, D., and Roberts, W., 1977, Effects of sample duration and spaced repetition upon delayed matching-to-sample in monkeys (*Macaca arctoides* and *Saimiri sciureus*), *Anim. Learn. Behav.* **5**:347–354.
Hill, C. O., 1960, *Primates*, Volume IV, *Cebidae*, Edinburgh University Press, Edinburgh.
Hodos, W., 1970, Evolutionary interpretations of neural and behavioral studies of living vertebrates, in: *Neurosciences*, Volume 12, (F. O. Schimdt, ed.), Rockefeller Univeristy Press, New York, pp. 26–39.

Hornbuckle, P. A., 1972, Delayed-response performance as a function of sensory stimulation in the squirrel and owl monkey, *J. Comp. Physiol. Psychol.* **79**:99–104.

Humphrey, N. K., 1972, "Interest" and "pleasure": Two determinants of a monkey's visual preferences, *Perception* **1**:395–416.

Humphrey, N. K., and Keeble, G. R., 1974, The reaction of monkeys to "fearsome" pictures, *Nature* **251**:500–502.

Hunt, E., 1983, On the nature of intelligence, *Science* **219**:141–146.

Hupfer, K., and Marus, M., 1976, Operant conditioning of the squirrel monkey with social reinforcement, *Naturwissenschaften* **62**:42–43.

Isaac, W., 1969, The influence of illumination upon the temporal patterning of responses in squirrel and owl monkeys, *Psychon. Sci.* **14**:243–244.

Izawa, K., 1979, Foods and feeding behavior of wild black-capped capuchins (*Cebus apella*), *Primates* **20**:57–76.

Jacobs, G. H., 1972, Increment-threshold spectral sensitivity in the squirrel monkey, *J. Comp. Physiol. Psychol.* **79**:425–431.

Jacobs, G. H., 1973, Visual sensitivity in the squirrel monkey, *Am. J. Phys. Anthropol.* **38**:371–376.

Jacobs, G. H., 1977, Visual sensitivity: Significant within-species variations in a non-human primate, *Science* **197**:499–500.

Jarosch, E., 1969, Social contacts and "task-directed" activity in the squirrel monkey, in: *Proceedings of the Second International Congress of Primatology*, Volume 1 (C. R. Carpenter, ed.), Karger, Basel, pp. 137–142.

Johnston, T. D., 1981, Contrasting approaches to a theory of learning, *Behav. Brain Sci.* **4**:125–173.

Johnston, T. D., 1982, Learning and the evolution of developmental systems, in: *Learning, Development and Culture* (H. C. Plotkin, ed.), Wiley, New York, pp. 411–442.

Kahneman, D., 1973, *Attention and Effort*, Prentice-Hall, Englewood Cliffs, New Jersey.

Kamil, A. C., 1978, Systematic foraging by a nectar-feeding bird, the amakihi (*Loxops virens*), *J. Comp. Physiol. Psychol.* **92**(3):388–396.

Kandel, E. R., 1976, *Cellular Basis of Behavior*, Freeman, San Francisco.

King, J. E., 1971, Determinants of serial discrimination learning by squirrel monkeys, *Learn. Motiv.* **2**:246–254.

King, J. E., and Goodman, R. R., 1966, Successive and concurrent discrimination by rock squirrels and squirrel monkeys, *Percept. Mot. Skills* **23**:703–710.

King, J. E., Flaningham, M., and Rees, W., 1968, Delayed response with different delay conditions by squirrel monkeys and fox squirrels, *Anim. Behav.* **116**:271–275.

Kintz, B. L., Foster, M. S., Hart, J. O., O'Malley, J. J., Palmer, E. L., and Sullivan, S. L., 1969, A comparison of learning sets in humans, primates, and subprimates, *J. Gen. Psychol.* **80**:189–204.

Leger, D., Mason, W. A., and Fragaszy, D. M., 1981, Sexual segregation, cliques, and social power in squirrel monkeys (*Saimiri*) groups, *Behaviour* **76**:163–181.

Mackintosh, N. J., 1974, *The Psychology of Animal Learning*, Academic Press, London.

Mahan, J. L., Jr., and Rumbaugh, D. M., 1963, Observational learning in the squirrel monkey, *Percept. Motor Skills* **17**:686.

Marriott, B. M., 1976, Picture Perception in Squirrel Monkeys (*Saimiri sciureus*), Ph.D. Dissertation University of Aberdeen (University Microfilms International #7810247).

Mason, W. A., 1968, Use of space by *Callicebus* groups, in: *Primates: Studies in Adapatation and Variability*, Holt, Rinehart and Winston, New York, pp. 200–216.

Mason, W. A., 1971, Field and laboratory studies of social organization in *Saimiri* and *Callicebus*, in: *Primate Behavior: Developments in Field and Laboratory Research*, Volume 2 (L. A. Rosenblum, ed.), Academic Press, New York, pp. 107–137.

Mason, W. A., 1974, Comparative studies of social behavior in *Callicebus* and *Saimiri*: Behavior of male–female pairs, *Folia Primatol.* **22**:1–8.

Mason, W. A., 1975, Comparative studies of social behavior in *Callicebus* and *Saimiri*: Strength and specificity of attraction between male–female cagemates, *Folia Primatol.* **12**:113–123.

Mason, W. A., 1982, Primate social intelligence: Contributions from the laboratory, in: *Animal Mind–Human Mind, Report of the Dahlem Workshop Konferenzen* (D. R. Griffin, ed.), Springer-Verlag, Berlin, pp. 131–143.

McGonigle, B., and Chalmers, M., 1977, Are monkeys logical?, *Nature* **267**:694–696.

McGonigle, B., and Flook, J., 1978, The learning of hand preferences by squirrel monkeys, *Psychol. Res.* **40**:93–98.

McGonigle, B., and Jones, B. T., 1978, Levels of stimulus processing by the squirrel monkey: Relative and absolute judgments compared, *Perception* **7**:635–659.

Menzel, E. W., Jr., 1973, Chimpanzee spatial memory, *Science* **182**:943–945.

Menzel, E. W., Jr., and Juno, C., 1982, Marmosets (*Saguinus fuscicollis*): Are learning sets learned?, *Science* **217**:750–752.

Milton, K., 1980, *The Foraging Strategy of Howler Monkeys*, Columbia University Press, New York.

Mitchell, G., 1972, Looking behavior in the rhesus monkey, *J. Phenom. Psychol.* **3**:53–67.

Murray, S. G., and King, J. E., 1973, Snake avoidance in feral and laboratory reared squirrel monkeys, *Behaviour* **47**:281–289.

Neisser, U., 1976, *Cognition and Reality*, Freeman, San Francisco.

Nitsch, F. M., and Jarosch, E., 1972, Individual learning strategies for discrimination between visual sequences in squirrel monkeys, *Psychol. Forsch.* **35**:335–354.

Noble, C. S., and Thomas, R. K., 1970, Oddity learning in the squirrel monkey, *Psychon. Sci.* **19**:305–307.

Nunnally, J. C., and Lemond, L. C., 1973, Exploratory behavior and human development, in: *Advances in Child Development and Behavior* (H. Reese, ed.), Academic Press, New York, pp. 60–109.

Parker, C. E., 1966, Total darkness as an aversive stimulus condition for the squirrel monkey, *Psychon. Sci.* **6**:111–112.

Parker, S. T. and Gibson, K. R., 1979, A developmental model of the evolution of language and intelligence in early hominids, *Behav. Brain Sci.* **2**:367–408.

Passingham, R. E., 1981, Primate specialization in brain and intelligence, *Symp. Zool. Soc. Lond.* **46**:361–368.

Petrinovich, L., and Patterson, T. L., 1981, Field studies of habituation: IV. Sensitization as a function of the distribution and novelty of song playback to white-crowned sparrows, *J. Comp. Physiol. Psychol.* **95**:805–812.

Phillips, M. J., and Mason, W. A., 1976, Comparative studies of social behavior in *Callicebus* and *Saimiri*: Social looking in male–female pairs, *Bull. Psychon. Soc.* **7**:55–56.

Plotkin, H. C., and Odling-Smee, F. J., 1979, Learning, change, and evolution: An inquiry into the teleonomy of learning, in: *Advances in the Study of Behavior*, Volume 10 (J. S. Rosenblatt, R. A. Hinde, C. Beer, and M. Busnel, eds.), Academic Press, New York, pp. 1–41.

Plotkin, H. C., and Odling-Smee, F. J., 1982, Learning in the context of a hierarchy of knowledge-gaining processes, in: *Learning, Development and Culture* (H. C. Plotkin, ed.), Wiley, New York, pp. 443–471.

Richard, A. F., 1981, Changing assumptions in primate ecology, *Am. Anthropol.* **83**:517–533.

Riopelle, A. J., 1967, *Animal Problem Solving*, Penguin Books, Hammondsworth, Middlesex, England.

Roiblat, H. L., 1982, The meaning of representation in animal memory, *Behav. Brain Sci.* **5**:353–406.

Rumbaugh, D. M., 1968, The learning and sensory capacities of the squirrel monkey in phylogenetic perspective, in: *The Squirrel Monkey* (L. A. Rosenblum and R. W. Cooper, eds.), Academic Press, New York, pp. 255–317.

Rumbaugh, D., 1974, Comparative primate learning and its contributions to understanding development, play, intelligence and language, in: *Perspectives in Primate Biology*, Volume 9 (B. Chiarelli, ed.), Plenum Press, New York, pp. 253–287.

Rumbaugh, D., and McCormack, C., 1967, The learning skills of primates: A comparative study of apes and monkeys, in: *Neue Ergebnisse der Primatologie* (D. Stark, R. Schnieder, and H. Kunh, eds.), Gustav Fisher, Stuttgart, pp. 289–306.

Rumbaugh, D., and McCormack, C., 1969, Attentional skills of great apes compared to those of gibbons and monkeys, in: *Proceedings of the Second International Congress of Primatology*, Volume 1 (C. R., Carpenter, ed.), Karger, Basel, pp. 167–172.

Schott, D., 1975, Quantitative analysis of the vocal repertoire of squirrel monkeys (*Saimiri sciureus*), *Z. Tierpsychol.* **38**:225–250.

Schwam, E., and Gamzu E., 1975, Constraints on autoshaping in the squirrel monkey: Stimulus and response factors, *Bull. Psychon. Soc.* **5**:369–372.

Smith, H. J., King, J. E., and Newberry, P., 1976, Facilitation of discrimination learning-set in squirrel monkeys by colored food stimuli, *Bull. Psychon. Soc.* **7**:5–8.

Sternberg, R. J., 1977, *Intelligence, Information Processing, and Analogical Reasoning*, Lawrence Erlbaum, Hillsdale, New Jersey.

Sternberg, R. J., and Salter, W., 1982, Conceptions of intelligence, in: *Handbook of Human Intelligence* (R. J. Sternberg, ed.), Cambridge University Press, Cambridge, pp. 3–28.

Thomas, R. K., 1980, Evolution of intelligence: An approach to its assessment, *Brain Behav. Evol.* **17**:454–472.

Thomas, R. K., 1982, The assessment of primate intelligence, *J. Hum. Evol.* **11**:247–255.

Thomas, R. K., and Boyd, M. G., 1973, A comparison of *Cebus albifrons* and *Saimiri sciureus* on oddity performance, *Anim. Learn. Behav.* **1**:151–153.

Thomas, R. K., and Chase, L., 1980, Relative numerousness judgments by squirrel monkeys, *Bull. Psychon. Soc.* **16**:79–82.

Thomas, R. K., and Crosby, T. N., 1977, Absolute versus relative class conceptual behaviour in squirrel monkeys (*Saimiri sciureus*), *Anim. Learn. Behav.* **5**:265–271.

Thomas, R. K., and Ingram, D. K., 1979, Conceptual volume judgments by squirrel monkeys, *Am. J. Psychol.* **92**:33–43.

Thomas, R. K., and Kerr, R. S., 1976, Conceptual conditional discrimination in *Saimiri sciureus*, *Anim. Learn. Behav.* **4**:333–336.

Thomas, R. K., and Peay, L., 1976, Length judgments by squirrel monkeys: Evidence for conservation?, *Dev. Psychol.* **12**:349–352.

Thomas, R. K., Fowlkes, D., and Vickery, J. D., 1980. Conceptual numerousness judgments by squirrel monkeys, *Am. J. Psychol.* **93**:247–257.

Thorington, R. W., 1967, Feeding activity of *Cebus* and *Saimiri* in a Colombian rain forest, in: *Neue Ergebnisse der Primatologie* (D. Starck, R. Schnieder, and H. Kuhn, eds.), Gustav Fisher, Stuttgart, pp. 180–189.

Thorington, R. W., 1968, Observations of squirrel monkeys in a Colombian forest, in: *The Squirrel Monkey* (R. W. Cooper and L. A. Rosenblum eds.), Academic Press, New York, pp. 69–85.

Treichler, F. R., Conner, J. B., and Ricciardi, A. M., 1970, Probability learning and reversal by experimentally naive rhesus and squirrel monkeys, *Learn. Motiv.* **1**:411–419.

Visalberghi, E., and Mason, W. A., 1983, Determinants of success in problem-solving situations in *Saimiri* and *Callicebus*, *Primates*. **24**:385–396.

Warren, J. M., 1965, Primate learning in comparative perspective, in: *Behavior of Nonhuman Primates*, Volume 1 (A. Schrier, H. Harlow, and F. Stollnitz, eds.), Academic Press, New York, pp. 249–281.

Warren, J. M., 1974, Possibly unique characteristics of learning by primates, *J. Hum. Evol.* **3**:445–454.

Welles, J. F., 1972, The Anthropoid Hand: A Comparative Study of Prehension, Ph.D. Dissertation, Tulane University (University Microfilms International #7325309).

Welles, J. F., 1976, A comparative study of manual prehension in anthropoids, *Sauget. Mitt.* **24**:26–38.

Williams, R. B., Lane, J. D., Kuhn, C. M., Melosh, W., White, A. D., and Schanberg, S. M., 1982, Type A behavior and elevated physiological and neuroendocrine responses to cognitive tasks, *Science* **218**:483–485.

Winter, P., 1968, Social communication in the squirrel monkey, in: *The Squirrel Monkey* (L. A. Rosenblum and R. W. Cooper, eds.), Academic Press, New York, pp. 235–253.

Wolman, B. D., 1973, *Dictionary of the Behavioral Sciences*, Van Nostrand, New York.

Worthington, C. S., 1972, Sensory input and circadian effects upon cage activity in two species of cebid monkey, *Psychon. Sci.* **28**:165–166.

4

Squirrel Monkey Communication

John D. Newman

1. INTRODUCTION

Communication in the squirrel monkey, *Saimiri sciureus* has been the subject of numerous analyses over the past two decades. One of the interesting aspects of having participated in some of these studies has been the realization that, even in this well-studied species, there are significant gaps in understanding the functional significance, ontogeny, and intraspecific variability of many common communication patterns. The causal mechanisms and evolutionary history of *Saimiri* communication are also areas of study for which there is a wealth of opportunity for further investigation.

Comparisons of the communication signals of various primate species permit an evaluation of where the squirrel monkey repertoire falls in the spectrum of repertoire complexity. From what is now known, the squirrel monkey is quite high in the spectrum in terms of the variety of *vocal* signals, but quite low in terms of the variety of *visual* signals. This is perhaps understandable in an arboreal species, where auditory signals would be more effectively transmitted.

This chapter will outline the communicative signals of the squirrel monkey, describing their characteristics and social significance. Major emphasis will be placed on vocalizations. The chapter will also touch upon a number of issues relating to communication in this species, including subspecies and sex differences in vocal behavior, an analysis of the structural classes of vocalizations, the roles of genetic and experiential influences on vocal behavior, and the structural variability of vocalizations.

2. OVERVIEW OF THE VOCAL REPERTOIRE

As is the case with most primates, squirrel monkeys employ vocalizations in a wide range of contexts, foremost being interactions with conspecifics.

John D. Newman • Laboratory of Comparative Ethology, National Institute of Child Health and Human Development, National Institutes of Health, Bethesda, Maryland 20205.

Vocal signals of the squirrel monkey occur over a large part of the audible frequency range, starting at about 0.1 kHz and ranging to at least 16 kHz. A few calls or parts of calls extend higher. However, while harmonics may occur well above this frequency, there is no evidence that squirrel monkeys ever produce vocalizations that are completely inaudible to humans with normal hearing. As a general rule, lower pitched vocalizations are similar to the grunts of some Old World monkeys, while most calls with a fundamental above 2 kHz are free of noisy components, some sounding like the contact calls of birds. A few calls in the repertoire, notably the Chucks, are rapidly frequency-modulated, frequency changing at a rate of up to 500 kHz/sec over three or four octaves. A further group of calls consists of clicklike pulses, the number of pulses and uniformity of their timing serving to separate variations within the group. Table I presents a summary of the vocal repertoire, indicating the main diagnostic structural characteristics for each group of calls, and the contexts, names, and major vocalizers for the variants within each group.

2.1. Previous Studies of the Squirrel Monkey Vocal Repertoire

The vocal repertoire of the squirrel monkey was first described by Winter *et al.* (1966). In its general outline, much of what was reported then has held up under subsequent investigation. Only a few additional adult vocalizations have been added to the species' repertoire since then: Vit [first described by Winter (1969a)]; Err Chuck [first illustrated by Winter (1972)]; Girren [peculiar to Roman-arch individuals and illustrated by Winter (1969b)]; Traller [named by Talmage-Riggs *et al.* (1972) and more specifically defined by Newman *et al.* (1983)]; Harsh Display Call (described in the present chapter). Calls given mainly by infants were described by Ploog *et al.* (1967), Hopf (1972), and Winter (1968). Most of the studies of *Saimiri* vocal behavior since 1966 have focused on issues relating to classification (grouping and naming of the sounds according to various criteria), to structural details (individual, contextual, and subspecific differences), to ontogeny, and to more detailed analyses of the behavioral significance of the sounds. As originally conceived by Winter *et al.*, the squirrel monkey vocal repertoire divided rather naturally into some 25–30 call, types which were further divided into six groups on the basis of shared structural features. The constituencies of these six groups have not held up as well as the identities of the individual call types. The reason for this is because of the occurrence of vocalizations that fall into more than one of the original call groups. Thus, e.g., Twitters generally contain elements ("syllables") from both the original "Peep" and "Twit" groups (Newman *et al.*, 1978), Chucks generally contain elements from the Peep, Twit, Ha, and Arr groups (Newman *et al.*, 1983; Smith *et al.*, 1982a,b), and, as recognized earlier (Winter *et al.*, 1966; Winter, 1968), calls with elements from both Ha and Arr groups commonly occurred. Thus, many calls fall, by the original classification scheme, into the "Combined Calls" group. This group, however, is just a catchall and has neither functional nor structural validity. A further problem with the original scheme arises from the authors' attempt to ascribe a single global function to all of the calls in a group; this fails due to the diversity of behavioral contexts

Table I
Vocal Repertoire of the Squirrel Monkey[a]

Call group	Main contexts	Variants	Main sources
Peeps (tonal, without periodically changing elements)			
	Isolation	Isolation Peep	All
	Play[b]	Play Peep	Under 3 years
	Exploration[b]	Peep, Vit	All
	Genital display	Variable Peeps	Male
	Mild disturbance	Baby-Piepen	Infant
	Rapidly moving stimulus	Alarm Peep	All
	Frustration	Squeal	Adult male
Twitters (tonal, comprised of periodically changing elements)			
	Feeding[b]	Twitter	Adult
	Partial Isolation[b]	Twitter	Adult
	Exploration	Twit	Infant
	Greeting[b]	Twitter	Adult
Chucks (rapidly descending frequency-modulated element)			
	Affiliation	Chuck, Traller	All
	Maternal behavior	Chuck, Err Chuck	Adult female
	Mild arousal, search for mother's nipple	Tuck	Infant
	Contact mother	Location Trill	Infant
	Sexual encounters	Err Chuck	Breeding adult
	Excitement, group disputes	Kecker	Adult (mostly males)
	Alarm, novel stimuli	Yap	All
Cackles (fundamental below 1 kHz)			
	Intragroup agonism, mild disturbance	Cackle	All
	Mild disturbance, nursing attempt	Oink	Infant
Pulsed calls (rapidly repeated pulses)			
	Solicit nursing	Mutter, Grumble	Infant
	Extended nursing bout	Milk Purr	Mother, infant
	Solicit contact with breeding male	Purr	Estrous female
	Solicit contact with partner	Err	Adult
	Mild hostility	Churr	Displaying male, mother threatening other females
	Agonism involving head-grasping	Spit	Adult (especially adult female to older infant)
	Maximum intensity display behavior	Harsh Display Call, Girren	Adult male
Noisy calls (nonharmonic acoustic energy over a broad frequency range)			
	Defeated, captured	Shriek	All
	Extreme distress	Noisy Infantile Isolation Call	Infant

[a] See Table II for references to call nomenclature.
[b] Usually more than one individual calling at the same time.

encompassed by the sounds of one group. Thus, for example, the Isolation Peep and the Squeal are both members of the "Peep" group, but the former occurs when contact with the group has been severed, whereas the latter occurs during close physical contact (Smith *et al.*, 1982a). Another example pertains to the Twitter and Chuck, both of which belonged to the "Twit" group, but whereas the former virtually never occurs during close affiliative contact, the latter is clearly associated with affiliative behavior (Winter *et al.*, 1966; Smith *et al.*, 1982a; Newman *et al.*, 1983). A third example pertains to the Churr and Purr, both of which were members of the original "Arr" group, but which have essentially opposite behavioral significance, the former occurring as a hostile gesture, whereas the latter occurs in positive situations such as sexual solicitation and nursing (Winter *et al.*, 1966; Schott, 1975). The overview presented in Table I takes these factors into consideration, and is meant to serve primarily as a convenient framework for grouping calls on the basis of shared structural features. One should bear in mind that grouping calls on the basis of *functional* affinities is also a useful endeavor, particularly where it can be shown that diverse sounds can be linked by structural intermediates to form a behaviorally integrated system. An overview of the main functional categories of squirrel monkey vocalizations is presented further on in this chapter.

Names applied to specific vocalizations have value, so long as it is made clear that structural features are being associated with which name.* The view that the entire vocal repertoire of this species represents a structural continuum, and therefore that named call types have no validity, is unfounded (Schott, 1975). At the other extreme, applying different names to each of many structural variants (Jürgens, 1979) is justified only where the structural and behavioral differences between variants are well documented. It is apparent that much of the squirrel monkey vocal repertoire can be linked by certain sounds containing elements common to two or more structurally disparate call groups in the scheme presented in Table I. Some examples are: Err Chuck (containing elements from the "Err," "Chuck," and "Cackle" groups, and from the "Peep" group as well, since the flag meets the definition of a Peep); noisy infant isolation calls (having the same duration and initial structure of the Isolation Peep, but containing broadband noise instead of a single tone over much of the call) Twitter syllables, all of which are tonal, but may either be periodic themselves (in some instances, a single periodic syllable making up the entire Twitter) or, like Peeps, lack periodicity but are combined with other periodic syllables to make a periodically repeating sound (in this way, resembling many bird songs); Errs (comprised of pulses) and Cackles (containing a low-frequency fundamental) are combined in some agonistic situations to produce the "Yow." Overall, then, the vocal repertoire of the squirrel monkey is in reality a complex lexicon, many components of which consist of structurally disparate lexical

* The nomenclature for certain vocalizations originally described by Winter *et al.* (1966 and subsequent papers) has been changed. In some cases, names originally designated for one call type have been assigned to completely different vocalizations, leading to some confusion in the literature as to which vocalization is being discussed. Table II is an attempt at summarizing these nomenclatural differences for the squirrel monkey vocal repertoire, referring in each case to the call type in the original description.

Table II
Nomenclature of Squirrel Monkey Call Types

Vocalization and first citation[a]	Equivalent names and first citation
Isolation Peep	Part of Chirping calls (Jürgens and Ploog, 1970)
Peep	Part of Chirping calls (Jürgens and Ploog, 1970); infant examples were termed Baby-Piepen (Ploog *et al.*, 1967)
Vit (Winter, 1969a)	None
Squeal	Shrieking (in part) (Jürgens and Ploog, 1970)
Alarm Peep	Jürgens (1982) includes with Clucking and Yapping
Chirp	Included with Twitters by Newman *et al.* (1978)
Twit	Chirp (Jürgens, 1979, 1982)
Chuck	Tschack-Rufe (Ploog *et al.*, 1967); Clucking (Jürgens, 1982)
Traller (Talmage-Riggs *et al.*, 1972)	Part of Chuck group (Newman *et al.*, 1983); Chatter (Jürgens, 1979)
Err Chuck (Winter, 1972)	None
Tuck (Winter, 1968)	Considered to be an infantile Chuck by Newman *et al.* (1983)
Location Trill (Winter, 1968)	Part of Chuck group (Newman *et al.*, 1983)
Kecker	Cackling (Jürgens and Ploog, 1970); part of Twittering/Chattering/Cackling group (Jürgens, 1982)
Yap	Restricted to calls with a pronounced mast by Newman *et al.* (1983); Bark (Schott, 1975)
Cackle	Quacking (Jürgens and Ploog, 1970); Caw (Jürgens, 1979)
Oink	Part of Cackling group (Jürgens and Ploog, 1970)
Grumble (Winter, 1968)	Calls with more pronounced initial pulses were termed Err-Grumbles (Winter *et al.*, 1973)
Mutter (this work)	Churr-Oink (Hopf, 1970)
Milk Purr	Trink-Quarren (Ploog *et al.*, 1967)
Purr	Part of Growling call group (Jürgens and Ploog, 1970)
Err	Part of Growling call group (Jürgens and Ploog, 1970)
Churr	Part of Growling call group (Jürgens and Ploog, 1970)
Girren (Winter, 1969b)	Harsh Growl (Jürgens, 1979)
Harsh Display Call (this work)	None
Chomp	Part of Shrieking call group (Jürgens and Ploog, 1970)
Groan	None
Shriek	Infant examples were called Baby-Schreien (Ploog *et al.*, 1967)
Scream (Winter, 1968)	Schrei-Piepen (Ploog *et al.*, 1967); Noisy Infantile Isolation Call (Newman, 1985)

[a] Names without citations originate from Winter *et al.*, (1966).

elements connected together to form functionally complete vocalizations. This is quite different from the picture drawn by some authors (e.g., Schott, 1975) of a continuously intergraded vocal system. The repertoire is more accurately viewed as consisting of subsets of sounds, with each subset containing sounds that vary over one or more dimensions. Some of this variability undoubtedly reflects variation in the mood of the vocalizer (Jürgens, 1979) or the presence of competing stimuli (Schott, 1975). However, other factors ignored by many workers are equally important in causing variability in the structure of vocalizations. Among these are differences between geographically separate populations or subspecies [see Newman and Symmes (1982) for examples], individual differences (Symmes *et al.*, 1979; Smith *et al.*, 1982b; Newman *et al.*, 1978); or differences in contextual details [see Newman *et al.*, (1983) for examples].

2.2. Major Functional Classes of Vocalizations

2.2.1. Alarm Calls

Two vocalizations, the Alarm Peep and Yap, are restricted to situations in which novel or alarming stimuli are present. The Alarm Peep is considered to be the aerial alarm call, its occurrence producing a sudden arrest of activity in all group members within hearing range. Animals foraging on the ground or sitting on a low perch will spring instantly to a higher location, then stop activity. This reaction lasts only a few seconds; normal activity quickly resumes. Stimuli eliciting the Alarm Peep appear to have rather low specificity. Hawks with *Accipiter*-like flight and of a size threatening to squirrel monkeys as well as vultures soaring overhead will produce similar alarm reactions in groups of captive squirrel monkeys. Also, Alarm Peeps may be given to a rapidly moving shadow or to a wad of paper tossed in front of or over an individual or group. In a number of instances, I have observed infants about 1 year old emit this vocalization, leap to another perch, and briefly stop activity in the absence of any obvious stimulus; adult group members in these instances scarcely reacted at all to this behavior. The Alarm Peep has not been subjected to extensive analysis for structural variability. However, it appears to be a relatively stereotyped call, with a duration of 50–100 msec and a peak frequency of 14–16 kHz, and it is sometimes is terminated by a rapidly descending FM component (Fig. 1). Usually a single Alarm Peep is produced, but several may be given in rapid succession. Experiments with single individuals indicate that the same object can produce both the Alarm Peep and the Yap. A large furry toy animal suddenly appearing in a monkey's visual field often results in a single Alarm Peep, followed by repeated Yaps as the toy remains in view. Yaps are readily elicited by a variety of genuinely threatening, or merely novel, stimuli. Such stimulus situations include: the attempted capture by humans wielding a net or, in a more confined area, extending a gloved hand; presenting a real animal (dog, cat) or a stuffed model; and entrance of a person wearing a fur coat or hat. Less threatening objects lose their effectiveness quickly; a small windup mechanical toy elicited Yaps from a group on the first four or five presentations, but subsequently only the two infants in the group showed any alarm. One wild-caught male in our colony regularly produced Yaps whenever a net bag of cabbages was brought into the room housing the colony. The Yap is part of a behavioral complex equivalent to the "mobbing" exhibited by many birds and mammals: the object eliciting this behavior, such as a snake, carnivore, or perched raptor in the wild, is slowly surrounded by troop members, who

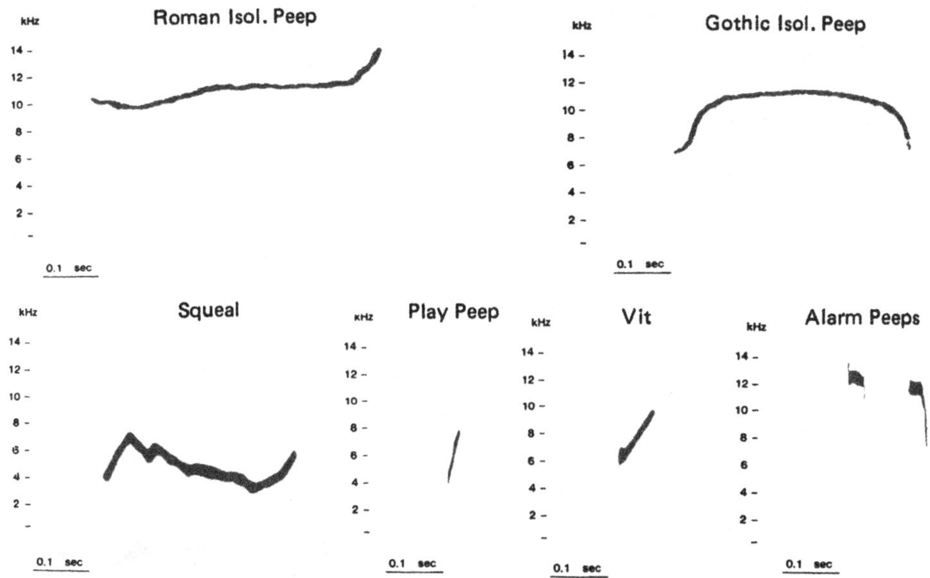

Fig. 1. Sound spectrograms of representative Peeps. Spectrograms in this and subsequent figures have been retouched to improve photographic quality. Spectrograms were made using either a 100 Hz or 600 Hz filter. Faint horizontal lines appearing in some spectrographic illustrations are frequency calibration marks generated along with the spectrograms. All examples in this figure were recorded from Gothic-arch individuals except the Roman Isolation Peep.

vocalize continuously while fixating on the stimulus object. This group behavior might lead one to regard the Yap as a threat signal; however, in instances where only infants in a group produce Yaps to a novel stimulus, it is difficult to consider their behavior as threatening to the stimulus object. The structure of the Yap is more complex and more variable than that of the Alarm Peep. The main features are a frequency-modulated (FM) component (the "mast") that descends rapidly from a frequency typically above 12 kHz, and a loud, low-frequency (<500 Hz) component (the "cackle") with prominent harmonic structure following the mast. The mast may be preceded by a tonal element (the "flag") that resembles the main tonal component of the Alarm Peep. Yaps are uttered persistently, at about one per second, as long as the eliciting stimulus is present. The variable structure of this vocalization is apparent from inspecting a sequence of repeated calls, and is most evident in the cackle component. Indeed, calls without a mast (i.e., Cackles) may be given to the same stimuli instead of Yaps. Other studies have suggested that Roman-arch individuals produce Yaps about four times more frequently than Cackles to alarming stimuli (Ploog *et al.*, 1975). Regardless of whether or not a mast is present, the cackles are usually loud, with prominent harmonics, and begin with a brief noisy component. Highly aroused animals may produce calls that are almost Shrieklike or, instead, have a longer duration flag preceding the mast (this latter being more common in Roman-arch individuals).

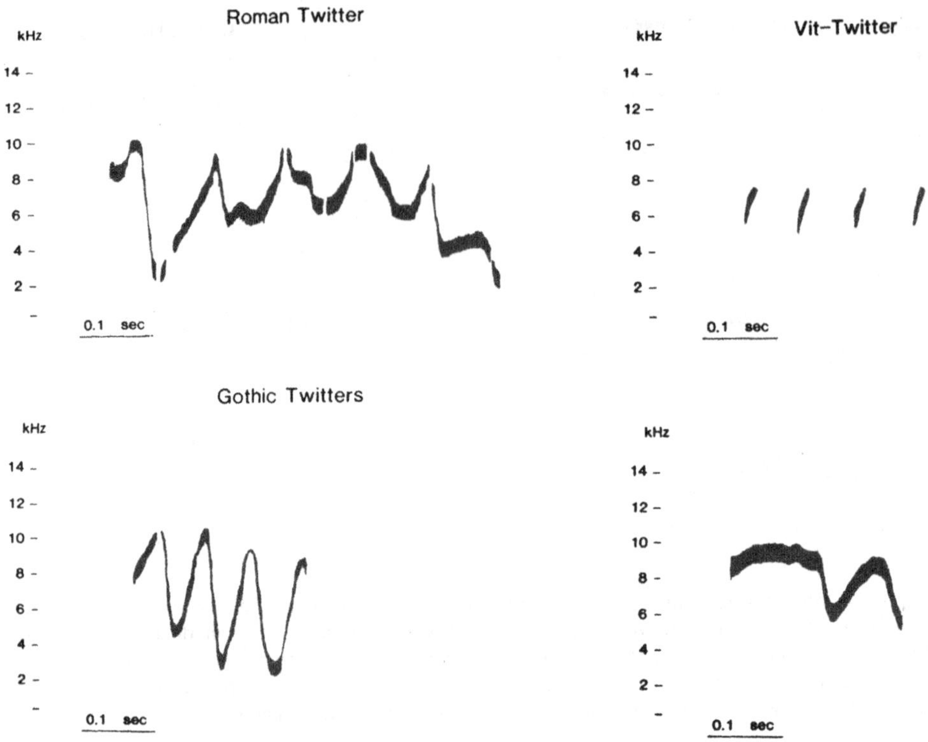

Fig. 2. Sound spectrograms of representative Twitters. The Vit-Twitter was recorded from a Gothic-arch individual; Roman-arch individuals produce similar calls.

2.2.2. Calls Associated with Excitement or General Disturbance

On occasion, a squabble within the group or the sudden appearance of a strange individual will cause a general increase in activity, accompanied by certain vocalizations. The most characteristic vocalization in such cases is the Kecker. Keckers are loud sounds most frequently given by adult males. In the case of family groups, personal observation indicates that the oldest adult male will give most or all of these sounds. Keckers are comprised of a series of uniform structural units, each of which consists of a flag, mast, and cackle. In the Kecker, the flag of each unit is as prominent as the mast, starts at a low frequency (usually below 2 kHz), and the first flag is preceded by a cackle (Fig. 3). Although associated more with Gothic-arch animals (Winter *et al.*, 1966), Roman-arch individuals give a similar call in the same circumstances. In Roman-arch individuals, however, the call is less distinctive, being merely an extended, louder version of the Chucks used in other situations. A 2-month-old Gothic-arch male made Keckers in response to the approach of an adult male who rarely interacted with the infant or its mother. The Kecker appears to function as a threat. Some captive males produce this sound to strange

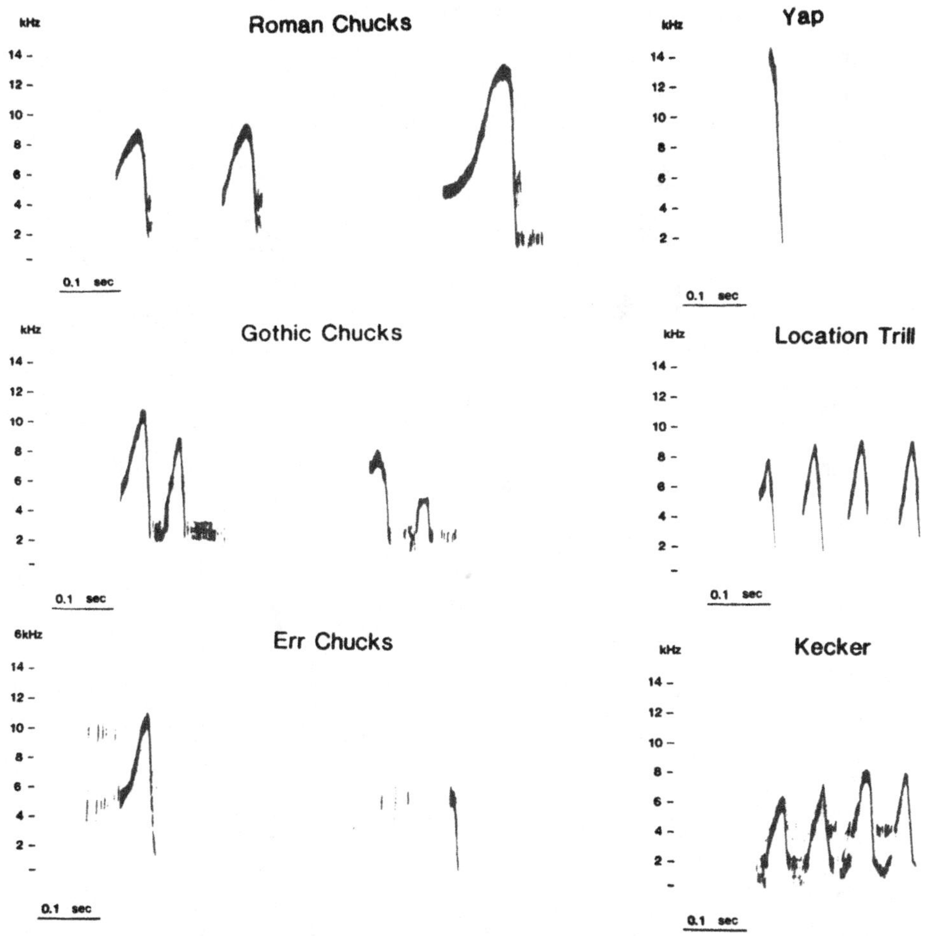

Fig. 3. Sound spectrograms of representative Chucks. All examples in this figure were recordea from Gothic-arch individuals except the Roman Chucks. The Yap was recorded from an infant less than 1 year of age, and lacks the proment "cackle" seen in many Yaps from older animals (see Fig. 4 for an example of a Yap with a prominent cackle).

humans entering their colony room, accompanying their vocalizations with a kind of intimidation display consisting of grasping the wire-mesh sides of their cage and bouncing up and down.

2.2.3. *Calls Associated with Intragroup Agonistic Encounters*

The most common agonistic sound is the Cackle (Fig. 4). Cackles are produced in a wide variety of contexts, but this call appears to express annoyance in all of them. The loudness, duration, and number of Cackles over time are useful measures of the strength of the vocalizer's annoyance. Situations in which Cackles have been recorded include: food stealing (uttered by the victim); jostling by other animals; during rough-and-tumble play (especially

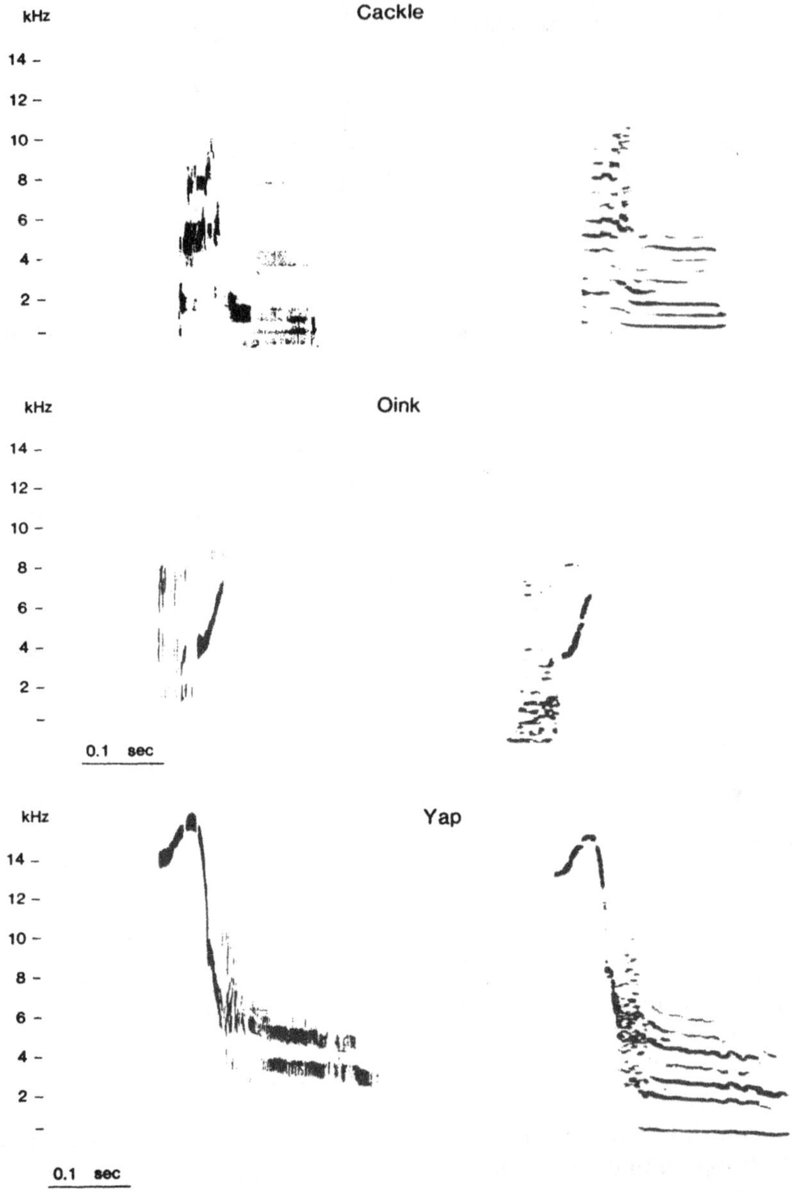

Fig. 4. Sound spectrograms of representative Cackles. In each of the 3 examples, spectrograms made using a 600 Hz filter (left side) and a 100 Hz filter are shown. The Yap is not, by definition, a member of the Cackle group, but is shown here because of its prominent cackle component for comparison purposes. This Yap was recorded from a Roman-arch female, the Cackle and Oink from Gothic-arch females.

Fig. 5. Sound spectograms of representative Pulsed Calls. All examples were recorded from Gothic-arch individuals. The Mutter illustrated on the left was recorded from a mother approaching her infant whereas the one on the right was from an infant soliciting access to its mother's nipple. Err Chuck's belong to the chuck group by definition, but the examples shown each have a particularly prominent "err" (series of pulses) preceding the chuck. The Err Chuck on the left is from an adult female, whereas the one on the right is from an infant

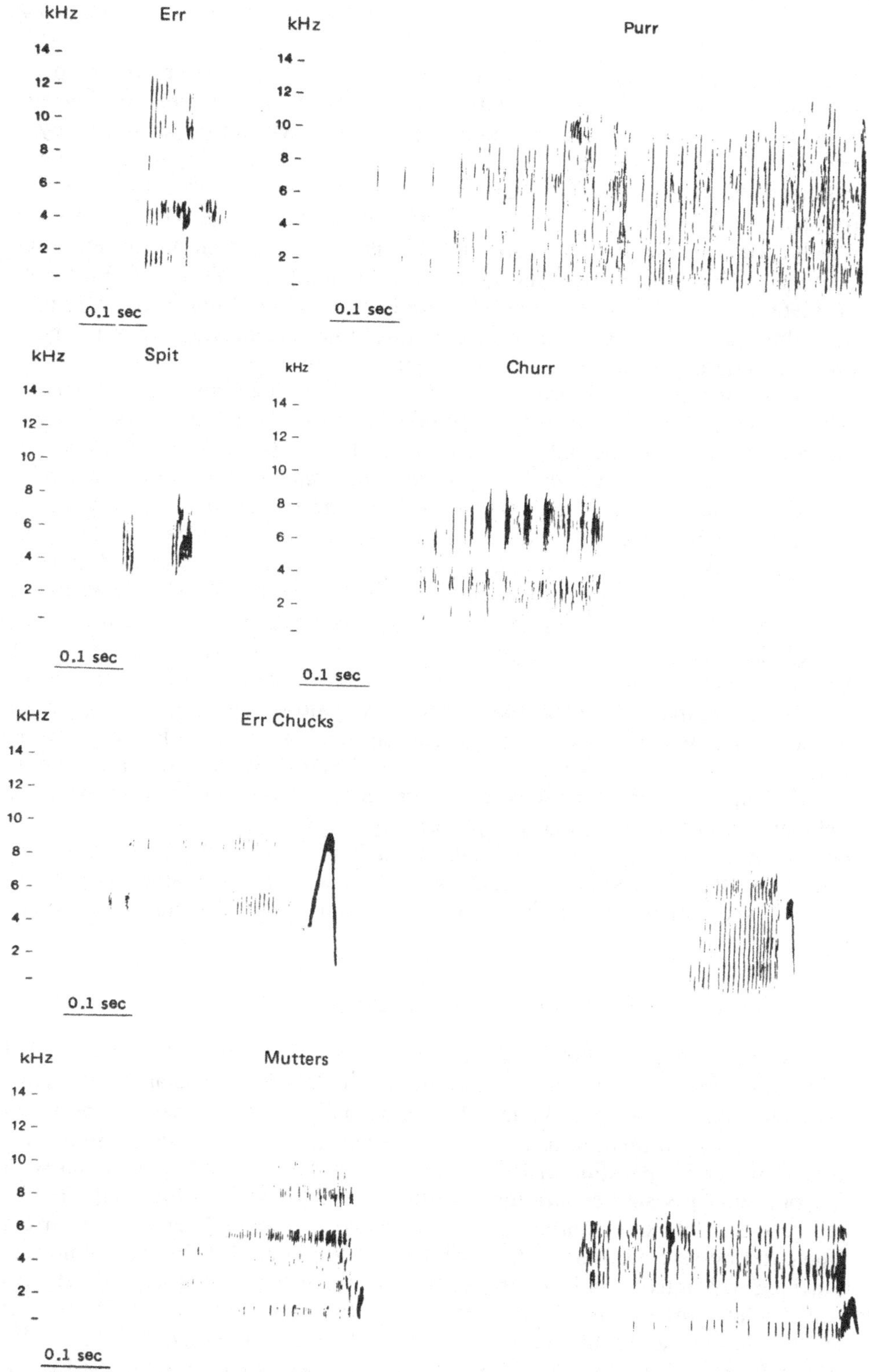
kHz
Err
14
12
10
8
6
4
2
0.1 sec
kHz
Purr
14
12
10
8
6
4
2
0.1 sec
kHz
Spit
14
12
10
8
6
4
2
0.1 sec
kHz
Churr
14
12
10
8
6
4
2
0.1 sec
kHz
Err Chucks
14
12
10
8
6
4
2
0.1 sec
kHz
Mutters
14
12
10
8
6
4
2
0.1 sec

by an adult play partner trying to fend off a juvenile); and presenting a novel stimulus object. Cackles occur with Yaps, Keckers, and Shrieks during encounters with, respectively, potential predators, newly introduced monkeys, and aggressive conspecifics. Cackle structure is highly variable, and as a result has been difficult to quantify. Operationally, any sound with a fundamental frequency lower than 1 kHz can be considered to be a Cackle. Duration varies widely, from less than 10 to more than 200 msec. Calls at the longer extreme contain both low-frequency and fricative elements [the "Yow" of Winter *et al.* (1966)]. Some Cackles are so obscured by broadband noise that a fundamental frequency cannot be determined. Two other vocalizations given in hostile interactions are the Spit and the Churr. In my experience, based on observations of Gothic-arch groups, the Spit and accompanying head grasp are used primarily by adult females to admonish obstreperous juveniles, and the Churr is given by an older infant or juvenile male as it displays to or approaches an adult male. These two calls are structurally related, both consisting of a series of rapidly repeated pulses. In the Spit, the pulses are 2–4 in number, while in the Churr the pulses are considerably greater in number and give the call a growl-like character (Fig. 5). In both, the pulses tend to occur as doublets, most with a clear peak of spectral energy. A fourth call, the Shriek, is associated with strong agonism and is an expression of extreme distress or fear. It is given by individuals who are being beaten in a fight, or are being restrained (as when held by a human following capture). Fights that break out in a group are invariably accompanied by Shrieks. This vocalization may be given *before* an actual attack, when the recipient expects to lose any ensuing battle (such as when a subordinate individual is being chased and harrassed by a more dominant group member). Shrieks are also given by group members observing a fight or another monkey being captured, the volume of sound approaching a deafening roar in the case of an infant being handled by a human. Shrieks are little more than emissions of noise over a broad frequency range. However, peaks of spectral energy within the broadband noise may sometimes be evident (Fig. 6).

2.2.4. Sounds Associated with Group Cohesion

A wide range of sounds fall into this general category, many associated with rather specific situations. The contextual categories associated with these vocalizations include: group contact when visually separated; contact between partners during affiliative interactions; sexual interactions; mother–infant interactions; and separation or isolation from the mother or troop. Contact with a group during visual separation is effected by Peeps, Twitters, and Chucks (Figs. 1–3). Published studies have established that short Peeps and Twitters [the latter being equivalent to the "Chirp" of Winter *et al.* (1966)] are common during group activity in free-ranging troops, being most often associated with travel while foraging for food (Thorington, 1968; Winter, 1972). Twitters of longer duration are common when food is present in abundance. Longer duration Peeps ("Isolation Peep"; see below) are restricted in adults to loss of contact with the troop or troop answers to a lost individual, but in infants the

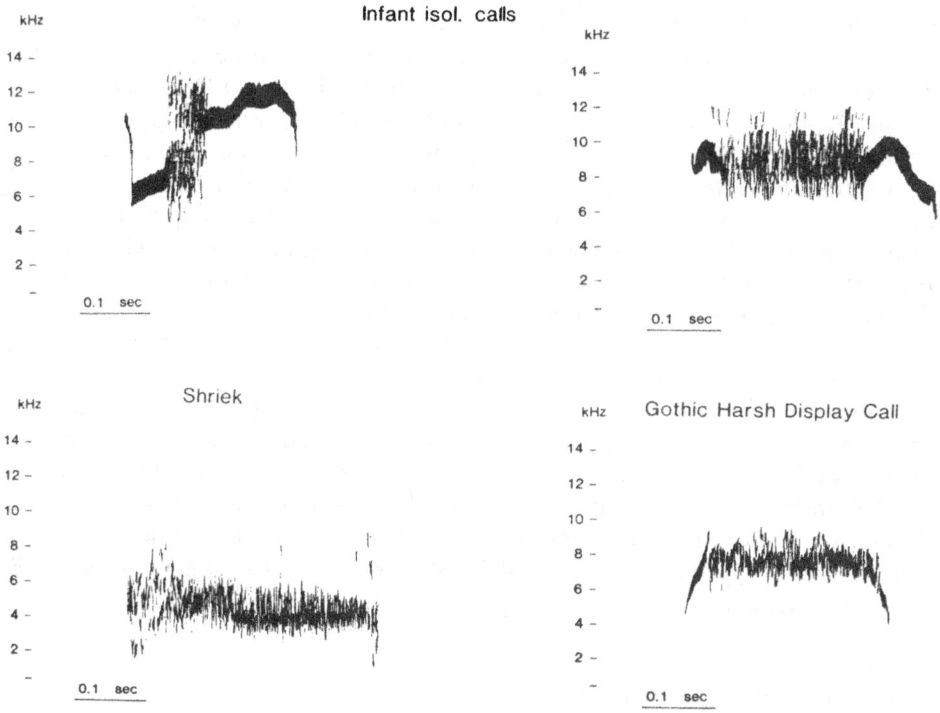

Fig. 6. Sound spectrograms of representative Noisy Calls. The infant isolation call in the upper left is from a Roman-arch individual; the other examples are from Gothic-arch animals.

same calls are frequently produced when the infant experiences some distress. Chucks are associated with closer contact between individuals, being common when group members are quietly sitting. Contact between affiliative partners is most closely associated with the Chuck (Smith *et al.*, 1982a). The Err and Err Chuck are common during sexual solicitation (Schott, 1975; Smith *et al.*, 1983). Errs in this context often are long series of soft, rapid pulses, giving them a purrlike character, and are therefore referred to as Purrs. Err Chucks are a combination of an Err (usually of shorter duration than a Purr), followed within 0.1 sec or less by a Chuck; they are given most commonly by females interacting with their own or another female's infant (Hopf, 1972; Schott, 1975). The foregoing suggests that seasonal factors enter into the relative frequency of Err, Chuck, and Err Chuck occurrence. As shown in Fig. 12, Errs are predominantly associated with the estrus period and Err Chucks with the period following the birth of infants, while Chucks are evenly distributed among nonreproductive, estrus, and postnatal phases of the annual cycle.

One of the most basic of all group-cohesion calls is the isolation call (sometimes referred to as the "separation" or "lost" call), since this sound aids the separated individual in regaining contact with other conspecifics. Of greatest importance to the infant that depends on its mother for survival, the isolation call nevertheless retains its functional significance into adulthood, as witnessed by the nearly automatic production of this vocalization whenever a captive adult is experimentally separated from other colony members. Isolated adult squirrel monkeys produce a characteristic call, the Isolation Peep. This call is a long whistle, with most of its energy above 6 kHz (cf. Fig. 1). Upon hearing an Isolation Peep, group members will answer with this same call. In captivity, an isolated individual can often be induced to call more vigorously by playing a tape recording of Isolation Peeps. Young squirrel monkeys, including newborn infants, produce shorter versions of the adult Isolation Peep. (Lieblich *et al.*, 1980). Separated infants usually call at higher rates than adults (personal observation), and when extremely distressed produce more variable calls with frequency-modulated and noisy sections (''infantile isolation calls''; Fig. 6), as well as the more stereotyped Isolation Peep.

The contact calls used by mother and infant are similar to those used between adults. Short Peeps are frequently emitted by exploring infants; Chucks and Err Chucks are used by both mothers and infants during close-range communication and after a separated infant has regained contact with the mother's body. Both the mother and infant sometimes emit Purrs while the infant is nursing [see Ploog *et al.* (1967), Winter *et al.* (1966), and Winter (1968) for further details].

2.2.5. *Display Vocalizations*

Squirrel monkeys are well known for a characteristic display involving exposure of their genital area. This ''genital display'' is best known from studies of adult Gothic-arch males, who will consistently display to their reflection in a mirror. The vocalizations accompanying this visual display are less well known. Vocalizations from displaying adult Gothic-arch males include Chucks (or Err Chucks), Peeps, and a type of call not otherwise produced by adults. This last, which I will name the ''Harsh Display Call'' (HDC), is quite variable but is distinguishable from all other adult calls by the presence of fricative noise separating briefer beginning and ending tonal components (Fig. 7). The HDC may occur as a series of closely spaced sounds or as a continual sound, in both cases lasting a total of 0.5–1.0 sec. Overall, the HDC most closely resembles noisy infantile isolation calls, otherwise never given by adults in isolation or any other context. Roman-arch males generally cannot be induced to display to their reflections reliably (MacLean, 1964). However, Roman-arch males with the dark crowns characteristic of the Bolivian *Saimiri sciureus boliviensis** do display to their reflection, but instead of exposing their genitals, they rapidly arch their back. In displaying Roman-arch males the Girren (Winter, 1969b) is

* In a recent reappraisal of squirrel monkey taxonomy, Hershkovitz (1984) concludes that Roman-arch and Gothic-arch squirrel monkeys are separable at the species level. The Roman-arch type is referred to the *Saimiri boliviensis* group (one species, two subspecies), while the Gothic-arch type is referred to the *Saimiri sciureus* group (three species). Behavioral differences summarized in the present chapter support the view that Roman- and Gothic-arch individuals represent different species. However, the taxonomic arrangement presented in Chapter 1 of this volume will be followed in the interest of maintaining consistency.

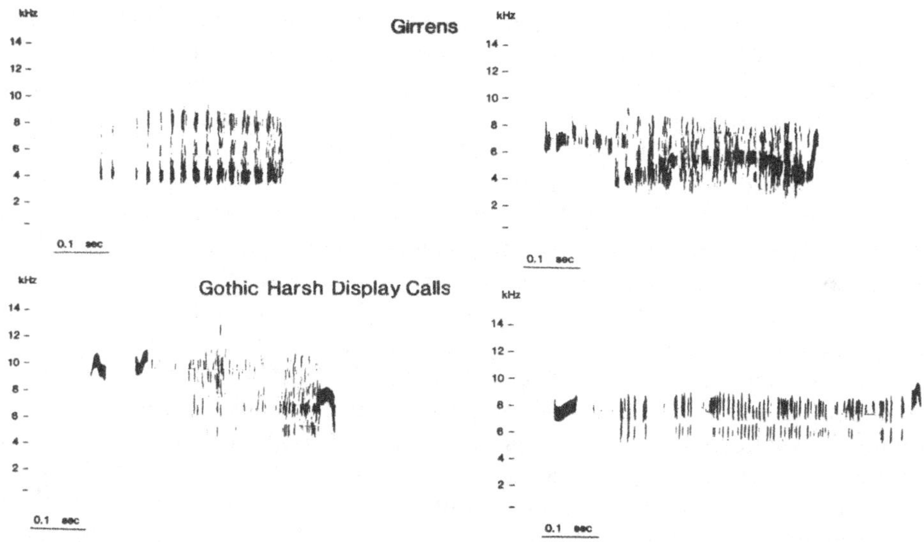

Fig. 7. Sound spectrograms of representative Harsh Display Calls (HDC). Above: Two examples of the Roman-arch HDC, the "Girren" of Winter (1969b), showing its conspicuous pulsed pattern. The Girren on the right was produced by an adult male with the "Bolivian" phenotype, accompanying a rapid arching of the back as it saw its reflection in a mirror. Below: Gothic-arch HDCs from two adult Gothic-arch males, during genital displays to their reflected image from a mirror. The example on the right shows the pulsed pattern particularly clearly. See Fig. 6 for a Gothic-arch HDC in which the pulses are replaced by continuous broadband noise.

produced (Fig. 7), a vocalization with similarities to both the Gothic HDC and to the Churr.

2.3. *Infant Vocalizations*

As is generally true for other primates, the vocal behavior of squirrel monkey infants is less well known than is the case for the adults of this species. Infant squirrel monkeys are not very vocal for their first few weeks of life, spending much of their time sleeping on their mother. During this neonatal period, vocalizations occur primarily in the contexts of nursing, moving around the mother's body, or when disturbed. Infants seeking their mother's nipple produce Grumbles and Tucks (Winter, 1968). During bouts of undisturbed nursing, infants may emit Purrs ["Milk Purr" of Winter *et al.* (1966)]. Disturbed infants will produce a soft Chuck or Peep (Ploog *et al.*, 1967). As has been demonstrated by previous work (Winter *et al.*, 1973), neonatal infants are actually capable of a wide range of vocalizations. It would appear that under normal social circumstances much of this vocal capability is not used until the infant begins its first tentative explorations off the mother. The brief Peeps heard from the disturbed neonate are the most common infant sounds at this point, being regularly emitted as the infant explores novel objects or comes into contact with other group members. During this period, the infant frequently interacts with other individuals, in particular adult females who are huddling partners of the mother. These interactions involve both adult and infant, each touching the other with the nose and, in the infant's case, reaching out with the hands as well. The infant vocalizes continuously during these interactions, making Peeps and a variety of variable tonal and noisy calls. Occasionally, such interactions lead to the infant climbing onto its "aunt." This initially appears to distress the infant, however, who emits isolation calls. The mother either retrieves the infant at this point, or the infant returns to its mother of its own accord. As the infant develops and its contact with other group members increases, the range of vocalizations increases as well. New types of social interactions for the older infant include play, displaying to the adult males in the group (more commonly shown by male infants), grabbing small bits of food from group members and soliciting contact with its mother. Infant vocalizations frequently associated with these contexts are, in respective order: Play Peep; infantile isolation calls, short Peep and Churr; Cackle and short Peep; and Mutter. The Mutter (Fig. 5) is a pulsed sound terminated by a Vit or Twit and sounding more emphatic than the Grumble. It is particularly characteristic of the weaning period, during which the infant follows its mother around for many seconds vocalizing repeatedly before being allowed to climb on her and nurse; the "churr-oink" of Hopf (1970) appears to be equivalent to this vocalization.

3. *INHERITED AND EXPERIENTIAL INFLUENCES ON SQUIRREL MONKEY VOCALIZATION*

3.1. *Introduction*

One of the major issues surrounding vocal communication in squirrel monkeys is that of the relative roles of inheritance and experience in determining the form and context of vocal behavior in this species. This issue is a general one, pervading all of primate communication, that goes well beyond the scope of this chapter. A recent paper (Newman and Symmes, 1982) treats this issue in some detail. At the present time, the prevailing evidence supports a dominant role for genetic factors in determining the basic structure of many vocalizations. This evidence derives from analysis of the vocalizations of newborn infants (Winter *et al.*, 1973; Ploog *et al.*, 1967; Lieblich *et al.*, 1980) and of infants reared by muted mothers or apart from other conspecifics (Winter *et al.*, 1973). Further, analysis of the vocalizations of deafened adults indicates that auditory feedback is not necessary for producing structurally normal vocalizations (Talmage-Riggs *et al.*, 1972). However, since some aspects of adult vocal behavior suggest a role for learning [e.g., answering the Chucks of affiliative partners (Smith *et al.*, 1982a)], it would be premature to eliminate a role for experiential factors in influencing squirrel monkey vocal behavior.

3.2. *Inherited Characteristics of the Squirrel Monkey Isolation Call*

The characteristic isolation call of the squirrel monkey—the Isolation Peep (IP)—has received particular attention in previous studies because of its relatively well-defined behavioral context and the ease of obtaining high-quality recordings in captive animals. Its simple, stereotyped structure has facilitated quantitative analysis, permitting the investigation of IP structural differences with age and between individuals of the same and different geographic origin. One of the most striking findings to have emerged from these studies is the difference in IP morphology between Roman-arch and Gothic-arch individuals. Originally discovered by Winter (1969b), the morphological differences were subsequently isolated to two related parameters, the point at which a call's highest frequency occurs (peak frequency location, PFL) and the slope of the terminal part (tail slope, TL) (Symmes *et al.*, 1979). In Roman-arch IPs, PFL is restricted to the last 15% of the call, whereas in the IPs from different Gothic-arch individuals PFL values range over much of the call (but never occur in the terminal 5%). Related to their late PFL is the fact that most Roman-arch IPs terminate with an upgoing change in frequency, giving the TL a strongly positive value. Some Roman-arch IPs, however, reverse direction at the very end of the call and terminate with a sudden, sharp descent, a trait more prominent in some individuals than in others. These differences between the IPs of Roman- and Gothic-arch individuals are present even in the IPs of newborn individuals, suggesting that genetic factors provide the major determinants in shaping these differences in IP morphology (Newman and Symmes, 1982). Changes in IP morphology observable in both Roman-arch and Gothic-arch individuals during development are restricted primarily to increased call duration and a diminished frequency change in the middle of the call (Lieblich *et al.*, 1980). The major differences between adult and infant IPs, then, are in call duration (DUR) and in the steepness of the slope of the call's midsection (MSL). By measuring parameters in adult and infant IPs of both subspecies and subjecting them to linear discriminant function analysis, a two-dimensional plot showing the clustered distribution of IP discriminant scores from Roman-arch adults and infants and Gothic-arch adults and infants can be made (Fig. 8). This plot shows that the discriminant scores of Roman- and Gothic-arch IPs, regardless of the age of the vocalizer, hardly overlap (not at all in adults) along the axis represented by canonical discriminant function 1, which is based primarily on PFL and TL. The plot also shows that the discriminant scores of IPs of 1- or 2-day-old infants are separated from those of adults along the axis represented by function 2, which is based primarily on DUR and MSL.

This difference in Roman- versus Gothic-arch IP morphology leads to questions regarding the inheritance and evolution of IP phenotype. Efforts to demonstrate inheritance patterns of *individualistic* IP traits by comparing IP morphology of infants with that of their parents have led to equivocal results (Lieblich *et al.*, 1980). On the other hand, efforts to demonstrate subspecific patterns of IP phenotype inheritance by cross-breeding Roman-arch individuals from Peru and Gothic-arch individuals from Colombia or Peru has led to some more consistent findings (Newman and Symmes, 1982). The general result of the Roman-arch × Gothic-arch crossbreeding studies is that the Roman-arch IP phenotype is more common than the Gothic-arch in F_1 "hybrid" offspring. That is most evident where the mother is the Roman-arch type (Fig. 9). The periorbital pattern in most hybrids also tends toward the Roman-arch type.

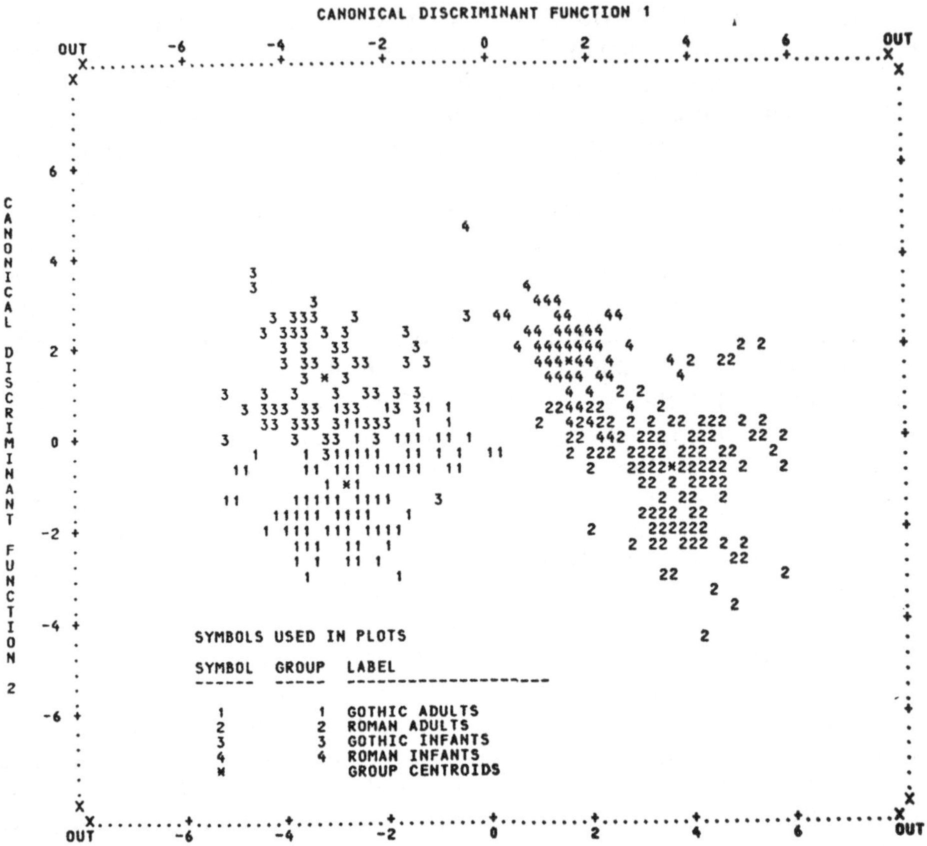

Fig. 8. Two-dimensional plot of Isolation Peep (IP) discriminant scores resulting from linear discriminant function analysis. The analysis was performed on eight measured parameters from the IPs of Roman-arch adults, Gothic-arch adults, and 1- or 2-day-old infants of both types. The most important variables comprising canonical discriminant function 1 in this plot are peak frequency location and tail slope. The most important variables in function 2 are duration and mid-slope. Note the lack of significant overlap in Roman- and Gothic-arch scores along the function 1 axis, indicating the reliable separation of the two types of IPs using the variables in function 1.

The IP and periorbital phenotypes of hybrid offspring with Gothic-arch mothers are more variable. These results suggest that the genes determining the Roman-arch IP phenotype and Roman-arch periorbital pattern may have greater penetrance compared to those determining the Gothic-arch IP and Gothic-arch patterns. The analysis of different geographic populations has been instituted to supplement laboratory studies of the genetics of IP phenotypic differences. With the cooperation of field biologists, importers, and the individuals actually capturing the monkeys, the geographic origins of captive squirrel monkeys can be determined. In such cases, correlation of IP structural details with particular geographic localities is feasible. The readily distinguishable differences in IP structure of Roman-arch and Gothic-arch monkeys can be demonstrated over a large part of their ranges. As shown in Fig. 10, Gothic-arch IPs are produced

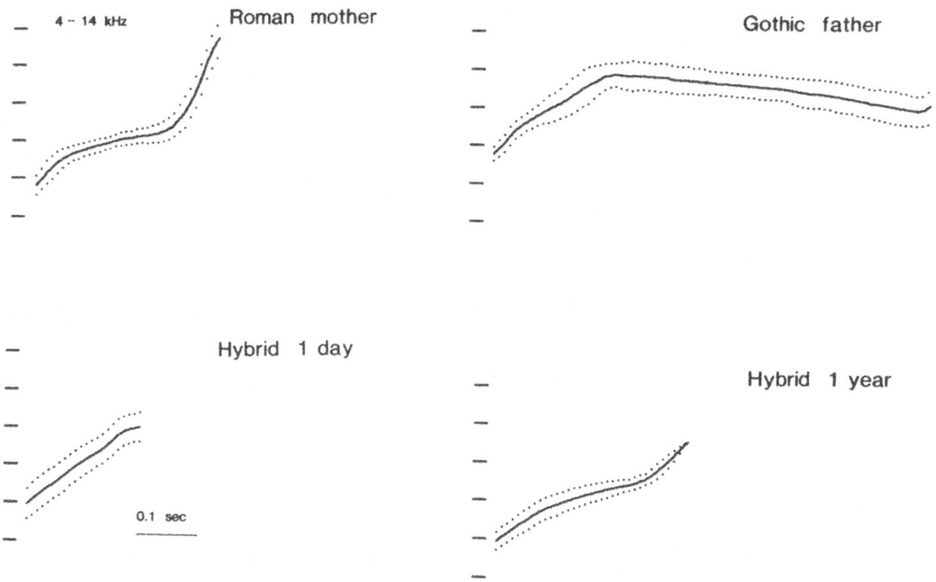

Fig. 9. Computer-generated tracings of the Isolation Peeps (IPs) of a Roman-arch adult female (upper left), a Gothic-arch adult male (upper right), and their hybrid offspring (below). The lower tracings are from the same hybrid, taken at two different ages. Each tracing depicts the average of 20 IPs recorded in succession. The dotted lines above and below each tracing show the variance (1 SD) of the sample. Note that the hybrid infant's IPs have the characteristic Roman-arch structure (see text for discussion of hybrid IPs).

by Gothic-arch individuals from Guyana and Peru (as well as those from Colombia; see Fig. 9), while Roman-arch IPs are produced by Roman-arch individuals from Bolivia and Peru. Costa Rican squirrel monkeys (*Saimiri sciureus oerstedii*) have a prominent Gothic-arch facial pattern and their IPs are likewise characteristic of Gothic-arch individuals (the author's unpublished observations in Corcovado National Park, Costa Rica). Data from Brazilian populations are unavailable at the present time. Whether there are differences in IP phenotype between different populations of the *same* type (i.e., Roman or Gothic-arch) has yet to be demonstrated. We have shown that Roman-arch populations living within 90 km of each other can overlap considerably in the details of IP structure (Newman and Symmes, 1982).

3.2.1. The IP Compared to Isolation Calls of Other Primates

Growing evidence suggests that most if not all primates produce a characteristic vocalization following separation from conspecifics. Analysis of the structure of various species' isolation calls indicates a widespread similarity in their long-duration tonal character, with no abrupt frequency changes except at the beginning and end of the calls (Newman, 1985). Many species even use the same frequency band. The IPs of squirrel monkeys are higher in pitch than the isolation calls of larger primates, but share the same frequency band with marmosets (Epple, 1968; Pola and Snowdon, 1975) and have the same

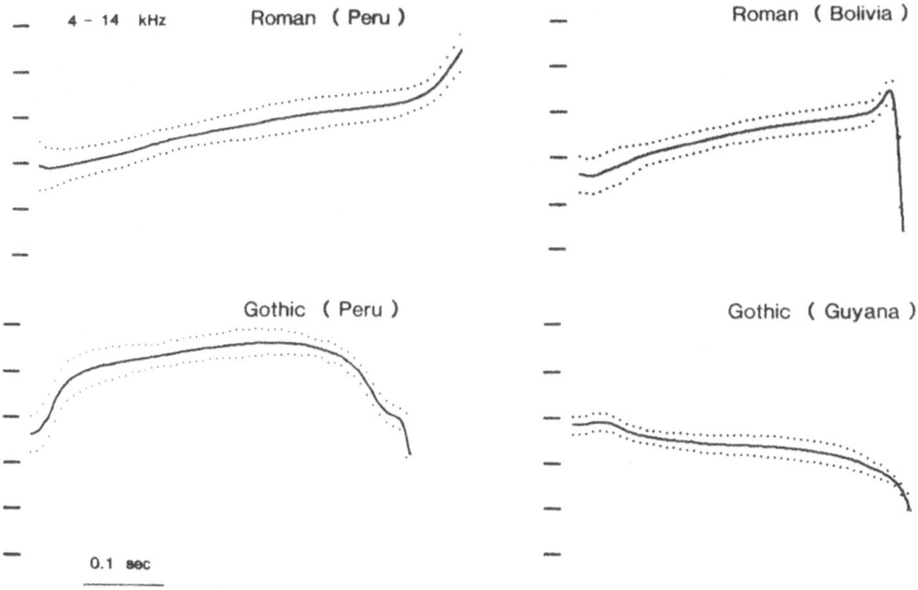

Fig. 10. Computer-generated tracings of the Isolation Peeps (IPs) of four adult squirrel monkeys, each from a different part of the species' South American range. Upper left: Roman-arch from Peru. Upper right: Roman-arch from Bolivia. Lower left: Gothic-arch from Peru. Lower right: Gothic-arch from Guyana. Each tracing and pair of dotted lines represent the average and variance, respectively, of 20 successive IPs, as in Fig. 9. The illustrated examples do not necessarily typify IPs from all individuals in a given geographic area, but demonstrate the correlation between facial type (Roman-arch or Gothic-arch) and IP morphology over a wide geographic area.

basic shape common to other primates. The widespread similarity in overall isolation call morphology across primate species suggests that those elements of the primate genome controlling isolation call phenotype have changed little over the course of primate evolution.

3.2.2. *Physiological Controls of the IP*

The widespread structural similarities of primate isolation calls suggests that the neural mechanisms underlying production of the isolation call may also be similarly shared by many primate species. However, at present these neural mechanisms are poorly understood. A first step in understanding such mechanisms is to identify pathways in the brain critical to the production of normal IPs. Some progress in that direction has been made in the finding that certain lesions restricted to parts of the thalamic tegmentum and adjacent core gray matter selectively disrupt production of normal IPs, but leave other vocalizations intact (Newman and MacLean, 1982). An earlier study found that IPs were among the call types elicited by electrically stimulating the basal anterior

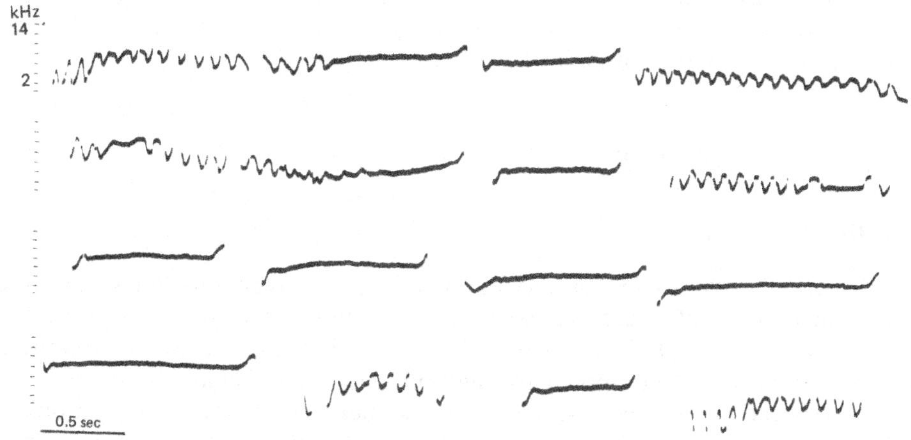

Fig. 11. A continuous 40-sec sequence of vocalizations from a "fatted" Roman-arch male. The male was in vocal contact but visually isolated from other members of its group. The time interval between successive calls has been compressed. The long Twitter–Peep combinations (first calls in the top and second rows), each more than 2 sec, in duration, have only been recorded from this one individual.

midline limbic cortex and midline thalamus (Jürgens and Ploog, 1970). Further progress can be expected in determining the nerve cell populations and their functional interactions essential for the expression of the IP. Of related interest is the question of neurochemically mediated changes in IP production. As cited above, behavioral changes occur during the squirrel monkey's mating season, a time when sex hormone levels are elevated. In particular, the otherwise very quiet Roman-arch males become more active and vocal (Baldwin, 1968; DuMond, 1968). It seems likely there would be a greater tendency for males to utter the IP and other contact-related calls during this season, as they attempt to maintain closer contact with the females forming the central core of the troop. In a study of adult Roman-arch males, Winter (1969b) described some calls given during visual isolation as being transitional in structure between the IP and Twitters. There was no suggestion in his study that these transitional calls were related to the hormonal state of the vocalizer. However, during recording attempts made by D. Symmes and myself of a group of captive Roman-arch individuals at the end of the mating season in Iquitos, Peru, the only adult male from which significant numbers of IPs could be obtained was still in the "fatted" condition. The records from this "fatted" male were particularly interesting, in that he produced numerous long vocal sequences (Fig. 11). There is the possibility, therefore, that endogenous hormone levels may influence the vocal behavior of male squirrel monkeys, a suggestion already made for the vocal behavior of females (Smith *et al.*, 1983). Aside from seasonally related changes in endogenous hormone levels, there is the question of what endogenous neurochemical events mediate the emission of isolation calls upon separation. Studies with rodents (Panksepp *et al.*, 1978) and with squirrel monkeys (Newman *et al.*, 1982) suggest that brain opioids play a role in mediating production of isolation calls.

4. VISUAL DISPLAYS

4.1. Introduction

Squirrel monkeys use a variety of postures and movements that often have communicative significance. These are covered in the review of squirrel monkey behavior by Hopf *et al.* (1974). Here, I will focus on some prominent visuomotor patterns that appear to function mainly as displays.

Undoubtedly the most distinctive display behavior of the squirrel monkey is the "genital display". In its essence, the display involves abduction of one or both thighs, exposing the genital area. Both sexes are known to perform this act, but it is much more commonly observed in males. The most effective stimulus appears to be introduction of a strange male (Winter, 1969b; Baldwin, 1967) and, in Gothic-arch males, seeing their own reflection (MacLean, 1964). There is evidence that incidence of the genital display in adult males is under the influence of testosterone. Baldwin (1967, 1968) observed a significant increase in the display by Roman-arch males during their "fatted" state, a seasonal condition coincident with active spermatogenesis (DuMond, 1968), which is dependent upon a high testosterone titer (Nadler and Rosenblum, 1972). Seasonal cyclicity in the incidence of the genital display is also observable in groups of Gothic-arch males (Mendoza *et al.*, 1978). A similar, though less pronounced, seasonal cyclicity is observable in the mirror-directed genital displays of individually housed Gothic-arch males (MacLean, 1978). This display was recognized as distinctive and described in detail by Ploog and MacLean (1963). Three types of display were recognized: the *distant display*; the *mirror display*; the *communal display*. The distant display was observed during the introduction of a new monkey or, less often, among familiar males. During this display, the monkey inclines the head to one side, abducts one thigh while the penis gradually erects (at which time a few spurts of urine may be seen), and emits a high-pitched, soft vocalization with the mouth retracted. With each vocalization there may be a thrusting forward of the fully erect penis. Typically, there is also scratching of the body with a hand or foot. The mirror display is practically identical with the distant display; full erection may be achieved more rapidly, as the reflected image of the displayer appears to facilitate the response. The communal display may be performed either in the *open* position (several inches from the target animal), or the *closed* position. In the open position, the display may sometimes consist only of thigh-spreading. In the closed position, the male approaches either a male or a female head on, places one or both hands on its back, thrusts the erect penis toward the face, thigh-spreads, and directs its gaze away from the recipient. The animal receiving the display sits in a cowered position with its head lowered. Females were observed to make open displays in the communal situation, thigh-spreading during enlargement of the clitoris. Subsequent to that study, it became apparent that the physical appearance of the males correlated with their tendency to display to their reflection. This led to the establishment of the now well-known "Gothic-arch" and "Roman-arch" distinction (MacLean, 1964), with the distant display and mirror display peculiar to Gothic-arch males. I have observed Gothic-arch males consistently display with their *backs* to the recipient. In one case, the animal displayed to its reflection in a mirror, in the other, to an older, more dominant male. In both cases, the vocalization accompanying the display was a soft Chuck. These displays were otherwise conventional, with an erect penis and spread thigh, the displayer performing while looking over its shoulder at the recipient. Roman-arch males either ignore their reflection in a mirror, or react differently, by bouncing and shaking the cage, and producing a stereotyped kicking movement of one hind foot (MacLean, 1964). DuMond (1968) observed this same stereotyped kick in free-ranging Roman males, given just before or after genital displays or sexual encounters. The kicking behavior is now described as a "kickwash" and is classified as a scent-marking behavior, since it involves the dispersion of urine on the ventral surface, occurs exclusively in males, and is exhibited more frequently during the mating season. Personal observation of Roman-arch males with the "Bolivian" phenotype (cf. Ariga *et al.*, 1978; Thorington, this volume, Chapter 1) indicates that their response to a mirror includes wiggling the ears, rapidly arching the back, and uttering Girren vocalizations.

The genital display appears to be instinctive, having been observed in an infant 2 days old (Ploog *et al.*, 1967). During the first few weeks of life, the display is directed to other group members while the infant remains on the mother. When the infant is several months old and more independent, displays are mostly directed to the mother and to the alpha male in the group. Displays by infants at that age appear to be triggered either by the mother's rejection during nursing attempts or by the adult male's threats during the infant's attempts to gain closer physical contact (Ploog *et al.*, 1967). In a free-ranging Roman-arch troop, Baldwin (1969) observed that infants 8–16 weeks of age would join their mothers in threatening approaching adult males, the infants' threats consisting of "shrieks" and genital displays.

A number of authors have raised the question of what function the genital display serves. Ploog (1967) concluded that a number of behavioral "messages" were being sent by displaying individuals: in adult males, open displays communicated a demanding self-assertion, and a desire to make closer contact, especially toward receptive females, while closed displays were more an expression of aggressive dominance; open displays by females and young animals indicated frustration and defense. It would appear, however, that the function of *appeasement* fits in a more general way most contexts in which displays are produced. The descriptions by Baldwin (1967) of 1- to 4-month-old Roman infants displaying whenever active adult males or other boisterous, excited animals came too near would suggest that the display serves as an appeasing gesture in those situations, reducing the excitability of the recipients. Similarly, displays between adult males and females during the breeding season would fit within the framework of an appeasement function, since a placated animal would be more receptive to close approach by the signaller, an obvious prerequisite for successful copulation. However, an appeasement function seems less likely where the incidence of displaying is highest in the most aggressive and dominant animals, as appears to be the case with Roman-arch males. The display in this context may best viewed as a "derived" ritual, related to increased intermale competition and sexual selection (Baldwin, 1967).

4.2. *Facial Expressions*

Squirrel monkeys do not have the well-developed facial musculature of the Old World monkeys and great apes. Consequently, they lack many of the communicative gestures employing the face seen in their Old World counterparts, such as elevating the eyebrows. However, facial gestures are not entirely lacking. As noted earlier, ear-wiggling is a part of the response of Roman-arch males to their reflection in a mirror, and is observed in close-range encounters of freely ranging Roman-arch individuals (DuMond, 1968). Ear-wiggling has also been observed in Gothic-arch animals during highly aggressive encounters (Marriott and Salzen, 1978). Marriott and Salzen (1978) have described a number of other facial gestures in Gothic-arch individuals. One, the "protruding-lips face," is restricted to males pursuing females during the mating season. Another, the "bared-teeth face," is associated with the genital display or aggressive encounters. Other facial expressions are associated with varying degrees of fear or aggression.

5. *OLFACTORY COMMUNICATION*

Olfaction evidently is important to male monkeys in determining the estrous condition of females. During the mating season, males make persistent approaches to females and sniff their genital region. Since the genital region of female squirrel monkeys changes much less than is the case with many Old World monkeys, males presumably must resort to other cues for judging the receptivity of females. Incidence of anogenital olfactory investigation by males increases significantly during the mating season (Hennessy *et al.*, 1978; Mendoza *et al.*, 1978). There is also an increased incidence of anogenital rubbing by females during this season. Whether urine or a vaginal secretion is the primary source of the scent is unknown. Squirrel monkeys are known to discriminate between the urine odors of familiar and foreign individuals (Candland *et al.*, 1980). Another source from which olfactory signals may arise is the nasal passage. In this case, arousal level and reaction to novel stimuli appear to be related to the incidence of nasal marking (Schwartz and Rosenblum, 1980). Body odors may also be useful in providing communicative information, particular the individual identity of infants (which are frequently sniffed by adult females), as well as promoting early attachment to the mother by infants (Kaplan *et al.*, 1977).

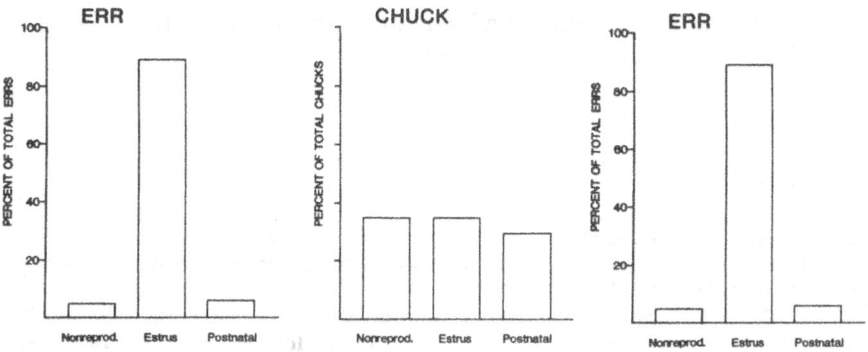

Fig. 12. Seasonal differences in the frequency of occurrence of Errs, Chucks, and Err Chucks. Vocalizers were four adult females living together with an adult male as a social group. Percentages were adjusted for equal observation times during each seasonal sample.

6. *CONCLUSION*

The squirrel monkey continues to be an important primate for behavioral research in the laboratory. It is understandable, therefore, that our knowledge of communication in this species in captive settings has reached a significant level. However, much remains unknown about the mechanisms underlying communication in the squirrel monkey and in primates in general. There is a need for studies addressing the genetic and developmental processes mediating communication-related brain circuits. The importance of experience and its role in neural development needs to be explored, particularly where infant vocal patterns come to be employed in new contexts with adulthood. In addition, the neural mechanisms underlying perception of species-specific communication signals are still poorly known.

In contrast to the abundance of studies of squirrel monkey communication in captive groups, there is a large gap in our knowledge of communication in freely living troops in their natural habitat. The contexts in which some patterns, such as the Isolation Peep, occur can be adequately reproduced in a laboratory setting. However, other patterns, particularly those influenced by cultural factors or involving interactions of several individuals, may occur only in undisturbed populations in nature.

Finally, the complex lexicon out of which the squirrel monkey vocal repertoire is constructed needs to be examined in finer detail. Understanding the behavioral significance of the lexical elements requires a close match in the analytic grain size applied to these lexical elements and concomitant behavior. At a coarse-grained level, the tendency to use particular lexical elements alone or in combination with other elements may vary with seasonal cycles (e.g., Err and Chuck; cf. Fig. 12). Under finer-grained analysis, the incorporation of specific lexical elements into more complex vocalizations may correlate with motivational state, age, or social status. Still more detailed analysis may reveal that parametric variations in particular lexical elements correlate with different contexts, independent of seasonal or motivational factors. The end result of these analyses will be a better understanding of the set of rules that govern communicative behavior in this species.

REFERENCES

Ariga, S., Dukelow, W. R., Emley, G. S., and Hutchinson, R. R., 1978, Possible errors in identification of squirrel monkeys (*Saimiri sciureus*) from different South American points of export, *J. Med. Primatol.* **7**:129–135.

Baldwin, J. D., 1967, A Study of the Social Behavior of a Semifree-Ranging Colony of Squirrel Monkeys (*Saimiri sciureus*), Ph.D. Dissertation, Johns Hopkins University, Baltimore.

Baldwin, J. D., 1968, The social behavior of adult male squirrel monkeys (*Saimiri sciureus*) in a seminatural environment, *Folia Primatol.* **9**:281–314.

Baldwin, J. D., 1969, The ontogeny of social behaviour of squirrel monkeys (*Saimiri sciureus*) in a seminatural environment, *Folia Primatol* **11**:35–79.

Candland, D. K., Blumer, E. S., and Mumford, M. D., 1980, Urine as a communicator in a New World primate, *Saimiri sciureus*, *Anim. Learn. Behav.* **8**:468–480.

DuMond, F. V., 1968, The squirrel monkey in a seminatural environment, in: *The Squirrel Monkey* (L. A. Rosenblum and R. W. Cooper, eds.), Academic Press, New York, pp. 87–145.

Epple, G., 1968, Comparative studies on vocalization in marmoset monkeys (Hapalidae), *Folia Primatol.* **8**:1–40.

Hennessy, M. B., Coe, C. L., Mendoza, S. P., Lowe, E. L., and Levine, S., 1978, Scent-marking and olfactory investigatory behavior in the squirrel monkey (*Saimiri sciureus*), *Behav. Biol.* **24**:57–67.

Hershkovitz, P., 1984, Taxonomy of squirrel monkeys genus *Saimiri* (Cebidae, Platyrrhini), *Am. J. Primatol.* **6**:257–312.

Hopf, S., 1970, Report on a hand-reared squirrel monkey (*Saimiri sciureus*), *Z. Tierpsychol.* **27**:610–621.

Hopf, S., 1972, Sozialpsychologische Untersuchungen zur Verhaltensentwicklung des Totenkopfaffen, Doctoral Dissertation, Philipps-Universitat, Marburg/Lahn.

Hopf, S., Hartmann-Wiesner, E., Kuhlmorgen, B., and Mayer, S., 1974, The behavioral repertoire of the squirrel monkey (*Saimiri*), *Folia Primatol.* **21**:225–249.

Jürgens, U., 1979, Vocalization as an emotional indicator: A neuroethological study in the squirrel monkey, *Behaviour* **69**:88–117.

Jürgens, U., 1982, A neuroethological approach to the classification of vocalization in the squirrel monkey, in: *Primate Communication* (C. T. Snowdon, C. H. Brown, and M. R. Petersen, eds.), Cambridge University Press, Cambridge, pp. 50–62.

Jürgens, U. and Ploog, D., 1970, Cerebral representation of vocalization in the squirrel monkey, *Exp. Brain Res.* **10**:532–554.

Kaplan, J. N., Cubicciotti, D., III, and Redican, W. K., 1977, Olfactory discrimination of squirrel monkey mothers by their infants, *Dev. Psychobiol.* **10**:447–453.

Lieblich, A. K., Symmes, D., Newman, J. D., and Shapiro, M., 1980, Development of the isolation peep in laboratory-bred squirrel monkeys, *Anim. Behav.* **28**:1–9.

MacLean, P. D., 1964, Mirror display in the squirrel monkey, Saimiri sciureus, *Science* **146**:950–952.

MacLean, P. D., 1978, Effects of lesions of globus pallidus on species-typical display behavior of squirrel monkeys, *Brain Res.* **149**:175–196.

Marriott, B. M., and Salzen, E. A., 1978, Facial expressions in captive squirrel monkeys (*Saimiri sciureus*), *Folia Primatol.* **29**:1–18.

Mendoza, S. P., Lowe, E. L., Resko, J. A., and Levine, S., 1978, Seasonal variations in gonadal hormones and social behavior in squirrel monkeys, *Physiol. Behav.* **20**:515–522.

Nadler, R. D., and Rosenblum, L. A., 1972, Hormonal regulation of the "fatted" phenomenon in squirrel monkeys, *Anat. Rec.* **173**:181–188.

Newman, J. D., 1985, The infant cry of primates: An evolutionary perspective, in: *Infant Crying: Theoretical and Research Perspectives* (B. M. Lester and C. F. Z. Boukydis, eds.), Plenum Press, New York pp. 307–323.

Newman, J. D., and MacLean, P. D., 1982, Effects of tegmental lesions on the isolation call of squirrel monkeys, *Brain Res.* **232:**317–329.

Newman, J. D., and Symmes, D., 1982, Inheritance and experience in the acquisition of primate acoustic behavior, in: *Primate Communication* (C. T. Snowdon, C. H. Brown, and M. R. Petersen, eds.), Cambridge University Press, Cambridge, pp. 259–278.

Newman, J. D., Lieblich, A. K., Talmage-Riggs, G., and Symmes, D., 1978, Syllable classification and sequencing in twitter calls of squirrel monkeys (*Saimiri sciureus*), *Z. Tierpsychol.* **47:**77–88.

Newman, J. D., Murphy, and Harbaugh, C. R., 1982, Naloxone-reversible suppression of isolation call production after morphine injections in squirrel monkeys, *Soc. Neurosci. Abstr.* **8:**940.

Newman, J. D., Smith, J. H., and Talmage-Riggs, G., 1983, Structural variability in primate vocalizations and its functional significance: An analysis of squirrel monkey chuck calls, *Folia Primatol.* **40:**114–124.

Panksepp, J., Herman, B. H., Vilberg, T., Bishop, P., and DeEskinazi, F. G., 1978, Endogenous opioids and social behavior, *Neurosci. Biobehav. Rev.* **4:**473–487.

Ploog, D. W., 1967, The behavior of squirrel monkeys (*Saimiri sciureus*) as revealed by sociometry, bioacoustics, and brain stimulation, in: *Social Communication among Primates* (S. A. Altmann, ed.), University of Chicago Press, Chicago, pp. 149–184.

Ploog, D. W., and MacLean, P. D., 1963, Display of penile erection in squirrel monkey (*Saimiri sciureus*), *Anim. Behav.* **11:**32–39.

Ploog, D. W., Hopf, S., and Winter, P., 1967, Ontogenese des Verhaltens von Totenkopf-affen (*Saimiri sciureus*), *Psychol. Forsch.* **31:**1–41.

Ploog, D., Hupfer, K., Jurgens, U., and Newman, J. D., 1975, Neuroethological studies of vocalization in squirrel monkeys with special reference to genetic differences of calling in two subspecies, in: *Growth and Development of the Brain* (M. A. B. Brazier, ed.), Raven Press, New York, pp. 231–254.

Pola, Y. V., and Snowdon, C. T., 1975, The vocalizations of pygmy marmosets (*Cebuella pygmaea*), *Anim. Behav.* **23:**826–842.

Schott, D., 1975, Quantitative analysis of the vocal repertoire of squirrel monkeys (*Saimiri sciureus*), *Z. Tierpsychol.* **38:**225–250.

Schwartz, G. G., and Rosenblum, L. A., 1980, Novelty, arousal, and nasal marking in the squirrel monkey, *Behav. Neural Biol.* **28:**116–122.

Smith, H. J., Newman, J. D., and Symmes, D., 1982a, Vocal concomitants of affiliative behavior in squirrel monkeys, in: *Primate Communication* (C. T. Snowdon, C. H. Brown, and M. R. Petersen, eds.), Cambridge University Press, Cambridge, pp. 30–49.

Smith, H. J., Newman, J. D., Hoffman, H. J., and Fetterly, K., 1982b, Statistical discrimination among vocalizations of individual squirrel monkeys (*Saimiri sciureus*), *Folia Primatol.* **37:**267–279.

Smith, H. J., Newman, J. D., Bernhards, D. E., and Symmes, D., 1983, Effects of reproductive state on vocalizations in squirrel monkeys (*Saimiri sciureus*), *Folia Primatol.* **40:**233–246.

Symmes, D., Newman, J. D., Riggs, G., and Lieblich, A. K., 1979, Individuality and stability of isolation peeps in squirrel monkeys, *Anim. Behav.* **27:**1142–1152.

Talmage-Riggs, G., Winter, P., Ploog, D., and Mayer, W., 1972, Effect of deafening on the vocal behavior of the squirrel monkey (*Saimiri sciureus*), *Folia Primatol.* **17:**404–420.

Thorington, R. W., 1968, Observations of squirrel monkeys in a Colombian forest, in: *The Squirrel Monkey* (L. A. Rosenblum and R. W. Cooper, eds.), Academic Press, New York, pp. 69–85.

Winter, P., 1968, Social communication in the squirrel monkey, in: *The Squirrel Monkey* (L. A. Rosenblum and R. W. Cooper, eds.), Academic Press, New York, pp. 235–253.

Winter, P., 1969a, The variability of peep and twit calls in captive squirrel monkeys (*Saimiri sciureus*), *Folia Primatol.* **10:**204–215.

Winter, P., 1969b, Dialects in squirrel monkeys: Vocalizations of the Roman arch type, *Folia Primatol.* **10:**216–229.

Winter, P., 1972, Observations on the vocal behaviour of free-ranging squirrel monkeys (*Saimiri sciureus*), *Z. Tierpsychol.* **31**:1–7.
Winter, P., Ploog, D., and Latta, J., 1966, Vocal repertoire of the squirrel monkey (*Saimiri sciureus*), its analysis and significance, *Exp. Brain Res.* **1**:359–384.
Winter, P., Handley, P., Ploog, D., and Schott, D., 1973, Ontogeny of squirrel monkeys calls under normal conditions and under acoustic isolation, *Behaviour* **47**:230–239.

5

Physiological Consequences of Maternal Separation and Loss in the Squirrel Monkey

Christopher L. Coe, Sandra G. Wiener, Leon T. Rosenberg, and Seymour Levine

1. INTRODUCTION

As the chapters in this volume indicate, the squirrel monkey has been a valuable subject in many areas of research. One area in which the squirrel monkey has been particularly popular is in the investigation of developmental issues. Our laboratory has found it to be the subject of choice for two reasons. First, the specificity and duration of the emotional bond between mother and infant make the squirrel monkey a good model for the study of attachment processes. Second, the resiliency of the infant and its high adrenocortical hormone levels are extremely conducive to experiments on the effects of psychological and environmental perturbations that occur early in life. Here we will review a series of studies on the physiological consequences of maternal separation in infant squirrel monkeys, and also discuss some of the factors that can ameliorate the deleterious hormonal and immunological effects of maternal loss.

Before we describe the specific findings, a few words should be said about our theoretical perspective and general procedures. Early studies on the squirrel monkey, and in fact on most primates, interpreted the infant's response to separation in behavioral and affective terms (Rosenblum, 1968; Kaplan, 1970; Rosenblum, 1971; Jones and Clark, 1973). Based on the theoretical formulations

Christopher L. Coe, Sandra G. Wiener, and Seymour Levine • Department of Psychiatry and Behavioral Sciences, Stanford University School of Medicine, Stanford, California 94305. *Leon T. Rosenberg* • Department of Medical Microbiology, Stanford University School of Medicine, Stanford, California 94305.

of Bowlby (1973), the infant's agitated activity and vocal responses to removal of the mother were described as indicative of the "protest" phase. As the infant's behavioral distress subsided, it was described as entering a stage of "despair" or, in cases where depression was not observed, a stage of adaptation and acquiescence. This emphasis on the emotional basis of the reaction can be thought of as a ballistic model, in that the infant's behavior follows a fixed course following maternal loss (Rosenblum and Plimpton, 1983). For many reasons, we have found the emphasis on the affective components of the separation response to be inadequate, and also believe that the infant should not be viewed simply as a passive responder. Instead, we have tried to examine the infant's behavioral and physiological responses to separation within the context of stress and coping theory.

From this perspective, the infant is viewed as a more active participant in the mother–infant relationship, and its behavior following maternal loss can be described in functional and cognitive terms as active coping attempts to resolve the aversive nature of the situation. For example, the infant's isolation calls are not simply a reflection of emotional distress; they also represent a concerted effort to reestablish contact with the mother. We have argued in previous review papers that this approach permits a more parsimonious interpretation of some discrepancies between behavioral and physiological responses, and better enables one to understand the influence of environmental variables on the separation response (Coe and Levine, 1981; Coe *et al.*, 1983a). Further, we believe that the infant's interactions with its mother under nondisturbed conditions establishes the basis of the coping strategy it will use to deal with aversive situations. Initially, the young infant learns a pattern of responses to threatening situations while on the mother's back. As it matures, the mother may also become a source of stress as the infant encounters increased maternal rejection associated with the weaning process.

Interpretation of the infant's separation response within the perspective of the stress and coping literature also leads one naturally to investigate a number of physiological parameters. Initially, we chose to measure plasma levels of cortisol, based on extensive research that showed that adrenal activation provides a reliable reflection of internal state (Mason, 1968; Hennessy and Levine, 1979). This simultaneous assessment of behavioral and hormonal measures has proved to be extremely efficacious for interpreting the infant's separation response and the influence of environmental variables. In addition, the striking adrenocortical changes led us to investigate some of the immunological consequences of maternal separation. Since Hans Selye (1950) first conceptualized the stress syndrome, it has been known that increased endocrine activity can result in thymolymphatic involution. In this chapter we will report some recent findings that show that maternal separation can affect the infant's immunoglobulin and complement protein levels, as well as its capacity to mount an antibody response to antigen challenge. Finally, we will describe some preliminary, but exciting, data on changes in brain neurotransmitter activity following separation.

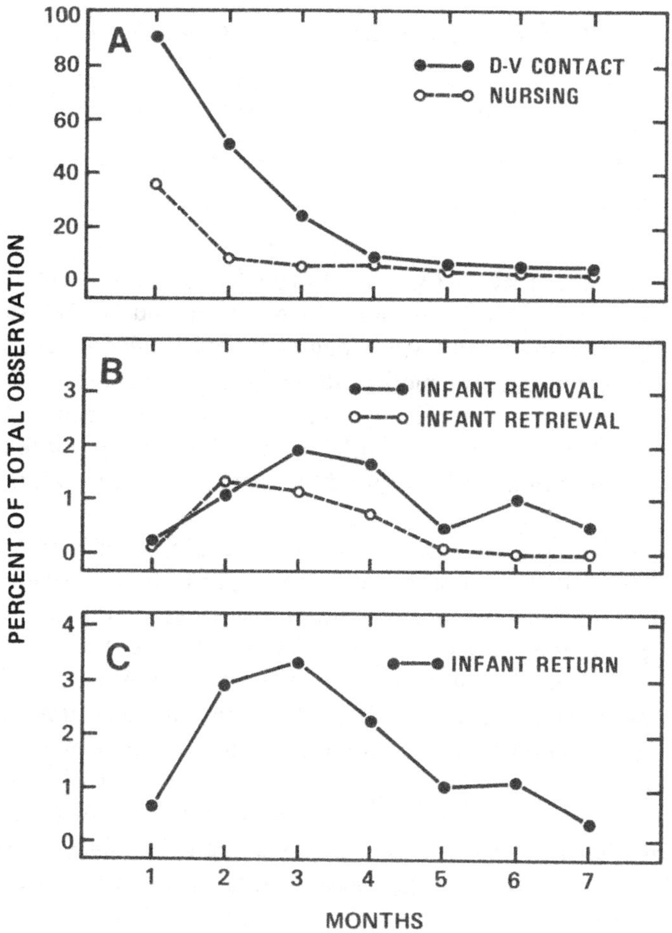

Fig. 1. Normal mother–infant behavior of the squirrel monkey. (A) Contact and nursing time, (B) changes in maternal retrieval and rejection, (C) loss of proximity-seeking by the increasingly independent infant. [Coe and Levine (1981).]

2. NORMAL DEVELOPMENT AND BASIC SEPARATION PROCEDURES

In order to evaluate the following results, it is important to understand the normal mother–infant relationship and the developmental stage at which the perturbations occur. A typical developmental profile is illustrated in Fig. 1 based on longitudinal observations of four male and four female infants (Rosenblum and Coe, 1977). As Rosenblum (1968) described in much greater detail, the infant squirrel monkey begins its life by climbing onto the mother's back, where it remains continuously for the first several weeks of life. The strength and specificity of the attachment bond emerges rapidly, for the mother within

Table I
Infant Body Temperature after 4 hr of Maternal Separation

Age (weeks)	14	16	18	20	22	24	26	28
Temperature (°C)	35.5	38.2	38.7	38.6	38.8	38.9	39.1	39.1
N	5	5	2	4	4	3	3	3

24 hr and for the infant by 1–2 weeks of age (Kaplan, 1979). Nevertheless, in laboratory conditions, where arboreal locomotion and predators are not a threat, the infant matures quickly and begins to make independent ventures by 1 month of age. At this age, the mother attempts to retrieve the departing infant, but by 3 months of age, her disposition shifts and she begins to reject the returning infant (Fig. 1). By 5–6 months of age, the nursing phase is complete and the mother now punitively responds to attempts by the weaned infant to maintain contact.

Our separation studies take the normal developmental pattern into account. For both humane and practical reasons, separation experiments are not initiated until the infants are at least 3 months of age. Our studies on the behavioral and physiological responses to brief separations (1–6 hr) are thus conducted between 3 and 5 months of age, when the infant is sufficiently independent to maintain itself. Experiments on the consequences of more prolonged separations are initiated when the infants are 6–7 months of age. This older age is when most laboratories wean infants into peer groups in order to breed the adult females again, and we have taken advantage of this normal feature of colony management to assess the physiological impact of the severed mother–infant bond. Although the infant is able to survive complete maternal loss at these older ages, we have nevertheless observed profound physiological changes after removal of the mother.

In the course of a study on the response to repeated brief separations that began at 3½ months of age, we were surprised to find that the infants showed a marked drop in body temperature. As can be seen in Table I, when infants were subjected to 4-hr separations, they showed an inability to maintain body temperature at the youngest ages. The inability to show independent thermoregulation at 14 weeks of age was unexpected, because infants normally spend most of the day away from the mother at this age. Although the magnitude of the drop at 14 weeks may have been slightly increased by some evaporative heat loss due to momentary exposure to ether prior to recording, the lower body temperature was already evident at the time of capture. These data suggest that the infant's periodic returns to the mother for nursing and clinging bouts still serve an extremely important function, and concur with a new realization about the importance of thermal factors in regulating ontogeny (Leon *et al.*, 1978; Schwartz and Rosenblum, 1983; Thiessen, 1983). From a pragmatic standpoint, these findings also provided empirical support for our heuristic feeling that separation studies should not begin before this age.

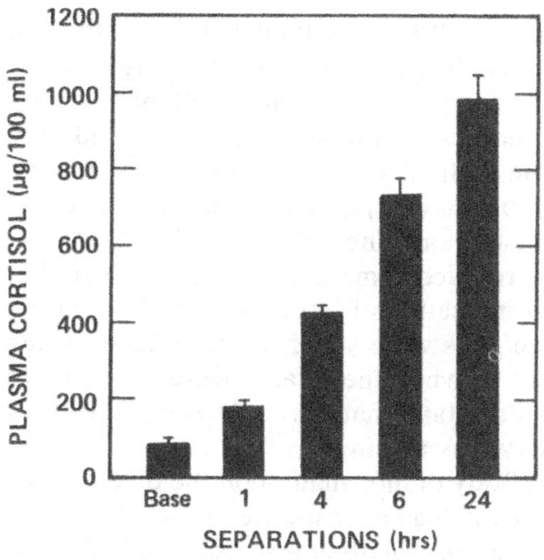

Fig. 2. Plasma cortisol response to maternal separations of various durations.

3. ADRENAL RESPONSES TO MATERNAL SEPARATION

Separation from the mother may be most life-threatening at a young age, but we have repeatedly found that separation leads to marked endocrine activation throughout the first year and even at older ages. In our early experiments we documented that both mother and infant squirrel monkeys show striking elevations of adrenal corticosteroids at 30 min following separation (Coe *et al.*, 1978; Levine *et al.*, 1978; Mendoza *et al.*, 1978). These studies established our basic experimental paradigm of simultaneously assessing behavioral and hormonal responses, and also revealed the unique adrenal physiology of the squirrel monkey. For reasons described in Chapter 8, this species secretes extremely high levels of adrenal hormones and has a remarkable capacity to produce cortisol when stressed (see Coe *et al.*, this volume, Chapter 8). The lability of the adrenocortical system has been of great utility for evaluating how the infant responds to the separation condition.

We have averaged the data from several recent studies to show the striking increase in adrenal corticoids that occurs during the first 24 hr after removal from the mother. As can be seen in Fig. 2, infants typically show basal levels of plasma cortisol between 75 and 150 μg/100 ml, and can undergo sustained increases in cortisol that may rise as high as 500–1000 μg % by 24 hr after separation. However, this adrenal response is not invariant, and in the following section we will describe how the magnitude and duration of the cortisol elevation are affected by the nature of the environment in which the infant is placed during separation.

4. EFFECT OF ENVIRONMENTAL CONDITIONS AND SOCIAL SUPPORT

In previously published papers, we demonstrated that both the behavioral and hormonal responses of the infant were influenced by proximity of the mother during the separation (Coe *et al.*, 1983a). Although infants show a marked cortisol response to all separations, the elevation is significantly reduced at 24 hr when the mother is proximal as compared to when the infant is totally separated. In this experiment on eight mother–infant pairs, the infant's adrenal activation plateaued after 6 hr when the mother was in an adjacent cage, even though physically and visually inaccessible, as compared to a continual increase in cortisol output during 24 hr of total isolation. The behavioral data from this study followed a different course and provided some of our first evidence that the intensity of calling was not a good indication of internal state. In contrast to the cortisol findings, infant isolation calls were significantly reduced when the mother was proximal as compared to when the infants were isolated. In fact, the calling response of the isolated infants had virtually subsided by 24 hr, at a time when their adrenal activity was maximal.

These data added to an increasing body of information on the discrepancy between the infant's calling and its physiological response, which led us to reevaluate the significance of the vocalizations. In other research, we had found that the frequency of calling declined progressively from the onset of separation as cortisol titers rose during the first 6 hr after removal of the mother. Moreover, a separated infant might not call at all if it was left in the presence of a care-giving female, even though it still showed an adrenal response to maternal loss (Coe *et al.*, 1978). We now believe that the level of calling is actually a poor measure of emotional disturbance. Instead, the vocal response may be better viewed as a malleable signaling behavior that functions primarily to elicit proximity-seeking and care-giving behavior from the mother. We shall return to this point in a later discussion of the behavioral and hormonal responses to repeated separation, but first let us consider the important issue of social support.

A number of investigators have reported that the behavioral disturbance of separated monkey infants is considerably reduced, and sometimes entirely ameliorated, when they are left in the presence of familiar conspecifics (Rosenblum and Kaufman, 1968; Dolhinow, 1980). These findings concur with studies on children, which have shown that their separation response is less marked in the home environment and, conversely, that the reaction is aggravated in unfamiliar places or when strangers are present (Robertson and Robertson, 1971; Kagan, 1974; Sroufe *et al.*, 1974; Weinraub and Lewis, 1977). Within the theoretical perspective of the stress and coping literature, it is also clear that familiarity with the environment and the presence of solicitous adults would also serve to reduce the novelty and uncertainty of separation.

We have now conducted several studies addressing the issue of social support and the degree to which it can buffer the behavioral and hormonal responses to separation. As mentioned above, in our first experiment we took advantage of the maternal disposition of a near-term pregnant female that readily engaged in care-giving behavior toward separated infants (Coe *et al.*, 1978). In keeping with the observations of Rosenblum (1971), the alloparenting by this female ameliorated any overt signs of behavioral distress during the 30-min separations. However, when we examined the infant's cortisol levels, we were struck by the high adrenal activation in the seemingly quiescent monkey.

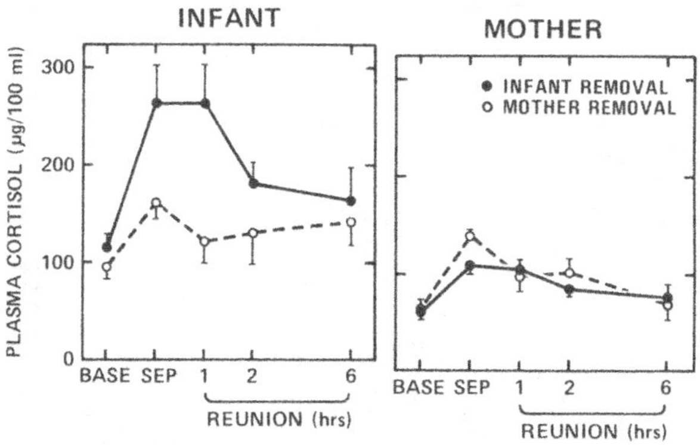

Fig. 3. Beneficial effect of social companions on the infant's adrenal response to 30-min separations in the home cage (mother removal) or a novel environment (infant removal).

In the published paper, we reported on the potentially misleading aspects of the behavioral response, at least during an acute separation.

In more extensive studies conducted since then, we have demonstrated that social support can reduce the adrenocortical stress response when longer time periods are examined. In one followup study designed to look at the protracted effects after reunion from different types of separation, there was an obvious beneficial effect of the home environment (Fig. 3). In this experiment, ten infants were separated for 30 min in a novel cage or the home cage, and then evaluated for 6 hr following reunion with the mother. The cortisol response to separation was significantly lower when the infant was allowed to remain in its social group and throughout the 6 hr after reunion with its mother. In addition, the infants showed a more rapid behavioral adjustment, in terms of their willingness to subsequently depart from the mother and resume play behavior, following separations in the home cage.

A similar effect of the home environment was found in another study that examined the hormonal and behavioral responses of infants after 1-, 2- or 4-hr separations in either the home or a novel cage. In this experiment, we also took advantage of a unique opportunity to examine the potential influence of prior experience with aunting behavior. As Thorington has described (this volume, Chapter 1), squirrel monkeys have speciated across their large range in South America. One of the many behavioral differences is a greater propensity of Peruvian squirrel monkeys to engage in aunting behavior under nondisturbed conditions as compared to Bolivian females, a difference that may be related to the different activity levels and separation responses already evident at birth (Kaplan, 1979; Roy *et al.*, 1981). Therefore, in this experiment, we compared the hormonal and behavioral responses of six Peruvian infants to that of four Bolivian infants.

Observations of the normative behavior of these infants indicated that Peruvian infants moved off their mothers more rapidly than did Bolivian infants, especially between 2 and 3 months of age, and that the Peruvian infants spent significantly more time with nonlactating females during the first 3 months of life. However, this difference in prior alloparenting experience proved to have

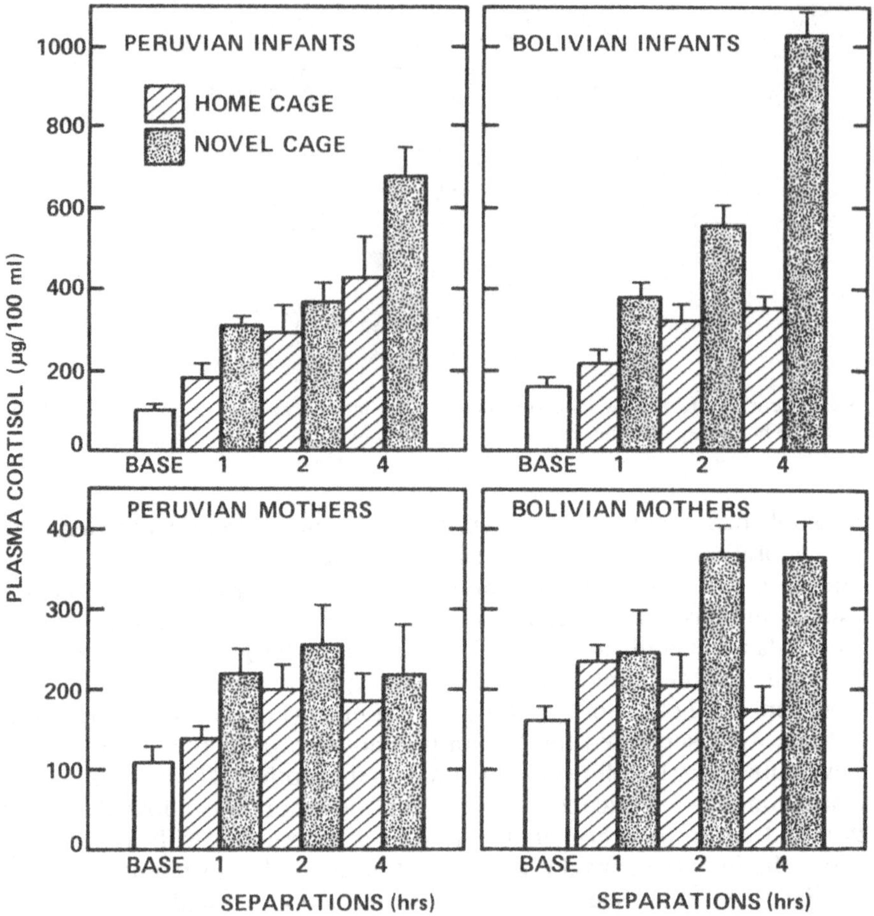

Fig. 4. Reduced cortisol response of mothers and infants separated for 1–4 hr in the home cage as compared to a novel environment. [Coe *et al.* (1983).]

little influence on either the infant's separation response or the beneficial influence of social companionship. All infants showed significant cortisol elevations during the separations, and smaller increases when left in the home environment after mother removal, compared to when isolated in a novel cage (Fig. 4). For both Peruvian and Bolivian infants, 4 hr of total isolation evoked the greatest cortisol response.* The infants' behavior also indicated that they were less distressed when left in the home environment. Isolation calls were emitted during only 20% of the 30-sec intervals in the home cage, as compared to 58% in the novel environment. Similarly, agitated and stereotyped movement patterns were exhibited less frequently in the home cage. The reduced disturbance in the home cage may have been partially attributable to the time

* Bolivian infants showed the greatest increment to 1029 μg %, in keeping with the greater adrenal responsiveness of this subspecies; see maternal data and this volume, Chapter 8.

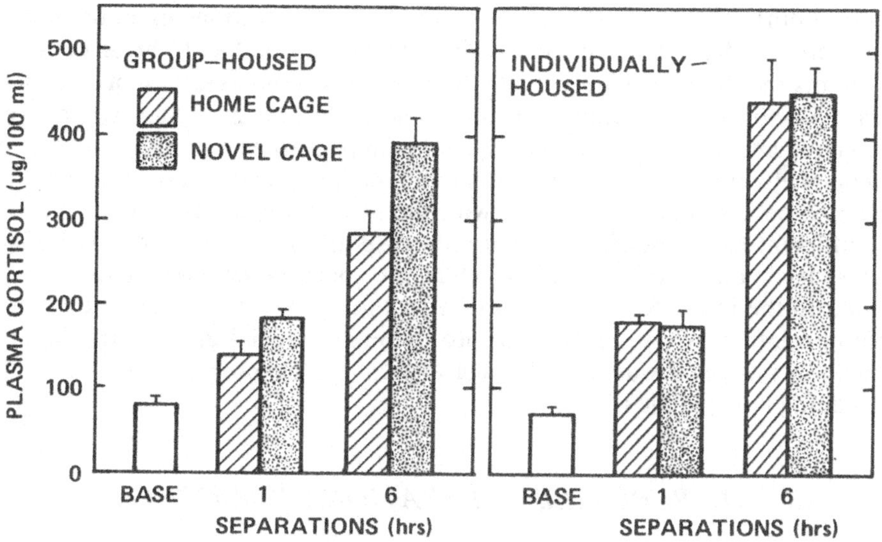

Fig. 5. Beneficial effect of the familiar environment on the separation response of group-reared, but not in individually housed, Guyanese infants.

spent in contact with other adult females, which averaged 20% of the observation. Of interest in this regard was the finding that Bolivian females spent an equivalent amount of time in care-giving behavior directed toward the separated infants. Thus, despite a difference in the propensity to aunt under nondisturbed conditions, both Peruvian and Bolivian females engaged in equal amounts of compensatory behavior toward a distressed infant.

Although both of the previous studies indicated a beneficial effect of social support for the separated infant, there also appeared to be a nonspecific effect of the home environment. The degree of benefit for each individual did not correlate directly with the level of aunting behavior observed; further, the separated mothers also showed a reduction in adrenal activation when they were left in the home environment during infant removal conditions. A recent study assessed more specifically the relative contributions of the social and physical properties of the home environment, and determined whether overt care-giving behavior is a necessary prerequisite for ameliorating the infant's separation response. In this study, 23 Guyanese infants were reared in two different environments. Twelve of the infants were reared in small social groups containing four mother–infant dyads (*group-housed*), and the remaining 11 infants were reared alone with only their mothers (*individually housed*). Thus, during the separations, there were no potential aunts for any of the infants, although group-reared infants had other mother–infant dyads available as social companions. Infants from both rearing conditions were subjected to 1- and 6-hr separations in the home or a novel environment (Fig. 5).

As can be seen, the group-housed infants showed the same beneficial effect of the home environment observed previously, even though no aunting behavior occurred during the separations. As compared to separations in a novel cage, there were significantly lower cortisol increments at both 1 and 6 hr. In contrast, the infants that had been individually housed showed no evidence of a reduced adrenal response to separation in the home environment. Thus, while overt care-giving and solicitous behavior does not appear to be a prerequisite for the beneficial effect of the home environment to occur, social companionship does appear to be essential. Familiarity with the physical features of the cage was not sufficient for reducing the separation response of the individually housed infants. It should also be mentioned that the social companions must also be familiar to the infant for the beneficial effect to occur. In another study where separated infants were introduced into unfamiliar social groups, the infant's separation response was actually aggravated by the presence of potentially threatening conspecifics.

5. *EFFECT OF REPEATED SEPARATION EXPERIENCES*

The previous experiments have shown how the infant's response can shift within a single separation and that environmental factors can differentially influence the behavioral and physiological reactions. However, it is in our studies on the effects of repeated separations that we have been most impressed with the changing relationship between behavioral and physiological measures. The dynamic nature of the infant's response across repeated separations has also provided the strongest evidence of its coping abilities. In several experiments, we have now evaluated the effects of prior experience with separation and the infant's ability to adapt to separation.

In our first experiment on the effect of repeated separations, we examined the behavioral and hormonal responses of five infants across six 1-hr separations conducted at 10-day intervals (Coe *et al.*, 1983b). On a theoretical level, one could have predicted a potentiated response across separations—a type of sensitization—or alternatively, a diminished response if the infant was able to habituate with repeated exposure. Prior studies on rhesus macaques, which evaluated longer and more traumatic separations than we impose in our squirrel monkey studies, had indicated that the infant's behavioral response becomes progressively worse with repeated separations (Mineka, 1981; Mineka *et al.*, 1981). However, these studies focused primarily on the onset of withdrawn-depressive behavior after several days, which occurs only in certain monkeys under select circumstances (Kraemer and McKinney, 1979). In contrast, our studies examined the potential change in the infant's initial response to repeated short separations.

Assessment of the monkeys' plasma cortisol levels indicated that the infants had a relatively constant adrenal response to the six separations, as compared to control infants that were blood-sampled while with their mothers on 6 equivalent days. However, in contrast to the stable adrenal response, the infants' behavioral response changed dramatically across the six separations. During each 1-hr separation, the infants responded with some vocalizations and agitated activity, but the level of calling decreased with each succeeding

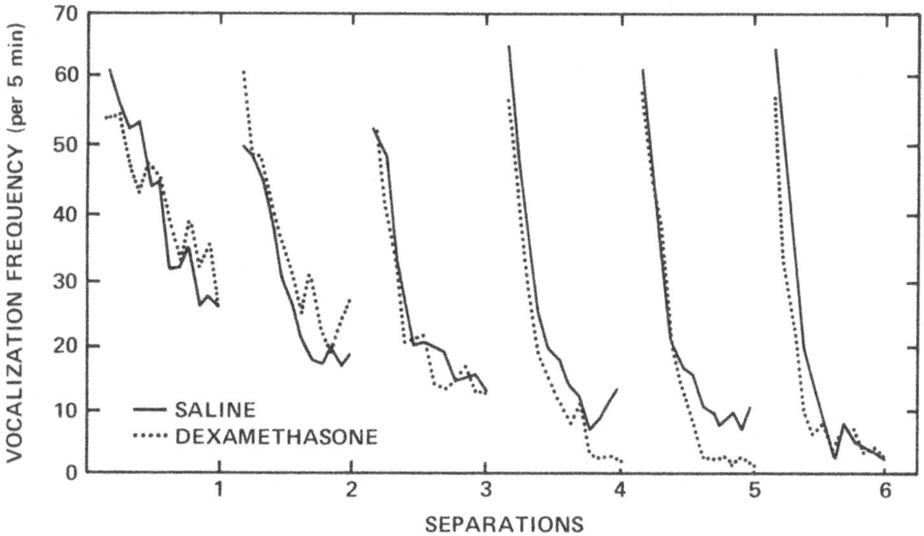

Fig. 6. More rapid reduction in the calling response of infants across repeated separations. Vocalizations were recorded during the first hour after separation, and the total frequency is portrayed at 5-min intervals.

separation. Thus, we were again confronted with a situation where the behavioral and hormonal data initially seemed to be in conflict. On the basis of the vocalization data alone, one might have concluded that the infants were less emotionally distressed after several separations. However, since the adrenal activation continued to occur, we believed that a more parsimonious explanation required a different interpretation of the vocalization data. As stated earlier, we now prefer a more functional interpretation of the isolation calling, which suggests that the decrease in vocalization is due to its ineffectiveness in eliciting the appropriate reinforcer, the mother. Thus, as reported in studies on the response to other stressors, such as electric shock or novelty, one would anticipate a more rapid behavioral adjustment as the organism attempts to cope with the noxious stimuli, and the physiological activation subsides only when behavioral or cognitive processes ultimately permit the individual to adjust to the aversive situation (Ader, 1969; Natelson *et al.*, 1976; Hennessy and Levine, 1979).

If one extends the separations to 4 hr as we did in a second experiment, the rate of learning is more rapid and the infants' behavioral adjustment is more complete. Thus, by the fourth separation, the infants virtually cease to emit the isolation call before the first hour of separation has elapsed (Fig. 6). In spite of this diminished behavioral reaction to separation, it is striking that the infant's adrenal response to separation after six 4-hr separations continues to be as great as on the first separation. In fact, pharmacologically blocking the cortisol response during prior separations has no effect on the basic tendency of the squirrel monkey infant to show adrenal activation to a subsequent separation. In this study, we dexamethasone-blocked (2 mg) the cortisol response

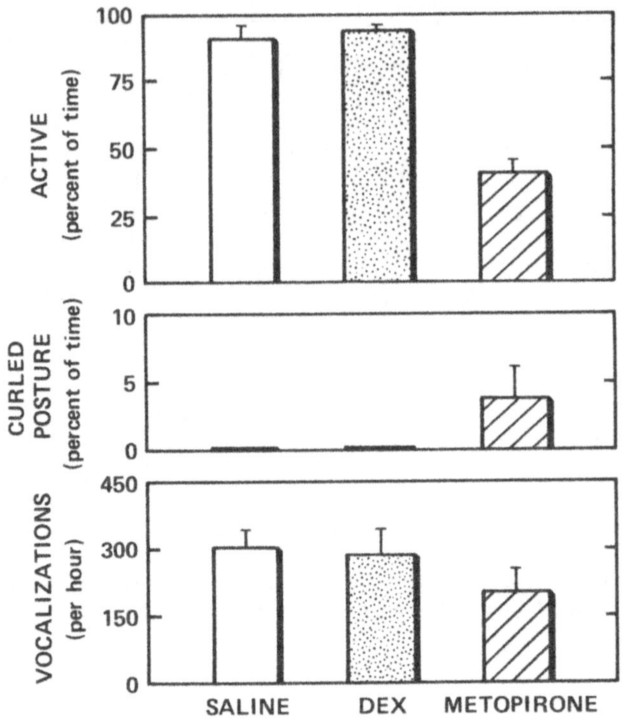

Fig. 7. Reduced activity and calling in separated infants following pretreatment with metopirone.

on the first six separations and then subjected them to a seventh separation with only saline pretreatment. The adrenal response of these infants on the seventh separation was of the normal magnitude, although the cortisol values were slightly lower than those of infants that received only saline injections prior to all previous separations.

6. *IMPORTANCE OF ADRENAL ACTIVATION DURING SEPARATION*

During the course of our investigation of how certain drugs might affect the infant's response to separation, we obtained a demonstration of the functional significance of increased cortisol output during stressful situations. In another phase of the repeated separation paradigm, we examined how metopirone pretreatment would effect the infants. Unlike dexamethasone, which blocks the adrenal release of cortisol but has potent glucocorticoid effects itself, metopirone blocks cortisol synthesis by inhibiting an important synthetic enzyme. Thus, the stressed organism is unable to benefit from the diverse actions of cortisol, and instead, is in a state of chronic ACTH hypersecretion. Four infants were pretreated with 25 mg of metopirone for 3 hr before each of six 4-hr separations that occurred in the usual manner. Figure 7 portrays the debilitating effect of adrenal blockade on the infant's behavioral response to separation. The infants did not engage in the typical pacing and stereotyped movements. Instead, they remained inactive during an average 61% of the observation. Moreover, for the first time in all of our studies, we observed evidence of depressive response. Three of the four infants showed the curled body posture that is usually interpreted as indicative of withdrawal or depression, and this posture was scored on 56% of the separations. The infants continued to emit the isolation call, although vocalizations also tended to be reduced in the more lethargic infants.

In addition to demonstrating an effect of pituitary–adrenal hormones on separation behavior, this finding has important implications for how we view the pituitary–adrenal response. As we will see in the next section, one tends to emphasize the deleterious physiological effects of adrenal activation. However, these negative effects usually occur only after sustained cortisol elevations or at pharmacological levels. Short-term activation may be extremely important for mobilizing the organism to deal with a stressful situation. Corticoids have diverse effects on body physiology, serving to enhance energy mobilization by affecting glucose, fat, water, and ion regulation, and by generally slowing down growth and digestive processes (Lorenzen, 1969; Cope, 1972).

7. *EFFECT OF SEPARATION ON THE IMMUNE SYSTEM*

While the acute adrenal response may be adaptive and necessary for the separated infant, we have also found that the prolonged cortisol elevations induced by sustained separations can adversely affect the infant's immune competence. The idea that stress can affect the immune system is not a new one. As mentioned earlier, Selye (1950) included changes in eosinophil and lymphocyte function, as well as thymolymphatic involution, in his early description of the stress syndrome. More recent work has shown that a variety of psychological and environmental perturbations can affect the immune system, although there is still considerable controversy over the degree of stress required and the neuroendocrine mediators (Ader, 1981). Of particular pertinence to the current work is a prior study indicating that an 11-day separation can affect one measure of immune function in macaques, the proliferative response of their lymphocytes to mitogen stimulation (Reite *et al.*, 1981). Examples of pertinent human studies include: (1) immunological changes following loss of a spouse (Bartrop *et al.*, 1977); (2) the interaction of a sense of loneliness with stressful events in mediating a reduction of natural killer cell activity (Kiecolt-Glaser *et al.*, 1984); (3) a higher prevalence of parental deaths, divorces, and relocations preceding the onset of leukemia in children (Jacobs and Charles, 1980).

We felt that the squirrel monkey might provide a good model for examining how maternal loss would affect the infant's immune system and for delineating the role of corticoids in mediating some of these effects. We will briefly describe two studies: one that found a change in circulating levels of complement proteins and immunoglobulins; and a second that demonstrated a diminution in

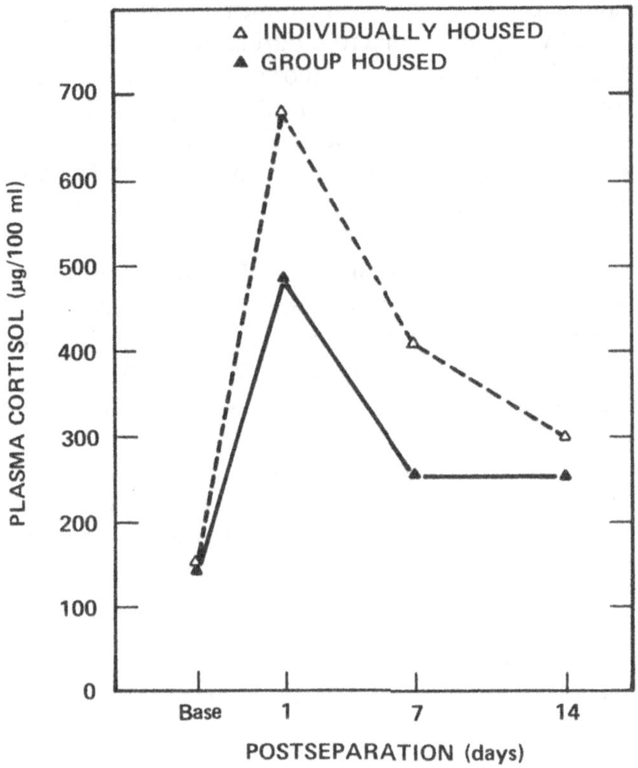

Fig. 8. Adrenal response to a 2-week separation in infants that were housed as a group or individually.

the separated infant's capacity to mount an antibody response to antigen challenge.

In the first study, we monitored 6-month-old infants during two types of separation. In one condition, the separated infants were housed alone in an unfamiliar environment within sight of other separated infants (*individual separation*, $N = 8$). In the second condition, the seven separated infants were allowed to remain as a group in their home cage after removal of their mothers (*group separation*, $N = 4$ and 3). We purposely waited until the infants were completely weaned before imposing the 2-week-long separations, in order not to confound the psychologically induced effect with a nutritionally or thermally induced one. Blood samples were obtained 1 week prior to separation, on the morning after separation, and after 7 and 14 days (Fig. 8).

Both types of separation resulted in significant increases in adrenal secretion throughout the first day, and plasma cortisol levels remained elevated over basal levels for the 2-week period. However, in accord with our prior work, those infants that were allowed to remain in the home cage with other infants had significantly lower cortisol levels at 24 hr and throughout the separation. In keeping with the beneficial effect of familiarity and social compan-

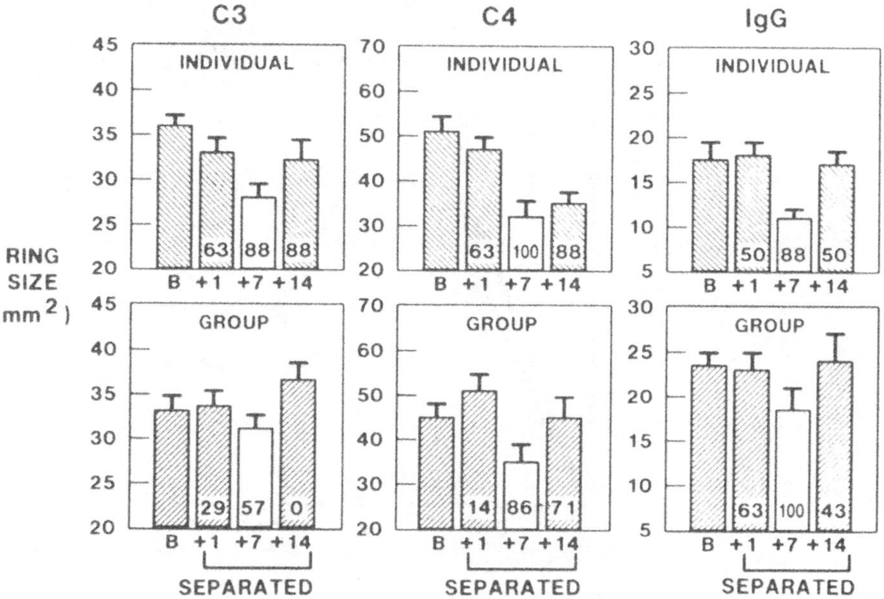

Fig. 9. Effect of a 2-week separation on complement proteins (C3 and C4) and immunoglobulins in infant monkeys. The percent of subjects showing a decrease from basal levels is shown within each bar.

ionship, there was also a smaller reduction in serum levels of complement proteins after separation in the home environment (Fig. 9).

Complement protein levels were measured in the sera of our monkeys by radial immunodiffusion tests (Kallstad Laboratories). The values shown in Fig. 9 are based on the squared diameters of the precipitation ring that forms when sera are added to wells in an agarose gel containing a monospecific antiserum for C3 or C4. As the principal effector in humoral immune reactivity, the complement system is an important object of study. It is involved in immune responses to bacteria, facilitates the formation of antigen–antibody complexes, and assists in localized immune responses by affecting vascular permeability. Decreased levels of C3 and C4 are found in a number of disorders, including systemic lupus erythematosis, glomerulonephritis, rheumatoid arthritis, and chronic liver diseases.

Infants separated in both types of housing conditions showed a significant decrease in C3 and C4 levels by 7 days after maternal loss. However, the decline in the individual separation condition was significantly greater, sustained longer, and more consistently shown by the isolated infants. As can be seen in Fig. 9, 88% of the infants housed alone still showed reduced levels of C3 and C4 by 14 days after the onset of separation, whereas infants housed as a group had returned to normal levels or had rebounded above normal levels by 14 days.

Separation also caused a significant decrease in serum levels of immunoglobulins, although there was no differential effect of housing conditions on

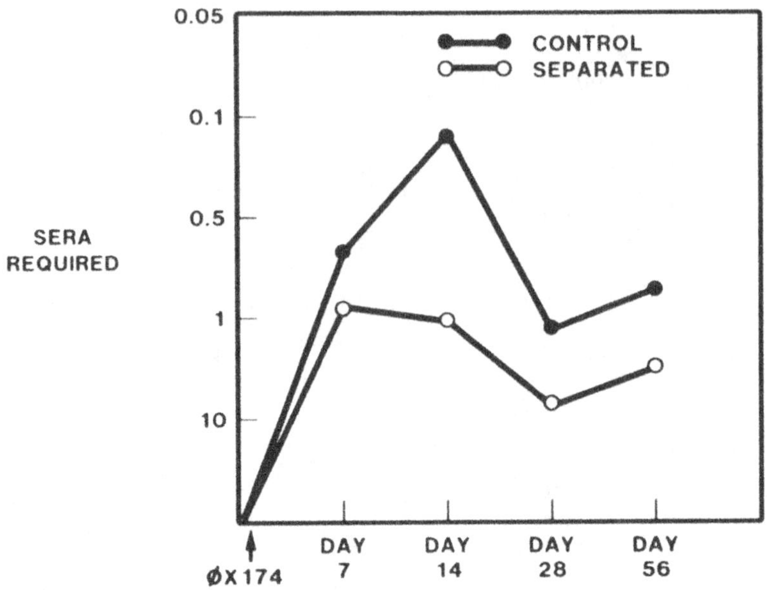

Fig. 10. Antibody response to a bacteriophage in control and separated infants. A higher antibody titer in control subjects is evident by the smaller amount of serum required for 50% neutralization of the phage. Serum volumes are presented in μl required to neutralize an expected plaque count.

this measure. The levels of IgG, which account for approximately 80% of the immunoglobulin titer in circulation, are also portrayed in Fig. 9. No effect was apparent on the first day of separation, but a significant decrease in IgG had occurred after 7 days in both conditions. Unlike the prolonged effect on complement, however, the infants returned to normal levels by 14 days.

Evidence for the functional significance of these immune changes after separation was obtained in a second experiment that assessed the antibody response to antigen challenge. Based on a number of human and animal studies, we evaluated the separated infant's response to a benign bacteriophage that evokes a reliable T-dependent response (Stashak *et al.*, 1970; Peacock *et al.*, 1973). We utilized the phage ϕX174, which does not cause any clinical symptoms, is passed out of circulation quickly, and requires a specific strain of *Escherichia coli* to subsist. The production of antibody in the monkey was measured with a phage neutralization assay. Antibody titers were assessed by determining the volume of sera required to neutralize 50% of a known quantity of virus incubated for 3 hr on *E. coli* C (Fig. 10). Sera were collected at 1, 2, 4, and 8 weeks after IV administration of ϕX174 (10^8 pfu).

We have now compared the response of six undisturbed infants to that of seven infants inoculated after separation. The six undisturbed infants remained with their mothers before and after inoculation; four infants received the phage after a 7-day separation and then remained separated from their mothers; three infants were inoculated after a 7-day separation, but were then reunited with their mothers following administration of the phage. Although our sample size is still relatively small, the four continuously separated infants showed lower levels of antibody at 1 and 2 weeks after receiving the phage than did all of the six control monkeys (Fig. 10). The neutralizing capacity of the sera from separated infants was also lower at 4 and 8 weeks, but this difference was primarily a continued reflection of the initial difference in the magnitude of the primary response during the first 2 weeks of antibody production. It is also of interest that the three infants reunited after phage administration showed an intermediate response on day 14 when the tenfold difference in antibody titer was observed between undisturbed and separated infants. Thus, it appears that sustained stress evoked by separation before and after administration had a strong effect on antibody production, while terminating the stress at the onset of the primary response reduced the inhibitory effect of separation.

8. NEUROTRANSMITTER ACTIVITY DURING SEPARATION

The prior studies have indicated that maternal separation evokes a number of behavioral and physiological changes in the young monkey. We would like to conclude our review with one more set of stress-induced changes that we have observed. These exciting data derive from a preliminary study that has been conducted in collaboration with Drs. K. Faull, R. King, and J. Barchas (Department of Psychiatry, Stanford University School of Medicine). In this work, we examined changes in monoamine metabolites in cerebrospinal fluid (CSF). These metabolites provide an opportunity to assess the activity of brain dopaminergic, serotonergic, and noradrenergic neurons: homovanillic acid (HVA) and dihydroxyphenylacetic acid (DOPAC); 5-hydroxyindoleacetic acid (5HIAA); and 3-methoxy-4-hydroxyphenylglycol (MHPG); respectively. A large literature already exists on stress-induced changes in rodents (Dunn and File, 1983; Swenson and Vogel, 1983), and Kraemer and McKinney (1979) have pioneered research on infant primates. The latter investigators have suggested that norepinephrine activity, in particular, may be predictive of which infant rhesus monkeys will lapse into the depressive response following separation.

In our initial study, we chose to examine the response of yearling monkeys in order to establish the feasibility of the research procedures. Thus, the separation paradigm involved separating juvenile monkeys from their peer group for either 1, 6, or 24 hr into a novel, isolated condition. Eighteen monkeys were studied, six at each time point, and both a basal and a stress sample were taken on each monkey in a counterbalanced order. The preliminary findings are shown in Table II illustrating the change in the various CSF metabolites over time. Based on all of the metabolites, there was clear evidence of increased cerebral monoamine activity after separation. For the metabolites of dopamine and serotonin, the effect was greatest during the first 6 hr after removal from the social group, while the norepinephrine metabolite MHPG remained elevated throughout the 24-hr period. Beyond the general effect of separation on brain activity, two correlations warrant mention. First, basal levels of the dopamine metabolites HVA and DOPAC showed a strong positive correlation with the initial level of agitated activity shown by the separated juveniles ($r =$

Table II
Monoamine Metabolite Levels (ng/ml) in CSF of Juvenile Monkeys after Social Separation[a]

	Base (N = 18)	1 Hr (N = 6)	6 Hr (N = 6)	24 Hr (N = 6)
HVA	204	260	342	256
DOPAC	4	7	10	2
5HIAA	111	141	150	125
MHPG	24	38	37	39

[a] HVA, Homovanillic acid; DOPAC, dihydroxyphenylacetic acid; 5HIAA, 5-hydroxyindoleacetic acid; MHPG, 3-methoxy-4-hydroxyphenylglycol.

0.51, n = 18). Second, the first 30 min of activity shown by the separated juveniles had a strong negative correlation with the final levels of MHPG manifested the next day at 24 hr (r = −0.94, n = 6). The source of these correlations remains to be determined. Investigators have reported that stress-induced increases in dopamine metabolism may derive from the frontal cortex (Thierry *et al.*, 1976; Dunn and File, 1983), although dopamine clearly is involved in nigrostriatal areas (Schofield and Dixson, 1982). In turn, the MHPG finding has several interpretations. First, the lack of norepinephrine depletion over 24 hr may partially account for the capacity of squirrel monkeys to maintain high levels of vigilance and activity across sustained separations. However, the inverse correlation between initial activity of individual monkeys and their final MHPG values after 24 hr suggests that there may ultimately be reduced norepinephrine if the high activity induced by separation is sustained.

9. GENERAL CONSIDERATIONS

In this chapter, we have tried to provide a broad review of many different studies on the effects of maternal separation in the squirrel monkey. These studies clearly demonstrate the profound behavioral and physiological consequences of maternal loss for the infant monkey. The initial behavioral response is best described as one of intense agitation involving high levels of rapid movement and frequent isolation calls. These behaviors are certainly designed to resolve the aversive event by effecting reunion with the mother. As Rosenblum and Plimpton (1983) have suggested, the "protest" behavior is probably the primary adaptive response, since the isolated infant would have little chance of survival under natural circumstances. As a consequence, a sustained separation becomes a source of severe stress for the infant if its behavioral repertoire is insufficient for establishing reunion. We have delineated some of these stressful physiological changes involving the endocrine, immune, and nervous systems. The potency and pervasive nature of these responses make mother–infant separation an excellent paradigm for investigators interested in studying the consequences of psychologically induced stress in the young organism.

It should be emphasized, however, that the infant monkey does show considerable resiliency in responding and adapting to this threatening situation. In spite of the physiological changes, or more probably because of them, the infant continues to show high levels of agitated movement for several days, which would provide the only likelihood of reunion once calling had failed to elicit retrieval by the mother. Our statements about the serious physiological consequences of involuntary separation should also be tempered by the fact that all of the changes do ultimately subside as the infants acquiesce to maternal loss. Even the monkeys that showed the sustained immunological changes did in a few weeks become healthy, functioning members of a large group of juveniles. It is possible, however, that these findings suggest that researchers should wean infants at a somewhat older age if it is a practical feature of colony management.

With regard to the resiliency of the squirrel monkey infant, it is of interest that one rarely observes the withdrawn-depressive behavior described from some macaque species following separation (Rosenblum and Kaufman, 1967; Mineka, 1981). It is possible that some of the infants' hardiness lies in the fact that all of our subjects are normally reared by their mothers prior to the studies. Recent research on the rhesus macaque has indicated that only about 15% of infants will lapse into the depressive phase unless one subjects them to traumatic rearing or separation procedures (Kaufman and Stynes, 1978; Kraemer and McKinney, 1979; Suomi *et al.*, 1983). In particular, peer-rearing instead of normal mother-rearing seems to predispose the infant monkey to an inability to cope with psychogenic stress. In our studies, which have involved more than 100 infants, we have observed the depressive posture in only three instances: (1) when the normal adrenal response to separation was blocked with metopirone; (2) when a physically weak offspring or "runt" was weaned; and (3) when a separated infant was overwhelmed by placement in an unfamiliar adult group.

Another example of the infant's resiliency is its capacity to take advantage of other aspects of the social environment during the separation in lieu of its attachment figure. In the presence of familiar conspecifics, and especially in the case of solicitous adult females, the physiological changes are markedly reduced. Indeed, we have recently had an exemplary case study in our squirrel monkey colony. An infant, barely over 3 months of age, lost its mother to illness. Under experimental conditions, an isolated monkey of this age would certainly not survive. Yet, initially due to the presence of familiar mothers and infants, and subsequently due to the "aunting" behavior of a subadult female, this infant continues to thrive. This capacity of the infant to shift its attachment bond and to derive benefit from other social companions speaks both to the will of organisms to survive and to the many biological advantages gained by a gregarious species. It is probably this early experience with the rewarding aspects of social relations that establishes the motivational basis of sociality in monkey infants and the reliance on group membership in adulthood.

ACKNOWLEDGMENTS. This research was supported by HD-02881 from NICH&HD; MH-23645 from NIMH; and Research Scientist Award MH-19936

from NIMH to S. Levine. The immunological studies were supported by MH-38126 from NIMH; and BRSG Grant RR5353 awarded by the Biomedical Research Support Grant Program, Division of Research Resources, NIH, to C. Coe. The authors would like to acknowledge the active participation of J. Glass, C. Gonzalez, H. Hu, A. Jacks, D. Johnson, E. Lowe, G. Mathews, and A. Vickers in this research.

REFERENCES

Ader, R., 1969, Adrenocortical function and the measurement of emotionality, *Ann. N.Y. Acad. Sci.* **159**:791–805.

Ader, R., 1981, *Psychoneuroimmunology*, Academic Press, New York.

Bartrop, R. W., Lazarus, L., Luckhurst, E., Kiloh, L. G., and Penny, R., 1977, Depressed lymphocyte functions after bereavement, *Lancet* **1**(B016):834–836.

Bowlby, J., 1973, *Attachment and Loss*, Volume 2, *Anxiety and Anger*, Basic Books, New York.

Coe, C. L., and Levine, S., 1981, Normal responses to mother–infant separation in nonhuman primates, in: *Anxiety: New Research and Changing Concepts* (D. F. Klein and J. G. Rabkin, eds.), Raven Press, New York, pp. 155–177.

Coe, C. L., Mendoza, S. P., Smotherman, W. P., and Levine, S., 1978, Mother–infant attachment in the squirrel monkey: Adrenal response to separation, *Behav. Biol.* **22**:256–263.

Coe, C. L., Wiener, S. G., and Levine, S., 1983a, Psychoendocrine responses of mother and infant monkeys to disturbance and separation, in: *Symbiosis in Parent–Offspring Interactions* (L. A. Rosenblum and H. Moltz, eds.), Plenum Press, New York, pp. 189–214.

Coe, C. L., Glass, J. C., Wiener, S. G., and Levine, S., 1983b, Behavioral, but not physiological, adaptation to repeated separation in mother and infant primates, *Psychoneuroendocrinology* **8**(4):401–409.

Cope, C. L., 1972, Adrenal function and stress, in: *Adrenal Steroids and Disease* (C. L. Cope, ed.), J. P. Lippincott, Philadelphia, pp. 199–225.

Dolhinow, P., 1980, An experimental study of mother loss in the Indian langur monkey (*Presbytis entellus*), *Folia Primatol.* **33**:77–128.

Dunn, A. J., and File, S. E., 1983, Cold restraint alters dopamine metabolism in frontal cortex, nucleus accumbens and neostratium, *Physiol. Behav.* **31**:511–513.

Hennessy, J. W., and Levine, S., 1979, Stress, arousal, and the pituitary-adrenal system: A psychoendocrine hypothesis, in: *Progress in Psychobiology and Physiological Psychology*, Volume 8 (J. M. Sprague and A. N. Epstein, eds.), Academic Press, New York, pp. 133–178.

Jacobs, T. J., and Charles, E., 1980, Life events and the occurrence of cancer in children, *Psychosom. Med.* **42**:11–24.

Jones, B. C., and Clark, D. L., 1973, Mother–infant separation in squirrel monkeys living in a group, *Dev. Psychobiol.* **6**:259–269.

Kagan, J., 1974, Discrepancy, temperament and infant distress, in: *The Origins of Fear* (M. Lewis and L. A. Rosenblum, eds.), Wiley, New York, pp. 229–248.

Kaplan, J., 1970, The effects of separation and reunion on the behavior of mother and infant squirrel monkeys, *Dev. Psychobiol.* **3**:43–52.

Kaplan, J. N., 1979, Growth and development of infant squirrel monkeys during the first six months of life, in: *Nursery Care of Nonhuman Primates* (G. C. Ruppenthal and D. J. Reese, eds.), Plenum Press, New York, pp. 153–164.

Kaufman, I. C., and Stynes, A. J., 1978, Depression can be induced in a bonnet macaque infant, *Psychosom. Med.* **40**:71–75.

Kiecolt-Glaser, J. K., Garner, W., Speicher, C., Pean, G. M., Holiday, J., and Glaser, R. 1984, Psychosocial modifiers of immunocompetence in medical students, *Psychosom. Med.* **46**:7–14.

Kraemer, G. W., and McKinney, W. T., 1979, Interactions of pharmacological agents which alter biogenic amine metabolism and depression, *J. Affect. Dis.* **1**:33–54.

Leon, M., Croskerry, P., and Smith, G., 1978, Thermal control of mother–infant contact in rats, *Physiol. Behav.* **21**:793–811.

Levine, S., Coe, C. L., Smotherman, W. P., and Kaplan, J. N., 1978, Prolonged cortisol elevation in the infant squirrel monkey after reunion with mother, *Physiol Behav.* **20**:7–10.

Lorenzen, I., 1969, Glucocorticoids in internal medicine, *Acta Med. Scand.* (Suppl. 500):1–83.

Mason, J. W., 1968, A review of psychoendocrine research on the pituitary-adrenal cortical system, *Psychosom. Med.* **30**:576–607.

Mendoza, S. P., Smotherman, W. P., Miner, M. T., Kaplan, J. N., and Levine, S., 1978, Pituitary-adrenal response to separation in mother and infant squirrel monkeys, *Dev. Psychobiol.* **11**:169–175.

Mineka, S., 1981, Depression and helplessness in primates, in: *Primate Behavior and Child Nurturance*, Volume 3 (H. Fitzgerald, J. Mullins, and P. Gage, eds.), Plenum Press, New York.

Mineka, S., Suomi, S. J., and DeLizio, R., 1981, Multiple separations in adolescent monkeys: An opponent-process interpretation, *J. Exp. Psychol. Gen.* **110**:56–85.

Natelson, B. H., Krasnegor, N., and Holaday, J. W., 1976, Relations between behavioral arousal and plasma cortisol levels in monkeys performing repeated free-operant avoidance sessions, *J. Comp. Physiol. Psychol.* **90**:958–969.

Peacock, D. B., Jones, J. V., and Gough, M., 1973, The immune response to ϕX174 in man, *Clin. Exp. Immunol.* **13**:497–513.

Reite, M. L., Harbeck, R., and Hoffman, A., 1981, Altered cellular immune response following peer separation, *Life Sci.* **29**:1133–1136.

Robertson, J., and Robertson, J., 1971, Young children in brief separation: A fresh look, *Psychoanal. Study Child* **26**:264–315.

Rosenblum, L. A., 1968, Mother–infant relations and early behavioral development in the squirrel monkey, in: *The Squirrel Monkey* (L. A. Rosenblum and R. W. Cooper, eds.), Academic Press, New York, pp. 207–233.

Rosenblum, L. A., 1971, Infant attachment in monkeys, in: *The Origins of Human Social Relations* (R. Schaffer, ed.), Academic Press, New York, pp. 85–113.

Rosenblum, L. A., and Coe, C. L., 1977, The influence of social structure on squirrel monkey socialization, in: *Primate Bio-Social Development* (S. Chevalier-Skolnikoff and F. E. Poirier, eds.), Garland, New York, pp. 479–500.

Rosenblum, L. A., and Kaufman, I. C., 1967, Laboratory observations of early mother–infant relations in pigtail and bonnet macaques, in: *Social Communication among Primates* (S. A. Altmann, ed.), University of Chicago Press, Chicago, pp. 33–41.

Rosenblum, L. A., and Kaufman, I. C., 1968, Variations in infant development and response to maternal loss in monkeys, *Am. J. Orthopsychiatr.* **38**:418–426.

Rosenblum, L. A., and Plimpton, E. H., 1981, Adaptation to separation: The infant's effort to cope with an altered environment, in: *The Uncommon Child: Genesis of Behavior*, Volume 3 (M. Lewis and L. A. Rosenblum, eds.), Plenum Press, New York, pp. 225–257.

Roy, M. A., Hennig, C. W., Hill, E. M., and Walker, M. L., 1981, Subspecies differences in curiosity and general activity for developing squirrel monkeys, *Dev. Psychobiol.* **14**:55–58.

Schofield, S. P. M., and Dixson, A. F., 1982, Comparative anatomy of brain monoaminergic neurons in New World and Old World monkeys, *Am. J. Primatol.* **2**:3–19.

Schwartz, G. G., and Rosenblum, L. A., 1983, Allometric influences on primate mothers and infants, in: *Symbiosis in Parent–Offspring Interactions* (L. A. Rosenblum and H. Moltz, eds.), Plenum Press, New York, pp. 215–248.

Selye, H., 1950, *Stress*, Acta, Montreal.

Sroufe, L. A., Waters, E., and Matas, L., 1974, Contextual determinants of infant affective response, in: *The Origins of Fear* (M. Lewis and L. A. Rosenblum, eds.), Wiley, New York, pp. 49–72.

Stashak, P. W., Baker, P. J., and Roberson, B. S., 1970, The serum antibody response to bacteriophage ϕX174 in germ-free and conventionally reared mice, *Immunology* **18**:295–305.

Suomi, S. J., Mineka, S., and DeLizio, R. D., 1983, Short- and long-term effects of repetitive mother–infant separations on social development in rhesus monkeys, *Dev. Psychobiol.* **19**:770–786.

Swenson, R. M., and Vogel, W. H., 1983, Plasma catecholamine changes during coping in rats exposed to stressful footshock, *Physiol. Biochem. Behav.* **18**:689–693.

Thierry, A. M., Tassier, J. P., Blanc, G., and Glowinski, J., 1976, Selective activation of the mesocortical DA system by stress, *Nature* **263**:242–243.

Thiessen, D. D., 1983, The thermoenergetics of communication and social interaction among Mongolian gerbils, in: *Symbiosis in Parent–Offspring Interactions* (L. A. Rosenblum and H. Moltz, eds.), Plenum Press, New York, pp. 113–144.

Weinraub, M., and Lewis, M., 1977, The determinants of children's responses to separation, *Monogr. Soc. Res. Child Dev.* **42**(172):1–78.

6

Effects of Surrogate-Rearing on the Infant Squirrel Monkey

Michael B. Hennessy

1. INTRODUCTION

At the time of this book's preceding companion volume, there were no pubished accounts of surrogate-rearing in the squirrel monkey. However, in the intervening years, this situation has changed dramatically. Young *Saimiri* have been raised on maternal surrogates in a number of laboratories, and more is now known about the effects of surrogate-rearing in this primate species than in any other, with the exception of the rhesus macaque. Most of the work with surrogate-reared *Saimiri* has focused on the infant's responsiveness and emotional attachment to its rearing figure. In addition, a substantial amount of effort has been directed at understanding the general behavior development of these animals.

Of course, the primary impetus for raising infants on surrogates is not to determine the effects of artificial rearing *per se*. Rather, it is to use the surrogate model as a tool, which in conjunction with the study of mother-reared animals, can help us to sort out the various factors that normally shape infant development. Because features of the surrogate can be readily modified, the impact of specific maternal characteristics can be assessed by comparing infants raised on surrogates with and without the particular characteristic in question. This approach is probably best exemplified in Harlow's (1958) original work on the importance of tactual cues in eliciting clinging and that of Mason and Berkson (1975) on the effect of rearing-figure mobility on the development of stereotypic body-rocking in the rhesus macaque. The inanimate surrogate [as distinguished from animate surrogates, such as dogs (W. A. Mason and Kenney, 1974)] also provides a means for holding maternal variables constant. This can be useful in determining whether an experimental treatment "directly" influences the infant or does so indirectly, by altering the mother's behavior or physiology,

Michael B. Hennessy • Department of Psychology, Wright State University, Dayton, Ohio 42435.

which in turn affects the infant. Similarly, in tests using the rearing figure as a stimulus or choice object, one can be assured that the response of the infant is not influenced by the rearing figure's reaction to the presence of the infant, if the rearing figure is an inanimate surrogate. Cross-species comparisons of the effects of surrogate-rearing can also be of value if these differences can be related in a meaningful way to species differences in normal patterns of mother–infant interactions, the capacities of the infants, or the environmental demands the animals normally encounter.

The present chapter will review what is known about the effects of surrogate-rearing in *Saimiri*. I will first discuss the infant's responsiveness to the surrogate, and next the effects of separation from the surrogate and exposure to novelty. The social and "atypical" behaviors of surrogate-reared squirrel monkeys will then be addressed. Finally, comparisons of the effects of surrogate-rearing in squirrel monkeys and macaques will be summarized. In the course of these discussions, implications for understanding the normal development of mother-reared animals will be considered.

2. RESPONSIVENESS TO THE SURROGATE

Figure 1 shows an infant squirrel monkey with a surrogate of the type used in our laboratory. The surrogate consists of a plastic cylinder (25 cm long × 5 cm in diameter) covered with soft fabric and mounted at about a 30° angle to the cage floor. An internal heat source is used to warm the surrogate for about the first month of life. A small plastic nursing bottle is fitted in the surrogate so that a plastic nipple protrudes from its undersurface, from which the infant can nurse in the typical (ventral–ventral) position. At about 6 weeks of age, this bottle is replaced with a larger bottle mounted on the front of the cage. Human infant formula is used. Monkeys are ordinarily taken from the mother and housed individually with their surrogate at about 1 week of age. The infant behaves in an agitated manner for 1 hr or so after removal from the mother, and then starts to cling to the surrogate in the usual position. However, infants do not spontaneously exhibit nursing behavior, and they must be trained to climb around the surrogate to the nipple. Once nursing is established, the behavior of the infant toward its surrogate is remarkably similar to that of an infant toward the biological mother.

Infants quickly begin to recognize their own surrogate and to prefer it to similar surrogates. At 2 weeks of age—1 week following initial presentation of their surrogate—infants placed on an unfamiliar surrogate (i.e., having a cover that either has been taken from an age-mate's surrogate, or has been freshly laundered) vocalize more, sniff the surrogate more, and climb off the surrogate more quickly than they do if placed on their own familiar surrogate (Kaplan and Cubicciotti, 1980). This discrimination is almost certainly mediated by olfactory cues. At 4 weeks of age, infants allowed to choose between their own surrogate and a surrogate with a clean cover spend significantly more time

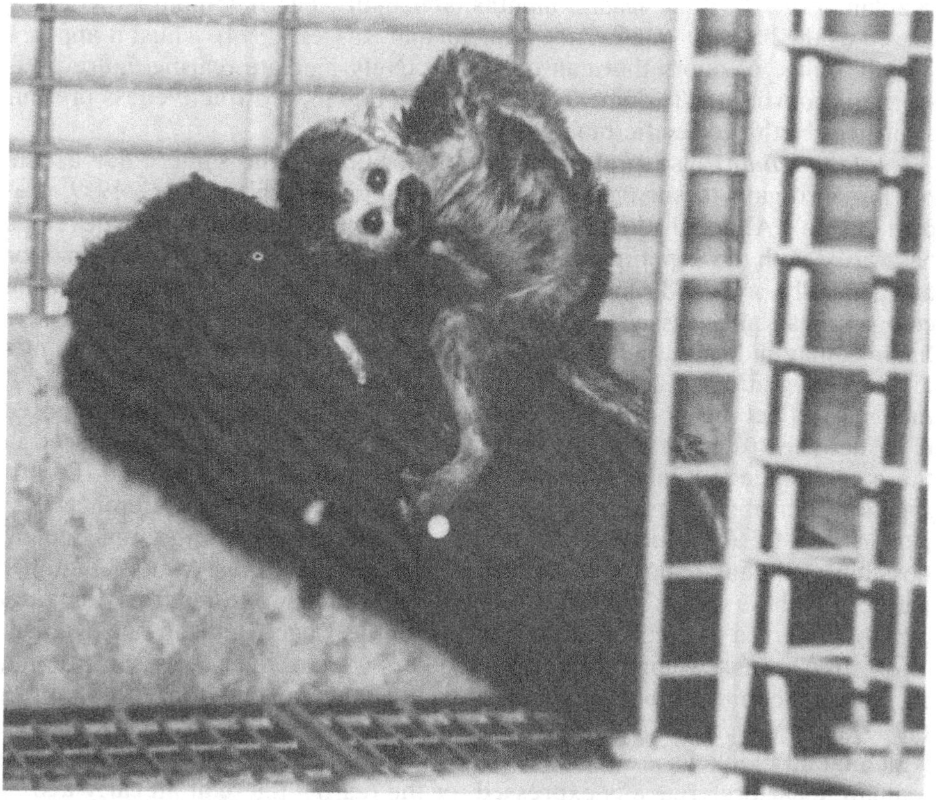

Fig. 1. Infant squirrel monkey on surrogate in our laboratory. This infant is scratching while clinging to the surrogate in the same manner as mother-reared infants perform this behavior while clinging to the mother. Note the nipple of the nursing bottle protruding from the undersurface of the surrogate.

on their own surrogate (Kaplan and Russell, 1974). This preference persists through at least 24 weeks of age (Kaplan, 1977a). *Saimiri* raised with their natural mother can discriminate the odor of the mother's fur from that of another lactating female (Kaplan *et al.*, 1977). The data from surrogate-reared animals suggest that this discrimination may be based, at least in part, on the infant recognizing its own odors deposited on the mother's fur, although, of course, the infant may also recognize individual odors of the mother herself.

From the above findings one might suppose that any distinctive odor associated with the rearing figure would come to be preferred to unfamiliar odors, and thereby mediate infant attraction to the rearing figure. However, not all distinctive odors are equally effective in this regard. Infant squirrel monkeys raised on synthetically scented surrogates did not show a preference for surrogates of the synthetic rearing odor until the end of the fourth month of life

(Kaplan *et al.*, 1979), at least 3 months later than when the infant's own odor served as the basis for preference (Kaplan and Russell, 1974). Thus, it appears that the range of odors that can most effectively mediate rearing-figure preference is constrained to some degree, perhaps due to an attractiveness present at or near birth to specific biological odors.

Infant *Saimiri* also prefer surrogates that are of the same color as the rearing surrogate to surrogates of different colors (Kaplan *et al.*, 1979; Kaplan, 1977a). Although the speed at which this preference behavior develops has varied with the experimental protocol employed, preference based on odor has generally preceded, and has been more robust, than preference based on color (Kaplan and Russell, 1974; Kaplan, 1977a; Kaplan *et al.*, 1979). These findings may indicate a more primary role for olfaction than for vision in mediating recognition of, and preference for, the rearing figure. Alternatively, they may reflect the more natural olfactory cues (infant's own odor) as opposed to visual cues (the color of the surrogate) that have most often been used in these experiments. [Only when synthetic rearing odors were used did the surrogate's odor appear less important than color in determining preference (Kaplan *et al.*, 1979).] Studies with mother-reared animals support this second possibility. Infants exposed to "natural" visual cues of their own and another mother (mothers viewed through one-way glass while awake and freely moving) clearly preferred their own mother by 1 month of age (Redican and Kaplan, 1978). In contrast, when anesthetized mothers were used as the visual stimuli, no preference for the mother was observed in infants ranging from 2 to 6 months old (Kaplan *et al.*, 1977). It may be that if visual cues approximate those to which infant monkeys are typically exposed (those of an awake, freely moving animal), clear preference is expressed for the rearing figure at an early age. If more arbitrary or artificial visual stimuli are used (an anesthetized mother or a surrogate of a particular color) preference is not expressed, or is retarded in its development.

Overall, the studies reviewed in this section indicate that the squirrel monkey infant is selectively responsive to its rearing figure from a very young age. Olfactory cues appear to play a prominent role, and visual cues are also utilized to some degree. This picture of the infant as precocious in its attentiveness and responsiveness to specific maternal characteristics accords well with the infant's major role in maintaining contact with the mother from the time of birth, as well as with the need for infants undergoing normal separations from the mother [e.g., while "aunted" by a nonlactating female (Rosenblum, 1971)] to identify and return to the mother for nourishment. Moreover, the recognition and preference that surrogate-reared (Kaplan and Russell, 1974; Kaplan *et al.*, 1979) and mother-reared (Kaplan *et al.*, 1977) squirrel monkeys exhibit for their rearing figure indicates that a specific attachment to the rearing figure is formed by infants of this species.

3. RESPONSES TO SEPARATION AND NOVELTY

Further evidence for attachment is provided by the infant's response to involuntary separation from the rearing figure. Removal of either the biological mother or rearing surrogate results in an immediate increase in behavioral signs of distress, particularly repetitive, high-pitched vocalizations (Coe *et al.*, 1983, M. B. Hennessy *et al.*, 1979; Kaplan, 1970). Thus, infants raised on either natural or artificial mothers exhibit three classical indicators of filial attachment (Ainsworth, 1972): they recognize the rearing figure, they prefer it to other similar figures, and they show behavioral signs of distress during separation.

If separation from the rearing figure is a truly traumatic event for the young squirrel monkey, then one would expect separation to affect not only behavior, but also physiological measures of emotionality. One physiological system that has been studied during separation is that of the hypothalamus–pituitary–adrenal axis. Activation of this system has been observed in a wide range of stressful or arousing situations, and it appears to be particularly responsive to psychological as opposed to physical events (J. W. Hennessy and Levine, 1979; Mason, 1975). Thus, measures of pituitary–adrenal activity seem well suited for assessing separation distress. As might be expected, removal of the biological mother from the home cage results in a clear and rapid increase in the infant's plasma levels of cortisol (Coe *et al.*, 1978), the primary glucocorticoid secreted from the adrenal gland of the squirrel monkey. In contrast, removal of the rearing surrogate from the home cage has no effect on the infant's plasma cortisol levels 30 min later (M. B. Hennessy *et al.*, 1979; M. B. Hennessy and Kaplan, 1982). The surrogate-reared infant is capable of a normal cortisol response because infants placed in a novel environment during separation show as great an elevation of plasma cortisol levels 30 min later as do mother-reared infants separated under the same conditions (Mendoza *et al.*, 1978b). It appears, therefore, that separation from the surrogate is not as arousing as is separation from the biological mother, and consequently, the attachment to the surrogate is diminished or in some way deficient relative to that for the natural mother. This interpretation is consistent with the conclusions of studies examining both physiological and behavioral responses to separation in surrogate-reared macaques (Meyer *et al.*, 1975; Reite *et al.*, 1978).

Another method often used to assess attachment is to examine the impact of the rearing figure in a novel or otherwise threatening environment. The capacity of the rearing figure to reduce distress in this environment, or to promote exploration of it, is taken as evidence of filial attachment (Ainsworth and Wittig, 1969; Harlow and Zimmerman, 1959). As noted above, infant squirrel monkeys placed in a novel room without their surrogate show a clear elevation of plasma cortisol levels 30 min following separation (Mendoza *et al.*, 1978b) along with behavioral disturbance (M. B. Hennessy and Kaplan, 1982). If, however, the infant is exposed to this novel environment in the presence of its surrogate, both the behavioral and cortisol responses are greatly diminished. Behavioral disturbance is reduced immediately upon exposure, whereas the effect on cortisol levels takes longer to develop. That is, 30 min following placement in the room, the infant's cortisol levels are as elevated when with the rearing surrogate as when in the room alone. However, with continued exposure to the room, cortisol levels in these two conditions clearly diverge, increasing significantly by 4 hr only when the infant is exposed to the room alone. The presence of the surrogate prevents this further increase (M. B. Hennessy and Kaplan, 1982).

Thus, in contrast to the results of separation experiments conducted in the home cage (M. B. Hennessy *et al.*, 1979), these findings indicate that the presence or absence of the rearing surrogate can affect the infant's pituitary–adrenal activity. However, before attributing this effect to the surrogate's role as the infant's attachment figure, it would be necessary to demonstrate that this moderating influence of the surrogate on cortisol levels is specific to the infant's own individual surrogate, and not the result of some property or properties of surrogates in general. Therefore, I compared the effects of the rearing surrogate with those of discriminable substitute surrogates on the infant's plasma cortisol levels, as well as vocalizations, during exposure to a novel room.

The subjects for this experiment were eight surrogate-reared *Saimiri* infants (two males, six females) of mixed subtype. Beginning at about 4 months of age, each subject was tested in five conditions, spaced at ~1-week intervals, in a counterbalanced order. The five conditions were: (1) base, the infant remained undisturbed in the home cage; (2) own surrogate, the infant was placed in a cage in a novel room with its surrogate [as in M. B. Hennessy and Kaplan (1982)]; (3) clean surrogate/different color, the infant was placed in the novel room with a surrogate having a freshly laundered cover that was green, rather than the rearing color of black; (4) scented surrogate/different color, the same as the previous condition except that the surrogate cover was sprayed with commercial air freshener (Glade, flower blossoms scent) just prior to being placed in the room; and (5), no surrogate, the infant was placed in an empty cage in the novel room. In conditions 2–5, infants remained in the novel room for a 4-hr period. During each condition, behavior was monitored from behind one-way glass, and at the end of each condition (~1400 hr), a blood sample was collected for analysis of plasma cortisol levels. Statistical significance was evaluated with analysis of variance procedures, followed by *post hoc* paired comparisons (Duncan new multiple range test).

As shown in Fig. 2, plasma cortisol levels were significantly greater when infants were placed in the novel room alone than when they were placed in the room with either their own or either of the two substitute surrogates ($p < 0.05$). Further, cortisol levels in these three surrogate conditions did not differ among themselves. These data indicate that the moderating influence of the infant's surrogate on plasma cortisol levels in a novel environment (Hennessy and Kaplan, 1982) is not a function of the infant's attachment to its own specific surrogate, but is instead due to some characteristic common to all three surrogates used in this experiment.

Placing infants in the novel room with no surrogate also resulted in a high level of vocalizations, particularly during the initial 15 min of exposure (Fig. 3). As expected, significantly fewer vocalizations were recorded when the infants were placed in the room with their own rearing surrogate ($p < 0.01$). Provision of a clean surrogate of a different color also reduced vocalizations relative to the no-surrogate condition ($p < 0.05$). The number of vocalizations in the clean surrogate/different color and own-surrogate conditions were not significantly different. However, when the infants were provided with a scented surrogate of the unfamiliar color, the number of vocalizations did not differ from that in the no-surrogate condition, and was significantly greater than in the own-surrogate condition ($p < 0.05$).

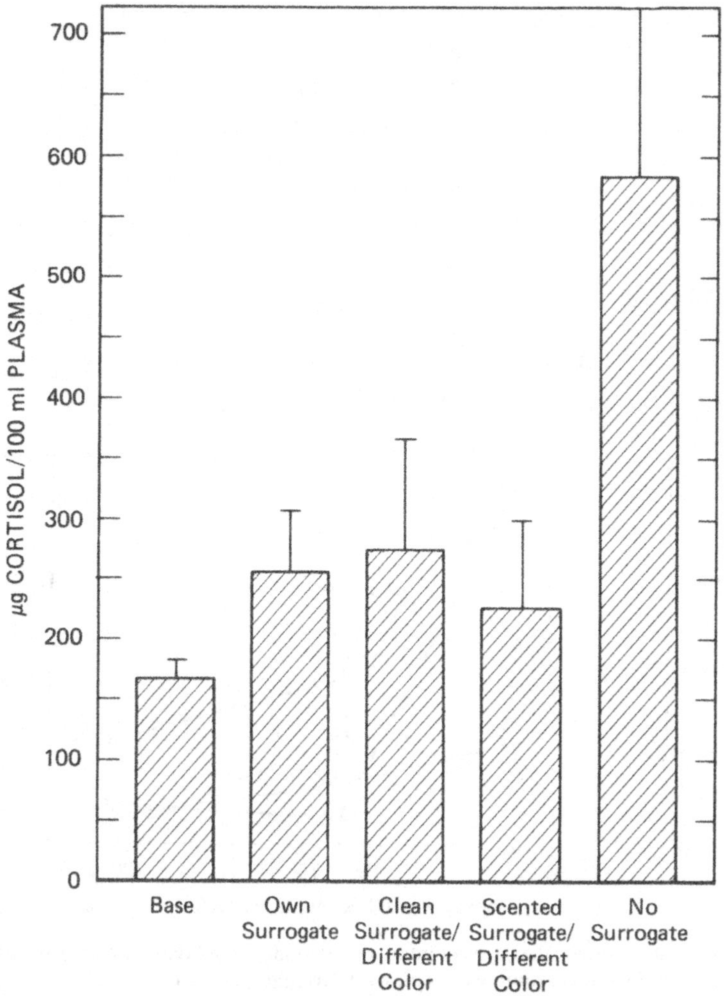

Fig. 2. Mean plasma cortisol levels of infants when undisturbed (base) and 4 hr following placement in a novel room with their own surrogate, either of two substitute surrogates, or no surrogate. Vertical lines indicate standard errors of the means.

Overall, the vocalization data must be considered less definitive than the cortisol results. The patterns of significant differences in vocalizations were not the same for the two substitute surrogate conditions. Further, as inspection of Fig. 3 reveals, the level of vocalizations in the two substitute surrogate conditions were, numerically at least, intermediate to those in the comparison conditions (own surrogate, no surrogate). Yet, these data are consistent with the cortisol results in indicating that a surrogate need not be the specific attachment object to reduce infant responsiveness in a novel environment, since one of the substitute surrogates (clean surrogate/different color) significantly diminished infant vocalizations. Not all surrogates were effective, though, because the scented surrogate of an unfamiliar color did not reduce vocalizations.

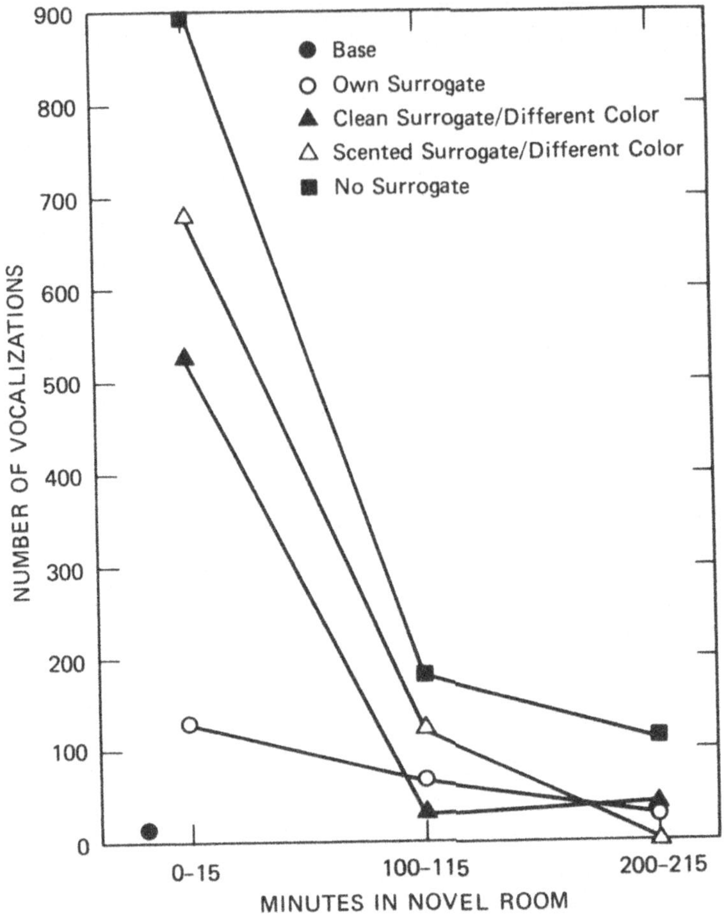

Fig. 3. Mean numbers of vocalizations of infants when undisturbed (base) and at various intervals following placement in a novel room with their own surrogate, either of two substitute surrogates, or no surrogate.

The difference in influence of the two substitute surrogates may represent an effect of relative familiarity. Although both of these surrogates were discriminable from the attachment object via both visual and olfactory cues, as indicated by previous work (M. B. Hennessy *et al.*, 1979; Kaplan, 1977a; Kaplan, *et al.*, 1979), the freshly laundered, unscented surrogate probably had a more familiar odor than did the scented surrogate. All infants had their own surrogate's cover changed at weekly intervals during rearing, and so experienced a freshly laundered cover for at least brief periods, i.e., until the infant imparted its own odor to the cover. Thus, the surrogate did not need to be the attachment object to reduce vocalizations, but may have needed to possess some degree of familiarity to be effective.

If the surrogate's role as the infant's attachment object cannot account for its moderating influence on infant responses, to what can we attribute this effect? The act of clinging has been hypothesized to reduce the arousal of infant primates (W. A. Mason, 1970). One might therefore suppose that the substitute surrogates reduced signs of increased arousal because they permitted the infant to cling. This suggestion is not substantiated, however, because infants were observed to cling more than twice as frequently to their own surrogate ($\bar{X}$ = 1335 sec) than to either the clean ($\bar{X}$ = 495 sec) or scented ($\bar{X}$ = 589 sec) surrogates. Further, the effects of the surrogates cannot be accounted for by the warmth or nourishment they might have provided, because no surrogate was warmed or contained a nursing bottle during testing. Perhaps the resolution of this issue lies in developmental considerations. It may be that at younger ages factors such as clinging, the receipt of warmth or nourishment, or even the presence of the specific attachment object are necessary to moderate infant arousal, but that by the age at which we tested subjects (~4–5 months), the range of effective stimuli has broadened or generalized so that merely the presence of a surrogate is sufficient to reduce arousal in a novel environment.

These findings with surrogate-reared *Saimiri* agree well with the results of studies with surrogate-reared macaques. Mason and colleagues (Hill *et al.*, 1973; W. A. Mason *et al.*, 1971; W. A. Mason, 1979) found that the presence of the rearing surrogate reduced the vocalizations, heart rate, and cortisol levels of rhesus infants during exposure to novelty, and that signs of arousal were also reduced, though generally to a lesser extent, when the infant was provided with a substitute surrogate.

One aspect of the studies reviewed in this section that deserves attention is the sometimes discrepant results for behavioral and hormonal measures of arousal. In particular, when the surrogate is removed briefly from the home cage, vocalizations but not cortisol levels increase (M. B. Hennessy *et al.*, 1979; M. B. Hennessy and Kaplan, 1982). There are several plausible explanations for this sort of dissociation. One possibility (Vogt and Hennessy, 1982) is that although both behavior and pituitary–adrenal activity reflect arousal, behavioral measures are also influenced by the stimulus situation impinging on the young monkey. That is, behaviors such as vocalizing and locomoting, in addition to being evoked by increases in arousal, may represent species-typical behaviors that have evolved to occur in response to a specific stimulus event (absence of the mother) because they increase the probability that mother and infant will be reunited. According to this hypothesis, the infant's attachment to its surrogate is not of sufficient magnitude for the removal of the surrogate to elicit an increase in arousal, or the consequent arousal-induced changes in cortisol levels or vocalizations. However, vocalizations do increase in this situation because they are a species-typical behavior elicited by the absence of maternal stimuli. This notion is not incompatible with other recent positions emphasizing the removal of maternal stimulation (Hofer, 1981) and the adaptiveness of infant behaviors (e.g., Rosenblum and Plimpton, 1981) in explaining maternal separation effects (see also Coe *et al.*, this volume, Chapter 5). Another possibility is that increased vocalizing and pituitary–adrenal activation may reflect changes in separate but closely related underlying states that, in our present state of knowledge, are undifferentiated and referred to broadly as "arousal" or "stress." If so, further studies delineating the precise variables that affect behavioral and pituitary–adrenal measures is needed to define the parameters of the true underlying states.

4. SOCIAL BEHAVIOR

In the rhesus macaque, some of the most dramatic effects of surrogate-rearing are on later social behavior. Being reared on an inanimate surrogate can completely disrupt the sexual behavior of both male and female rhesus monkeys (Harlow and Harlow, 1969). Similarly, surrogate-reared rhesus females much more frequently become indifferent or abusive mothers than do females who are themselves mother-reared (Ruppenthal *et al.*, 1976).

These findings contrast sharply with those for *Saimiri*. Surrogate-reared male and female squirrel monkeys have been observed to show all major components of sexual behavior, males exhibit the typical fatting phenomenon, and females show interest in, and care for, their young (Roy *et al.*, 1978b). Although it is clear that surrogate-reared *Saimiri* can exhibit typical sexual and maternal responses, the relative performance of surrogate- and mother-reared squirrel monkeys has not been directly compared. Quantitative differences in the social behavior of surrogate- and mother-reared animals could well exist.

If the maternal behavior of *Saimiri* is impaired by surrogate-rearing, one might expect these effects to be most evident in the treatment of a female's first infant. In the rhesus, females raised without a biological mother are more likely to ignore their first infant than they are subsequent infants (Ruppenthal *et al.*, 1976). Experience with infants appears to reverse the effects produced by the mother's own deprived rearing history. Experience with young may also be an important factor in determining the quality of infant care provided by squirrel monkey mothers (Taub *et al.*, 1978). In our own colony, the success of females raising their first infants is generally poor.

As a first step in directly comparing the maternal competence of surrogate- and mother-reared animals, I have reviewed records from our colony and examined success rates in raising first-born infants. I identified all females that were born in our colony and subsequently bore at least one live infant of their own. Adequate records were available for 26 such females—15 raised with their biological mother, and 11 with a maternal surrogate. My comparisons are limited to the mothers' success in raising infants through the first week of life, because some infants were removed from their mother at this time for surrogate-rearing. However, the great majority of infant rejections (infant not permitted to cling) occur within the first few days of life, and infants that die while with the mother commonly do so during the first week. The fate of the first-born infants of the mother-reared and surrogate-reared mothers is depicted in Table I. As expected, the overall success of mothers raising their first infant was poor. Fourteen of the 26 infants were rejected, and another four died during the first week. However, as is clear in the table, there is no indication that surrogate-rearing reduced later maternal success. Four of 11 surrogate-reared (36%) and four of 15 mother-reared (27%) animals maintained their infant through the first week. It should be emphasized that these data were collected

Table I
Fate of First-Born Infant of Mothers That Were Themselves Mother- or Surrogate-Reared

Mother's rearing history	*N*	Number rejected	Percent	Number dying while with mother <1 week	Percent	Number successfully raised >1 week	Percent
Mother-reared	15	8	53	3	20	4	27
Surrogate-reared	11	6	55	1	9	4	36

retrospectively and are intended to be considered only as preliminary. Yet, the data support the notion that maternal competence is little affected by surrogate-rearing in *Saimiri*.

With regard to measures of sociability other than those involved in copulation and maternal care, the most striking feature of the surrogate-reared squirrel monkey's behavior is its relative normalcy. These animals have been found to emit a wide range of complex and age-appropriate social behaviors, including play, vocal communication, presumed scent-marking activity, inter-animal approaches, withdrawals, and grooming, as well as aggressive and dominance behaviors (Kaplan, 1977b; Roy *et al.*, 1978; Herzog and Hopf, 1983; Hopf *et al.*, in press; M. B. Hennessy, unpublished observations). These behaviors are typically well coordinated and oriented appropriately toward the environment. When mother-reared and surrogate-reared infants have been directly compared, no differences have been found on measures of various social behaviors, or the relative preference these animals exhibit for cage-mates and strangers (Kaplan, 1977b; Kaplan and Cubicciotti, 1980). In most laboratories, surrogate-reared animals are maintained with continual visual access to other monkeys caged in the rearing room. However, exposure to freely-behaving conspecifics does not appear to be required for the development of social competence in surrogate-reared subjects, because a variety of species-typical social behaviors and aspects of social responsiveness to artifical monkey models have been observed in infants reared, from several hours following birth in complete visual isolation from other conspecifics (Herzog and Hopf, 1983).

Conditions of surrogate-rearing do appear to encourage infants to exhibit social responses toward humans. Roy (1980) found that the genital display that is commonly observed during dominance interactions and other interanimal encounters (Latta *et al.*, 1967; Mendoza *et al.*, 1978a) was frequently directed at an approaching human by surrogate-reared *Saimiri*. A group of feral-reared animals never showed this response. This finding is consistent with informal observations in our own colony. Whereas mother-reared infants are only infrequently observed to direct genital displays toward humans, this behavior is extremely common in surrogate-reared animals, and has been noted as early as the second week of life. However, mother-reared *Saimiri* in another colony that have had very extensive experience with humans have been observed to direct genital displays toward them with regularity (S. P. Mendoza, personal communication). Thus, it seems likely that the tendency to direct genital displays toward humans is related to the surrogate-reared infant's regular exposure to caretakers. It is noteworthy, however, that high levels of human-directed genital displays have also been observed in juveniles that had relatively restricted experience with human caretakers during surrogate-rearing (Hopf *et al.*, in press).

Overall, the above findings demonstrate that various aspects of squirrel monkey social behavior and responsiveness will develop in the absence of social stimuli provided by the biological mother. However, this should not be taken to imply that the rearing figure has no significant impact on social development. In most cases, surrogate-reared monkeys have not been directly compared with mother-reared controls, so that it has not been possible to look for more subtle or quantitative effects on social behavior. Possible maternal influences acting prenatally have not been examined. Moreover, factors that the natural mother and surrogate have in common (e.g., tactual qualities, shape that permits clinging) might also influence the development of social behavior. An example of the type of subtle influence that surrogate-rearing might have on *Saimiri* social behavior is provided by findings in macaques. Rhesus monkeys raised on inanimate surrogates were found to be capable of engaging in dyadic interactions with social partners, but unlike animals raised with animate partners, they were unable to manipulate other dyads to their own advantage [i.e., to gain access to a water spout (Anderson and Mason, 1978; Capitanio, 1982)].

Further, it has been demonstrated that, under certain conditions, the social choices of *Saimiri* can be influenced by surrogate-rearing (Kaplan and Cubicciotti, 1980). Squirrel monkeys raised on distinctively colored surrogates (either black or green) were housed for 2 months following weaning with peers wearing cloth jackets of the same color. In subsequent preference tests, the monkeys showed greater approach to both familiar and unfamiliar peers that wore jackets of the rearing color than they did to peers that either wore jackets of the other color or wore no jackets. Exposure to the color prior to weaning appeared to be critical for this effect, because infants raised with their mother in the usual fashion and then exposed to peers of a particular jacket color after weaning showed no preference based on jacket color. These findings indicate that certain preceptual qualities of the maternal figure may influence the animal's choice of social partners following weaning. Now that it is clear that the social behavior of surrogate-reared squirrel monkeys is relatively normal, further studies varying specific surrogate features to examine the ontogeny of social choice behavior would certainly be of interest.

5. ATYPICAL BEHAVIORS

Squirrel monkeys raised on surrogates have been found to engage in unusually high levels of self-oral behaviors [i.e., mouthing or sucking of body parts, including digits, genitals, and tail (Kaplan, 1974, 1979; Roy, 1980, 1981; King and King, 1970)]. The most ubiquitous form of self-orality is thumbsucking, a behavior only rarely observed in mother-reared animals. Thumbsucking has been reported to occur in surrogate-reared *Saimiri* from the first week of life until about the time of sexual maturity (King and King, 1970; Kaplan, 1979). As animals raised on surrogates in our colony have grown older, it has become clear that thumbsucking can persist well into adulthood. *Figure 4* shows two surrogate-reared females, 6.5 and 7 years of age, sucking their thumbs in their home pen. Neither of these females display other obvious behavioral anomalies, and each has successfully reared offspring of her own.

The prevalence of thumbsucking appears to be related to the availability of a nipple during rearing. When infants have been presented with a nipple only during scheduled feedings, more thumbsucking has been observed than when a nursing bottle with nipple is continuously present (Kaplan, 1974). However, even when a nipple is continuously present during surrogate-rearing, an appreciable number of infants suck their thumbs (Kaplan, 1974; Kaplan, 1979). Because it takes several days to train these infants to locate the nipple, they are, in effect, deprived of free nipple access during the training period. Thus, their thumbsucking might stem from the brief period of deprivation when the nipple is inaccessible during training. This possibility is consistent with the view that thumbsucking arises from the more immediate availability of digits than nipple when rooting and placing reflexes are strong (Benjamin, 1967). Moreover, the formula given infants often is totally consumed or spoils before the nursing bottle can be changed. If the infant tends to avoid the nipple when this occurs, thumbsucking might be further encouraged.

A variety of atypical behaviors that do not involve self-orality have also been reported in surrogate-reared *Saimiri*. The degree of behavioral disturbance observed has varied greatly among laboratories and seems to be related to the severity of the early deprivation. In our own laboratory, using the surrogate-rearing procedures described above, few clear behavioral anomalies have been observed. Kaplan (1977b) reported that one animal made unusual rubbing movements with his hands and arms over his face and head. The behavior was sometimes accompanied by squealing vocalizations and self-biting. A surrogate-reared juvenile male presently in our colony frequently grasps his hindquarters with his hands in an unusual manner, often while jumping excitedly or walking bipedally (Fig. 5). A variation of this behavior, in which the animal sits and curls his body so that his head is tucked between the legs, has also been observed (Fig. 6). Yet, overall, such anomalous behaviors are rare among our surrogate-reared monkeys. Apparently, only those individuals that are for some reason most vulnerable to conditions of deprived rearing exhibit these aberrant behaviors. Large subtype differences in the development of abnormal behaviors have been reported in surrogate-reared *Saimiri* (Roy *et al.*, 1981), suggesting that substantial genetic variation in susceptibility to these effects may exist.

Investigators who have used a surrogate similar in design to our own have also found few instances of unusual behavior patterns (Herzog and Hopf, 1983; Hopf *et al.*, in press). However, Roy (1981), who provided his infants only with a rolled cloth towel, observed a variety of atypical behaviors. In large part, the behavioral anomalies found in his squirrel monkeys resembled the

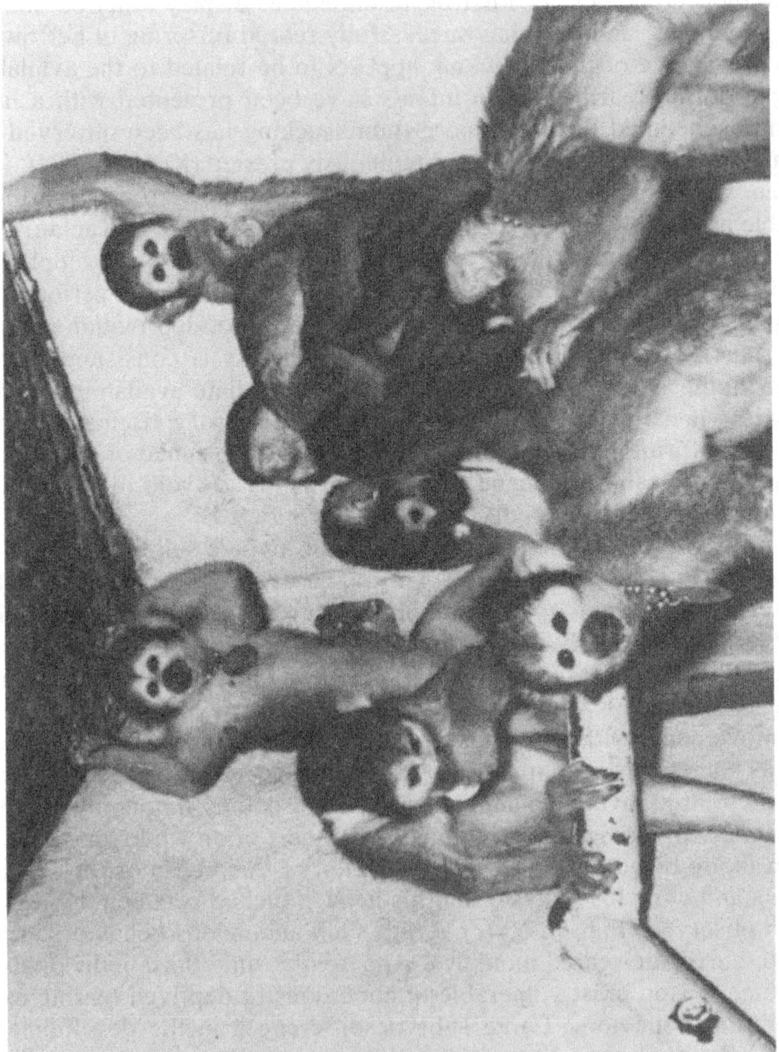

Fig. 4. Squirrel monkey breeding group. The two females sucking their thumbs (far left and upper right) were both surrogate-reared, and are now 6.5 and 7 years old.

Fig. 5. Surrogate-reared juvenile male in our colony performing atypical behavior in which he grasps his own hindquarters while locomoting.

severe atypical behaviors observed in rhesus macaques raised on inanimate surrogates (W. A. Mason and Berkson, 1975). Roy's (1981) animals exhibited extreme self-clasping, stereotypic rocking, various forms of oral self-manipulation, head grasping, and behavior involving clutching of objects or their own tail in a manner reminiscent of the clinging response usually directed toward the rearing figure (Fig. 7). Stereotypic body-rocking in a nursery-reared squirrel monkey was also reported by King and King (1970). This infant was raised with a yarn doll. The behavior of the infant toward the doll is not described,

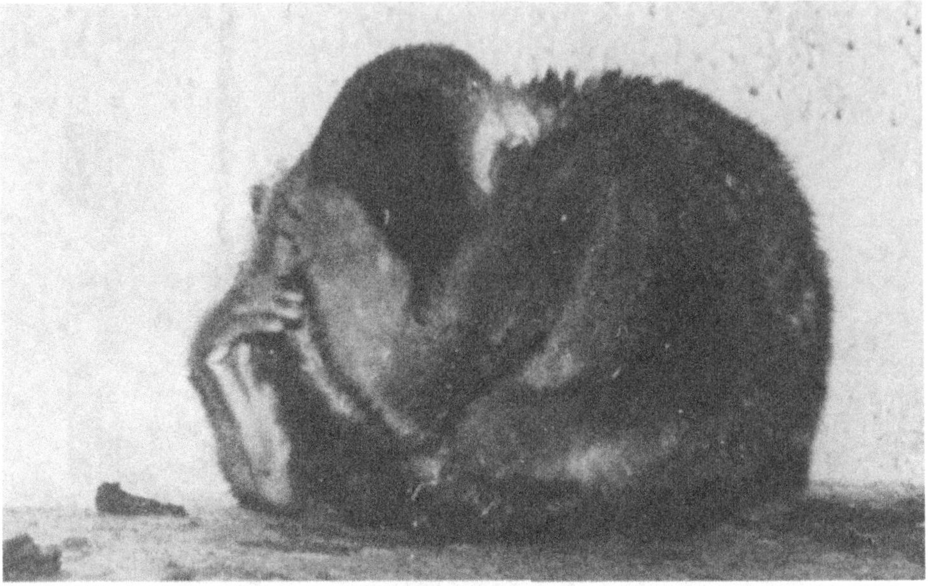

Fig. 6. A variation of the atypical behavior shown in Fig. 5. Here the male infant is curling his body forward and grasping his hindquarters while sitting.

but this object would seemingly afford a rearing figure not unlike Roy's cloth surrogate.

There are a number of differences between Roy's and our rearing procedures, several of which appear likely to have contributed to the differences observed in abnormal behavioral development. First, Roy's simple cloth surrogate did not support a normal clinging response. If the infant attempted to cling in the typical "dorsal–ventral" position, the surrogate would roll over and the infant would fall on its side (M. A. Roy, personal communication). Second, Roy permanently removed the surrogate from the home cage at 3.5 months of age, rather than at the 6 months or older that is standard in our laboratory. Deprivation of normal clinging, perhaps combined with early weaning, may have been critical, because some of the atypical behaviors Roy observed appear to involve a redirection of the clinging response to objects or the infant's own body (e.g., see Fig. 7). Third, Roy's infants, unlike ours, did not have a nipple continuously present for the first 35 days of life. The overall level of social stimulation does not appear to have been a factor. Roy's animals were reared in wire cages permitting continual visual and vocal communication between infants from 1 month of age until testing began at about 7 months. In contrast, Hopf and colleagues (Herzog and Hopf, 1983); Hopf *et al.*, in press) raised infants in complete visual isolation from conspecifics, with surrogates similar in design to our own, and observed no major behavioral abnormalities.

Several investigators have noted that the atypical behaviors of surrogate-reared *Saimiri* tend to occur following disturbance or during other times of

Fig. 7. Squirrel monkey that was reared on a rolled-cloth surrogate shown clinging to its own tail.

apparent arousal (Kaplan, 1977b; King and King, 1970; Roy, 1981). The nature of this relationship between atypical behaviors and arousal is unclear. Perhaps the normal behavior patterns of surrogate-reared *Saimiri* simply begin to deteriorate during heightened arousal. On the other hand, the monkeys may perform atypical behaviors when aroused because these behaviors have a calming effect. In line with this second possibility, W. A. Mason (1978) has proposed that a fundamental effect of surrogate-rearing in rhesus monkeys is to alter the animal's behavioral strategy for "coping" with novel or challenging situations.

6. SUMMARY OF COMPARISONS WITH MACAQUES

In comparing the effects of surrogate-rearing in macaques and squirrel monkeys, one is struck by the extreme disruption of sexual and maternal behavior observed in the rhesus but not *Saimiri*. However, there are also many similarities in the effects of surrogate-rearing, and as the data base for species comparisons continues to grow, more similarities are emerging. In both *Macaca* and *Saimiri*, infants cling and show attachment to their surrogate. Yet, in both macaques and squirrel monkeys, separation studies indicate that this attachment is not as great as shown for the biological mother. During exposure to a novel environment, the rearing surrogate reduces signs of infant arousal in both rhesus and squirrel monkeys, and in both species, discriminable substitute surrogates can also reduce signs of arousal in this situation. Rhesus monkeys raised on inanimate surrogates exhibit many aberrant behavior patterns, such as self-clasping and body-rocking. Although such effects are not commonly obtained in surrogate-reared squirrel monkeys, they have been observed in a group of *Saimiri* exposed to particularly restrictive surrogate-rearing procedures.

As these latter findings suggest, an important consideration in evaluating species differences is the relative amount of deprivation that comparable surrogate-rearing procedures pose for rhesus and squirrel monkey infants. From birth, the ventrally carried rhesus infant receives more direct maternal stimulation (e.g., cradling, grooming) than does the squirrel monkey infant, who remains largely out of sight on the mother's back (Rosenblum, 1968; Hinde and Spencer-Booth, 1967). Thus, comparable conditions of surrogate-rearing probably constitute a relatively greater deprivation for the young rhesus, and some species differences in surrogate-rearing effects may be a function of the relative degree of early deprivation.

Note added in proof. The reader is also referred to a recently published paper by Huebner and King (1984).

ACKNOWLEDGMENTS. I would like to thank Drs. Sigrid Hopf and M. Aaron Roy for their ideas and cooperation in providing unpublished data and photographs (Fig. 7 is courtesy of Dr. Roy), Deborah Mallis for her photography (Fig. 4–6) and assistance in data collection, and Drs. Joel N. Kaplan and Sally P. Mendoza for their thoughtful comments on the manuscript. The author's original work was supported by Grant HD 14591 from NIH.

REFERENCES

Ainsworth, M. D. S., 1972, Attachment and dependency: A comparison, in: *Attachment and Dependency* (J. L. Gewirtz, ed.), V. H. Winston and Sons, Washington, pp. 97–137.

Ainsworth, M. D. S., and Wittig, B. A., 1969, Attachment and exploratory behavior of one-year-olds in a strange situation, in: *Determinants of Infant Behaviour IV* (B. M. Foss, ed.), Methuen, London, pp. 111–136.

Anderson, C. O., and Mason, W. A., 1978, Competitive social strategies in groups of deprived and experienced rhesus monkeys, *Dev. Psychobiol.* **11**:289–299.

Benjamin, L. S., 1967, The beginning of thumbsucking, *Child Dev.* **38**:1065–1078.

Capitanio, J. P., 1982, Early Experience, Behavior and Organization in Rhesus Macaques (*Macaca mulatta*) Social Groups, Ph.D. Dissertation, University of California, at Davis (unpublished).

Coe, C. L. Mendoza, S. P., Smotherman, W. P., and Levine, S., 1978, Mother–infant attachment in the squirrel monkey: Adrenal response to separation, *Behav. Biol.* **22:**256–263.

Coe, C. L., Wiener, S. G., and Levine, S., 1983, Psychoendocrine responses of mother and infant monkeys to disturbance and separation, in: *Symbiosis in Parent–Young Interactions* (L. A. Rosenblum and H. Moltz, eds.), Plenum Press, New York, pp. 189–214.

Harlow, H. F., 1958, The nature of love, *Am. Psychol.* **13:**673–685.

Harlow, H. F., and Harlow, M. K., 1969, Effects of various mother–infant relationships on rhesus monkey behaviors, in: *Determinants of Infant Behaviour IV* (B. M. Foss, ed.), London, pp. 15–36.

Harlow, H. F., and Zimmerman, R. R., 1959, Affectional responses in the infant monkey, *Science* **130:**421–432.

Hennessy, J. W., and Levine, S., 1979, The pituitary–adrenal system: A psychoendocrine hypothesis, in: *Progress in Psychobiology and Physiological Psychology* (J. Sprague and A. Epstein, eds.), Volume 8, Academic Press, New York.

Hennessy, M. B., and Kaplan, J. N., 1982, Influence of the maternal surrogate on pituitary–adrenal activity and behavior of infant squirrel monkeys, *Dev. Psychobiol.* **15:**423–431.

Hennessy, M. B., Kaplan, J. N., Mendoza, S. P., Lowe, E. L., and Levine, S., 1979, Separation distress and attachment in surrogate-reared squirrel monkeys, *Physiol. Behav.* **23:**1017–1023.

Herzog, M., and Hopf, S., 1983, Effects of species-specific vocalizations on the behaviour of surrogate-reared squirrel monkeys, *Behaviour* **86:**197–214.

Hill, S. D., McCormack, S. A., and Mason, W. A., 1973, Effects of artificial mothers and visual experience on adrenal responsiveness of infant monkeys, *Dev. Psychobiol.* **6:**421–429.

Hinde, R. A., and Spencer-Booth, Y., 1967, The behavior of socially living rhesus monkeys in their first two and a half years, *Anim. Behav.* **15:**169–196.

Hofer, M. A., 1981, Parental contributions to the development of their offspring, in: *Parental Care in Mammals* (D. J. Gubernick and P. H. Klopfer, eds.), Plenum Press, New York, pp. 77–115.

Hopf, S., Herzog, M., and Vogel-Köhler, C., in press, Social integration of surrogate-reared infant squirrel monkeys to captive groups, *Acta Paedopsychiat.*

Huebner, D. K., and King, J. E., 1984, Kittens as therapists: Social behavior sequences in isolated squirrel monkeys after exposure to young nonconspecifics, *Dev. Psychobiol.* **17:**233–242.

Kaplan, J., 1970, The effects of separation and reunion on the behavior of mother and infant squirrel monkeys, *Dev. Psychobiol.* **3:**43–52.

Kaplan, J., 1974, Growth and behavior of surrogate-reared squirrel monkeys, *Dev. Psychobiol.* **7:**7–13.

Kaplan, J., 1977a, Perceptual properties of attachment in surrogate-reared and mother-reared squirrel monkeys, in: *Primate Bio-Social Development: Biological, Social, and Ecological Determinants* (S. Chevalier-Skolnikoff and F. E. Poirier, eds.), Garland, New York, pp. 225–234.

Kaplan, J., 1977b, Some behavioral observations of surrogate- and mother-reared squirrel monkeys, in: *Primate Bio-Social Development: Biological, Social, and Ecological Determinants* (S. Chevalier-Skolnikoff and F. E. Poirier, eds.), Garland, New York, pp. 501–511.

Kaplan, J., and Russell, M., 1974, Olfactory recognition in the infant squirrel monkey, *Dev. Psychobiol.* **7:**15–19.

Kaplan, J. N., 1979, Growth and development of infant squirrel monkeys during the first six months of life, in: *Nursery Care of Nonhuman Primates* (G. C. Ruppenthal and D. J. Reese, eds.), Plenum Press, New York, pp. 153–164.

Kaplan, J. N., and Cubicciotti, D. D., 1980, Early perceptual experience and the development of social preferences in squirrel monkeys, in: *Maternal Influences and Early Behavior* (R. W. Bell and W. P. Smotherman, eds.), Spectrum Press, New York, pp. 253–270.

Kaplan, J. N., Cubicciotti, D., and Redican, W. K., 1977, Olfactory discrimination of squirrel monkey mothers by their infants, *Dev. Psychobiol.* **10:**447–454.

Kaplan, J. N., Cubicciotti, D. D., and Redican, W. K., 1979, Olfactory and visual differentiation of synthetically scented surrogates by infant squirrel monkeys, *Dev. Psychobiol.* **12:**1–10.

King, J. E., and King, P. A., 1970, Early behaviors in hand-reared squirrel monkeys *Saimiri sciureus*), *Dev. Psychobiol.* **2:**251–256.

Latta, J., Hopf, S., and Ploog, D., 1967, Observation on mating behavior and sexual play in the squirrel monkey (*Saimiri sciureus*), *Primates* **8:**229–246.

Mason, J. W., 1975, A historical view of the stress field, part 2, *J. Hum. Stress* **1**:22–36.

Mason, W. A., 1970, Motivational factors in psychosocial development, in: *Nebraska Symposium on Motivation*, (W. J. Arnold and M. M. Page, eds.), University of Nebraska Press, Lincoln, pp. 35–67.

Mason, W. A., 1978, Social experience and primate cognitive ability, in: *The Development of Behavior* (G. Burghardt and M. Bekoff, eds.), Garland, New York, pp. 233–251.

Mason, W. A., 1979, Wanting and knowing: A biological perspective on maternal deprivation, in: *Origins of the Infant's Social Responsiveness* (E. B. Thoman, ed.), Lawrence Erlbaum, New Jersey, pp. 225–249.

Mason, W. A., and Berkson, G., 1975, Effects of maternal mobility on the development of rocking and other behaviors in rhesus monkeys: A study with artificial mothers, *Dev. Psychobiol.* **8**:197–211.

Mason, W. A., and Kenney, M. D., 1974, Redirection of filial attachments in rhesus monkeys: Dogs as mother surrogates, *Science* **183**:1209–1211.

Mason, W. A., Hill, S. D., and Thomsen, C. E., 1971, Perceptual factors in the development of filial attachment, in: *Proceedings of the 3rd International Congress of Primatology*. Volume 3 *Behavior* (H. Kummer, ed.), Karger, Basel, pp. 125–133.

Mendoza, S. P., Lowe, E. L., and Levine, S., 1978a, Social organization and social behavior in two subspecies of squirrel monkeys, *Folia Primatol.* **30**:126–144.

Mendoza, S. P., Smotherman, W. P., Miner, M. T., Kaplan, J., and Levine, S., 1978b, Pituitary-adrenal response to separation in mother and infant squirrel monkeys, *Dev. Psychobiol.* **11**:169–175.

Meyer, J. S., Novak, M. A., Bowman, R. E., and Harlow, H. F., 1975, Behavioral and hormonal effects of attachment object separation in surrogate-peer-reared and mother-reared infant rhesus monkeys, *Dev. Psychobiol.* **8**:425–435.

Redican, W. K., and Kaplan, J. N., 1978, Effects of synthetic odors on filial attachment in infant squirrel monkeys, *Physiol. Behav.* **20**:79–85.

Reite, M., Short, R., and Seiler, C., 1978, Physiological correlates of maternal separation in surrogate-reared infants: A study in altered attachment bonds, *Dev. Psychobiol.* **11**:427–435.

Rosenblum, L. A., 1968, Mother–infant relations and early behavioral development in the squirrel monkey, in: *The Squirrel Monkey* (L. A. Rosenblum and R. W. Cooper, eds.), Academic Press, New York, pp. 207–233.

Rosenblum, L. A., 1971, Infant attachment in monkeys, in *The Origins of Human Social Relations* (R. Schaffer, ed.), Academic Press, New York, pp. 85–113.

Rosenblum, L. A., and Plimpton, E. H., 1981, The infant's effort to cope with separation, in: *The Uncommon Child* (M. Lewis and L. Rosenblum, eds.), Plenum Press, New York, pp. 225–257.

Roy, M. A., 1980, Atypical behaviors and social perferences in two New World primate species, in: *Species Identity and Attachment* (M. A. Roy, ed.), Garland, New York, pp. 167–180.

Roy, M. A., 1981, Abnormal behaviors in nursery-reared squirrel monkeys (*Saimiri sciureus*), *Am. J. Primatol.* **1**:35–42.

Roy, M. A., Wolf, R. H., Martin, L. N. Rangan, S. R. S., and Allen, W. P., 1978, Social and reproductive behaviors in surrogate-reared squirred monkeys (*Saimiri sciureus*), *Lab. Anim. Sci.* **28**:417–421.

Roy, M. A. Hennig, C. W., Hill, E. M., and Walker, M. L., 1981, Subspecies differences in curiosity and general activity for developing squirrel monkeys, *Dev. Psychobiol.* **14**:55–58.

Ruppenthal, G. C., Arling, G. L., Harlow, H. F., Sackett, G. P., and Suomi, S. J., 1976, A 10-year perspective of motherless-mother monkey behavior, *J. Abnormal Psychol.* **85**:341–349.

Taub, D. M., Adams, M. R., and Auerbach, K. G., 1978, Reproductive performance in a breeding colony of Brazilian squirrel monkeys (*Saimiri sciureus*) *Lab. Anim. Sci.* **28**:562–566.

Vogt, J. L., and Hennessy, M. B., 1982, Infant separation in monkeys: Studies on social figures other than the mother, in: *Child Nurturance Series*, Volume III, *Studies of Development in Nonhuman Primates* (H. E. Fitzgerald, J. Mullins, and P. Gage, eds.), Plenum Press, New York, pp. 109–133.

7

Reproductive Cyclicity and Breeding in the Squirrel Monkey

W. Richard Dukelow

1. INTRODUCTION

Some behavioral research was done with squirrel monkeys in a captive environment in the 1930s, but the first captive births of this species did not occur until 1940, and only more recently has extensive use been made of it. Its value in various nutritional and cardiovascular studies has led to increased efforts to breed this animal in captivity. Its small size, ease of handling, and relatively simple management requirements make it an attractive animal for research laboratories, and in fact it is the second most commonly used species for the purpose.

Although the basics of squirrel monkey reproduction have been known for some time, several problems exist with this nonmenstruating short-cycle animal. There have been great variances in reported estrous cycle length, and vaginal cytology has been less than satisfactory for predicting the stage of the cycle. Although a reasonable conception rate can occur in captivity, the animal is very susceptible to stress, and reproduction is characterized by a high rate of abortion and neonatal death. The large size of the infant relative to its mother contributes to the latter problem. Until recently very little was known about implantation and placentation of the squirrel monkey, but increasing research in this area is under way. The results of these studies, coupled with studies of reproductive behavior and the relationship of stress and adrenal hormones to reproduction, bode well for the future breeding and research use of this fascinating primate. The objective of this chapter is to outline the research that has been done on reproduction, to define the present state of knowledge of reproductive physiology of the squirrel monkey, and to suggest future areas where research is needed and where progress most likely will be made.

W. Richard Dukelow • Endocrine Research Center, Michigan State University, East Lansing, Michigan 48824.

2. THE REPRODUCTIVE CYCLE

2.1. Cycle Length

It is now recognized that the cycle length of the squirrel monkey ranges from 7 to 12 days, with most cycles falling between 8 and 10 days. Earlier reports (Denniston, 1963; Lang, 1967; Srivastava *et al.*, 1970a; Richter, 1976) reported cycle lengths ranging as high as 25 days. These reports used various cytological techniques to examine the vaginal cornification cycle, and probably reflect differences in the types of stress to which the monkeys were exposed. In fact Hutchinson (1970) described different squirrel monkey cycle lengths depending on the type of housing and whether the animals were housed as single animals or in groups. Rosenblum *et al.* (1967) described a carefully designed study of vaginal cornification cycles and the presence of sperm in vaginal smears from females housed with males and reported a cycle length of 7–8 days. This was confirmed by similar studies by Gould *et al.* (1973) and Travis and Holmes (1974) the latter reporting a 10.9-day cycle.

More recent work has utilized activity cycles and behavioral characteristics (Richter, 1976; Latta *et al.*, 1967; Jarosz *et al.*, 1977; Wilson, 1977), and confirm the 8- to 10-day period of cycle. Additionally, studies on steroid hormone variation in the squirrel monkey (R. C. Wolf *et al.*, 1977; Wilson, 1977; Ghosh *et al.*, 1982) show an 8- to 10-day cycle. Three reviews have appeared regarding reproductive cycle patterns (Butler, 1974; R. H. Wolf *et al.*, 1975; Dukelow, 1978).

Although the sexual cycle of higher primates is normally referred to as the "menstrual cycle," the squirrel monkey does not menstruate. Accordingly, in the present chapter I will use the words "estrous cycle" to describe this activity, recognizing that others have questioned this terminology, pointing out that a clearcut indication of sexual receptivity for the male is not a characteristic of the squirrel monkey. This distinction has been pointed out by Lang (1967).

2.2. Cycle Endocrinology

It is only in recent years that an accurate picture of the variation of steroid and gonadotropic hormones has been evident with the squirrel monkey, but early publications indicated estrogenic effects on the cervical canal of the squirrel monkey and the effects of the estrogenic stimulation on the cervix and various cervical enzymes (Graham, 1969a,b; Graham and Manocha, 1971). Because of the difficulty of drawing continuous samples of blood from the small squirrel monkey, early studies were conducted on urinary steroids as a measure of cycle length. Travis and Holmes (1974), working with animals showing a 10.9-day cycle length, divided the cycle into follicular and luteal phases based on a karyopycnotic index that peaked on day 5. They found that the daily rate of pregnanediol excretion increased abruptly on the sixth day of the cycle and thereafter the daily rate of pregnanediol excretion was slightly higher, and both the mean rate of daily excretion and the total amount excreted throughout the luteal phase was significantly greater than the values found for the follicular phase.

The development of radioimmunoassay techniques for use on small blood samples led to examination of the circulating steroid levels in squirrel monkey serum during the cycle. The most significant discovery was that such levels are strikingly higher than are found in humans or Old World primates. These have been observed in other New World species as well. The estradiol maximum and minimum values have been found to be 503 ± 57.5 and 81.0 ± 8.2 pg/ml, respectively. Serum progestins reached peak concentrations of 399 ± 27.7 ng/ml, 3–4 days after the estradiol peak (Wolf *et al.*, 1977; Wilson, 1977). In a study of estradiol 17β, progesterone, and LH levels from 14 animals over a 16-day period, Ghosh *et al.* (1982) found that serum estradiol and LH levels increased significantly on day 4 of the cycle. Progesterone levels peaked on day 6, suggesting ovulation on day 5 or early day 6 of the cycle. Based on hormonal peak intervals, these workers reported the reproductive cycle of the squirrel monkey to be 8–9 days in length.

Gonadal steroids are not the only hormones showing high serum levels in the squirrel monkey. Resting levels of cortisol can approach 405 μg/100 ml (Brown *et al.*, 1970). This high adrenal output can be confirmed by high levels of urinary free cortisone (Setchell *et al.* 1977). It should be kept in mind that reproductive behavior and environmental influences can exert strong influences on steroid cyclicity in the squirrel monkey (Coe and Levine, 1981). Some workers have used immunofluorescence techniques to study the hypothalamo-infundibular LRH tract in the squirrel monkey. Preliminary data show that serum FSH and LH levels progressively decrease from the luteal phases to the late follicular phase. In the case of FSH this decrease is from an average of 16 to about 5 mIU/ml in the late follicular phase (Barry and Croix, 1978; Barry, 1979).

2.3. Seasonal Influences and Captivity Adaptation

The seasonality of reproductive response is well established for the squirrel monkey and has been the subject of several reviews (Butler, 1974; Wolf *et al.*, 1975; Dukelow, 1978a). This seasonality and, in particular, a shift in the breeding season when squirrel monkeys are moved from South America to the northern hemisphere were first described by DuMond (1968). With squirrel monkeys housed under seminatural conditions in Miami, Florida, over a 6-year period, all births occurred within a discrete season of 8–12 weeks each year. Over the 6-year period there was a marked shift of the mating and birth seasons and this was attributed to the degree of rainfall. It was suggested that the start of the dry season triggered mating whether it was in the Amazon or Miami area. Later work by Baldwin and Baldwin (1971) further substantiated the humidity effect. This seasonality effect is further observed in female response to ovulation induction regimens and also appears to be correlated with humidity even in temperature- and light-controlled environments (Harrison and Dukelow, 1973; Dukelow, 1983). This response has been attributed to alterations in the sensitivity of the hypothalamic–hypophyseal axis relative to FSH secretion.

While evidence suggests that there may be a humidity relationship, this is difficult to confirm. There is little evidence for the existence of humidity receptors in most mammalian species, and in fact some squirrel monkey breeding colonies do not show a seasonality relative to the environmental precipitation in the area (Kaplan, 1977). Other workers have suggested that photoperiod may be a controlling factor, particularly if the animals are removed from their equatorial environment and moved to a more northern latitude (Coe and Levine, 1981). These workers also emphasize the complex environmental and sociological interactions that could trigger the breeding season. It is well accepted in many mammalian species that the presence of receptive females triggers male behavior, and similarly, the presence of males can alter female behavior. In a study of the onset of increased progesterone secretion in females, Kaplan *et al.* (1981) found elevations occurring at about 2½ years of age in February and May and continuing to increase throughout the fall months. There also were differences between various subtypes of squirrel monkeys.

In summary, distinct seasonal patterns of reproduction are observed in the wild, in a seminatural environment, and in the laboratory for the squirrel monkey. The exact cause of these variations is unknown, but probably reflects a complex interaction of sociological and environmental cues that may exert their influence at different locations relative to the natural habitat of the squirrel monkey. Additional research is needed in this vital area.

Adaptation to captivity is important in establishing a breeding colony. While some animals will adapt by 9 months of captivity (Harrison and Dukelow, 1973), at least three breeding seasons is required for the female to adapt (Lorenz *et al.*, 1973). If the female is established and the male is not, reproductive success is as would be expected if both members were fully adapted to captivity, i. e., female adaptation is more essential to effective breeding than is male adaptation. Seasonality is also observed in the male, as evidenced by the "fatted male" phenomenon (DuMond and Hutchinson, 1967). While pronounced changes in body size and male sexual activity occur, the squirrel monkey can produce sperm throughout the year and these sperm are capable of fertilization both *in vivo* and *in vitro*.

3. OVULATION AND FERTILIZATION

3.1. Normal Follicular Morphology

Because of the difficulty in detecting overt signs of the ovarian cycle in the squirrel monkey, few studies have been done on normal follicular development and ovulation. Srivastava *et al.* (1970b) performed ovariectomies on 13 female squirrel monkeys and determined that masses of luteal tissue were present in all cases and located centrally in the ovary. These workers described the luteal tissue as being present in two distinct stages of activity. In the first, where the corpus luteum was active, the theca interna did not seem to contribute to the structure of the corpus luteum, but rather the corpus luteum came totally from the granulosa cells. Follicular growth during this stage of activity

Table I
Time Sequence of Morphological Changes in the Ovary of the Squirrel Monkey

Mean morphological finding at ovary	Hours after HCG injection[a]
Clear	8–15 (11)
Probably time of ovulation	8–16 (12)
Hemorrhagic appearance	14–20 (16)
Indentation of dome	19–40 (34)
Luteinization, slight	14–38 (34)
Corpus luteum	36+

[a] Range of time sequence when finding was noted, with most frequent time in parentheses.

was very limited. On the other hand, when luteal activity was low, the corpus luteum seemed to be composed of well-defined granulosal and theca interna areas. The cytoplasmic vacuolation was less obvious in this case. Where the theca interna component of the luteal tissue appeared active, follicular development was much more advanced. Both anovular and polyovular follicles of various sizes were present and some follicles appeared to be undergoing luteinization without ovulation. Atresia of follicles was associated with the formation of cysts of various sizes and the cavity of atretic follicles was often filled with a pink, watery fluid.

The relative number and size of developing follicles have been reported by Jarosz *et al.* (1977) from control animals housed in an outdoor environment during the spring and summer seasons. During the spring season there were 3.9–4 small follicles (1 mm or smaller) on each ovary, 1.7–2.7 medium and large follicles (2–4 mm) per ovary, and 0.3–0.6 very large follicles and 0.1–0.3 corpora lutea per ovary. During the summer months the numbers of follicles present were nearly identical, except for the very large (preovulatory) follicles, which numbered from 0.1 to 0.3 per ovary. This undoubtedly reflects the anovulatory season of the squirrel monkey in this environment.

Laparoscopy has been used extensively to study follicular morphology in the squirrel monkey (Dukelow, 1978b; Dukelow and Ariga, 1976). Based on over 200 observations (some with induced follicular growth, some with natural follicular growth), it is evident that the follicle becomes very transparent and at times a vascular pattern is observed on the surface similar to what has been observed in Old World species. Harrison and Dukelow (1974) timed the sequence of morphological changes in the ovary of the squirrel monkey following an FSH-stimulated follicular growth and a single HCG injection. This time sequence is illustrated in Table I.

At the time of ovulation in the squirrel monkey there is extensive hemorrhaging and less clearly defined vascular patterns on the follicular wall than is observed in Old World species (Dukelow *et al.*, 1973). Due to this hemorrhaging, postovulatory follicles in the squirrel monkey are generally more easily discernible than in other species, but the formation of clear areas (stigma) is not as evident. Under normal circumstances a single oocyte occurs in each follicle, but treatment with diethylstilbestrol can cause polyovular follicles to occur in 30% of all ovarian follicles (Graham and Bradley, 1971).

3.2. Ovulation Induction

Ovulation induction can occur in the squirrel monkey under a wide variety of regimens. Much of the original work in this area derived from studies with other common laboratory and domestic species and utilized pregnant mares serum (PMS) supplemented with human chorionic gonadotropin (HCG) to induce ovulation. Previous work had demonstrated that in nonhuman primates the requirement for an FSH source was not species-specific but that the LH source (to cause ovulation) must be of primate origin. Accordingly, HCG has been used almost exclusively to trigger the ovulatory response. Major studies in the late 1960s and early 1970s using PMS and HCG in the squirrel monkey indicated that follicular growth and ovulation can be induced by this method, and furthermore, "superovulation" (that is, the production of more than a single egg at each ovulation) can be achieved with higher doses of PMS. More recently it has been found in a number of laboratory and domestic species, as well as primate species, that repeated subjection of the animal to a PMS regimen results in a refractoriness to the PMS that greatly lessens the effectiveness of the regimen. This is probably due to the antigenicity of the PMS.

Because of the necessity of subjecting captive squirrel monkeys held for biomedical research to several regimens of ovulation induction, studies were undertaken to develop techniques to induce single or double ovulations in the squirrel monkey (Dukelow, 1970). The initial work of Bennett (1967c) was repeated using PMS and HCG and again superovulation was observed. To mimic the luteal phase of the 8- to 10-day cycle, 5 mg of progesterone was given daily for each of 5 days. Following this optional pretreatment various treatments of FSH, human menopausal gonadotropin (HMG), and PMS were given. The optimal treatment was found to consist of 4 days of 1 mg FSH (Armour Laboratories, Omaha, Nebraska) followed, on the last day, by a single injection of 250–500 IU of HCG. This treatment resulted in a mean number of ovulation points of 1.3 per animal with 56.6% of the animals ovulating.This regimen has been used over the past 13 years with virtually no change in efficacy. Its use has been confirmed in the squirrel monkey by others (Travis and Holmes, 1974), as well as being used in a variety of other laboratory, domestic, and zoo animals. A summary of a 7-year period of use with this regimen has been published (Dukelow *et al.*, 1981). While the progesterone pretreatment is not essential, a significant difference was noted in the percentage of animals ovulating when the pretreatment was used. There did appear to be a slight suppressive effect of the progesterone pretreatment on follicular growth but not on percentage of animals ovulating or the number of ovulations per animal. Of particular interest is the effect of the ovulation induction regimen when used over an extended period of time, and data have been presented (Dukelow *et al.*, 1981) showing the effectiveness when used 20 or more times in a single squirrel monkey over a 7-year period. Differences in the percentage,

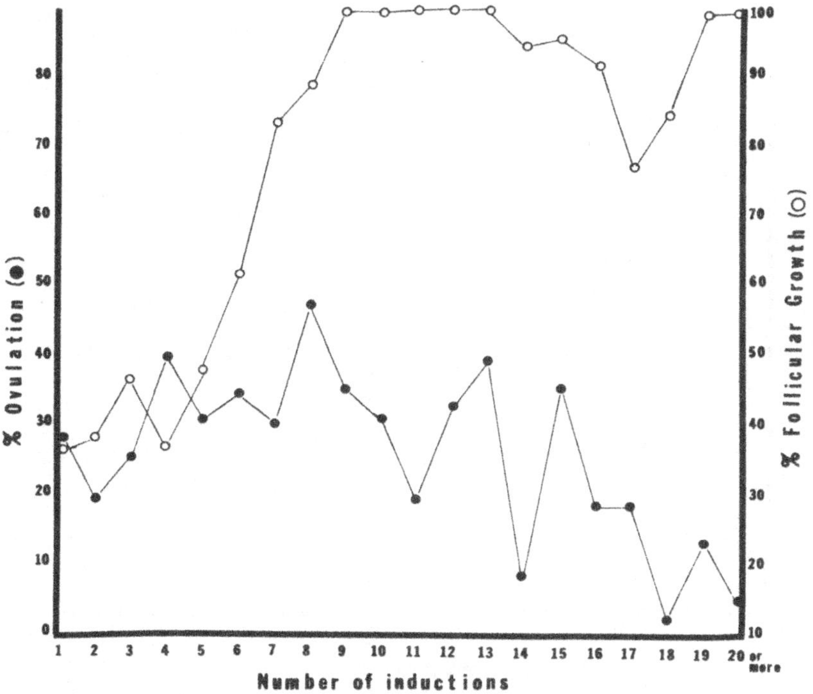

Fig. 1. Ovulatory and follicular responses of squirrel monkeys subjected to multiple ovulation inductions over an extended period.

of animals ovulating or the number of ovulations per animal were not significant although there was a slight trend toward decreasing effectiveness. There was a highly significant effect on the effectiveness to induce follicular growth, with an increase beginning after the fifth induction and a response of 70% or higher with every induction after the seventh. These data are indicated in Fig. 1. The fact that the HCG caused less refractoriness was not surprising, but the tendency of the animals to continue to ovulate even after repeated treatments was encouraging. The very slight decrease over time probably does represent the production of antibodies to the HCG; however, the low levels of HCG employed probably keep this response to a minimum. Since FSH is not of high antigenicity, one would not expect refractoriness with regard to follicular growth and this was indeed the case.

The level of HCG used to induce ovulation has been a matter of concern to scientists and clinicians with many species of animals. It is known that a rabbit will ovulate in response to as little as 25 IU of HCG, and yet reports occur in the literature where standard levels of 1000–2000 IU are used to ovulate many domestic species. In the human, 5000–10,000 IU is commonly used clinically, and as much as 20,000 IU has been reported in the literature. Because of concern over possible antigenicity of high levels of HCG and some work that indicated that HCG administered to rabbits to induce ovulation resulted in

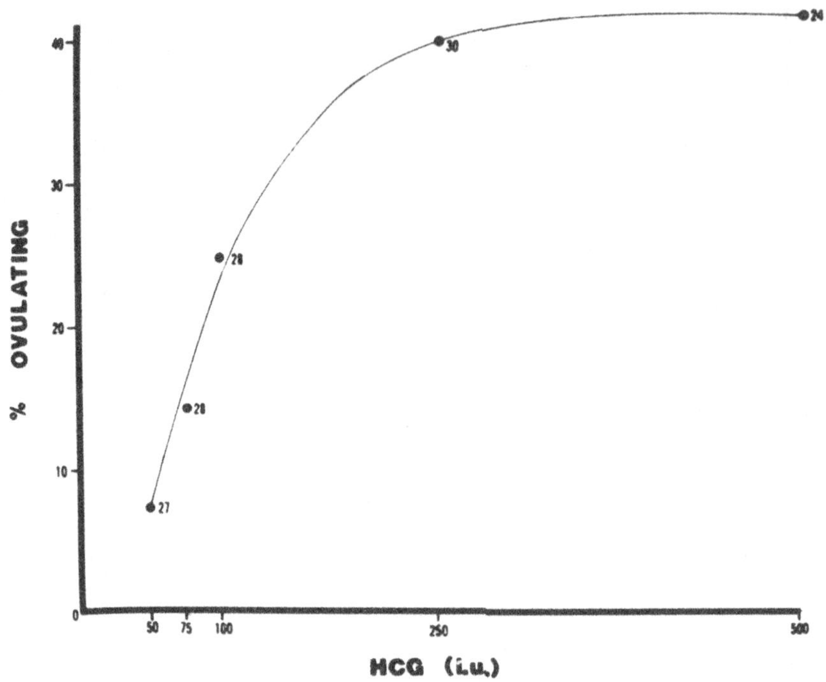

Fig. 2. Percent of squirrel monkeys ovulating in response to a given dose of HCG after treatment with FSH for 4 days. Values at each point indicate the numbers of animals tested.

9.7% chromosomally defective 6-day old blastocysts (Fujimoto *et al.*, 1974), a series of experiments was carried out to determine the minimum effective dose of HCG to induce ovulation in the squirrel monkey (Dukelow, 1979). The results of these experiments are indicated in Fig. 2. It is clearly evident that the minimum effective dose of HCG lies between 100 and 250 IU in this species.

It has already been indicated that reproduction in the squirrel monkey in the wild is seasonal, and this seasonality extends also to its response to an ovulation induction regimen. Harrison and Dukelow (1973) reported a seasonal response to the FSH–HCG regimen, with peak ovulation occurring in late winter and early spring and very low levels of ovulation in July, August, and September in a northern United States environment. Later work by Kuehl and Dukelow (1975a) indicated that the anovulatory response during the summer months could be overcome by a variety of techniques, including sequential estradiol and progesterone pretreatments. The chief limiting factor, however, appeared to be the FSH level, and satisfactory levels of ovulation could be obtained merely by doubling the amount of FSH given daily from 1 to 2 mg or, alternatively, by administering FSH for 5 days rather than 4 days during the anovulatory season. There was no effect of increasing the level of HCG during the summer months (Dukelow, 1978a).

While distinct differences occur in squirrel monkeys from different geographic areas (Jones *et al.*, 1973; Ariga *et al.*, 1978) and some of these differences are manifested in captivity by behavioral and endocrine characteristics (Coe and Levine, 1981), there has been no indication that a difference in ovulatory response to an induction regimen is evident between different subspecies of squirrel monkey (Dukelow, 1970; Dukelow *et al.*, 1981).

The ultimate test, of course, of any ovulation induction regimen is whether the oocytes produced are capable of fertilization. That oocytes produced by FSH–HCG ovulation induction can be fertilized *in vitro* has been demonstrated (Kuehl and Dukelow, 1975b, 1979; Dukelow and Kuehl, 1975; Chan *et al.*, 1982; Asakawa and Dukelow, 1982). Jarosz *et al.* (1977) demonstrated that oocytes resulting from ovulation induction were capable of being fertilized *in vivo,* and further, that the products of such conception carried to term with the birth of normal offspring.

Some modification of standard ovulation induction regimens have been utilized on an experimental basis. Rawson and Dukelow (1978) used laparoscopic direct administration of 0.5 IU of HCG onto the matured ovarian follicles of the squirrel monkey. This resulted in ovulation in 45% of the animals treated. Interestingly, when purified ovine LH was administered by the same technique, ovulation occurred in 55% of the animals, whereas intrafollicular injection of saline had no effect. The use of ovine LH was interesting considering the previously mentioned fact of a species-specific requirement for the LH source for primates. This suggests that the species specificity seen in primates might be at a suprafollicular level, possibly relating to binding in the circulatory system or a reactivity at the blood–follicle barrier. Additionally, some clinical applications of the ovulation induction regimens have demonstrated that the squirrel monkey can be a useful model for various contraceptive and drug testing. Harrison and Dukelow (1971) showed a dose–response effect of megestrol acetate on blocking ovulation that had been induced in the squirrel monkey. Furthermore, this effect could be shown using silastic implants filled with megestrol acetate (Harrison *et al.*, 1974). A synthetic polypeptide (TPAL) reported to occur naturally in the hamster and capable of blocking ovulation was tested at doses of 3 or 18 mg/day for antiovulatory activity in squirrel monkeys induced to ovulate. The treatment was shown to be ineffective in the squirrel monkey (Kuehl and Dukelow, 1978).

3.3. Fertilization

The only successful report of artificial insemination in the squirrel monkey was that of Bennett (1967b). Following ovulation induction the animals were artificially inseminated on the first or second day after the cessation of gonadotropin treatment. Of 16 eggs recovered from three monkeys, ten showed evidence of fertilization or cleavage up to the four-cell stage. Two additional squirrel monkeys, similarly treated, were allowed to go to term, but pregnancy did not occur. The author suggested that the PMS–HCG treatment resulted in increased transport of the embryos down the oviduct to the uterus, where they disintegrated. The only other report of fertilization *in vivo* following ovulation induction is that of Jarosz *et al.* (1977) previously quoted. Other reports relate mainly to breeding colonies in captive conditions.

The technique of *in vitro* fertilization has given investigators a rich opportunity to examine fertilization in the squirrel monkey, its timing sequence, and the normality of the embryos produced. The entire field of ovulation induction, *in vitro* fertilization, and embryo transfer in the nonhuman primate has recently been reviewed (Dukelow, 1982). While a preliminary report appeared in 1972 (D. C. Kraemer, personal communication) regarding the successful *in vitro* fertilization of a baboon embryo, in that same year two reports appeared simultaneously that described successful *in vitro* fertilization in squirrel monkey oocytes (Johnson *et al.*, 1972; Cline *et al.*, 1972). In an expanded report of one of these studies (Gould *et al.*, 1973), 22 oocytes were recovered, of which 11 had sperm in the perivitelline space, extrusion of the second polar body, or pronuclear formation. Six of these ova cleaved to the two-cell stage. From the other original report, extended studies were reported in 1975 (Kuehl and Dukelow, 1975b; Dukelow and Kuehl, 1975) and 1979 (Kuehl and Dukelow, 1979). By 1979 a level of 35% *in vitro* fertilization had been achieved and embryos developed to the eight-cell stage. The optimal medium for *in vitro* fertilization was found to be TC-199 with 20% agamma fetal calf serum and 72 mg pyruvate/ml as an energy source. By 1982 a report from this same laboratory (Chan *et al.*, 1982) reported a total of 1788 oocytes laparoscopically recovered from squirrel monkey follicles with 669 matured and 332 fertilized, for a fertilization rate of 49.6%. Oocyte maturation was found to be highest from March to June, again reflecting a seasonal response, and among the oocytes, 50% showed the first polar body extrusion by 21 hr after recovery. The presence of corona radiata cells about the cells resulted in a significantly higher rate of maturation compared with denuded oocytes, and the addition of 2.6–5.2 mM caffeine increased motility over controls for the first 5 hr. More significant was the effect of addition of 1 or 10 μM dibutyryl cyclic adenosine 3′, 5′-monophosphate (dbcAMP), which enhanced fertilization to over 90% over controls. This increase reflected the effect of cAMP on sperm capacitation, sperm acrosome reaction, and sperm motility. The time required for squirrel monkey sperm capacitation had previously been reported to average 4.7 hr with a minimum time of 2 hr (Kuehl and Dukelow, 1982), but the addition of cAMP reduced this sperm capacitation time requirement to 45 min or less. This phenomenon has also been observed in other laboratory species.

Various studies have been carried out to determine the normality of squirrel monkey embryos produced by the *in vitro* fertilization system, and these have generally involved two approaches: cytogenetic markers, and biochemical analysis of the embryo. Such studies are complicated by the necessity of using microtechniques for the very small numbers of oocytes that are available. In 1981 a technique was developed (Mizoguchi and Dukelow, 1981) for the gradual fixation of chromosomes within squirrel monkey oocytes recovered after gonadotropic treatment. This was based on earlier work with the Chinese hamster by workers in Japan. Utilizing this procedure, Asakawa *et al.* (1982) characterized the chromosomal normality in metaphase I and metaphase II squirrel monkey oocytes recovered from the follicle. Of 145 squirrel monkey oocytes, seven (4.8%) of 50 oocytes found in the metaphase II stage showed aneuploidy. After *in vitro* fertilization and chromosome analysis of first cleavage metaphases, 88% had the normal diploid number of chromosomes and aneuploidy occured 12% of the time, with five cases of triploidy noted (Asakawa and Dukelow, 1982). These indications of chromosomal abnormality should not be viewed with alarm, since the values represent the approximate normal values found after natural fertilization in a variety of species. In other words, it is doubtful if such embryos would have survived to term had they been allowed to proceed. The biochemical normality of *in vitro* fertilized squirrel monkey embryos has been reported by Chan *et al.* (1982) and Hutz *et al.* (1983). Squirrel monkey ova and embryos were incubated with either [^{3}H]-leucine [^{3}H]-estradiol-17β (E_2) or [^{3}H]-progesterone (P) supplemented medium and assayed for radioactivity or fixed and processed for autoradiographic analysis. Further, oxygen consumption was measured with polarographic O_2 electrode. The results demonstrate a decrease in leucine incorporation with ovum maturation and subsequent *in vitro* fertilization as assessed autoradiographically, whereas E_2 and progesterone uptake increased with fertilization in the squirrel monkey embryo. It was clear that detectable biochemical changes occur in the primate ovum with *in vitro* fertilization, including diminished relative protein synthesis and augmented steroid uptake. These results indicated normal metabolic development of *in vitro* fertilized primate embryos that was comparable with preimplantation development in other species.

For several decades scientists have known that the pseudopregnant rabbit oviduct can be used for temporary incubation and development of embryos of various species. This procedure has been used, by shipment of the rabbit, to transport embryos of both domestic and laboratory species. *In vitro* fertilized squirrel monkey embryos have been transferred to the oviduct of the pseudopregnant rabbit and developed to the 16-cell stage; these types of studies suggested the possibility that it might be possible to fertilize the eggs of the squirrel monkey with squirrel monkey sperm in the oviduct of the pseudopregnant rabbit (xenogenous fertilization), and indeed that has been accomplished (DeMayo *et al.*, 1980). These workers recovered 22 out of 40 embryos placed in the pseudopregnant rabbit oviduct, and of these, eight were fertilized, for a fertilization rate of 36%. At the time of this writing several preliminary reports have appeared relating to further development of early preimplantation of squirrel monkey embryos. These include techniques for the freezing and subsequent fertilization of the embryos (DeMayo, 1983) and for the separation of blastomeres from developing embryos to produce identical twins (Rawlins, 1983). Dynamic advances can be expected in the next few years.

4. CAPTIVITY BREEDING AND ARTIFICIAL INSEMINATION

A conception rate of 50–70% can be expected after squirrel monkeys have been acclimated to their captive environment, but this is coupled with a high incidence of stillborns or abortions (16.8%) and of neonatal death (34.3%) (Kaplan, 1977; Dukelow, 1982). Over the past 15 years a number of laboratories have reported their management procedures for the breeding of squirrel monkeys with varying degrees of success (Bantin, 1969; Hupp, 1972; Kaplan, 1977;

Taub *et al.*, 1978; Rasmussen *et al.*, 1980a; Tsuji and Tatsumi, 1981). Rather than reviewing all of the variables from these laboratories, the following synopsis provides a general indication of modern management techniques for optimum breeding performance in the squirrel monkey.

Males and females are normally housed in group cages at a ratio of 1:10 or in smaller cages at a 1:3 ratio. The male need not be removed after mating and will generally not bother mothers or newborn infants. Single housing of squirrel monkeys should be avoided unless absolutely essential to experimental design, because of social requirements of this animal. In caging, vertical height is more important with this species, and perches should be provided that will allow the animal to rest with its tail hanging and not touching the bottom of the cage. A minimum of 0.15 m^2 of floor space should be provided for each monkey, and, more importantly, they should have access to at least 51 cm of vertical space. Normal standards of cleanliness for caging, bedding, feeders, and water containers should be maintained. Individual water bottles for automatic water dispensing can be used. An adult squirrel monkey will consume 50–150 ml of water/day.

Animal rooms should be maintained between 21 and 24°C at a relative humidity of about 55%. When the latter is less than 50%, sneezing and sniffling may be observed and lead to serious respiratory problems, which in turn affect breeding capabilities. It is questionable if alteration of the photoperiod alters squirrel monkey cyclicity in captivity. However, most modern animal rooms do have controlled lighting on a 12:12 or 14:10 (light–dark) cycle. The nutritional requirements of the squirrel monkey are normally met by providing freshly prepared commercial feed, but emphasis must be placed on the fact that a folic acid deficiency can occur with squirrel monkeys, which has been shown to be detrimental to pregnancy. Similarly, ascorbic acid deficiency is sometimes observed in colonies. This vitamin is normally dusted on commercial pellets and removed if the pellets are soaked prior to feeding. Maintenance of protein levels is crucial for successful reproduction. High abortion rates can be attributed to protein deficiency. For greater detail on proper management of housing and nutritional factors, and for consideration of social behavioral factors relating to reproduction, readers are referred to other appropriate chapters in this volume.

Artificial insemination is often considered as an aid for normal reproduction when dealing with wild species of animals. Unfortunately, many times the causes of infertility in such species held in captivity cannot be simply corrected by the artificial placement of semen in the female reproductive tract. To the date of this writing and the knowledge of this author there has never been a birth of a live squirrel monkey resulting from artificial insemination. In work previously referenced, Bennett (1967a,b) collected semen and artificially inseminated a small number of females. Subsequently some of the animals were sacrificed and embryos recovered from the reproductive tract, indicating that fertilization did occur. In two animals where pregnancy was allowed to continue, no births occurred, nor was there evidence that fertilization had occurred. Denis *et al.* (1976) provided procedures for the freeze preservation of sperm in the hope that it could be utilized in artificial insemination for timed mating pregnancies. Unfortunately, these studies report only the motility of sperm after thawing and there is no evidence that the semen was used for artificial insemination to ascertain its fertility. One interesting attempt to better evaluate the fertility of captive squirrel monkeys was that of Stolzenberg *et al.* (1979), who used a continuous mating procedure coupled with daily swabbing of the vagina to determine when mating occurred. By this timed mating procedure, 277 females (of 373) were considered inseminated, for a rate of 74%. Forty-six full-term progeny were produced, and with continuous cohabitation the mean time to insemination was 5 days, with 75% of the matings occurring within 8 days. Pregnancy evaluation by mouse bioassay indicated a very high degree of resorptions within the first 50 days of gestation, in accord with reports of other workers.

Another factor for consideration for mating is the age at which pregnancy first occurs. Since it is difficult to precisely age animals from the wild, one must rely on dental formulas and other procedures. With captive-born animals it has been reported that 40% of all pregnancies occur in females between 3.5 and 4 years of age, with 57% of all pregnancies occurring between 3.5 and 5 years of age (Taub, 1980).

In recent years there has been increased emphasis on captive breeding of all species of nonhuman primates, including the squirrel monkey. Government interest in developing breeding colonies not only to supply biomedical needs but to preserve species in the wild has led to expanded breeding programs in the southern United States and Caribbean. Progress from these breeding colonies in the next few years should contribute greatly to our knowledge of basic breeding parameters of the squirrel monkey.

5. *PREGNANCY*

5.1. *Diagnosis*

A variety of techniques have been utilized to diagnose pregnancy in the squirrel monkey (Table II). Of the methods listed, uterine size increase and chorionic gonadotropin concentration occur early in the first trimester at a time when the embryo is in the primitive stage. Since it is desirable to detect pregnancy as early as possible, these procedures would appear to have the greatest promise. A pregnancy test kit has been developed by the National Institutes of Health, which can be used for squirrel monkey pregnancy detection between 40 and 60 days of pregnancy. This has an inherent risk of 10% for false-negative responses (Hodgen *et al.*, 1978). Any increase of uterine width, whether detected by laparotomy, laparoscopy, or palpation, to greater than 20 ml indicates pregnancy. The infant itself can be palpated by 70 days of gestation.

5.2. *Gestation Length*

Because of the difficulties of precisely timing the date of conception, it has been difficult to determine the gestation length in the squirrel monkey.

Table II
Pregnancy Detection in the Squirrel Monkey

Characteristic	Days after conception	Reference
Uterine size increase (by laparotomy)	20–25	Goss *et al.* (1968)
Chorionic gonadotropin		
Plasma	20–105	Nathan *et al.* (1966)
Urine	40–60	Hodgen *et al.* (1978)
Uterine size increase (by palpation)	42–56	Rosenblum (1968)
	61	Nathan *et al.* (1966)
Diabetes insipidus	60	Clewe (1969)
	60–135	Travis and Holmes (1974)
Maternal weight gain	60–95	Hopf (1967)
	67–147	Travis and Holmes (1974)
Visual abdominal enlargement	74–102	Hopf (1967)
Fetal skeleton by X ray	81	Nathan *et al.* (1966)

Earlier reports gave estimates ranging from 165 to 170 days (Goss *et al.*, 1968; Wiswell, 1965; Beischer and Furry, 1964). Lorenz *et al.* (1973) reported a shorter mean gestation length of 162.6 days, with a range of 152–168 days after pairing. Jarosz *et al.* (1977), using induced ovulation and laparoscopic examination of ovulation, reported on five pregnancies of Bolivian squirrel monkeys where the mean gestation length was 155 days, with a range of 137–175 days. Two of the five deliveries resulted in stillbirth or neonatal death, and thus prematurity cannot be ruled out for the lowest values. Kerber *et al.* (1977) reported on ten pregnancies where the length of gestation period was found to range from 141 to 154 days, with a mean of 146.9, a median of 146.5, and a modal length of 146 days. It would appear, based on recent evidence using sophisticated techniques, that the gestation length of the squirrel monkey ranges between 145 and 155 days in length.

5.3. *Time of Implantation*

In most mammalian species transport of the embryo throughout the oviduct or fallopian tube requires 3–4 days. As the developing embryo reaches the uterus, dissolution of the zona pellucida occurs and expansion of the blastocyst begins. While embryos have been recovered from the squirrel monkey oviduct at various preimplantation stages of development, the only uterine embryos recovered from this species were obtained utilizing a laparoscopic technique (Ariga and Dukelow, 1977a). Using this procedure, two 5-day preimplantation blastocysts were recovered. The first was estimated to be 93–109 hr after fertilization and clearly showed the early stages of an expanding blastocyst. The second 5-day preimplantation blastocyst was more difficult to exactly stage, but was estimated to have been recovered at 125.7 hr after HCG administration. This blastocyst was more spherical and expansion had not yet begun. In the human the zona pellucida disappears between day 4 and day 5 at a cell stage of approximately 107 cells. These results indicate that the zona pellucida of the squirrel monkey ova might be lost at a slightly younger age than human or some other nonhuman primate embryos. This could imply earlier ovum entry into the uterus and earlier embryonic development in the squirrel monkey but with implantation occuring shortly after the fifth day following fertilization (Ariga and Dukelow, 1977b).

5.4. *Placental Physiology*

In a monumental description of placental development and endocrinology of nonhuman primates (Luckett, 1974), little mention is made of the squirrel monkey, but extensive descriptions are made of the placental development of related New World primate species. Benirschke and Miller (1982) describe two placentas from squirrel monkeys that were available from zoo material. One was serially sectioned *in situ* and one was a term organ composed of two lobes and weighing 22 g. The hemochorial placenta of the squirrel monkey has been demonstrated to be a trabecular structure similar to that of other playtrrhine primates. It shares with these other features such structures as an extensive proliferation of syncytium and particularly the enclosure of large maternal vessels by the trophoblast as described by Luckett (1974) for other primates. These large vessels (arteries) may reach to the chorionic plate, become interlaced with extensive fibrin deposits, and eventually open into the pseudolabyrinthine intervillous space. Later in development these large vascular spaces are surrounded by very large syncytial cells. The villi contain more connective tissue than is found in Callithricids, but villous erythropoiesis is very much reduced. The amount of decidual tissue shed at delivery is very extensive, as is found in other Ceboids. The number of umbilical vessels in the squirrel monkey has recently benn examined from 40 nonsurviving squirrel monkey fetuses and neonates (Baker and Hendrickx, 1977). Four umbilical vessels were present in 72.5% of the cords and 27.5% contained three vessels. Coincident fetal malformations were observed.

Pregnancy-associated plasma proteins (PAPPs) have been studied in a number of species of nonhuman primates and in chimpanzees and orangutans. These appear to be immunologically identical to human PAPPs (Lin and Halbert, 1978). In macaque species, equivalent, partially crossreacted PAPPs are found, but in the squirrel monkey only faint traces of these proteins unique to the pregnant female are observed.

5.5. *Stillbirth and Abortion*

Some years ago Wolf *et al.* (1975) reported on the reproductive statistics for 2 years of breeding at the Delta Regional Primate Research Center. Of the 113 pregnancies, 75.2% resulted in live births, with a stillborn rate of 20.4% and an abortion rate of 4.4%. A more recent summary of 512 squirrel monkey pregnancies over a 9-year period (Rasmussen *et al.*, 1980b) showed that a majority (84%) of pregnancies resulted in live births. Of those born alive, 4% died within 24 hr and an additional 7% died in the first week of life. Of

Table III
Temporal Development of Squirrel Monkey Embryos in Vitro

Hours after insemination	Event	Reference
2–4.7	Sperm capacitation	Kuehl and Dukelow (1982)
6	Condensation and swelling of sperm	Asakawa and Dukelow (1982)
10	Nucleate stage	Asakawa and Dukelow (1982)
16	First cleavage metaphase	Asakawa and Dukelow (1982)
46–52	Second cleavage	Kuehl and Dukelow (1979)
52–72	Third cleavage	Kuehl and Dukelow (1979)
96–120	Blastocyst stage	Ariga and Dukelow (1977b)

those not delivered, 26% were aborted and 74% were stillborn. During the 9-year period 30 of 63 female deaths were reproduction-related. The most common cause (nine cases) was hemorrhage at the time of delivery, with an additional seven deaths due to dystocia or poor fetal presentation. Six died as a result of complications following a retained placenta. There was a significant correlation between birth rate and infant survival, and mortality was especially high for infants weighing less than 80 g at birth. These figures serve to further emphasize earlier statements in this chapter relating to the high incidence of abortion and stillbirths in the squirrel monkey as a major problem in management and reproductive study.

One case of an ectopic pregnancy in the squirrel monkey has been reported (McClure and Chang, 1975).

6. EMBRYONIC DEVELOPMENT

With the exception of the two *in vivo* fertilized uterine blastocysts that were recovered 5 days after fertilization, there are no other studies of normal embryological development in the squirrel monkey. This undoubtedly reflects problems in establishing pregnancy, in determining the exact time of conception, and in the availability of material for embryological studies.

From studies on *in vitro* fertilization and comparative studies of the temporal stages of development with other nonhuman primates and humans, it is possible to determine the sequence of events that occur following fertilization. The initial stages are indicated in Table III.The time required for sperm capacitation with *in vitro* fertilization has been determined to be 4.7 ± 2.6 hr with the earliest possible penetration after mixing of sperm and eggs at 2 hr (Kuehl and Dukelow, 1982). These values are comparable with values obtained from other primate species, including humans. Table IV shows the comparative development through the cleavage stages for the squirrel monkey in comparison with data on the rhesus monkey, baboon, and human (Kuehl and Dukelow, 1979). It can be seen that these stages of development are quite comparable and this gives credibility to the use of the squirrel monkey as a model for studying human *in vitro* fertilization.

Table IV
Comparative Time Course of Early Embryonic Development in Primates

	Hours after fertilization			
	Squirrel monkey	Rhesus monkey	Baboon	Human
First cleavage	16–40	24–36	—	30–38
Second cleavage	46–52	36–48	48	38–46
Third cleavage	52–73	48–72	48–72	51–72
Morula	96	—	120–148	96–135
Blastocyst	96–120	—	120	123–147

7. *PREGNANCY OUTCOME*

The high rate of stillbirth and abortions in the squirrel monkey has already been alluded to. There is also a high neonatal death loss. Price *et al.* (1972) reported that only three of 18 newborn survived the neonatal period, with 12 of the 15 deaths occurring about the time of parturition. All deaths occurred with infants with body weights from 52 to 100 g and the majority of these occurred within 2 days of parturition. These workers reported that newborns of body weight of 90 g or more showed no evidence of renal glomerulogenesis to suggest tissue immaturity. On the other hand, four of six lighter newborn animals had glomerulogenic zones. In other studies there was no association between the outcome of pregnancy and the sex of the infant (Rasmussen *et al.*, 1980b) but there were large differences between years and from month to month. The proportion of neonatal death loss (including stillborns and abortions) was highest from January through March. There also was a significant relationship between the occurrence of aborted or delivered stillborn infants and the age of the mother, with older animals (average 9.7 years) showing greater difficulty. Interestingly, the highest proportion of adverse outcomes was observed in infants born to mothers that had been in the colony for 6 years or more.

An interesting series of studies has been conducted on the effect of folic acid supplementation and deficiency in squirrel monkey diets during pregnancy (Rasmussen *et al.*, 1979, 1980a,b). These workers found that more than 75 mg of total folacin/kg of body weight per day may be needed to assure growth and normal hematological parameters and bone marrow cytology. Supplementation in the diet of pregnant females resulted in better maternal weight gain during pregnancy and higher infant birth weight. These workers have suggested the use of the squirrel monkey as a model for this problem in human pregnancies.

After the first 7 days the weight gain of infants is at a steady pace. By 6 weeks of age both sexes are ~150 g in body weight. By 14 weeks females are nearly 200 g and males about 250 g. By 30 weeks the weights for females and males are 350 and 400 g, respectively. A comprehensive study of reflexes and behavioral characteristics of developing squirrel monkeys has been published (Kaack *et al.*, 1979).

8. SUMMARY

Reproductively the squirrel monkey is a very interesting nonhuman primate. Having a nonmenstrual cycle of very short duration with no clearly evident vaginal cytological or behavioral characteristics, and coupled with a very strong seasonality, this small, stress-susceptible primate has been an enigma to those wishing to breed it in captivity. The high stillbirth, abortion, and neonatal death rates further complicate reproduction in this species. Increasing research, however, is providing additional clues in the endocrinology and behavioral aspects of squirrel monkey reproduction. Overcoming the obstacles that occur in natural reproduction under natural or seminatural conditions represents the clue to successful breeding of the squirrel monkey in captive conditions. A great deal of progress has been made in recent years relative to artificial control of the reproductive cycle using hormones to induce estrus and ovulation, collection of sperm from the males, artificial insemination, and *in vitro* fertilization. All of these techniques, however, are geared to application for biomedical research, and should not be looked upon as alternatives to natural fertilization. Such studies do yield findings that relate to the normal embryological development of the squirrel monkey *in vitro*. As such, these studies are valuable in contributing to a better understanding of squirrel monkey reproduction.

REFERENCES

Ariga, S., and Dukelow, W. R., 1977a, Non-surgical (laparoscopic) uterine flushing and egg recovery techniques in the squirrel monkey (*Saimiri sciureus*), *Primates* **18**(2):453–457.

Ariga, S., and Dukelow, W. R., 1977b, Recovery of preimplantation blastocysts in the squirrel monkey by a laparoscopic technique, *Fertil. Steril.* **28**(5):577–580.

Ariga, S., Dukelow, W. R., Emley, G. S., and Hutchinson, R. R., 1978, Possible errors in identification of squirrel monkeys (*Saimiri sciureus*) from different South American points of export, *J. Med. Primatol.* **7**:129–135.

Asakawa, T., and Dukelow, W. R., 1982, Chromosomal analyses after *in vitro* fertilization of squirrel monkey (*Saimiri sciureus*) oocytes, *Biol. Reprod.* **26**:579–583.

Asakawa, T., Chan, P. J., and Dukelow, W. R., 1982, Time sequence of *in vitro* maturation and chromosomal normality in metaphase I and metaphase II of the squirrel monkey (*Saimiri sciureus*) oocyte, *Biol. Reprod.* **27**:118–124.

Baker, C. A., and Hendrickx, A. G., 1977, The number of umbilical vessels in the squirrel monkey (*Saimiri sciureus*), *Folia Primatol.* **27**:230–233.

Baldwin, J. D., and Baldwin, J. I., 1971, Squirrel monkeys (*Saimiri*) in natural habitats in Panama, Columbia, Brazil and Peru, *Primates* **12**:45–61.

Bantin, G. C., 1969, Reproduction in the squirrel monkey (*Saimiri sciureus*), *J. Inst. Anim. Tech.* **20**(3):83–90.

Barry, J., 1979, Immunofluorescence study of the preoptico-terminal LRH tract in the female squirrel monkey during the estrous cycle, *Cell Tissue Res.* **198**:1–13.

Barry, J., and Croix, D., 1978, Immunofluorescence study of the hypothalamoinfundibular LRH tract and serum gonadotropin levels in the female squirrel monkey during the estrous cycle, *Cell Tissue Res.* **192**:215–226.

Beischer, D. E., and Furry, D. E., 1964, *Saimiri sciureus* as an experimental animal, *Anat. Rec.* **148**:615–624.

Benirschke, K., and Miller, C. J., 1982, Anatomical and functional differences in the placenta of primates, *Biol. Reprod.* **26**:29–53.

Bennett, J. P., 1967a, Semen collection in the squirrel monkey, *J. Reprod. Fertil.* **13**:353–355.

Bennett, J. P., 1967b, Artificial insemination of the squirrel monkey *J. Endocrinal.* **37**:473–474.

Bennett, J. P., 1967c, The induction of ovulation in the squirrel monkey (*Saimiri sciureus*) with pregnant mares serum (PMS) and human chorionic gonadotrophin (HCG), *J. Reprod. Fertil.* **13**:357–359.

Brown, G. M., Grota, L. J., Penney, D. P., and Reichlin, S., 1970, Pituitary–adrenal function in the squirrel monkey, *Endocrinology* **86**:519–529.

Butler, H., 1974, Evolutionary trends in primate sex cycles, in: *Contributions to Primatology,* Volume 3 (H. Kuhn, ed.), S. Karger, Basel, pp. 2–35.

Chan, P. J., Hutz, R. J., and Dukelow, W. R., 1982, Nonhuman primate *in vitro* fertilization: Seasonality cumulus cells, cyclic nucleotides, ribonucleic acid and viability assays, *Fertil. Steril.* **38**:609–615.

Clewe, T. H., 1969, Observations on reproduction of squirrel monkeys in captivity, *J. Reprod. Fertil.* **6**(Suppl.):151–156.

Cline, E. M., 1972, Investigations on *in vitro* fertilization for three mammalian species: Golden hamster, squirrel monkey and New Zealand White rabbit, M. S. Thesis, Department of Biochemistry, University of Georgia, Athens, Georgia.

Cline, E. M., Gould, K. G., and Foley, C. W., 1972, Regulation of ovulation, recovery of mature ova and fertilization *in vitro* of mature ova of the squirrel monkey (*Saimiri sciureus*), *Fed. Proc.* **31**:177.

Coe, C. L., and Levine, S., 1981, Psychoendocrine relationships underlying reproductive behavior in the squirrel monkey, *Int. J. Ment. Health* **10**(2–3):22–42.

DeMayo, F. J., 1983, Cryopreservation and xenogenous fertilization of hamster and squirrel monkey ova, *Proc. Mich. Acad. Sci.*, p. 14.

DeMayo, F. J., Mizoguchi, H., and Dukelow, W. R., 1980, Fertilization of squirrel monkey and hamster ova in the rabbit oviduct (*Xenogenous fertilization*), *Science* **208**:1468–1469.

Denis, L. T., Poindexter, A. N., Ritter, M. B., Seager, S. W. J., and Deter, R. L., 1976, Freeze preservation of squirrel monkey sperm for use in timed fertilization studies, *Fertil. Steril.* **27**:723–729.

Denniston, R. H., 1963, Notes on the vaginal cornification cycle of captive squirrel monkeys, *J. Mammal.* **45**:471.

Dukelow, W. R., 1970, Induction and timing of single and multiple ovulations in the squirrel monkey (*Saimiri sciureus*), *J. Reprod. Fertil.* **22**:303–309.

Dukelow, W. R., 1978a, Reproduction in the squirrel monkey (*Saimiri sciureus*), *Rec. Adv. Primatol.* **2**:195–200.

Dukelow, W. R., 1978b, Laparoscopic research techniques in mammalian embryology, in: *Methods in Mammalian Reproduction* (J. C. Daniel, ed.), pp. 247–271, Academic Press, New York.

Dukelow, W. R., 1979, Human chorionic gonadotropin: Induction of ovulation in the squirrel monkey (*Saimiri sciureus*), *Science* **206**:234.

Dukelow, W. R., 1982, The squirrel monkey, in: *Universities Federation for Animal Welfare Handbook* (J. Remfry, ed.), MTP Press, London.

Dukelow, W. R., 1983, Ovum recovery and embryo transfer in primates, in: *Mammalian Egg Transfer* (C. E. Adams, ed.), CRC Press, Boca Raton, Florida, pp. 155–174.

Dukelow, W. R., and Ariga, S., 1976, Laparoscopic techniques for biomedical research, *J. Med. Primatol.* **5**:82–99.

Dukelow, W. R., and Kuehl, T. J., 1975, *In vitro* fertilization of non human primates, in: *Proc. 4th Eur. Congr. Fertil. Steril. — Extrait de la fecondation: Colloque de la societé nationale pour l'étude de la sterilité et de la fecondité,* Masson, Paris, pp. 67–80.

Dukelow, W. R., Jewett, D. A., and Rawson, J. M. R., 1973, Follicular development and ovulation in *Macaca fascicularis, Saimiri sciureus* and *Galago senegalensis, Am. J. Phys. Anthropol.* **38**:207–210.

Dukelow, W. R., Theodoran, C. G., Howe-Baughman, J., and Magee, W. T., 1981, Ovulatory patterns in the squirrel monkey (*Saimiri sciureus*), *Anim. Reprod. Sci.* **4**:55–63.

DuMond, F. V., and Hutchinson, T. C., 1967, Squirrel monkey reproduction: The "fatted" male phenomenon and seasonal spermatogenesis, *Science* **158**:1067–1070.

DuMond, F. V., 1968, The squirrel monkey in a seminatural environment, in: *The Squirrel Monkey* (L. A. Rosenblum and R. W. Cooper, eds.), Academic Press, New York, pp. 87–145.

Fujimoto, S., Pahlavan, N., and Dukelow, W. R., 1974, Chromosome abnormalities in rabbit preimplantation blastocysts induced by superovulation, *J. Reprod. Fertil.* **38**:97–103.

Ghosh, M., Hutz, R. J., and Dukelow, W. R., 1982, Serum estradiol 17β, progesterone and relative LH levels in *Saimiri sciureus*: Cyclic variations and the effect of laparoscopy and follicular aspiration, *J. Med. Primatol.*, **11**:312–318.

Goss, C. M., Popjoy, L. T., Fusiler, J. L., and Smith, T. M., 1968, Observations on the relationship between embryological development, time of conception and gestation, in: *The Squirrel Monkey* (L. A. Rosenblum and R. W. Cooper, eds.), pp. 171–192, Academic Press, New York.

Gould, K. G., Cline, E. M., and Williams, W. L., 1973, Observations on the induction of ovulation and fertilization *in vitro* in the squirrel monkey (*Saimiri sciureus*), *Fertil. Steril.* **24**(4):260–268.

Graham, C. E., 1969a, Glycogen and glycogen-metabolizing enzymes in the squirrel monkey uterine cervix, *Am. J. Obstet. Gynecol.* **102**(7):989–996.

Graham, C. E., 1969b, Estrogen-induced changes at the cervical squamocolumnar junction of the squirrel monkey, *Am. J. Obstet. Gynecol.* **104**(6):839–845.

Graham, C. E., and Bradley, C. F., 1971, Polyovular follicles in squirrel monkeys after prolonged diethylstilboestrol treatment, *J. Reprod. Fertil.* **27**:181–185.

Graham, C. E., and Manocha, S. L., 1971, The effect of chronic estrogenic stimulation on the squirrel monkey cervical epithelium, *Virchows Arch. Abt. B. Zellpathol.* **7**:147–156.

Harrison, R. M., and Dukelow, W. R., 1971, Megestrol acetate: Its effect on the inhibition of ovulation in squirrel monkeys, *Saimiri sciureus, J. Reprod. Fertil.* **25**:99–101.

Harrison, R. M., and Dukelow, W. R., 1973, Seasonal adaption of laboratory-maintained squirrel monkeys (*Saimiri sciureus*), *J. Med. Primatol.* **2**:277–283.

Harrison, R. M., and Dukelow, W. R., 1974, Morphological changes in the *Saimiri sciureus* ovarian follicles as detected by laparoscopy, *Primates* **15**(4):305–309.

Harrison, R. M., Rawson, J. M. R., and Dukelow, W. R., 1974, Megestrol acetate. II. Effects on ovulation in nonhuman primates as determined by laparoscopy, *Fertil. Steril.* **25**(1):51–56.

Hodgen, G. D., Stolzenberg, S. J., Jones, D. C. L., Hildebrand, D. F., and Turner, C. K., 1978, Pregnancy diagnosis in squirrel monkeys: Hemagglutination test, radioimmunoassay, and bioassay of chorionic gonadotropin, *J. Med. Primatol.* **7**:59–64.

Hopf, S., 1967, Notes on pregnancy, delivery, and infant survival in captive squirrel monkeys, *Primates* **8**:323–332.

Hupp, E. W., 1972, Husbandry procedures in an indoor breeding squirrel monkey colony, in: *Medical Primatology* (E. I. Goldsmith and J. Moor-Jankowski, eds.), Karger, Basel, Part I, pp. 100–104.

Hutchinson, T. C., 1970, Vaginal cytology and reproduction in the squirrel monkey (*Saimiri sciureus*), *Folia Primatol.* **12**:212–223.

Hutz, R. J., Ghosh, M., Dukelow, W. R., 1984, Steroid uptake and ^{3}H-uridine incorporation by early hamster and squirrel monkey (*in vitro* fertilized) embryos: Effects of ovulatory regimen, *Zool. Sci.*, **1**:771–776.

Jarosz, S. J., Kuehl, T. J., and Dukelow, W. R., 1977, Vaginal cytology, induced ovulation and gestation in the squirrel monkey (*Saimiri sciureus*), *Biol. Reprod.* **16**:97–103.

Johnson, M. J., Harrison, R. M., and Dukelow, W. R., 1972, Studies on oviductal fluid and *in vitro* fertilization in rabbits and nonhuman primates, *Fed. Proc* **31**:278.

Jones, T. C., Thorington, R. W., Hu, M. M., Adams, E., and Cooper, R. W., 1973, Karyotypes of squirrel monkeys (*Saimiri sciureus*) from different geographical regions, *Am. J. Phys. Anthropol.* **38**:269–277.

Kaack, B., Walker, L., and Brizzee, K. R., 1979, The growth and development of the squirrel monkey (*Saimiri sciureus*) *Growth* **43**:116–135.

Kaplan, J. N., 1977, Breeding and rearing squirrel monkeys (*Saimiri sciureus*) in captivity, *Lab. Anim. Sci.* **27**:557–567.

Kaplan, J. N., Chen, J., Smith, E., and Davidson, J., 1981, The development of seasonal variation in gonadal hormones and body weight in the maturing squirrel monkey, *Int. J. Primatol.* **2**:369–380.

Kerber, W. T., Conaway, C. H., and Moore Smith, D., 1977, The duration of gestation in the squirrel monkey (*Saimiri sciureus*), *Lab. Anim. Sci.* **27**:700–702.

Kuehl, T. J., and Dukelow, W. R., 1975a, Ovulation induction during the anovulatory season in *Saimiri sciureus, J. Med. Primatol.* **4**:23–31.

Kuehl, T. J., and Dukelow, W. R., 1975b, Fertilization *in vitro* of *Saimiri sciureus* follicular oocytes, *J. Med. Primatol.* **4**:209–216.

Kuehl, T. J., and Dukelow, W. R., 1978, The effect of a synthetic polypeptide, threonyl-prolyl-arginyl-lysine, on ovulation in the squirrel monkey (*Saimiri sciureus*) and hamster, *J. Reprod. Fertil.* **52**:25–28.

Kuehl, T. J., and Dukelow, W. R., 1979, Maturation and *in vitro* fertilization of follicular oocytes of the squirrel monkey (*Saimiri sciureus*), *Biol. Reprod.* **21**:545–556.

Kuehl, T. J., and Dukelow, W. R., 1982, Time relations of squirrel monkey (*Saimiri sciureus*) sperm capacitation and ovum maturation in an *in vitro* fertilization system, *J. Reprod. Fertil.* **64**:135–137.

Lang, C. M., 1967, The estrous cycle of the squirrel monkey (*Saimiri sciureus*), *Lab. Anim. Care* **17**(5):442–451.

Latta, J., Hopf, S., Ploog, D., 1969, Observations on mating behavior and sexual play in the squirrel monkey, *Primates* **8**:229–246.

Lin, T.-M., and Halbert, S. P., 1978, Immunological relationships of human and subhuman primate pregnancy-associated plasma proteins, *Int. Arch. Allergy Appl. Immunol.* **56**:207–223.

Lorenz, R., Anderson, C. O., and Mason, W. A., 1973, Notes on reproduction in captive squirrel monkeys (*Saimiri sciureus*), *Folia Primatol.* **19**:286–292.

Luckett, W. P., 1974, Comparative development and evolution of the placenta in primates, in: *Contributions to Primatology,* Volume 3, Karger, Basel, pp. 142–234.

McClure, H. M., and Chang, J., 1975, Ectopic pregnancy in a squirrel monkey, *J. Am. Vet. Med. Assoc.* **1975**:167–168.

Mizoguchi, H., and Dukelow, W. R., 1981, Gradual fixation method for chromosomal studies of squirrel monkey oocytes after gonadotropin treatment, *J. Med. Primatol.* **10**:180–186.

Nathan, T. S., Rosenblum, L. A., Limson, G., and Nelson, J. H., 1966, Diagnosis of pregnancy in the squirrel monkey, *Anat. Rec.* **155**:531–536.

Price, R. A., Anver, M. R., and Garcia, F. G., 1972, Simian neonatology I. Gestational maturity and extrauterine viability, *Vet. Pathol.* **9**:301–309.

Rasmussen, K. M., Thenen, S.W., and Hayes, K. C., 1979, Folacin deficiency and requirement in the squirrel monkey (*Saimiri sciureus*), *Am. J. Clin. Nutr.* **32**:2508–2518.

Rasmussen, K. M., Ausman, L. M., and Hayes, K. C., 1980a, Vital statistics from a laboratory breeding colony of squirrel monkeys (*Saimiri sciureus*), *Lab. Anim. Sci.* **30**:99–106.

Rasmussen, K. M., Thenen, S. W., and Hayes, K. C., 1980b, Effect of folic acid supplementation on pregnancy in the squirrel monkey (*Saimiri sciureus*), *J. Med. Primatol.* **9**:169–184.

Rawlins, R. G., 1983, Blastomere separation and culture techniques for hamster and squirrel monkey ova, *Proc. Mich. Acad. Sci.* p. 17.

Rawson, J. M. R., and Dukelow, W. R., 1978, Effects of intrafollicular administration of gonadotropins in two species of nonhuman primates using laparoscopy, *J. Med. Primatol.* **7**:223–227.

Richter, C. P., 1976, Artificial seven-day cycles in spontaneous activity in wild rodents and squirrel monkeys, *J. Comp. Physiol. Psychol.* **90**:572–582.

Rosenblum, L. A., 1968, Some aspects of female reproductive physiology in the squirrel monkey, in: *The Squirrel Monkey* (L. A. Rosenblum and R. W. Cooper, eds.), pp. 147–170, Academic Press, New York.

Rosenblum, L. A., Nathan, T., Nelson, J., and Kaufman, I. C., 1967, Vaginal cornification cycles in the squirrel monkey (*Saimiri sciureus*), *Folia Primatol.* **6**:83–91.

Setchell, K. D. R., Chua, K. S., and Himsworth, R. L., 1977, Urinary steroid excretion by the squirrel monkey (*Saimiri sciureus*), *J. Endocrinol.* **73**:365–375.

Srivastava, P. K., Cavazos, F., and Lucas, F. V., 1970a, Biology of reproduction in the squirrel monkey (*Saimiri sciureus*): I. The estrus cycle, *Primates* **11**:125–134.

Srivastava, P. K., Cavazos, F., and Lucas, F. V., 1970b, Biology of reproduction in the squirrel monkey (*Saimiri sciureus*): II. The ovary, *Primates* **11**:265–270.

Stolzenberg, S. J., Jones, D. C. L., Kaplan, J. N., Barth, R. A., Hodgen, G. D., and Madan, S. M., 1979, Studies with timed-pregnant squirrel monkeys (*Saimiri sciureus*), *J. Med. Primatol.* **8**:29–38.

Taub, D. M., 1980, Age at first pregnancy and reproductive outcome among colony-born squirrel monkeys (*Saimiri sciureus, Brazilian*), *Folia Primato.* **33**:262–272.

Taub, D. M., Adams, M. R., and Auerbach, K. G., 1978, Reproductive performance in a breeding colony of Brazilian squirrel monkeys (*Saimiri sciureus*), *Lab. Anim. Sci.* **28**:562–566.

Travis, J. C., and Holmes, W. N., 1974, Some physiological and behavioral changes associated with oestrus and pregnancy in the squirrel monkey (*Saimiri sciureus*), *J. Zool. Lond.* **174**:41–66.

Tsuji, K., and Tatsumi, T., 1981, Some experiences of captive breeding of squirrel monkeys (*Saimiri sciureus*), *Jpn. J. Med. Sci. Biol.* **34**:259–261.

Wilson, M. I., 1977, Characterization of the oestrous cycle and mating season of squirrel monkeys from copulatory behavior, *J. Reprod. Fertil.* **51**:57–63.

Wiswell, O. B., 1965, Gestation, parturition and maturation in the *Saimiri sciureus* (squirrel monkey), *Excerpta Med. Int. Congr. Ser.* **99**:E52.

Wolf, R. C., O'Connor, R. F., and Robinson, J. A., 1977, Cyclic changes in plasma progestins and estrogens in squirrel monkeys, *Biol. Reprod.* **17**:228–231.

Wolf, R. H., Harrison, R. M., and Martin, T. W., 1975, A review of reproductive patterns in new world monkeys, *Lab. Anim. Sci.* **25**:814–821.

8

The Endocrine System of the Squirrel Monkey

Christopher L. Coe, Erla R. Smith, and Seymour Levine

1. INTRODUCTION

In 1968, when the first treatise on the squirrel monkey appeared, it was already apparent that the physiology of New World monkeys differed from that of Old World primates, but few investigators were aware of the extensive differences in hormone metabolism and secretion (Rosenblum and Cooper, 1968). Since that time, it has become increasingly evident that the endocrine system of New World monkeys is extremely divergent from the general primate pattern in terms of both absolute hormone levels and the synthetic pathways utilized. We now know that most of the Callithricidae and smaller Cebidae have unusually high levels of adrenal and gonadal hormones, although typical values are difficult to describe because of the remarkable diversity in their reproductive physiology [for a species comparison, see Hearn (1983)]. In this chapter we will review a series of recent studies on the endocrine system of the squirrel monkey, which has some of the highest hormone levels observed so far. The level and magnitude of the hormone responses in the squirrel monkey have also provided a unique opportunity for assessing the influence of psychological and environmental variables, which may exert less overt effects in species with lower hormone levels. We will discuss the influence of seasonal and diurnal rhythms and the effect of genetic and maturational variables, and finally consider several of the important psychological processes that affect endocrine function.

Christopher L. Coe and Seymour Levine • Department of Psychiatry and Behavioral Sciences, Stanford University School of Medicine, Stanford, California 94305. *Erla R. Smith* • Department of Physiology, Stanford University School of Medicine, Stanford, California 94305.

Table I
General Hormone Levels in the Squirrel Monkey

	Mating season	Nonmating season	Typical primate value
Cortisol (μg/dl)	100–200	75–150	10–30
Testosterone (ng/dl)	125–500	10–30	5–50
Androstenedione (ng/ml)	86–525	—	1
Estrogen (pg/ml)	400–500	30–80	100–600
Progesterone (ng/ml)	200–400	20–30	2–25
Thyroxine (T_4) (μg/dl)	4	2	9
Triiodothyronine (T_3) (ng/dl)	95	40	120
ACTH (pg/ml)	—	15–30	10–80
Prolactin (ng/ml)	—	9–30	6–14
Growth hormone (ng/ml)	—	4–6	2–5

2. HORMONE PROFILE

The first inkling that New World monkeys differed from the typical endocrine profile described for Old World primates occurred in a number of disparate studies on squirrel monkeys and marmosets (Snipes *et al.*, 1969; Brown *et al.*, 1970; Yamamoto *et al.*, 1977; Coe *et al.*, 1978; Wilson *et al.*, 1978). These investigators reported unusually high levels of adrenal corticosteroids and gonadal hormones, ranging 10–50 times higher than human values, with tremendous lability in basal levels even within a single day. The full significance of the high hormone levels was not clear at the time, but these studies served to stimulate a great deal of interest in the endocrine system of New World monkeys. As can be seen in Table I, the high secretion pattern proved to be restricted to the steroid hormones; other hormones, the peptide hormones in particular, are usually in the normal range (Brown *et al.*, 1971; Wolf *et al.*, 1977; Kaack *et al.*, 1980; Wiebe *et al.*, 1984). Thus, the alteration in the endocrine system appears to be peripheral, occurring at the organ level, rather than emanating from excessive secretion at the hypothalamic or pituitary levels.

Some of the most insightful research on the significance of the high hormone levels has just been completed during the last few years (Cassorla *et al.*, 1982; Chrousos *et al.*, 1982a,b). These studies have suggested that an alteration in hormone receptors may be the causal event underlying the hypersecretion of steroid hormones because of either a low affinity or low concentration of receptors. According to this view, alterations in the receptors of target cells have reduced the efficacy of hormone action and necessitated a compensatory response on the part of the adrenals and gonads. While the factual basis for the receptor hypothesis is well substantiated, it is also true that other changes have occurred in the endocrine system of New World monkeys. In addition to the alteration in hormone receptors, a number of differences in hormone synthesis and metabolism have been reported (Preslock and Steinberger, 1977). With regard to gonadal hormone production in males, for example, squirrel

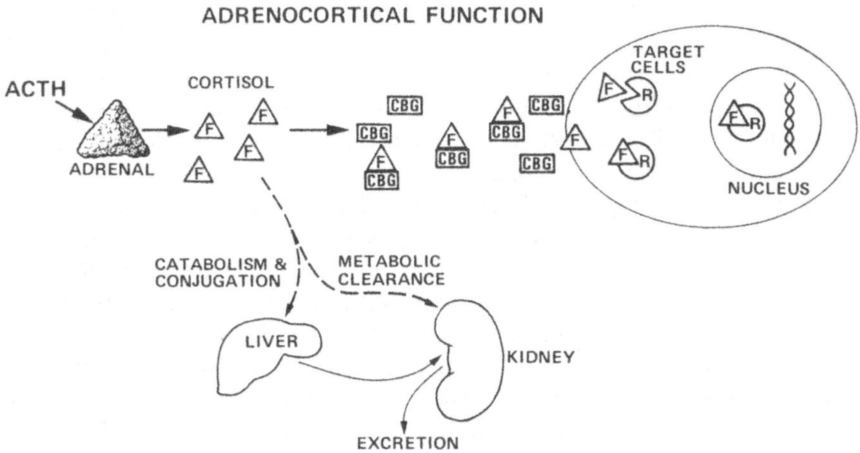

Fig. 1. Diagram of the pituitary–adrenal system illustrating the major factors that have been examined in the squirrel monkey.

monkeys have extremely high levels of androstenedione and relatively low levels of dihydrotestosterone. In fact, the plasma levels of androstenedione are even greater than the high testosterone titers (Wiebe *et al.*, 1984). The dihydrotestosterone levels also reflect an inability of peripheral cells to convert testosterone via the enzyme action of 5-α-reductase. Instead, based on tests of foreskin fibroblasts, it appears that the target cells of male squirrel monkeys have extremely high aromatase activity capable of converting testosterone to estrogen (Goldman *et al.*, 1983). A similarly complex scenario can be described for hormone secretion in females. For example, female squirrel monkeys have a high capacity for converting estradiol to sulfated forms, and thus have uniquely high levels of sulfated estrogens in both plasma and urine (Hodges *et al.*, 1981; Musey *et al.*, 1982). Suffice it to say for the moment that the alteration in gonadal hormone secretion involves changes not only at the receptor level, but also in hormone metabolism.

Perhaps the most effort has been expended in attempting to elucidate the factors accounting for the high adrenocortical output. Since many of the studies discussed later in the chapter focus on cortisol responses, we will briefly describe some of the pertinent information that has been obtained on the pituitary–adrenal axis. One hypothesis proposed to account for the high adrenal activity was that the pituitary of the squirrel monkey secreted equivalently high levels of ACTH (Brown *et al.*, 1970). However, in collaboration with Dr. M. Dallman (University of California at San Francisco) we determined that ACTH levels were not elevated (Coe *et al.*, 1978) and this finding has recently been confirmed by Cassorla *et al.* (1982). An alternative explanation would have been provided if there were any changes in the transport or binding of cortisol that would decrease its efficacy (Fig. 1). Brown *et al.* (1970) assessed the binding of tritiated cortisol and concluded that there is actually a relatively low level of binding of cortisol by blood proteins in the squirrel monkey (approximately 40% at physiological levels). Chrousos *et al.* (1982a) have reported that free cortisol accounts for 15% of the total hormone in circulation, while we have found in collaboration with Dr. P. Siiteri (University of California at San Francisco) that the free hormone can account for up to 95% of the total under certain conditions. Other peripheral factors that have been ruled out are (1) an unusual ratio of the different corticoids, (2) an exceptionally high level of conjugation, (3) a rapid clearance from circulation, and (4) an excessive excretion rate of corticoids (Seal and Doe, 1965; Kittinger and Beamer, 1968; Brown *et al.*, 1970; Setchell *et al.*, 1977; Cassorla *et al.*, 1982). The ability of the squirrel monkey to clear cortisol is actually half that of a macaque, while the half-life of cortisol (approximately 79 min) is similar to that for humans.

As discussed above, the first breakthrough in our understanding of the reason for the high cortisol levels was the demonstration of an alteration in receptor sensitivity (Cassorla *et al.*, 1982; Chrousos *et al.*, 1982a). It appears that the glucocorticoid receptors in the squirrel monkey, as well as in several other New World monkeys, have an extremely low affinity for cortisol (K_D = 49.5 × 10^9 M). At present, this finding provides the most plausible explanation for the high adrenal activity. In order to produce cortisol at such levels, the squirrel monkey adrenal has undergone a number of morphological and enzymatic changes, which have been described in detail elsewhere (Penney and Brown, 1971; Cassorla *et al.*, 1982). Further, the adrenal may be particularly sensitive to ACTH, and the pituitary relatively insensitive to the negative feedback of cortisol, since high doses of dexamethasone are required to inhibit the release of cortisol via the suppression of ACTH (Chrousos *et al.*, 1982a).

As this detailed information indicates, tremendous progress has been made in elucidating the physiological processes involved in the secretion of adrenal hormones. Nevertheless, much remains to be learned about the functional significance of the high hormone output. For example, we have no information about any natural process that would have selected for this type of endocrine profile. The maintenance of high hormone levels must have been particularly important, because the squirrel monkey appears to pay a high biological price for its adrenal activation. Squirrel monkeys in captivity are prone to a number of stress-related diseases, including glomerulonephritis and atherosclerosis; and aortitis is observed even in wild populations (see Strickland and Clarkson, this volume, Chapter 12). Moreover, our laboratory has found that the sustained adrenal activation that can occur over a 7-day period of psychological stress has a number of suppressive effects on the immune system (see Coe *et al.*, this volume, Chapter 5). One could offer some speculative explanations for the high cortisol output, based on the adrenal's involvement in glucose, water, fat, and ion regulation, but further work will be required to uncover the primary reason for the cortisol secretion pattern.

3. ANNUAL REPRODUCTIVE PATTERN

It is possible to provide a more complete description of the general biological factors and endocrine physiology that underlie the annual reproductive

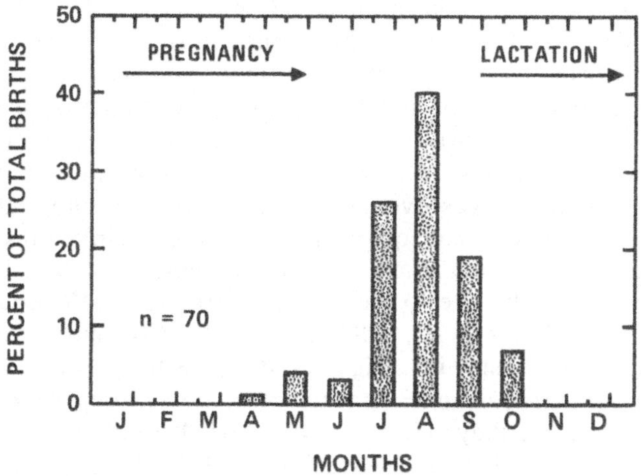

Fig. 2. Annual distribution of births across two seasons (1979–1980) in squirrel monkeys that had been acclimated to local environmental conditions [Coe and Levine (1981).]

pattern of the squirrel monkey. An understanding of the seasonal breeding pattern is essential for any investigator working with this species, because of the pronounced changes in behavior and physiology that occur across the year. Under natural conditions, squirrel monkeys are strictly seasonal breeders, with demarcated mating and birth periods occurring at approximately 6-month intervals (DuMond, 1968; Baldwin, 1970). The birth period lasts 3–4 months and usually coincides with the season of high rainfall. This temporal orientation appears to be to the advantage of the water-dependent pregnant females (Clewe, 1969) and may facilitate the initial foraging attempts of infants that are weaned during the first flush of new fruits and leaves at the end of the wet season (T. Rowell, University of California at Berkeley, personal communication). The timing of the mating period, which occurs during the dry season, is probably set by the fixed 5½-month gestation and lactation phases (Coe and Rosenblum, 1978).

While food availability and rainfall are extremely important *zeitgebers* in equatorial regions, the reproductive pattern of the squirrel monkey can be strongly influenced by light cycles when the animals are brought into temperate zones. In North America, squirrel monkeys typically shift their mating activity to our winter months between December and April. An early report on this phenomenon in Florida suggested that the temporal shift was due to an effect of humidity, because the monkeys oriented their birth period to the local season of high rainfall (DuMond, 1968). However, it may have also reflected an effect of photoperiod, since the mating activity shifted to the winter months. As can be seen in Fig. 2, a summary of the breeding records of acclimated monkeys in the Stanford colony showed that mating also occurred during our winter months. Thus, births occurred during the summer months—a period of extreme dryness in California. We believe that this temporal orientation is due to the fact that the squirrel monkey shows a strong propensity for being a short-day breeder, a basic trait in diurnal primates (Lancaster and Lee, 1965; Michael and Zumpe, 1976; Van Horn, 1980). More specifically, we have evidence that suggests that the high output of gonadal hormones during the long nights may be the critical endocrine factor in laboratory-maintained monkeys (see Section 4 on diurnal hormone rhythm). It is also of interest that squirrel monkeys from different geographic regions in South America show a differential sensitivity to the seasonal light changes even when housed in the same laboratory. In general, Bolivian and Peruvian squirrel monkeys engage in mating 1–3 months earlier than do Guyanese and Brazilian monkeys when maintained in captivity, a disposition that may be due to their origin further from the equator (Travis and Holmes, 1974; Kaack *et al.*, 1980; Taub, 1980; Coe and Levine, 1981).

Due to the work of many investigators, it is possible to provide an idealized picture of the endocrine changes that underlie the seasonal breeding of male and female squirrel monkeys (Latta *et al.*, 1967; Rosenblum, 1968; Baldwin, 1970; Wolf *et al.*, 1977; Coe and Rosenblum, 1978; Mendoza *et al.*, 1978a,b; Wilson *et al.*, 1978). As shown in Fig. 3, the seasonal occurrence of ovarian cycles in females is one of the essential features of the annual reproductive pattern. In stable social groups, females typically cycle only during the mating period, when they undergo several comparatively short cycles for a primate (Mendoza *et al.*, 1978b; Diamond *et al.*, 1984). Ovulatory cycles are usually 8–10 days in duration and are characterized by a high preovulatory surge of estrogen and a rising progesterone titer during the luteal phase, although no menses occur (Wolf *et al.*, 1977; Ghosh *et al.*, 1982). Females can conceive rapidly after the initiation of mating activity, and those that do not become pregnant will usually terminate cycling after the mating period. The occurrence of fertile mating is not the only factor that causes the cessation of sexual behavior, however. In a 16-month study of heterosexual groups with vasectomized males, Mendoza *et al.* (1978b) demonstrated that annual changes in reproductive physiology and behavior occurred even in the absence of conceptions. Assessment of female estrogen levels in this study indicated that their ovarian cycles occurred seasonally and that the males' hormonal changes were oriented to the estrous status of the females. In research on the response to induced ovulation regimens, Dukelow and co-workers have shown further that females may undergo an annual shift in the sensitivity to hormone action at the ovary (Kuehl and Dukelow, 1975; Jarosz *et al.*, 1977; Dukelow, this volume, Chapter 7).

These hormonal changes in females also affect their attractiveness and receptivity to males. A study of scent-marking and olfactory behavior in captive groups indicated that there may be a change in pheromonal cues, since males showed a significant increase in olfactory investigations of females during the mating period (Hennessy *et al.*, 1978). In addition, field studies have indicated that females become more receptive to the approaches of males, which results in more frequent heterosexual interactions and, ultimately, copulatory behavior (Baldwin, 1970; and this volume, Chapter 2). As illustrated in Fig. 3, assessment of the annual changes in male testosterone levels has shown that peak secretion coincides with the time of ovarian cycles in females (Mendoza *et al.*, 1978a; Coe *et al.*, 1981). Heightened spermatogenic activity also occurs during this

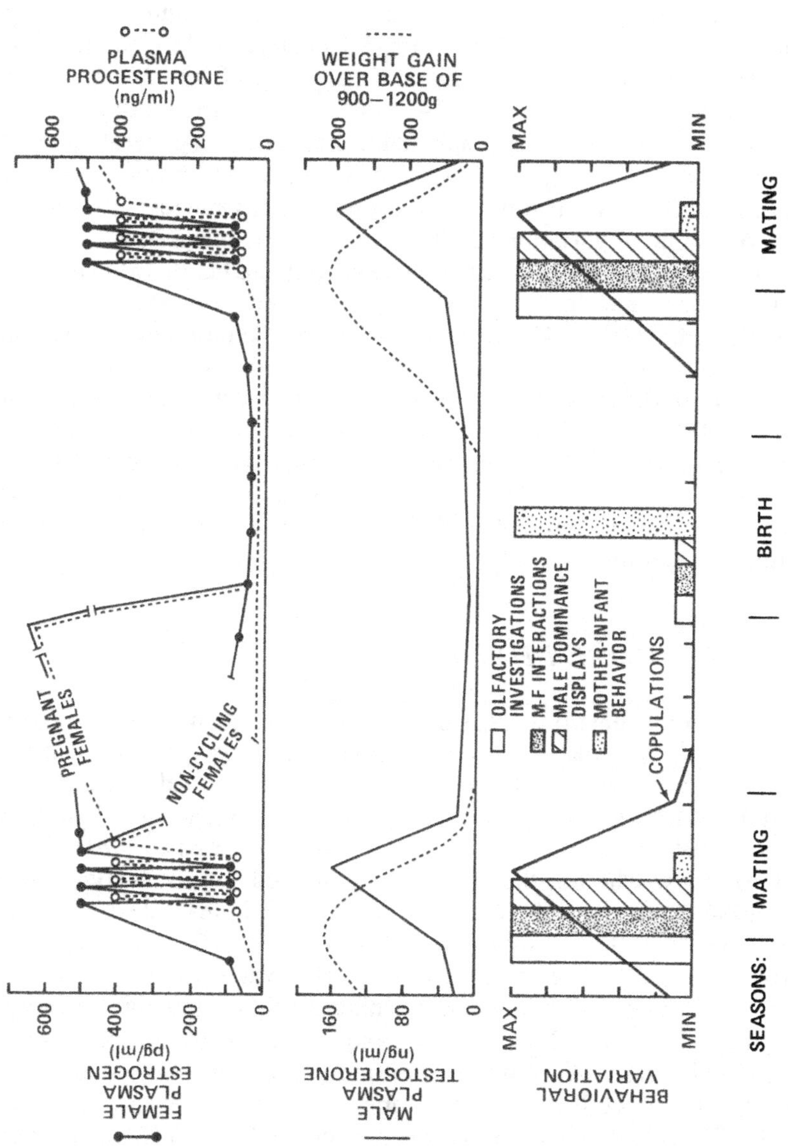

Fig. 3. Idealized model of the annual variation in reproductive behavior and physiology. (A) Profile of ovarian steroids indicates that females are seasonally polyestrus. (B) Body weight and testosterone changes in males are also oriented to the annual mating cycle. (C) Social interactions, dominance, and sexual behavior all increase during the delimited mating season. [Coe and Levine (1981).]

period, as evidenced by studies of testicular biopsies (DuMond, 1968; Belt and Cavazos, 1970) and electroejaculation (Chen *et al.*, 1981).

Although it might seem that the seasonal occurrence of reproductive activity could therefore be accounted for primarily by the responsiveness of males to female physiology, several findings led us to question this simple view. In particular, the occurrence of an annual weight cycle in the squirrel monkey suggests that males undergo some physiological changes independent of females. Most adult males will show an annual weight gain of 200–300 g, the so-called "fatting response," which begins 2–3 months before the mating activity (DuMond, 1968). At the onset of the weight gain, there is also a general increase in dominance contention and aggressivity, even in the absence of females (Baldwin, 1970; Coe *et al.*, 1983a). The timing of these changes suggests that males may undergo some annual variation that has an endogenous basis. There is no doubt that testosterone has a role in some of the weight and behavioral changes, since castration causes a weight loss (Nadler and Rosenblum, 1972), but it is definitely not the only factor. Mendoza *et al.* (1978a) demonstrated that the rate of change in body weight and testosterone secretion are dissociated; and castrated males do undergo a small increase in body weight each year (Coe and Rosenblum, 1978). Other likely hormones involved in these annual changes are thyroid hormones, growth hormone, prolactin, or adrenal androgens (Kaack *et al.*, 1980; Wiebe *et al.*, 1984). We will also show later that the initiation of aggressive interactions between males may actually facilitate endocrinological changes related to reproductive behavior.

4. DIURNAL HORMONE RHYTHM

One of the obvious questions regarding the seasonal recrudescence of gonadal function is the relationship between annual and diurnal changes in testosterone output. From prior work, we knew that male squirrel monkeys had higher testosterone levels during the mating season when the blood samples were collected at 0800 (Mendoza *et al.*, 1978a; Coe and Levine, 1981). However, we did not know whether this difference extended across the day, since Wiebe *et al.* (1984) had reported a clear diurnal rhythm in testosterone during the mating season.

In the following study, we examined the diurnal variation in cortisol and testosterone at two times of year when morning levels of gonadal hormones were either relatively high or low. In one group of males, the blood samples were collected beginning in April while hormone output was high; in a second group, the samples were collected in November, prior to the seasonal increase in testosterone. The two all-male groups were composed of 5–6 Guyanese subjects that were housed within visual and olfactory contact of a group of six females. All monkeys were wild-born adults of Guyanese origin. Blood samples from each male were collected at six different times of day at weekly intervals, with the temporal order of the samples balanced across subjects (Fig. 4).

As can be seen in Fig. 4, there was a significant difference in testosterone levels at the two times of year, as well as a shift in the pattern of daily variation.

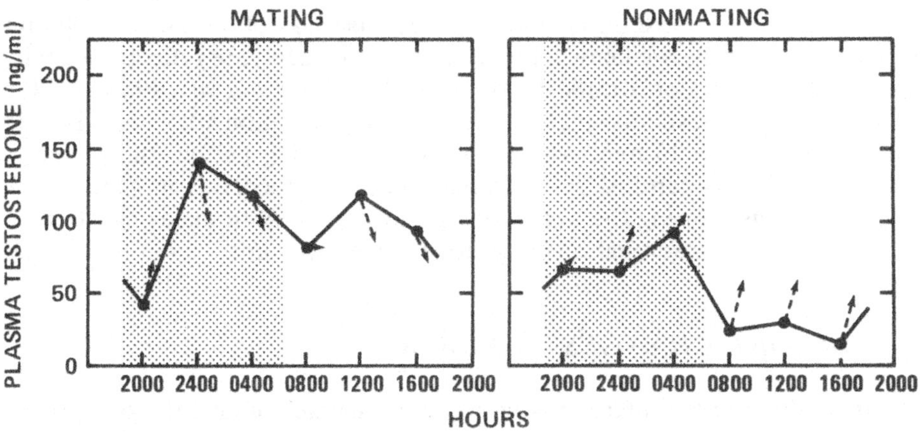

Fig. 4. Diurnal variation in male testosterone levels during the periods of high and low hormone secretion. Arrows indicate the direction and magnitude of the testosterone response 30 min after stress.

In keeping with our prior studies, we confirmed that male testosterone levels were greater at 0800 during the mating period. However, an examination of the complete daily rhythm provided more insight into the annual variation. In both time periods, testosterone secretion was greater during the night than during the day; but during April, the levels were strikingly higher at night, and the high secretion pattern was sustained throughout the day. This prolonged high output was in marked contrast to the sharp decline during the day in the nonmating period, especially in the afternoon values. Thus, these data establish the basis for our belief that it is the cumulative output of high testosterone, particularly during the long winter nights, that is one of the primary factors in the annual variation of male squirrel monkeys.

Two other findings from this study were of particular interest. The first was concerned with the effect of social dominance on these rhythms. We assessed the males' dominance hierarchy based on the occurrence of genital displays and manual grabbing, and examined the respective testosterone levels of dominant and subordinate monkeys. Not only was the effect of rank greater during the mating period when hormone levels were elevated, the interaction of rank and hormone level was evident in the daily rhythm as well. The effect of rank was considerably more pronounced at night, when high-ranking males showed the greatest nocturnal surges between 2400 and 0400. In contrast, there was no association between dominance rank and testosterone secretion when hormone levels were low during the afternoon in the nonmating period. A second finding of interest was the effect of these annual and diurnal changes in basal testosterone levels on stress responsiveness. At all time points, we obtained a second blood sample 30 min after the basal one to determine the stress response to the capture and brief ether anesthesia utilized in the initial blood collection. As shown in Fig. 4, when basal levels were relatively low, the testosterone titers were markedly elevated 30 min later, whereas when testosterone levels were initially high, a decrease was observed after 30 min. Based on prior work, we knew that the testosterone decrease reflected a ceiling effect and initiation of a rapid negative feedback (Coe *et al.*, 1983a). When basal testosterone levels are high, the testosterone elevation after stress is brief, usually lasting only 15 min, whereas in the nonmating period, the increased testosterone levels can last 30–240 min. Endocrine feedback in general may be more sluggish during the nonmating period, since both males and females also show a greater adrenal response to stress at this time.

5. HORMONE CHANGES AT PUBERTY

In addition to the interaction between diurnal and annual rhythms, one of the other factors that must be considered when discussing the endocrinology of squirrel monkeys is age. In longitudinal studies on developing squirrel monkeys, we have found that there is a marked change in hormone levels between 1 and 3 years of age, and also a strong biological predisposition for the emergence of an annual cycle in reproductive physiology (Coe *et al.*, 1981). These studies have now been conducted on wild-born, laboratory-reared monkeys and on laboratory-bred monkeys. In both cases, the monkeys were reared in the absence of adults after the first year of life. Nevertheless, in both populations, males and females showed the typical annual variation in reproductive physiology, and discrete mating and birth periods emerged.

Several important findings from the first study on wild-born monkeys warrant a brief summary in the context of this chapter. These data were based on nine male and 15 female juveniles that were imported from Iquitos, Peru at approximately 1½ years of age. For the next 20 months, plasma levels of gonadal steroids, physical growth, and behavioral maturation were assessed. It was immediately apparent that puberty in a seasonal breeder like the squirrel monkey does not develop in the progressive manner observed in humans and fast-maturing rodents. Instead, puberty emerged in a sudden, quantum fashion that resembled the physiology underlying the annual reproductive cycle.

Based on the monthly progesterone values and the occurrence of pregnancies, we determined that the females began to cycle in two synchronized clusters. Five of the heaviest females became pubescent at 2½ years of age, which was appropriate for their chronological age and their birth season in the wild. The slower growing females and the ones of smaller stature began to cycle in a later period, a phenomenon that revealed an interesting interaction between environmental and endogenous control. Typically, these remaining females would have delayed an entire year until the next mating period, when they were 3½ years of age (Taub, 1980), but because of their relocation to North America, they began to cycle only 6 months after the first group of females. This second period coincided with the laboratory mating season, indicating that pubertal processes can be influenced by environmental cues once a certain maturational stage has been reached. In this case, maturation was accelerated and the mating period oriented by either photoperiod or social facilitation. Figure 5 shows the progesterone profiles of three of these females,

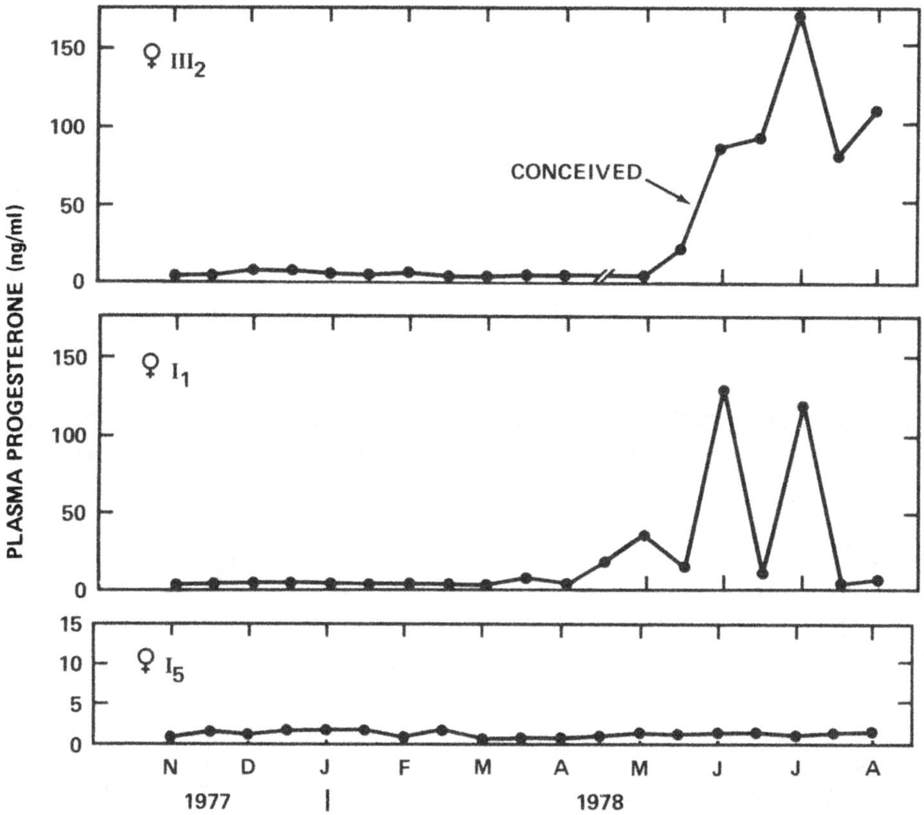

Fig. 5. Developmental changes in the progesterone levels of three juvenile females. Female III_2 reached puberty and conceived; female I_1 cycled for several months; female I_5 became pubescent 6 months later. [Coe *et al.*, (1981).]

two that began to cycle in the early group, and one that matured 6 months later, illustrating the utility of monthly progesterone values for assessing development in females.

Puberty in the developing males also showed a strong resemblance to the annual pattern, and the onset of the first testosterone surge was highly synchronized between males. The nine males in the prior study underwent their first seasonal increase during the laboratory mating period when the second group of females conceived (Fig. 6). Prior to puberty, their plasma testosterone levels at 0900 fluctuated around 3 ng/ml. However, in December all of the males showed testosterone surges between 7 and 40 ng/ml, which were sustained for 2–4 months before subsiding. The seasonal nature of these surges indicates that the annual rhythmicity has an endogenous basis, but that the temporal orientation can be influenced by photoperiod or social stimuli. In laboratory-born subjects, the first testosterone increases occur at the same time of year, although the timing of puberty is appropriate to the monkey's chronological age and birth period (Kaplan *et al.*, 1981). With regard to the attainment of

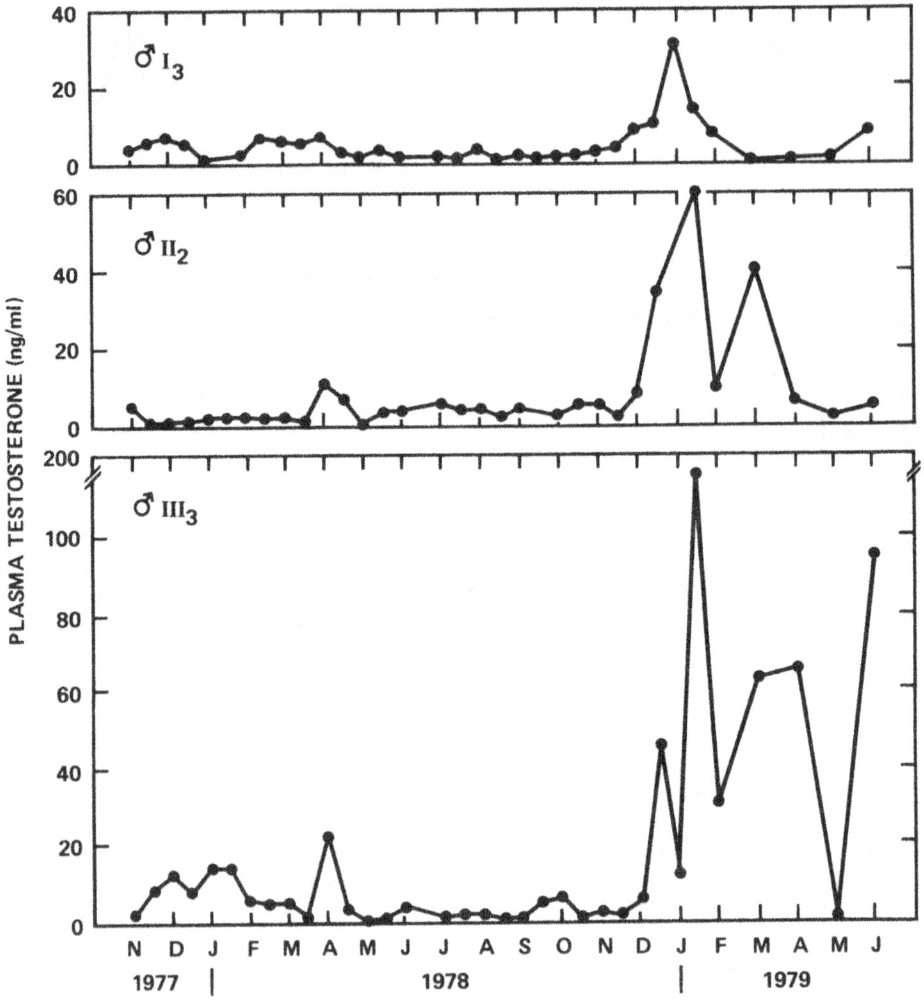

Fig. 6. Developmental changes in the testosterone levels of three juvenile males. The profiles illustrate the synchronous onset of the first testosterone surges as well as the individual differences in the magnitude of the hormone output. [Coe *et al.*, (1981).]

adult testosterone values, we know from the latter cross-sectional study that the seasonal pattern becomes established at this point, and that males show progressive, stepwise increments in their annual peaks and troughs for several more years.

6. *EFFECT OF BODY WEIGHT*

One other important finding that emerged from the puberty research was a pronounced effect of body weight on hormone secretion. Although the onset

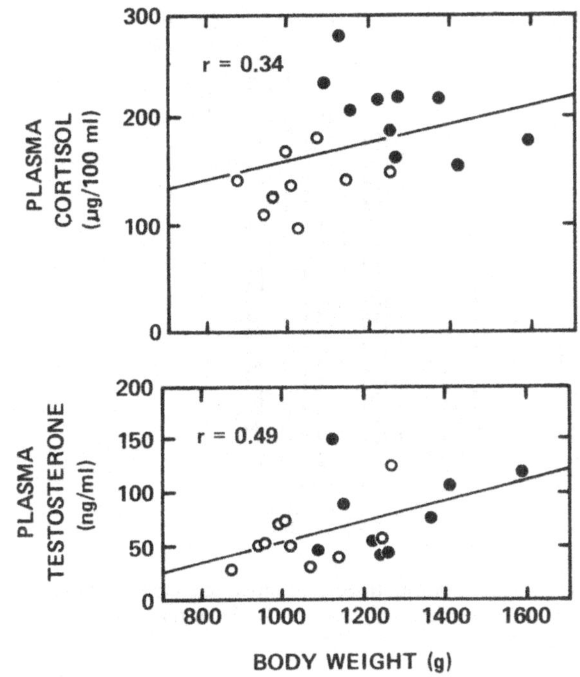

Fig. 7. Scatter plot of body weight and hormone levels from cross-sectional study of 19 male monkeys. (●) Bolivian; (○) Guyanese.

of puberty in developing females was constrained by age and seasonal factors, attainment of a certain body weight (>620 g) was critical for the occurrence of ovarian cycles. Body weight also influenced the first seasonal testosterone surge in males. The body weights of the nine males were highly correlated with the peak magnitude ($r = 0.68$) and monthly duration ($r = 0.92$) of the first testosterone surges.

In developing males, this correlation occurred only during the period of high testosterone secretion, and the relationship could not be attributed to prior growth rates or initial gonadal activity. However, in adult males, a similar effect of body weight on cortisol and testosterone levels can be demonstrated by making a cross-sectional assessment of laboratory subjects. In the following study, we determined basal hormone levels in 19 males and simply plotted their average cortisol and testosterone against their mean body weights (Fig. 7). Each hormone value was based on 6–7 samples collected over a 3-month period at 0800. The monkeys were housed in small, all-male groups composed of 4–6 subjects, and were either of the Guyanese or Bolivian variety. As can be seen in the scatter plot, body weight was positively correlated with both cortisol and testosterone levels, and the strength of this allometric relationship is quite similar to that observed in humans (Rose *et al.*, 1968; Nieschlag, 1979; Migeon, 1981).

Moreover, the effect of body weight appears to account for some of the differences in hormone levels that have been observed between Guyanese and Bolivian males. The larger Bolivian monkeys tended to have the highest hormone levels, and when Bolivian and Guyanese males overlapped in size, so did

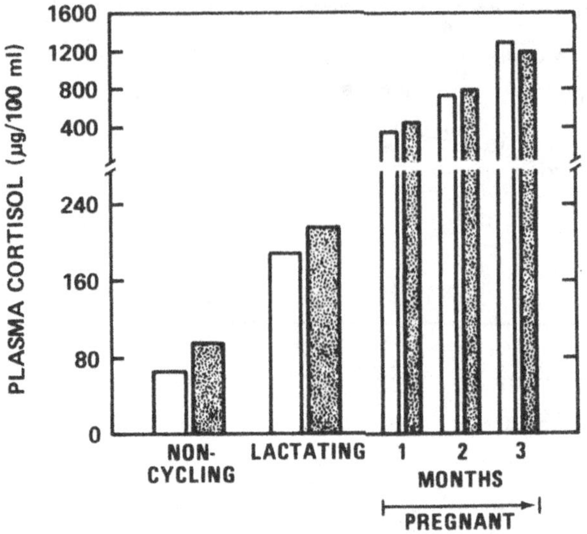

Fig. 8. Cortisol levels in female squirrel monkeys during three reproductive states. Bolivian females tend to have higher adrenal output. (□) Guyanese females; (▩) Bolivian females.

their hormone output. It should be mentioned that these males were not in the "fatted" condition. In fact, much greater differences in testosterone levels are observed during the mating season, when Bolivian males weigh 1300–1650 g and Guyanese males weigh 1000–1250 g. At this time, Bolivian males can show testosterone surges between 400 and 500 ng/ml, while Guyanese males typically peak between 100 and 150 ng/ml (Mendoza *et al.*, 1978a; Coe *et al.*, 1983a).

7. *HORMONE DIFFERENCES IN SAIMIRI FEMALES*

In spite of these findings from the last study, we believe that the hormonal difference between Guyanese and Bolivian squirrel monkeys is only partially accounted for by the differences in body weight. Many investigators have now reported important population differences in chromosomal (Jones *et al.*, 1973; Ariga *et al.*, 1978), physiological (L. N. Martin and McNease, 1982), and behavioral patterns (Mendoza *et al.*, 1978c). In our own studies, we have also observed a number of important differences in the stress physiology of Bolivian and Guyanese monkeys. To illustrate this point, we will turn to female squirrel monkeys and review two studies on their adrenal activity in different reproductive states and after acute stress.

The first study was based on a large cross-sectional comparison of basal cortisol levels in Bolivian and Guyanese females. The subjects included 24 noncycling females, ten pregnant females, and 19 lactating females. As seen in Fig. 8, the reproductive state of female squirrel monkeys has a marked effect on their level of adrenal secretion. Lactating females have higher cortisol levels than do noncycling females. In addition, following conception there is a large

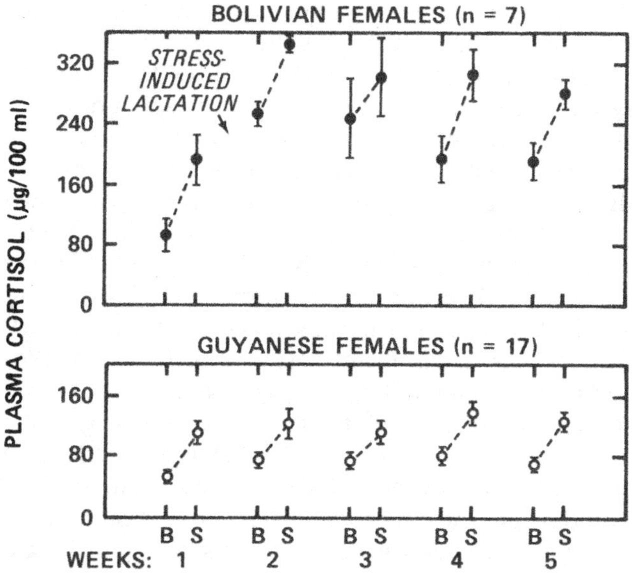

Fig. 9. Changes in the basal and 30-min stress levels of cortisol during 5 weeks of blood collection. The onset of lactation resulted in chronic adrenal activation in Bolivian females.

and progressive rise in cortisol secretion that continues for several months before stabilizing at a dramatically high level. In the data from this study, it is also possible to see that noncycling and lactating Bolivian females tend to have higher cortisol levels than do Guyanese females. This difference in adrenal secretion is not accounted for by an effect of body weight, since females do not show the variation in size that occurs in males from these two populations.

A more dramatic demonstration of the difference in the pituitary–adrenal systems of Bolivian and Guyanese squirrel monkeys was obtained when we examined the adrenal response of females to acute stress. The next two experiments evaluated a phenomenon that we have labeled "stress-induced lactation." When subjected to ether anesthesia, females of the Bolivian variety will often begin to lactate even after only one or two brief exposures. Figure 9 shows the hormonal sequelae following induction of lactation in Bolivian females. In this study, we collected blood samples from seven females at five weekly intervals. A 0.5-ml sample was collected at 0800, and then a second sample was collected at 0830 to assess the stress response to the capture and ether anesthesia involved in the basal sampling. As can be seen in Fig. 9, all of the Bolivian females began to lactate after the first week, and there was a concomitant increase in adrenal activation. Cortisol levels rose sharply by the second week of sampling and then remained chronically elevated for the duration of the study. This dynamic shift in basal levels was in marked contrast to the stable adrenal activity observed in Guyanese females subjected to the same experimental manipulations (Fig. 9).

At a later date we attempted to examine this phenomenon further in a second experiment on different Bolivian females. To our surprise, we found that the eight new subjects showed signs of lactation on the first day of sampling. Moreover, the cortisol levels of these females were already somewhat elevated on the first day (136 μg/dl), and they did not show the dynamic shift in basal activity during the 5 weeks of sampling. Thus, it appears that females of the Bolivian variety can continue to lactate for prolonged periods of time, and that the dramatic increases in adrenal activity subside from the high levels observed during the induction phase. The physiological mechanisms underlying the lactational state are still unclear. After 5 weeks of sampling in the second experiment, we attempted to inhibit lactation by reducing prolactin levels with bromocriptine. For 2 weeks, the eight females were given daily IM injections of 0.5 mg bromocriptine. No reduction in milk production was observed, and there was only a small decrease in basal cortisol levels to 118 μg/dl. We do not believe that this finding indicates a minimal role for prolactin in the induction of lactation. Rather, the failure to inhibit milk production may indicate that once lactation is established, it is no longer completely dependent on prolactin, as is observed in some other species (Martin and Bateson, 1982). An alternative explanation would be provided if bromocriptine treatment failed to substantially inhibit prolactin levels because the squirrel monkey requires high doses to exert a dopaminergic action.

At present, we cannot offer a good explanation for the stress-induced lactation, although it is certainly an important phenomenon for investigators to bear in mind. Most significantly, it may account for some of the differences in the adrenal responses of Bolivian and Guyanese females to psychosocial stimuli. Following the formation of new social groups, Bolivian females tend to show sustained increases in plasma cortisol levels for several weeks, whereas Guyanese females return to basal levels of adrenal activity more quickly (Gonzalez *et al.*, 1981). The difference in the endocrine responses of these females may therefore reflect both distinctive psychological and physiological reactions.

8. *PSYCHOLOGICAL INFLUENCES ON HORMONE SECRETION*

In this section, we turn to some of the psychological responses that can influence endocrine function in the squirrel monkey. Over the last 6 years, we have attempted to take advantage of the high and labile hormone levels in this species in order to assess how a variety of psychological and environmental perturbations affect hormone output. In particular, we have described the important influence of social dominance and aggression on cortisol and testosterone secretion in males (Coe *et al.*, 1983a). A brief review of these findings will serve to illustrate the importance of taking psychosocial variables into account, especially when the monkeys are living in small captive groups.

In one of our early papers on this topic, we reported that dominance rank affected basal and stress levels of hormones in males housed as pairs (Coe *et al.*, 1979). Cortisol and testosterone responses were assessed in eight pairs of Bolivian males at 0800 during the mating season. The dominance relationship in each pair was evaluated in a series of five behavioral tests, and rank was

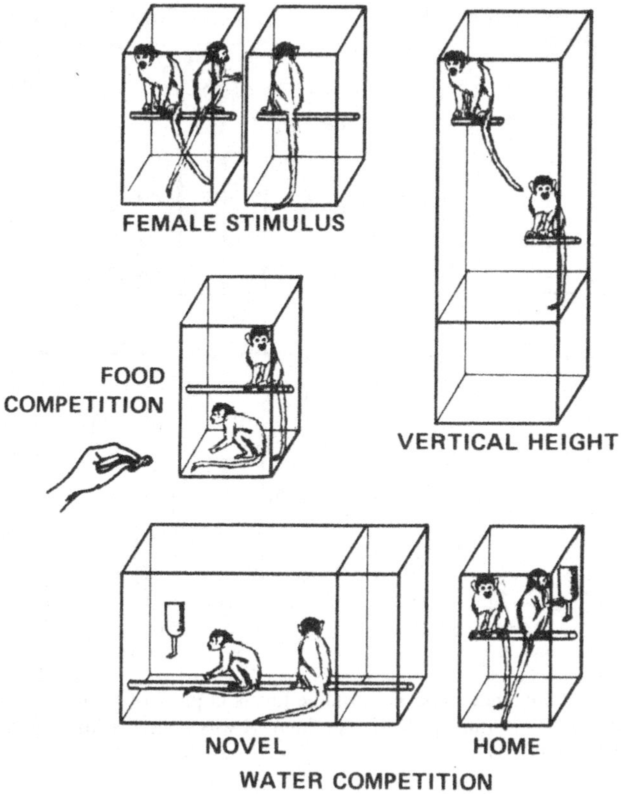

Fig. 10. Behavior tests used to assess dominance rank in pair-housed monkeys. [Coe *et al.* (1983a).]

assigned after we determined the outcome of induced competition between males (Fig. 10). In all pairs, dominant males had higher basal levels of cortisol and testosterone than did their subordinate partners (Fig. 11, experiment 1). The occurrence of higher cortisol levels in dominant males has been a fairly consistent finding in our research, although it does differ from reports on other species and an early paper on squirrel monkeys (Manogue *et al.*, 1975). Moreover, the difference in basal hormone levels can affect the male's hormonal response to stress. Subordinate males with lower basal levels showed a greater cortisol increment at 30 min after stress. In addition, they continued to show an elevation in testosterone levels at 30 min after stress, whereas the dominant males had returned to basal levels. In an extension of this research, we compared the effect of dominance during the mating season to its effect during the nonmating season when hormone levels are lower (Fig. 11, experiment 2). As can be seen in Fig. 11, the results from the four new pairs during the mating season replicated the initial data that had been collected at this time of year. However, when the hormone levels had declined during the nonmating season, an effect of dominance was no longer evident. The weaker influence of rank

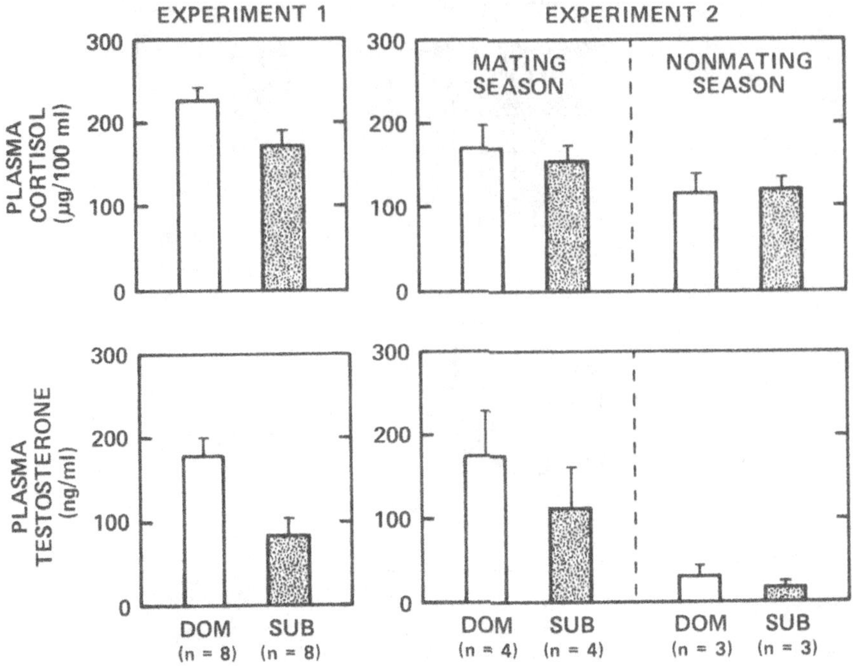

Fig. 11. Effects of dominance rank on hormone levels in pairs of Bolivian males. Experiment 1 was conducted during the mating season; experiment 2 was a replication at two different times of year. [Coe and Levine (1981).]

on hormone levels in the nonmating season is in keeping with our study on diurnal rhythms discussed earlier in this chapter, and it also concurs with research on macaques carried out in the nonmating season (Eaton and Resko, 1974).

This statement regarding the effect of dominance in the nonmating period must be qualified, however, when one considers the response to the formation of dominance relations. Even in the nonmating season, it is possible to detect a transient effect of rank on hormone levels during dominance contention. This effect is illustrated in Fig. 12, which shows the cortisol and testosterone responses of seven male pairs over a 2-week period. In this study, the hormonal response to pair formation was examined in 14 Guyanese males during October, when basal levels are relatively low (Coe *et al.*, 1982). As can be seen, dominant males showed a striking increase in testosterone secretion at 24 hr after pair formation, while subordinates either remained the same or showed lower levels. Thereafter, as the social relationship stabilized, testosterone values for dominant males returned to normal levels observed in the nonmating season. Adrenal activation was also increased during the first 24 hr of behavioral agitation, but both high- and low-ranking monkeys showed a similar pattern of response.

In another study, we demonstrated that the establishment of dominance relations can have a more protracted effect on testosterone secretion even when the groups are formed in the nonmating season. This study was conducted with Dr. L. Bromley (University of California at Berkeley) and involved the as-

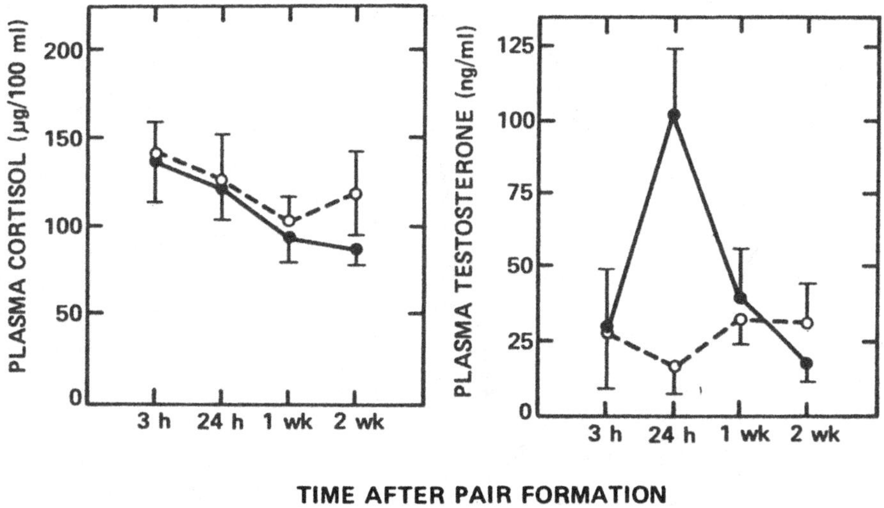

Fig. 12. Hormonal responses to the establishment of dominance relations in seven pairs of Guyanese males during the nonmating season. (●) Dominant; (○) Subordinate. [Coe *et al.* (1982).]

sessment of hormone levels following the formation of two four-male groups of Bolivian and Guyanese males. The groups were formed in October, and blood samples were collected until the onset of the first seasonal surges in testosterone. At 24 hr after the group formation, the monkey that emerged as the dominant male in each group showed the highest testosterone level (Fig. 13). This effect of rank was transient, however, and the hormone levels subsided and became inconsistent for the next 2 months. Nevertheless, with the onset of the seasonal testosterone surge, there was once again a strong influence of rank. In January, the dominance effect reemerged and we found a perfect delineation of hormone levels by hierarchy position in both groups.

An effect of dominance on the initiation and magnitude of seasonal testosterone increases has also been observed in long-term stable groups. In a longitudinal assessment of well-established group of four Bolivian males, we found that the alpha and beta males were the first to show the seasonal weight gain and were the only ones to show a large seasonal surge in testosterone. In other groups, the subordinate males do show a seasonal hormone surge, but it is often of a smaller magnitude and may occur 1–2 months after the testosterone increases in dominant males (e.g., Mendoza *et al.*, 1978a). These findings may have profound implications for the effect of dominance on reproductive success under natural conditions. Since testosterone titers can affect spermatogenesis and presumably sexual drive, these findings suggest that dominant males are in an advantageous position for impregnating more females early in the mating season. Field data on a natural troop in Costa Rica also indicate that females show a preference for copulating with males in the fatted condition (S. Boinski, University of Texas, personal communication). If the ability to undergo an early weight gain conveys reproductive advantage, the reemergence of aggression and dominance contention could serve an important function at

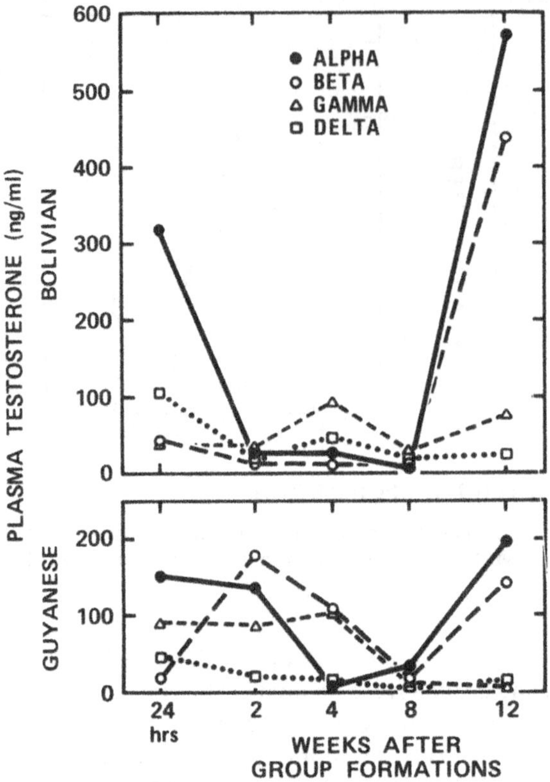

Fig. 13. Longitudinal assessment of testosterone levels in two four-male groups. The effect of rank was apparent at 24 hr after group formation and again after the onset of the seasonal surges.

this time of year. Males that attain a high position in the social hierarchy following the initial period of aggression would show an enhancement of testosterone and weight gain. Further, we also know that wounding can cause prolonged suppression of testosterone titers and a rapid loss of weight, indicating that there could be some inadvertent consequences of injuries in subordinate males.

To return from the speculative realm, we would like to end this chapter with two studies that provide some substantive support for these views on the important role of dominance behavior. This research was concerned with how changes in social relationships may directly affect the seasonal recrudescence of reproductive physiology. Several investigators have reported an intriguing phenomenon of induced sexual activity following the formation of new social groups of squirrel monkeys in the nonmating season (Travis and Holmes, 1974; Coe and Rosenblum, 1978). In captivity, at least, it is possible to rapidly induce fertile mating simply by forming new social groups even if the monkeys are in a state of reproductive quiescence. The next experiment examined the hormonal changes in males associated with induced breeding (Mendoza *et al.* 1979).

In June, after the normal mating season had ended, nine Bolivian males were initially housed alone to determine basal levels of cortisol and testosterone

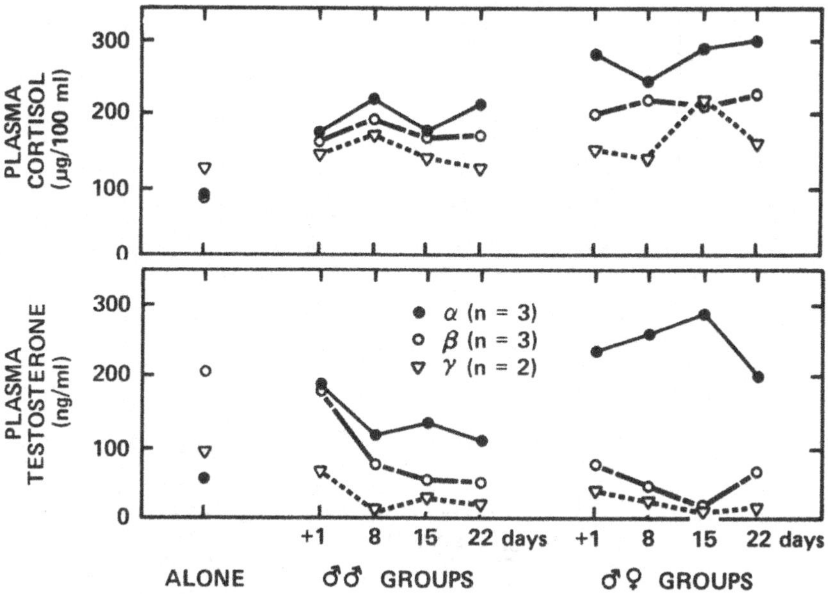

Fig. 14. Effect of dominance rank on the hormone levels in Bolivian males after the establishment of all-male groups and with the subsequent introduction of females. [Coe *et al.* (1983a).]

at 0800. These males were then housed together in three social groups, each consisting of three males. During the next month, four samples were collected from each subject at weekly intervals, and then five females were added to each group. Four more blood samples were collected from the males at weekly intervals during the heterosexual phase. The hormonal findings are illustrated in Fig. 14, although the data from one low-ranking male have not been included because he had to be removed during the male–female phase.

Behavioral observations of the three social groups indicated that linear dominance hierarchies emerged soon after group formation, and that a male's social position had a pervasive effect on his subsequent interactions and hormone changes. Analysis of the hormonal data indicated that individual testosterone levels while the males were housed alone were not predictive of rank, but attainment of high social rank caused a significant elevation of gonadal secretion. The higher testosterone levels in alpha males persisted throughout the all-male and male–female phases, while subordinate males showed progressive decreases in testosterone levels. This differential effect of rank became particularly pronounced during the male–female phase. By 24 hr after the introduction of females, mean testosterone levels in dominant males had risen to mating season values (237 ng/ml), whereas testosterone secretion in beta and gamma males either declined or did not respond to the female stimuli. The pattern of change in plasma cortisol also revealed strong adrenocortical responses to the three social conditions. Plasma cortisol levels increased in all males following group formation, but dominant males again showed the highest hormone levels, especially after the introduction of the females.

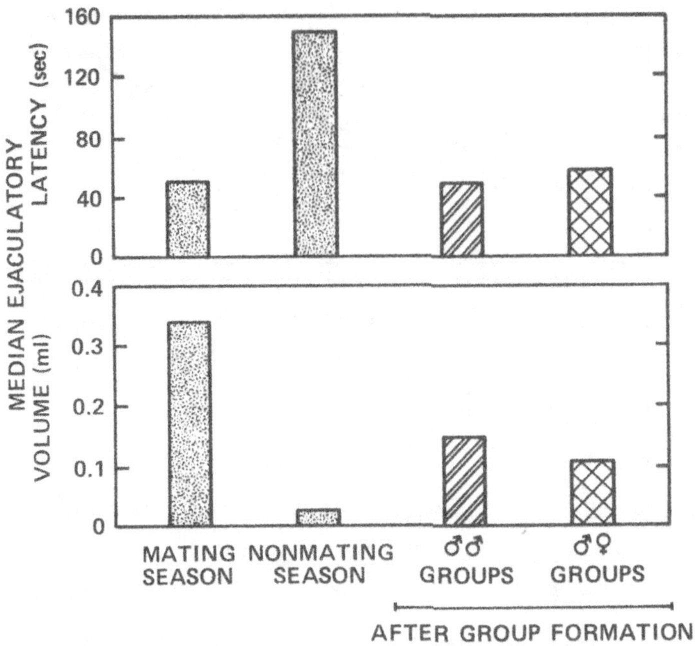

Fig. 15. Effects of season and group formation on the response to electroejaculation in four Bolivian males. [Coe and Levine (1981).]

The marked effect of social rank on hormone levels during the male–female phases seemed to be related to the induction of mating activity, which began soon after the females were added. At least one female in each group became pregnant, and the first births occurred 5½ months later, indicating that some conceptions had taken place almost immediately. Thus, as previous reports had indicated, the formation of new social groups did indeed induce fertile mating even in the nonmating season. Moreover, this induced mating was accompanied by hormonal changes in males that reflected a strong influence of dominance behavior on endocrine physiology.

It was not possible to determine the functional significance of the altered hormone levels in this experiment, so a followup study was designed to investigate whether specific benefits might accrue from increased testosterone secretion. This study was conducted in collaboration with Drs. E. Smith, J. Chen and J. Davidson (Stanford University). To see if reproductive ability was actually enhanced, the ejaculatory responses of males were evaluated before and after group formation in the nonmating season (Fig. 15). As previous research on electroejaculation had indicated (Chen *et al.*, 1981), the volume of ejaculatory fluid was decreased, and the latency to ejaculate was prolonged during the nonmating season. However, within 24 hr after the formation of either all-male or male–female groups, the males' ejaculatory responses were suddenly transformed, shifting toward mating season performance.

These findings indicated that the psychoendocrine changes induced by group formation can cause the testicular physiology of males to shift rapidly from a state of relative inactivity to one of fertile reproductive potential. Although this facilitation of ejaculation appeared to be associated with increased testosterone output, it should be noted that the change occurred as a general response to heightened behavioral arousal in both unisexual and heterosexual groups. Therefore, other hormonal or neural mechanisms may also be involved in the enhancement of ejaculation, especially since the change persisted in some males that showed only a transient rise in testosterone levels.

To date we do not have equivalent information on ovarian responses to group formation in females, although the occurrence of fertile mating is certainly indicative of similar activation of reproductive physiology. Given the evidence that females are typically acyclic during nonmating season, the rapidity of conception following group formation suggests that ovulatory cycles can be quickly initiated by means of social facilitation. This effect of social interactions on female reproduction is not the first demonstration of a psychological influence on female physiology, since it concurs with the finding that the onset and termination of cycling are highly synchronized between squirrel monkeys during the normal mating season (Coe and Rosenblum, 1978; Mendoza *et al.*, 1978b; Coe *et al.*, 1981). In addition, as with males, the actual manifestation of sexual behavior may be related to increased adrenocortical output. Evaluation of plasma cortisol levels in females has shown that they undergo marked increases in adrenocortical secretion following group formation and that the hormone levels attained reflect the degree of social agitation within the group (Gonzalez *et al.*, 1981). Similarly, plasma cortisol levels of cycling females tend to be higher than those of noncycling females, presumably because of the facilitory effect of estrogen on the adrenal that occurs in most species (Parkes and Deanesley, 1966).

9. GENERAL CONSIDERATIONS

To summarize the important points we have tried to review on the endocrine system of the squirrel monkey, we should certainly begin with the unusual hormone profile. Although the early assessment of plasma cortisol levels by Brown *et al.* (1970) has proved to be somewhat high, it is now well established that both adrenal and gonadal hormones are extremely elevated in the squirrel monkey. At present, the most plausible explanation to account for the hypersecretion of steroid hormones is the "altered receptor" hypothesis, but we believe that other factors related to hormone synthesis and degradation are also involved (Chrousos *et al.*, 1982a,b). The actual reduction in receptor affinity for cortisol, for example, is insufficient by itself to explain the magnitude of the increase in free and total hormones (Siiteri *et al.*, 1982).

It is important that investigators be cognizant of the high adrenal activity because of the diverse secondary effects on digestive, cardiovascular, and renal function that makes this species particularly prone to stress-related diseases (Strickland and Clarkson, this volume, Chapter 12). In addition, the labile nature of the hormone output makes it imperative to be aware of many procedural and biological variables even in research that does not focus on endocrinological issues. Biorhythms, both diurnal and annual, exert an especially strong influence on this species. The general pattern of hormone secretion is similar to that of other seasonally breeding primates, but the magnitude of the hormone levels permits large changes to occur even within a few hours. When housed alone or manipulated frequently, the fluctuations in hormone levels may yield only an inconsistent pattern (Wilson *et al.*, 1978). Similarly, variables that are usually considered of minor importance in other species, such as body weight, may have pronounced effects in the squirrel monkey. In adult males, the correlation between body weight and cortisol ($r = 0.34$) or testosterone ($r = 0.49$) may account for a substantial component of the hormone variation. In pubescent animals, body weight may be even more important, since it appears that the onset of cycling in females is affected by physical growth and that the weight itself may determine whether a developing female becomes pubescent at 2½ or 3½ years of age. We have also found that the body weight of pubescent males can affect the magnitude and duration of their first seasonal testosterone surge.

It is somewhat surprising, however, that the annual "fatting" response is not under the exclusive control of testosterone. We know that testosterone is involved, because castration of intact males causes a weight loss, and replacement of testosterone by injection causes a weight gain (Nadler and Rosenblum, 1972). Nevertheless, the annual weight and testosterone cycles may become dissociated in captivity (Mendoza *et al.*, 1978a). In addition, castrated males and females do show some annual weight variation, albeit less (Coe and Rosenblum, 1978), and pubescent males do not show the weight gain until 1–3 years after their first testosterone surge (Coe *et al.*, 1981). It is likely that testosterone works in a synergistic manner with one or more other hormones. The fact that injections of testosterone cause a weight loss when given to already "fatted" males suggests the possible involvement of other androgens, such as androstenedione, which are markedly elevated in the squirrel monkey (Wiebe *et al.*, 1984). Alternatively, there may be some involvement of nongonadal hormones, such as thyroid hormones, which are also elevated during the mating season (Kaack *et al.*, 1980). A third possibility is the involvement of estrogen, given the high capacity of peripheral cells to aromatize testosterone (Goldman *et al.*, 1983).

Beyond the endogenous variation in hormone levels, the endocrine system of the squirrel monkey is extremely sensitive to environmental and psychological perturbations. We have observed sustained elevations in adrenal activity for 1–2 months after the formation of new social groups (Mendoza *et al.*, 1979). Further testosterone, which is not usually considered to be a stress-sensitive hormone, can undergo marked elevations and suppressions within a 15–45 min period after stress (Coe *et al.*, 1983a). In this chapter we have reviewed a number of studies showing that housing conditions and social relations can have protracted effects on basal hormone levels, and that these changes, in turn, can affect stress responsiveness.

Perhaps the most interesting of these endocrine responses is the capacity of the squirrel monkey to shift within 24 hr from a physiological state of reproductive quiescence to mating capability. We have reported several studies that demonstrate that dominance contention and behavioral agitation will alter both testosterone levels and spermatogenic activity. Moreover, the rapid occurrence of conceptions following the formation of new social groups indicates the same capacity in females. There are several practical applications of these findings. We have taken advantage of the rapid hormonal response to override the normal seasonal breeding pattern of our colony and have spread the birth season to cover 6 months of the year. By selectively weaning 6- to 8-month-old infants in a synchronized manner and forming new social groups with unfamiliar males and females, we have been able to obtain births from June through January.

There are other interesting applications of the finding that males show a dramatic testosterone rise to the introduction of females. We have recently used this reliable hormone response to evaluate how Bolivian and Guyanese males perceive females from the two populations (Coe *et al.*, 1983b). It was possible to show that males responded preferentially to females from their own geographic area. This difference in a hormone system directly related to reproduction provides further support for Thorington's contention (this volume, Chapter 1) that we need a taxonomic distinction between Guyanese and Bolivian monkeys.

The importance of endogenous and psychological factors should not be taken to suggest a smaller role for environmental ones. In this chapter, we have discussed the importance of light cycles. Rosenblum (1968) was one of the first investigators to show that alterations in the daily light rhythms could affect the occurrence of ovarian cycles in female squirrel monkeys. We have briefly discussed two studies on pubescent and adult monkeys, indicating that photoperiod may influence the temporal orientation of the annual cycle. We believe that the timing of the mating season in temperate zones is determined by the cumulative output of gonadal hormones during the long winter nights. This interpretation concurs with the views of other investigators, which have suggested that diurnal primates tend to be short-day breeders (Lancaster and Lee, 1965; Michael and Zumpe, 1976; Van Horn, 1980). However, based on the current data, it may be more appropriate to describe the squirrel monkey as a long-night breeder when maintained in North America. A particularly interesting finding is the differential sensitivity of Bolivian and Peruvian monkeys which breed 1–2 months before Guyanese and Brazilian monkeys, although the mechanism remains to be elucidated.

In summary, these studies describe the complex synergism of physiological, psychological, and environmental factors in determining endocrine function. The high level of hormone secretion in the squirrel monkey makes this synergism more evident, although, clearly, the regulation of the endocrine system in all species is multidetermined. Such redundancy of control permits the stable continuance of hormone output and cycles over time, while permitting the capacity for change when a particular factor becomes salient enough. The great challenge of research with squirrel monkeys is to make sure that these factors and the unusual endocrine profile work to the advantage of the investigation, rather than serving to confound it.

ACKNOWLEDGMENTS. This research was supported by HD-02881 from NICH&HD; MH-23645 from NIMH; and Research Scientist Award MH-19936 from NIMH to S. Levine. Additional support was provided by MH-21178 to J. Davidson. The technical assistance of E. Lowe and H. Hu and the participation of S. Mendoza, L. Bromley, J. Chen, and C. Gonzalez in parts of this research are also acknowledged.

REFERENCES

Ariga, S., Dukelow, W. R., Emley, G. S., and Hutchinson, R. R., 1978, Possible errors in identification of squirrel monkeys from different South American points of export, *J. Med. Primatol.***7**:129–135.

Baldwin, J. D., 1970, Reproductive synchronization in squirrel monkeys (*Saimiri*), *Primates* **11**:317–326.

Belt, W. D., and Cavazos, L. F., 1970, Fine structures of the interstitial cells of Leydig in the squirrel monkey during seasonal regression, *Anat. Rec.* **169**:115–128.

Brown, G. M., Grota, L. J., Penney, D. P., and Reichlin, S., 1970, Pituitary-adrenal function in the squirrel monkey, *Endocrinology* **86**:519–529.

Brown, G. M., Schalch, D. S., and Reichlin, S., 1971, Patterns of growth hormone and cortisol responses to psychological stress in the squirrel monkey, *Endocrinology* **88**:956–963.

Cassorla, F. G., Albertson, B. D., Chrousos, G. P., Booth, J. D., Renquist, D., Lipsett, M. B., and Loriaux, D. L., 1982, The mechanisms of hypercortisolemia in the squirrel monkey, *Endocrinology* **111**(2):448–452.

Chen, J. J., Smith, E. R., Gray, G. D., and Davidson, J. M., 1981, Seasonal changes in plasma testosterone and ejaculatory capacity in squirrel monkeys (*Saimiri sciureus*), *Primates* **22**:253–260.

Chrousos, G. P., Renquist, D., Brandon, D., Eil, C., Pugeal, M., Vigersky, R., Cutler, G. B., Loriaux, D. L., and Lipsett, M. B., 1982a, Glucocorticoid hormone resistance during primate evolution: Receptor mediated mechanisms, *Proc. Natl. Acad. Sci. USA* **79**:2036–2040.

Chrousos, G. P., Renquist, D., Brandon, D., Barnard, D., Fowler, D., Loriaux, D. L., and Lipsett, M., 1982b, The squirrel monkey: Receptor-mediated end-organ resistance to progesterone? *J. Clin. Endocrinol. Metab.* **55**(2):364–368.

Clewe, T. H., 1969, Observations on reproduction of squirrel monkeys in captivity, *J. Reprod. Fertil.* **6**(Suppl.):151–156.

Coe, C. L., and Levine, S., 1981, Psychoendocrine relationships underlying reproductive behavior in the squirrel monkey, *Int. J. Ment. Health* **10**:22–42.

Coe, C. L., and Rosenblum, L. A., 1978, Annual reproductive strategy of the squirrel monkey (*Saimiri sciureus*), *Folia Primatol.* **29**:19–42.

Coe, C. L., Mendoza, S. P., Davidson, J. M., Smith, E. R., Dallman, M. F., and Levine, S., 1978, Hormonal response to stress in the squirrel monkey (*Saimiri sciureus*), *Neuroendocrinology* **26**:367–377.

Coe, C. L., Mendoza, S. P., and Levine, S., 1979, Social status constrains the stress response in the squirrel monkey, *Physiol. Behav.* **23**:633–638.

Coe, C. L., Chen, J. J., Lowe, E. L., Davidson, J. M., and Levine, S., 1981, Hormonal and behavioral changes at puberty in the squirrel monkey, *Horm. Behav.* **15**:36–53.

Coe, C. L., Franklin, D., Smith, E. R., and Levine, S., 1982, Hormonal responses accompanying fear and agitation in the squirrel monkey, *Physiol. Behav.* **29**:1051–1057.

Coe, C. L., Smith, E. R., Mendoza, S. P., and Levine, S., 1983a, Varying influence of social status on hormone levels in male squirrel monkeys, in: *Hormones, Drugs, and Social Behavior* (A. S. Kling and H. D. Steklis, eds.), Spectrum, New York, pp. 7–32.

Coe, C. L., Smith, E. R., Franklin, D., and Levine, S., 1983b, Male recognition of subspecies differences in female squirrel monkeys, *Am. J. Primatol.* **4**(4):340.

Diamond, E. J., Aksel, S., Hazelton, J. M., Jennings, R. A., and Abee, C. R., 1984, Seasonal changes of serum concentrations of estradiol and progesterone in Bolivian squirrel monkeys, *Am. J. Primatol.* **6**:103–113.

DuMond, F. V., 1968, The squirrel monkey in a seminatural environment, in: *The Squirrel Monkey* (L. A. Rosenblum and R. W. Cooper, eds.), Academic Press, New York, pp. 87–145.

Eaton, G. G., and Resko, J. A., 1974, Plasma testosterone and male dominance in a Japanese macaque (*Macaca fiscata*) troop compared with repeated measures of testosterone in laboratory males, *Horm. Behav.* **5**:251–259.

Ghosh, M., Hutz, R. J., and Dukelow, W. R., 1982, Serum estradiol 17β, progesterone and relative luteinizing hormone levels in *Saimiri sciureus:* Cyclic variations and the effect of laparoscopy and follicular aspiration. *J. Med. Primatol.* **11**:312–318.

Goldman, E. S., Coe, C. L., Murai, J. T., and Siiteri, P. K., 1983, Corticosteroid binding globulin (CBG) response to estrogens and aromatase (estrogen synthetase) activity in the squirrel monkey, in: *Endocrine Society Abstracts* (65th meeting, San Antonio, Texas, June 8–10).

Gonzalez, C. A., Hennessy, M. B., and Levine, S., 1981, Subspecies differences in hormonal and behavioral responses after group formation in squirrel monkeys, *Am. J. Primatol.* **1**:439–452.

Hearn, J., 1983, *Reproduction in New World Primates,* MTP Press, Lancaster.

Hennessy, M. B., Coe, C. L., Mendoza, S. P., Lowe, E. L., and Levine, S., 1978, Scent-marking and olfactory investigatory behavior in the squirrel monkey (*Saimiri sciureus*), *Behav. Biol.* **24**:57–67.

Hodges, J. K., Gulick, B. A., Czekala, N. M., and Lasley, B. L., 1981, Comparison of urinary estrogen excretion in South American primates, *J. Reprod. Fertil.* **61**(1):83–90.

Jarosz, S. J., Kuehl, T. J., and Dukelow, W. R., 1977, Vaginal cytology, induced ovulation and gestation in the squirrel monkey (*Saimiri sciureus*), *Biol. Reprod.* **16**:97–103.

Jones, T. C., Thorington, R. W., Hu, M. M., Adams, E., and Cooper, R. W., 1973, Karyotypes of squirrel monkeys (*Saimiri sciureus*) from different geographic regions, *Am. J. Phys. Anthropol.* **38**:269–277.

Kaack, B., Walker, M., and Walker, L., 1980, Seasonal changes in the thyroid hormones of the male squirrel monkey, *Arch. Androl.* **4**(2):133–136.

Kaplan, J. N., Chen, J. J., Smith, E. R., and Davidson, J. M., 1981, The development of seasonal variation in gonadal hormones and body weight in maturing squirrel monkeys, *Int. J. Primatol.* **2**(4):369–380.

Kittinger, G. W., and Beamer, N. B., 1968, Quantitative gas chromatography of squirrel monkey (*Saimiri sciureus*) corticosteroids, *Steroids* **12**(2):275–289.

Kuehl, T. J., and Dukelow, W. R., 1975, Ovulation induction during the anovulatory season in *Saimiri sciureus, J. Med. Primatol.* **4**:23–31.

Lancaster, J. B., and Lee, R. B., 1965, The annual reproductive cycle in monkeys and apes, in: *Primate Behavior: Field Studies on Monkeys and Apes* (I. DeVore, ed.), Holt, Rinehart and Winston, New York, pp. 486–513.

Latta, J., Hopf, S., and Ploog, D., 1967, Observation of mating behavior and sexual play in the squirrel monkey (*Saimiri sciureus*), *Primates* **8**:229–246.

Manogue, K. R., Leshner, A. I., and Candland, D. K., 1975, Dominance status and adrenocortical reactivity to stress in squirrel monkeys (*Saimiri sciureus*), *Primates* **16**:457–463.

Martin, L. N., and McNease, P. E., 1982, Genetically determined antigens of squirrel monkey (*Saimiri sciureus*) IgG, *J. Med. Primatol.* **11**:272–290.

Martin, P., and Bateson, P., 1982, The lactation-blocking drug bromocriptine and its application to studies of weaning and behavioral development, *Dev. Psychobiol.* **15**:139–157.

Mendoza, S. P., Lowe, E. L., Davidson, J. M., and Levine, S., 1978a, Annual cyclicity in the squirrel monkey (*Saimiri sciureus*): The relationship between testosterone, fatting, and sexual behavior, *Horm. Behav.* **11**:295–303.

Mendoza, S. P., Lowe, E. L., Resko, J. A., and Levine, S., 1978b, Seasonal variations in gonadal hormones and social behavior in squirrel monkeys, *Physiol. Behav.* **20**:515–522.

Mendoza, S. P., Lowe, E. L., and Levine, S., 1978c, Social organization and social behavior in two subspecies of squirrel monkeys (*Saimiri sciureus*), *Folia Primatol.* **30**:126–144.

Mendoza, S. P., Coe, C. L., Lowe, E. L., and Levine, S., 1979, The physiological response to group formation in adult male squirrel monkeys, *Psychoneuroendocrinology* **3**:221–229.

Michael, R. P., and Zumpe, D., 1976, Environmental and endocrine factors influencing annual changes in sexual potency in primates, *Psychoneuroendocrinology* **1**:303–313.
Migeon, D. J., 1981, Physiology and pathology of adrenocortical function in infancy and childhood, in: *Pediatric Endocrinology* (R. Collier, ed.), Raven Press, New York, pp. 465–522.
Musey, P. I., Collins, D. C., Gould, K. G., and Preedy, J. R. K., 1982, Estrogen metabolism in the New World monkeys: Urinary excretion of estradiol metabolites by squirrel monkeys, *Physiologist* **25**(4):248.
Nadler, R. D., and Rosenblum, L. A., 1972, Hormonal regulation of the "fatted" phenomenon in squirrel monkeys, *Anat. Rec.* **173**:181–187.
Nieschlag, E., 1979, The endocrine function of the human testis in regard to sexuality, in: *Sex, Hormones and Behavior* (Ciba Foundation Symposium, 1978), Excerpta Medica, Amsterdam, pp. 183–197.
Parkes, A. S. and Deanesley, R., 1966, Relation between the gonads and the adrenal glands, in: *Marshall's Physiology of Reproduction,* Volume 3 (A. S. Parkes, ed.), Little, Brown, Boston, pp. 1064–1111.
Penney, D. P., and Brown, G. M., 1971, The fine structural morphology of adrenal cortices of normal and stressed squirrel monkeys, *J. Morphol.* **134**:447–466.
Preslock, J. P., and Steinberger, E., 1977, Testicular steroidogenesis in the common marmoset, *Callithrix jacchus, Biol. Reprod.* **17**:289–293.
Rose, R. M., Poe, R. O., and Mason, J. W., 1968, Psychological state and body size as determinants of 17-OCHS excretion, *Arch. Intern. Med. (Chicago)* **121**:406.
Rosenblum, L. A., 1968, Some aspects of female reproductive physiology in the squirrel monkey, in: *The Squirrel Monkey* (L. A. Rosenblum and R. W. Cooper, eds.), Academic Press, New York, pp. 147–169.
Rosenblum, L. A., and Cooper, R. W., eds., 1968, *The Squirrel Monkey,* Academic Press, New York.
Seal, U. S., and Doe, R. P., 1965, Vertebrate distribution of corticosteroid-binding globulin and some endocrine effects on concentration, *Steroids* **8**:827–841.
Setchell, K. D. R., Chua, K. S., and Himsworth, R. L., 1977, Urinary steroid excretion by the squirrel monkey, *J. Endocrinol.* **73**:365–375.
Siiteri, P. K., Murai, J. T., Hammond, G. L., Nisker, J. A., Raymoure, W. J., and Kuhn, R. W., 1982, The serum transport of steroid hormones, *Recent Prog. Horm. Research* **38**:457–510.
Snipes, C. A., Forest, M. G., and Migeon, C. J., 1969, Plasma androgen concentrations in several species of Old and New World monkeys, *Endocrinology* **85**:941–945.
Taub, D. M., 1980, Age at first pregnancy and reproductive outcome among colony-born squirrel monkeys (*Saimiri sciureus,* Brazilian), *Folia Primatol.* **33**:262–272.
Travis, J. C., and Holmes, W. N., 1974, Some physiological and behavioral changes associated with oestrus and pregnancy in the squirrel monkey, *J. Zool. Lond.* **174**:41–66.
Van Horn, R. N., 1980, Seasonal reproductive patterns in primates, in: *Seasonal Reproduction in Higher Vertebrates* (R. J. Reiter and B. K. Follett, eds.), Karger, Basel, pp. 181–221.
Wiebe, R. H., Diamond, E., Aksel, S., Liu, M. D., and Abee, C. R., 1984, Diurnal variations of androgens in sexually mature Bolivian squirrel monkeys (*Saimiri sciureus*) during the breeding season, *Am. J. Primatol.*
Wilson, M. I., Brown, G. M., and Wilson, D., 1978, Annual and diurnal changes in plasma androgen and cortisol in adult male squirrel monkeys (*Saimiri sciureus*) studied longitudinally, *Acta Endocrinol.* **87**:424–433.
Wolf, R. C., O'Connor, R. F., and Robinson, J. A., 1977, Cyclic changes in plasma progestins and estrogens in squirrel monkeys, *Biol. Reprod.* **17**(2):228–231.
Yamamoto, S., Utsu, S., Yanioka, Y., and Ohsawa, N., 1977, Extremely high levels of corticosteroids and low levels of corticosteroid binding molecule in plasma of marmoset monkeys, *Acta Endocrinol.* **85**:398–405.

9

Thermoregulation in the Squirrel Monkey

Eleanor R. Adair

1. BASIC CONCEPTS

The squirrel monkey is an endotherm, an organism capable of maintaining a stable internal body temperature in the face of rather wide fluctuations in the temperature of the environment. Thermoregulation in endotherms is accomplished by fine adjustments in appropriate autonomic response systems, by which the body gains or loses heat, acting in concert with a wide range of behavioral maneuvers that provide a hospitable microclimate for the animal. Whenever possible, the behaviorally generated microclimate is thermally neutral, a situation that maximizes the economy of energy stores and water in the body, and minimizes the involvement of autonomic mechanisms. Thus, the description of thermoregulation in any endotherm involves detailed knowledge of thermoregulatory behavior, both instinctive and learned, and of individual autonomic processes of heat production and heat loss. As we shall see, the particular autonomic response that may be operative at a given time is dictated by the environmental temperature: i.e., squirrel monkeys shiver in the cold and sweat in the heat, but not the reverse, and they will avoid doing either one if an efficient behavioral maneuver is available to them.

The normal colonic temperature of an awake squirrel monkey restrained in a thermoneutral environment (33°C) is 39 ± 0.5°C (Stitt and Hardy, 1971). Deviations from this range have been reported by many investigators and result from many variables, including degree and method of restraint, time of day, age, sex, subspecies, ambient conditions, acute versus steady-state determination, locus of measurement, and nutritional state. Body temperatures *well* outside of this range must be related to disease states, unusual activity, or extraordinary environmental conditions, particularly hot environments. As a general rule, the temperature of the anterior hypothalamic/preoptic thermo-

---

Eleanor R. Adair • John B. Pierce Foundation and Yale University, New Haven, Connecticut 06519.

regulatory center in the brainstem runs ~0.5°C higher than that measured in the colon (see Section 3.3).

To aid the understanding of how thermoregulation is accomplished by the squirrel monkey, a few fundamental concepts from thermal physiology, including a description of the methods of partitional calorimetry, are necessary as background. The law of conservation of energy is the most important principle underlying the science of thermal physiology. In the steady state, the heat produced in the body of an endotherm is balanced by the heat lost to the environment such that storage of heat within the body is minimal. This concept can be expressed by a generalized heat balance equation of the form

$$M - W = R + C + E \pm S \tag{1}$$

(Bligh and Johnson, 1973). The thermal energy produced in the body by metabolic processes M will be modified by any work W produced by the animal on the environment. While W may be a significant factor for humans or beasts of burden, it may be considered negligible for other endothermic mammals, particularly in a laboratory setting. Thus, for practical purposes, M represents the heat production of the body.

The first three terms on the right side of the equation represent the different avenues by which heat is exchanged between the body and the environment: R represents radiation, C represents convection, and E is the heat lost through the evaporation of water from the skin surface and respiratory tract. No term appears in the equation for heat exchange by conduction since it is usually insignificant in most species. If the environmental temperature is higher than that of the body, the direction of heat transfer may be *into* the body and R and C may have a negative sign. Under such conditions, evaporation of fluid from the body surface is the only available avenue of heat loss. Man and certain of the nonhuman primates (e.g., *Erythrocebus patas*) sweat profusely and thus are able to thermoregulate efficiently even during exercise in hot environments (Elizondo, 1977). The squirrel monkey does not have this capability, possessing eccrine sweat glands only on the friction surfaces of the palms and soles (Machida *et al.*, 1967). Thus, this species must rely to a large extent on behavioral responses (e.g. spreading urine, creating a cool microclimate) to ensure maintenance of a stable body temperature in the heat.

The last term in the heat balance equation, S, represents the rate of heat storage in the body. If S is positive, body temperature rises; if S is negative, body temperature falls. Clearly, the goal of normal thermoregulation is the minimization of S and thus the achievement of a stable internal body temperature. The methods of partitional calorimetry are particularly useful to determine the values of individual terms in the heat balance equation at any given environmental temperature for a particular species. These methods are described in detail below as they have been used to study thermoregulation in the squirrel monkey.

It is pertinent to ask how individual thermoregulatory effector responses are mobilized. In other words, what is the nature of the control system for the regulation of the body temperature of an endotherm? The thermoregulatory system appears to function like a negative-feedback control system with a reference or "set" temperature. The controlled variable in this system is the internal body temperature. Temperature sensors are distributed around the body to provide information about local tissue temperatures. Neural signals from these sensors are integrated by a central controller, the integrated signal is compared with the internal reference, and an effector command is generated to energize appropriate responses whenever a load error occurs. A negative load error (i.e., body temperature lower than set point) will produce increased heat storage; a positive load error will produce increased heat loss.

As noted above, the particular effector response that is mobilized, as well as its vigor, depends upon the prevailing environmental temperature, and thus that of the skin. Therefore, it has been argued, thermal sensors must exist in the outer skin layers to detect the local temperature. Indeed, cutaneous nerve endings have been found whose activity (i.e., firing rate) depends on their own temperature. Some fibers are sensitive to warming, others to cooling, but none of these are sensitive to other stimuli such as mechanical deformation. Such temperature-sensitive cells have been found not only in the skin but also in many other regions within the central nervous system (CNS), such as the hypothalamus, midbrain, medulla, cortex, spinal cord, and deep abdominal structures. The most important CNS region for thermoregulation appears to be the medial preoptic/anterior hypothalamic (PO/AH) area of the brainstem. In all species studied so far, including the squirrel monkey, local temperature changes produced experimentally in the PO/AH will mobilize all the effector mechanisms of heat production and heat loss as well as appropriate thermoregulatory behavior. For example, local heating will induce vasodilation of the peripheral vasculature, sweating, and the behavioral selection of a cooler environment. Thus, the experimental evidence is compelling that this brainstem region, which contains many cold- and warm-sensitive neurons, is the site of the "central thermostat" and the control center for the regulation of the body temperature. Since the mid-1960s, the study of thermoregulation in the squirrel monkey has provided a substantial contribution to our knowledge of basic thermoregulatory processes in endotherms, including many of the operating characteristics of the PO/AH thermoregulatory center.

2. *AUTONOMIC THERMOREGULATORY RESPONSES*

2.1. *Method of Partitional Calorimetry*

The basis of this experimental method is the heat balance equation (1) described above. In general, the experimental animal is brought into thermal equilibrium (i.e., $S = 0$) with a particular environmental temperature and the steady-state heat production and heat loss responses are measured. A wide range of environmental temperature is studied, which encompasses the range to which the animal is normally exposed, a design that yields a thermoregulatory profile of the species in question.

Metabolic heat production M is calculated from oxygen consumption, often assuming a constant respiratory quotient or RQ (0.83 for the squirrel monkey).

Total evaporative water loss, which includes water lost through respiration, passive diffusion through the skin, and that evaporated as sweat, is determined from the total reduction in body weight during the experiment, urine and feces being trapped under oil. Evaporative heat loss E is then calculated assuming the latent heat of evaporation of water to be 0.72 W-hr/g. Dry heat exchanged with the environment through convection C and radiation R must be expressed in terms of the surface area of the body. For the squirrel monkey, Stitt *et al.* (1971a) have determined that a body mass of 1 kg is equivalent to a surface area of 0.108 m^2.

Heat exchange via radiation and convection is determined indirectly in the method of partitional calorimetry and requires measurement of the animal's average skin temperature $\bar{T}_{sk}$. Under steady-state conditions of rest at a constant air temperature T_a, Eq. (1) can be simplified to

$$M = R + C + E \tag{2}$$

In the steady state, when the walls of the environmental test chamber are at the same temperature as T_a, the heat losses from the body due to radiation and convection $R + C$ are a function of the thermal gradient between skin and air; thus

$$R + C = h(\bar{T}_{sk} - T_a) \tag{3}$$

where h is the coefficient of heat transfer to the environment. Substituting in Eq. (2) yields

$$M = h(\bar{T}_{sk} - T_a) + E \tag{4}$$

in which h is the only unknown. This coefficient can be determined at any given T_a from the relation

$$h = \frac{M - E}{\bar{T}_{sk} - T_a} \tag{5}$$

and a plot of $M - E$ versus $\bar{T}_{sk} - T_a$ yields a straight line of slope h that passes through the origin, where $\bar{T}_{sk} = T_a$. The units in which h is expressed are W/m^2°C.

2.2. Stitt and Hardy's Experiment

The autonomic thermoregulatory responses of five adult male squirrel monkeys to T_a ranging from 10 to 39°C were measured by Stitt and Hardy (1971). During the experimental tests, single animals were restrained in a Plexiglas chair (Adair, 1969a) inside a closely regulated environmental test chamber. Air movement within the chamber was <8 m/min (still air). A Plexiglas hood over the animal's head collected the expired air, which was drawn outside the chamber at 7–10 liters/min for analysis of oxygen content. The increase in relative humidity of the expired air was also measured so that respiratory evaporative heat loss E_{res} could be calculated. The restraining chair was mounted on a sensitive balance so that the reduction in body mass could be monitored

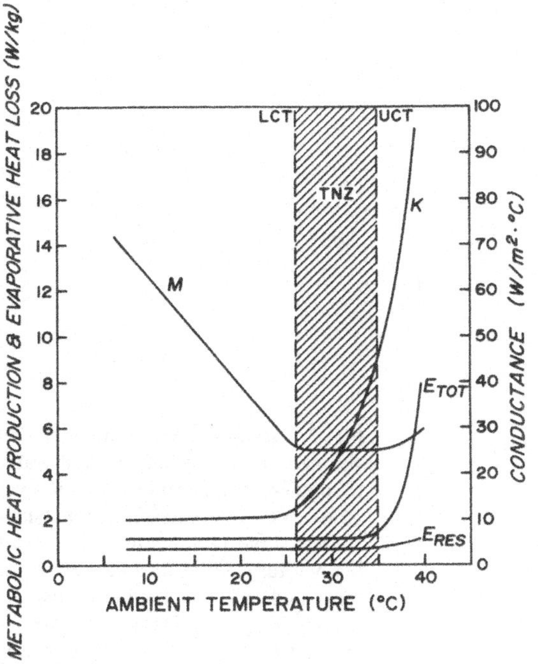

Fig. 1. Thermoregulatory profile for the restrained squirrel monkey equilibrated to ambient temperatures ranging from 10 to 39°C. Individual functions show metabolic heat production M, total evaporative heat loss E_{tot}, respiratory evaporative heat loss E_{res}, and tissue conductance K. The thermoneutral zone of vasomotor control (TNZ) encompasses ambient temperatures between the lower critical temperature (LCT) of 26°C and the upper critical temperature (UCT) of 35°C. [From data in Stitt and Hardy (1971).]

continuously. Deep colonic temperature, 10 cm beyond the anal sphincter, was measured with a polyethylene-encased copper–constantan thermocouple with a reference junction in melting ice and water. Four representative skin temperatures, taken from shaved areas on the abdomen, tail, leg, and foot, were also measured with copper–constantan thermocouples and used to calculate a weighted mean skin temperature as suggested by Stitt *et al.* (1971a):

$$\bar{T}_{sk} = 0.45T_{abd} + 0.37T_{leg} + 0.11T_{tail} + 0.07T_{foot} \tag{6}$$

Each animal underwent at least three exposures to each of the following T_a: 10, 15, 20, 25, 30, and 35°C. A few exposures were made at T_a = 39°C, but a steady state could not be achieved, since this T_a is equal to the animal's normal regulated body temperature. Figure 1 is the thermoregulatory profile of the squirrel monkey derived from the data of Stitt and Hardy. The figure shows the mean steady state levels of metabolic heat production M, respiratory evaporative heat loss E_{res}, total evaporative heat loss E_{tot}, and thermal conductance K as a function of the ambient temperature to which the animals were exposed. Conductance is a measure of heat flow from the body core to the skin and reflects the vasomotor tonus of the peripheral vasculature. Steady-state colonic temperature, weighted mean skin temperature, and the temperatures of tail and foot skin, measured in the same experiments, are shown in Fig. 2. The data in these two figures represent the fundamentals of autonomic thermoregulation in the squirrel monkey. Individual responses are discussed in detail below.

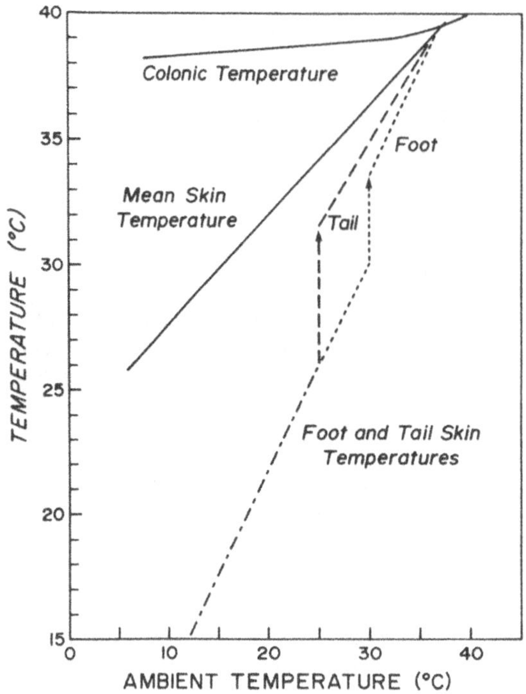

Fig. 2. Body temperatures of the restrained squirrel monkey equilibrated to ambient temperatures ranging from 10 to 39°C. Individual functions show deep colonic temperature, mean skin temperature based on four skin sites, foot, and tail skin temperatures. Arrows indicate vasodilation of vessels in tail and foot skin. [From data in Stitt and Hardy (1971).]

2.3. Basic Data on Autonomic Thermoregulation

2.3.1. Metabolic Heat Production

When squirrel monkeys are restrained in cool environments, body temperature is regulated by an increase in metabolic heat production *M*. Figure 1 shows that as the ambient temperature falls below ~26–27°C, *M* increases linearly at a rate of about 0.35 W/kg°C (Stitt and Hardy, 1971). Active shivering may be observed at T_a = 20°C and below and heart rate and mean arterial blood pressure may be elevated (Wasserstrum and Herd, 1977). Although Chaffee *et al.* (1966a,b) demonstrated that cold-acclimated squirrel monkeys possess brown fat, nonshivering thermogenesis is probably not involved in the heat production of nonacclimated animals exposed to cold environments.

At 26–27°C, designated the lower critical temperature (LCT), resting heat production reaches its nadir and remains at this low level throughout the range of T_a that is designated the thermoneutral zone or TNZ (T_a = 27–35°C). Stitt and Hardy found that the resting heat production of chaired animals ranged from 4.5 to 7 W/kg in the TNZ. Other data (Adair, 1976; Malinow and Wagner, 1966; Adair and Adams, 1982) confirm this general range, the particular values obtained depending on the air movement in the test chamber as well as the method used to measure oxygen consumption. It appears that resting heat production in unrestrained animals may range slightly lower than the values cited above (LeMaho *et al.*, 1981; Smoake, *et al.*, 1977) and may also vary with age and subspecies (E. R. Adair, unpublished observations). At any given ambient temperature, the metabolic heat production will be higher when the animal is in moving air than in still air conditions (Adair and Adams, 1982).

The resting metabolic rate RMR of an animal at thermoneutrality may be predicted from its body mass M using an allometric equation derived from data on many species by Kleiber (1961). This equation, $\text{RMR} = 3.35M^{0.75}$, predicts resting levels of heat production 10–75% lower than the range of values actually measured for the squirrel monkey. For this reason, it has been concluded that the squirrel monkey is hypermetabolic when compared with other laboratory animals of the same size (Smoake *et al.*, 1977). Anyone who has observed squirrel monkeys scampering wildly about cage or forest canopy would tend to agree with this conclusion.

Excessively warm environments, e.g., temperatures above 35°C, threaten effective thermoregulation in this species. An increase in metabolic heat production has been reported under such conditions that is presumably related to increased behavioral activity. The upturn in the M function in Fig. 1 was likely the result of struggling against the restraining chair. Hickey and Kelley (1972) have reported that an ambient temperature above 37.8°C at 100% relative humidity produced fatal heatstroke in several members of a colony of squirrel monkeys. Thus, escape from a thermally hostile environment is essential to the survival of this species.

2.3.2. Tissue Conductance and Vasomotion

Figure 2 illustrates the relationship between various body temperatures and ambient temperature as measured by Stitt and Hardy (1971). Mean colonic temperature (termed rectal temperature by most authors) was regulated between 38.5 and 39.8°C over the range of ambient temperatures 10–39°C, with a slightly higher rate of increase above the TNZ. The weighted mean skin temperature increased linearly with ambient temperature; the line of best fit to the data has the equation $\bar{T}_{sk} = 0.44T_a + 23.2$. It should be remembered however, that this specific relationship will hold only for still air conditions such as those employed by Stitt and Hardy. Lynch *et al.*, (1980), who conducted similar experiments on monkeys restrained in a moving airstream ($4.5 \geq V \geq 0.4$ m/sec), reported this equation to be $\bar{T}_{sk} = 0.61T_a + 15.9$.

Figure 2 also shows dramatic discontinuities in the temperature functions for foot and tail skin. These discontinuities occur at different discrete ambient temperatures and reflect vasodilation of the peripheral blood vessels of the tail and foot. Warm blood from deep in the body is brought close to the skin surface when these vessels vasodilate, aiding the transfer of metabolic heat to the environment. Changes in vasomotor tonus provide the means for efficient thermoregulation in the squirrel monkey across the entire TNZ.

Tissue conductance K is an indicant of vasomotor state and represents the flow of heat from the body core to the skin. Figure 1 shows that there is a sharp increase in conductance at or near the lower end of the TNZ, increasing dramatically as the environment becomes warmer. All heat generated in the body, with the exception of that lost from the respiratory tract, must be carried to the surface by the circulatory system, from whence it is transferred to the environment by radiation, convection, and evaporation. When the animal is in a steady state ($S = 0$), the heat leaving the body is equal to the heat production less that lost through respiration, $M - E_{res}$. Conductance is defined as this value divided by the temperature gradient between the body core and the skin:

$$K = \frac{M - E_{res}}{T_{co} - \bar{T}_{sk}} \tag{7}$$

Figure 1 shows that conductance is minimal at ambient temperature below the LCT; in other words, the peripheral vasculature of squirrel monkeys is maximally vasoconstricted in environments cooler than 26°C. In warmer environments, these animals exhibit a remarkable ability to increase tissue conductance, as first the tail and then the hands and feet vasodilate (cf. Fig. 2).

The precise ambient temperatures that produce vasodilation of tail and foot were pinpointed by Lynch *et al.* (1980), who measured steady-state skin temperatures of four monkeys restrained in many discrete ambient temperatures. Figure 3 shows that tail and foot skin temperatures were essentially at ambient levels for all ambient temperatures below the threshold for vasodilation. At an ambient temperature 26.5°C for the tail and 32.5°C for the foot, these skin temperatures varied widely, indicating a wide variation in local tissue blood flow. The corresponding skin temperatures at the threshold for vasodilation were 27°C for the tail and 32°C for the foot. Similar results have recently been reported by Adair and Adams (1980b).

The vasomotor response is truly thermoregulatory in nature: experimental warming of the anterior hypothalamic thermoregulatory center (see Section 3.3) in animals equilibrated to subthreshold ambient temperatures produces prompt vasodilation of both tail and foot vessels (Stitt and Hardy, 1971; Lynch and Adair, 1978; Lynch *et al.*, 1980). However, the tail appears to exhibit a much greater degree of vasodilation than does the foot. The range over which the tail skin temperature varies during dilation is about 7°C, while the corresponding range for the foot is only about 3°C. Since the surface area of the tail represents 11% of the total surface area of the body, this organ has the capability of dissipating significant amounts of metabolic heat to the environment. Two large superficial veins extend laterally along the tail just under the skin (Stitt and Hardy, 1971) and these vessels undoubtedly play a major role in heat loss in warm environments. This explains why squirrel monkeys in the heat are often observed to be perched or sprawled with the tail hanging below the perch or branch (DuMond, 1968; Thorington, 1968). This attitude takes full advantage of any air currents that would increase the ability of the tail to dissipate body heat.

2.3.3. Evaporative Heat Loss and Sudomotor Activity

Figure 1 displays two functions for evaporative heat loss, that from the body as a whole (E_{tot}) and that from the respiratory tract (E_{res}). The difference between the two represents evaporative heat loss through sweating (E_{sw}) and a minimal amount of passive diffusion of water through the skin (commonly

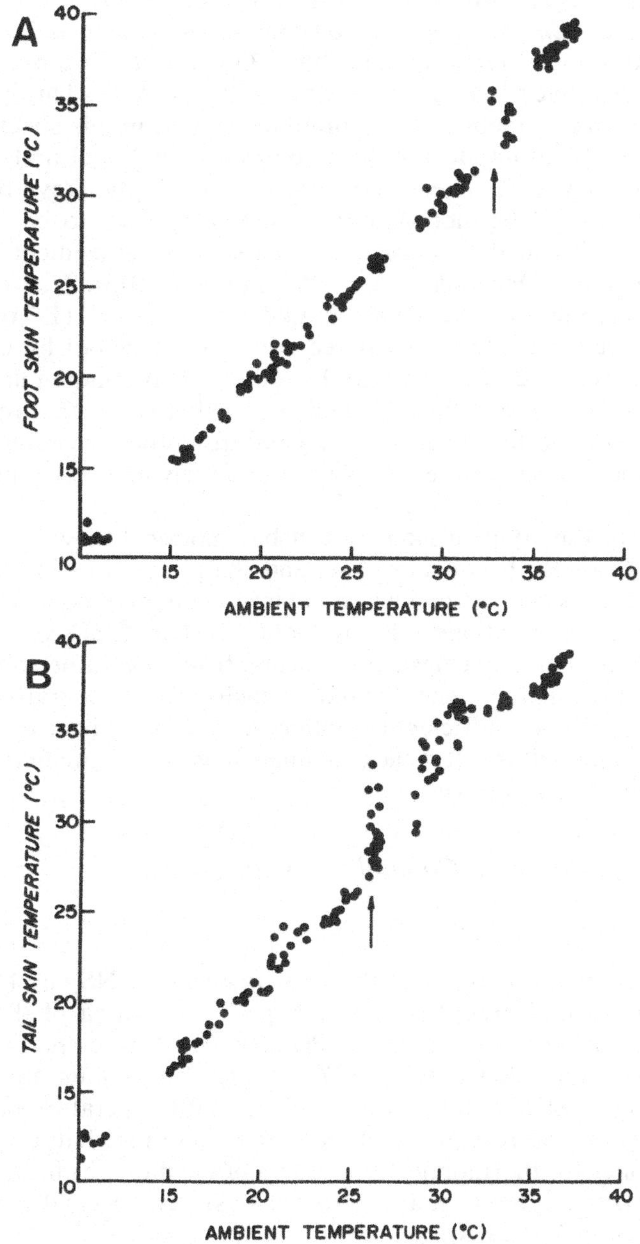

Fig. 3. (A) Foot skin temperature and (B) tail skin temperature of four squirrel monkeys as a function of prevailing ambient temperature. Each data point is mean of ten samples taken at 1-min intervals during steady state. Arrows indicate vasodilation of peripheral blood vessels. [Reproduced from Lynch *et al.* (1980), with permission of the publisher.]

designated "insensible perspiration"). The latter tends to be constant, whatever the ambient temperature, and is insignificant for thermoregulation.

Both E_{tot} and E_{res} are low and constant at ambient temperatures below about 30°C. At ambient temperatures above 35°C (UCT), E_{tot} becomes significantly elevated above the basal level, whereas E_{res} changes hardly at all. This implies that the major avenue of evaporative heat loss in this species is sweating, not panting. This implication was confirmed by Nadel and Stitt (1970), who recorded changes in sweat rate from the foot of monkeys restrained in warm environments. Thermoregulatory sweating appears to be restricted to the palms and soles in this species, a fact that is in agreement with earlier histological evidence (Machida *et al.*, 1967; Perkins, 1975). That the sweating is truly thermoregulatory was demonstrated by Stitt *et al.* (1971b); these investigators altered the rate of sweat secretion from the foot by changing the local tissue temperature of the anterior hypothalamic thermoregulatory center. Recent data of V. Candas (unpublished) show that the sweat response is a linear function of the hypothalamic temperature within the range 38.7–42°C and that the temperature of the skin plays a relatively minor role in the control of sweat rate.

Sweating in the squirrel monkey can be emotional as well as thermoregulatory, however. Both sweating onset and sharp increases in the sweat rate of animals in thermoneutral and warm environments may be accompanied by sudden increases in metabolic heat production (Adair, 1981). Such conditions might occur in an animal struggling to escape from uncomfortably warm surroundings. Unfortunately, the limited capacity for evaporative heat loss through sweating is not sufficient to counteract increased heat production and fatal hyperthermia will result if these animals must long remain in an ambient temperature of 38°C or above.

2.4. Probing Autonomic Capabilities with Thermodes

2.4.1. Fundamental Methodology

It has been stated above that the most important CNS site for thermoregulation is the medial preoptic/anterior hypothalamic area of the brainstem. This region is variously designated as PO/AH, PO, HY, or preoptic area and the temperature recorded therein as T_{po} or T_{hy} by various authors. Useful methods for the control and assessment of PO/AH temperature have been developed that allow the researcher to probe the operating characteristics of the central controller for thermoregulation. Most of the methods in current use are an outgrowth of a thermode and reentrant tube system originally conceived by Hammel (cf. Hammel, 1968).

Figure 4 diagrams an adaptation of Hammel's thermode system as made by the author for the squirrel monkey. Pairs of sealed stainless steel tubes are implanted in the brain under stereotaxic guidance with their tips positioned in the site of interest (e.g., PO/AH). The tubes may be closed by silver soldering and then polished or they may be welded shut in an inert atmosphere. The latter is the method of choice. The tube of larger diameter (18 ga) serves as

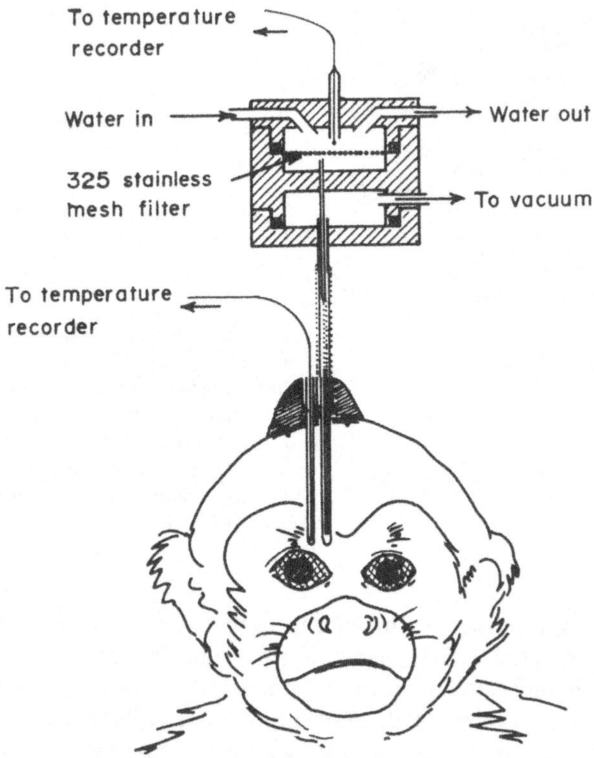

Fig. 4. Diagrammatic representation of the thermode method by which experimental manipulation and measurement of local brain temperature is accomplished. A sealed stainless steel thermode tube, chronically implanted in the brain, is perfused via a two–chamber circulator with temperature-controlled water. An adjacent reentrant tube allows the insertion of a fine thermocouple to measure changes in tissue temperature produced by the thermode.

the thermode and the second tube (21 ga) serves as a chronic orifice into which a 36-ga thermocouple or a thermistor may be inserted to measure local temperature changes produced by the thermode. The closer the tubes are to each other, the more accurate the assessment of tissue temperature in the vicinity of the thermode. To prevent vertical movement of the implanted tubes, steel washers (2 mm diameter × 0.5 mm thick) may be brazed to each tube 22–23 mm from the tip. After implantation of such a tube in the PO/AH, this washer should lie just above the skull surface. In recent years we have used instead a miniature adjustable clamp machined from stainless steel that can be pushed down to the skull surface and tightened with a setscrew against the implanted tube: dental acrylic packed around and under washer or clamp, as well as to stainless steel screws inserted in the calvarium, fixes the implant securely in position. An open well constructed from dental acrylic after the method of Adair (1969b) may be used to protect each protruding tube.

During experiments to study thermoregulatory responses, the thermode is perfused with temperature-controlled water from a circulator located above the

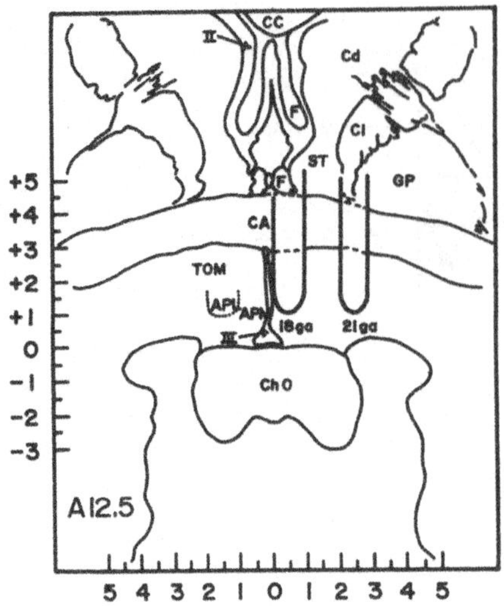

Fig. 5. Coronal section through the squirrel monkey brain at 12.5 mm anterior to the interaural line showing optimal placement of thermode and reentrant tube tips in medial preoptic area (APM) of the hypothalamus. Other major landmarks include optic chiasma (ChO), third ventricle (III), anterior commissure (CA), internal capsule (CI), and fornix (F).

animal's head. Water passes through the upper circulator chamber under pressure (15 psi) and is forced into inner thermode tubing (26 ga thin-wall connected to PE 20) that is inserted to within 0.5 mm of the bottom of the implanted tube. The water is then aspirated through concentric polyethylene tubing into the lower circulator chamber, to which a vacuum is applied. A flow rate of water through the thermode of 7–10 ml/min will produce a change in tissue temperature (measured by thermocouple in the adjacent reentrant tube) of 5–7°C/min. It is important that the circulator contain a 325-mesh stainless steel filter to prevent tiny dirt particles from clogging the thermode.

The most effective placement of a thermode–reentrant tube pair in the PO/AH thermoregulatory center of the squirrel monkey is shown in Fig. 5. The coordinates for the thermode in a 750- to 900-g animal, according to the atlas of Emmers and Akert (1963), are 12.5 mm anterior to the interaural line, 0.5 mm lateral to the midline, and 1.0 mm above the horizontal zero plane. The reentrant tube is centered 2.0 mm lateral to the thermode. For more effective thermal control of the PO/AH tissue, thermodes may be placed bilaterally or at AP 13.5 and 11.5 to bracket the medial preoptic area.

2.4.2. *Influence of Preoptic and Other CNS Temperatures on Autonomic Thermoregulatory Responses*

Neurophysiological studies have provided evidence for the existence of thermosensitive neural units not only in the PO/AH but also in several other CNS sites. Several of these have been investigated for their possible involvement in normal thermoregulation using the thermode technique. In the squirrel monkey, local temperature changes induced via thermode in the posterior hypothalamus (Adair, 1974), midbrain reticular formation (Adair and Stitt, 1971), lateral hypothalamus, or globus pallidus (E. R. Adair, unpublished observa-

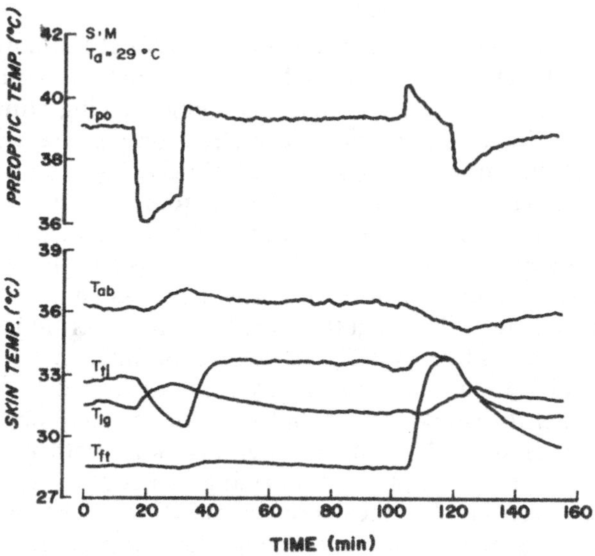

Fig. 6. Sample record of a daily session for one monkey in experiments to demonstrate the influence of preoptic temperature T_{po} on peripheral vasomotor state. Tracings show changes in abdominal T_{ab}, tail T_{tl}, leg T_{lg}, and foot T_{ft} skin temperatures that result from increases and decreases in T_{po} at an ambient temperature T_a of 29°C. [Reproduced from Lynch *et al.* (1980), with permission of the publisher.]

tions) produce *no change* in autonomic mechanisms of heat production or heat loss. In several endothermic species, the spinal cord is one area reported to exert control over these mechanisms that is comparable to the control exerted by the PO/AH (Chai and Lin, 1972). However, the small diameter of the spinal canal in the squirrel monkey (quite apart from the general fragility of the species) has conspired against any attempt to implant flexible thermodes therein to study the thermoregulatory effects of heating and cooling the spinal cord. Heat introduced into the abdominal cavity by an oversized rectal thermode will alter thermoregulatory behavior, but the effects on autonomic responses are equivocal (Adair, 1971).

All the autonomic mechanisms of heat production and heat loss are altered in predictable ways when the temperature of the PO/AH is changed by means of implanted thermodes. Stitt and Hardy (1971) demonstrated that hypothalamic warming could induce vasodilation in the tail and foot of awake squirrel monkeys, a finding confirmed by Adair (1974). Subsequently, the ambient–hypothalamic temperature interaction affecting vasomotion in this species was quantified by Lynch and his co-workers (Lynch and Adair, 1978; Lynch *et al.*, 1980). They studied six animals equilibrated to a wide range of ambient temperatures (0–39°C) and subjected to 20–min episodes of PO/AH warming or cooling. A sample experiment shown in Fig. 6 illustrates the various effects of increasing and decreasing the preoptic temperature T_{po}. Both vasodilation and vasoconstriction occurred at this ambient temperature (T_a = 29°C) from warming and cooling the hypothalamus, respectively. Further, the effect of a change in T_{po} depended on the state of the vasomotor system at the initiation of that T_{po} change. Thus, the already vasodilated tail skin constricted during preoptic cooling, whereas, later in the experiment, preoptic heating stimulated vasodilation in the constricted foot but had little effect on the already vasodilated tail. Although not shown in this figure, preoptic cooling stimulates a rise in rectal temperature, which is indicative of heat storage; conversely, preoptic warming stimulates a fall in rectal temperature, which is indicative of heat loss. During the experiment depicted in Fig. 6, these changes in the temperature of the body core were mirrored by drifts in T_{po}, which was controlled by a single thermode in this case. It is clear that the use of multiple thermodes would have allowed more precise control over T_{po} in these experiments.

The findings of Lynch *et al.* (1980) may be summarized as follows: (1) at T_a below the normal threshold (LCT), tail vasodilation may be stimulated by preoptic warming; (2) at T_a above the normal threshold, both vasodilation and vasoconstriction occur during preoptic warming and cooling, respectively; (3) for comparable increases in T_{po}, the degree of vasodilation above the T_a threshold is greater than it is below; and (4) the magnitude of peripheral vasodilation that results from preoptic warming varies as the magnitude of the preoptic temperature change. Similar results for vasomotion in the foot skin were reported by Lynch and Adair (1978). Both studies demonstrated that the threshold for vasodilation in the squirrel monkey varies in nearly linear fashion with changes in preoptic and mean skin temperatures over most of the range in which these temperatures can be changed without injuring the tissues involved. In other words, the cooler the skin, the higher the PO/AH temperature must be to initiate peripheral vasodilation. However, since the temperature of the hypothalamus varies little in the nonexperimental situation except under emergency conditions that accompany illness or extreme thermal stress, the temperature of the skin will play the predominant role in the control of peripheral vasomotion.

As we have seen, when squirrel monkeys are restrained in cool environments, body temperature is regulated by an increase in metabolic heat production. At any given T_a below the LCT, the level of heat production will be a linear function of the PO/AH temperature. Stitt *et al.* (1971b) found an additive interaction between mean skin temperature and preoptic temperature in the control of metabolism in one squirrel monkey equilibrated to a range of ambient temperatures from 10 to 30°C ($27 \leq \bar{T}_{sk} \leq 38$°C). At all T_a, preoptic warming reduced and preoptic cooling increased metabolic heat production. This finding was later confirmed in several animals by Adair and Stitt (1971) and by Lynch (1976, and unpublished data). The phenomenon can also be demonstrated in a behaving animal (Adair, 1976). When monkeys exert inefficient behavioral control over the temperature of their environment such that T_a is below the LCT, metabolic heat production will be elevated. Under these conditions, the level of heat production can be altered by warming and cooling the PO/AH. Thus, the picture is similar to that described above with regard to the alteration of vasomotor responses by PO/AH temperature change: preoptic warming drives the ongoing thermoregulatory effector response in the direction that would be produced by an environment warmer than that in which the animal finds itself, and preoptic cooling does just the reverse.

In a recent study at the Pierce Laboratory, V. Candas (unpublished) has examined the control of sweating from the foot of squirrel monkeys equilibrated to T_a between 28 and 35°C ($34.5 \leq \bar{T}_{sk} \leq 37.5$°C), when the PO/AH temperature was varied between 38.7 and 42°C. The local temperature of the foot skin T_{ft} from which sweat rate was measured was also closely regulated in these experiments. Candas found that the sweating response is once again a linear function of preoptic temperature, but that neither mean skin $\bar{T}_{sk}$ nor foot skin T_{ft} temperature provided any systematic facilitation of sweat secretion when these temperatures were manipulated within physiological limits. Candas concluded that because of a limited capacity for evaporative cooling and because of a nonsystematic involvement of peripheral afferent inputs to the sweating mechanism, at least within physiological limits, the squirrel monkey appears to be a poor model for studying the control of sweating in humans.

2.5. *Autonomic Thermoregulation during Fever*

In contrast, the squirrel monkey is an especially appropriate primate model for the study of the origins and control of febrile responses in man, because, unlike many other nonhuman primates, it is susceptible to both the systemic and central administration of agents that cause fever in humans. In general, fever is a regulated rise in body temperature that is caused by endogenous pyrogens released into the circulation in response to invading bacteria, viruses, and other agents. The endogenous pyrogen appears to alter the firing rate of temperature-sensitive neurons of the central thermoregulatory center in such a way that heat loss responses are curtailed and heat storage is increased.

Generally speaking, fevers may be produced in experimental animals in two ways: (1) by the intravenous injection of a pyrogenic substance such as *Salmonella typhosa* endotoxin, or (2) by the direct intracerebral introduction of endotoxin or other substances through injection cannulae that have been chronically implanted in the brain. Both methods have been used to study the production of fever in the squirrel monkey; much of this research has been conducted by J. M. Lipton and his colleagues.

Lipton and Fossler (1974) first demonstrated that squirrel monkeys would generate high monophasic fevers of many hours' duration in response to both intravenous and intracerebral injections of *S. typhosa* endotoxin. The course of thermogenesis was monitored by changes in rectal temperature, and the involvement of the peripheral vasomotor system was assessed from temperature changes on the tail and leg skin. The febrile response to intravenous endotoxin typically showed a latency of about 40 min and peaked 40 min later at a rectal temperature approximately 2°C above baseline levels. Some tolerance developed if systemic injections were given more frequently than once every 11 days.

Fevers were also produced when small doses (e.g., 500 ng) of endotoxin were injected directly into many sites in the forebrain; other sites were found to be endotoxin-insensitive. The most responsive region was located immediately ventrocaudal to the anterior commissure. While these fevers were similar to those produced by the intravenous route, they exhibited a longer latency and longer time to maximum. Bilateral injections produced higher fevers than unilateral injections. Using the same preparation, Lipton and Fossler also demonstrated brief fevers of rapid onset when they injected prostaglandin E_1 (PGE_1) intracerebrally, the magnitude of the fever being a direct function of the dose employed. On the other hand, even large doses of cAMP injected bilaterally failed to stimulate a rise in rectal temperature, indicating that PGE_1 probably does not produce fever by increasing the activity of the cAMP system in pyrogen-sensitive cells. A major finding in this study was that there is an upper limit to the febrile response, which is not exceeded even when the dose of endotoxin is very high. This upper thermal limit implies the existence of an antagonistic control mechanism whose function is to hold the body temperature below dangerously high levels.

More light was shed on the mechanisms involved in the production of fever during a study conducted by Crawshaw and Stitt (1975). They implanted injection cannulae in the PO/AH area of six squirrel monkeys so that PGE_1 in various doses could be injected directly into the preoptic thermoregulatory center. They found that when the monkeys were tested at room temperature, reliable, dose-dependent increases in rectal temperature were produced by PGE_1 injections of 20–500 ng. When the ambient temperature was below the LCT (T_a = 22°C), the fevers were produced entirely by increases in metabolic heat production. When the ambient temperature was near the upper end of the TNZ (32°C), the fevers were produced by vasoconstriction of the tail vasculature accompanied by smaller increases in metabolic heat production. Control injections of NaCl produced no fever.

Some of the animals were trained to regulate the temperature of their environment behaviorally (see Section 3). When 250 ng PGE_1 was injected into the PO/AH area of behaving animals, a fever was generated because the monkeys selected an environment several degrees warmer than they normally preferred. In all cases, the peak fever as well as its time course were similar for a given animal and a given dose, irrespective of whether autonomic or behavioral mechanisms were mobilized to increase heat storage in the animal's body.

Additional pyrogenic substances that cause regulated increases in the rectal temperature of squirrel monkeys when injected intracerebrally include sodium pentobarbital (Lipton and Fossler, 1974), leukocycte pyrogen (LP) derived from blood leukocyctes of the cat (Lipton and Trzcinka, 1976), and highly purified LP derived from human blood (Lipton *et al.*, 1979). Pretreatment with acetaminophen (5 mg/kg, IV) reduced but did not abolish the febrile response to human LP (100 μl/kg) in tests conducted on three monkeys. The ability to develop a fever in response to all these substances persists after the PO/AH area has been destroyed bilaterally and the pattern of responses that increase heat production and reduce heat loss remains essentially unchanged (Lipton and Trzcinka, 1976). These results imply that other CNS thermoregulatory sites in addition to the PO/AH area must be involved in the control of fever in primates. Followup experiments showed that the medullary thermosensitive area is not one of these; no fever follows microinjection of endotoxin into this area either before or after PO/AH lesions. On the other hand, heating or cooling the PO/AH with implanted thermodes when a monkey is febrile produces smaller changes in metabolic heat production and rectal temperature than when the animal is in an afebrile state (Lipton and Kennedy, 1979). Sodium salicylate injected into the PO/AH region has no effect on the thermosensitivity of the area, indicating that this agent probably does not exert its antipyretic effect by acting directly on CNS temperature-sensitive cells.

A recent study by Clark *et al.* (1980) probed the question of whether the aged may have a reduced capacity to generate fevers. Injection cannulae were chronically implanted in the lateral cerebral ventricle of 31 male and female squirrel monkeys of ages from 1 to 17+ years. The change in rectal temperature to intracerebroventricular (ICV) injections of *S. typhosa* endotoxin (0–50 ng) or PGE_2 (0–500 ng) or to intravenous endotoxin (0–10 μg/kg) were recorded for 7–9 hr after administration of the pyrogen. While the higher doses always produced greater fevers in any age group that responded, animals 14 years of age or older showed little or no febrile response to IV endotoxin, and animals over 10 years of age developed no fevers to ICV endotoxin. Young monkeys developed robust fevers to ICV PGE_2, but there was a progressive decrease with age in the magnitude of this febrile response. It is not yet clear whether this age-related deficiency is due to a reduced capacity to produce endogenous pyrogen, to an altered central action of the endogenous pyrogen on CNS thermosensitive cells, or to a reduced capacity to produce and store body heat. The resolution of these questions awaits further study.

2.6. *Effects of CNS Lesions and Chemical Agents Other Than Pyrogens*

As we have seen, the ability of the squirrel monkey to generate a fever in response to an endotoxin challenge is not abolished following bilateral destruction of the preoptic area, the region that is often considered to be the "central thermostat." Heat production in animals exposed to cold environments is also not compromised following low bilateral decerebration that transects the brain stem between the upper and lower thirds of the pontine tegmentum. However, if the initial decerebration is at a higher level (e.g., just caudal to the hypothalamus), all heat production responses in the animal's repertoire are lost. These responses will then reappear if a subsequent low decerebration is made (Liu, 1979). The implication of these results is that a tonic inhibitory mechanism may originate in the tegmentum of the upper pons and mesencephalon and this inhibition may be released when descending tracts from the hypothalamus are severed, suppressing thermoregulatory mechanisms that originate in the brain stem. These results confirm similar findings in the dog and cat, but still leave unanswered many questions about the locus and organization of CNS centers that exert control over the body temperature in health and disease.

Barbiturates, such as secobarbital and methohexital, administered in low doses will decrease heart rate, lower blood pressure, and cause a fall in the deep body temperature (Byrd, 1980). Dissociative anesthetics such as phencyclidine and ketamine, on the other hand, can exert substantial stimulant effects over a range of doses, as manifested by increased heart rate and blood pressure and by increases in schedule-controlled operant behavior (Byrd, 1982).

Substances that are reported to produce hypothermia, particularly in older primates, include taurine and diazapam (Clark and Lipton, 1981a,b). The two substances appear to act differently, however. Taurine, a putative inhibitory neurotransmitter, acts centrally and even small doses injected intracerebrally in aged squirrel monkeys produce uncontrolled hypothermia. On the other hand, diazepam injected intramuscularly produces a hypothermia that is potentiated by cold environments, but when injected intraventricularly produces hyperthermia. Thus, it would appear that the action of this drug takes place peripherally rather than centrally.

It should be clear from the above summary that information about the effects of drugs and lesions on the thermoregulatory system of the squirrel monkey is spotty. There is a definite need for a comprehensive program to investigate the thermoregulatory consequences of substances whose locus of action is known. Unraveling the complexities of the neural substrate subserving thermoregulation may be the reward of those who undertake such a program. One can safely say that the squirrel monkey appears to be an excellent subject for such researches.

2.7. Circadian Variations in Body Temperature

Like all endotherms, the squirrel monkey displays a prominent diurnal variation in body temperature. The range of rectal temperatures about the 24-hr mean is about 2°C. For animals entrained to a light–dark (LD) 12:12 cycle (12 hr light : 12 hr darkness), the body temperature is maintained well above the mean while the lights are on and then falls progressively during the night, reaching a nadir ~2 hr before the lights are scheduled to come on (Fuller *et al.*, 1979).

While changes in activity (i.e., metabolic rate) can exert considerable influence on the body temperature, it is important to recognize that activity rhythms *per se* do not generate the circadian rhythm of body temperature. Instead, activity rhythms, together with cycles of food intake and sleep, act to modulate the regulated temperature (Fuller and Sulzman, 1982).

The primary natural synchronizer of the circadian rhythm is the cycle of light and dark. Thus, not only the temperature rhythm, but other body rhythms maintain characteristic phase relationships with the environment. When a phase shift occurs in the LD cycle, resynchronization over a period of time occurs in all body rhythms, usually at different rates. Detailed descriptions of such phase shifts in the body temperature rhythm of the squirrel monkey have been given by Fuller *et al.* (1979, 1981b).

In general, the body temperature is higher when lights are on and lower when lights are off. This will be true even when lights go on and off every 2 hr, although in this case the *average* level of the body temperature will continue to follow an endogenous LD 12:12 cycle (Fuller *et al.*, 1981b). Curiously, if an imposed LD cycle (e.g., LD 9:9) is outside of the range of the entrained or endogenous rhythm, the circadian system will free-run and a complex pattern of changes in body temperature will result (Fuller *et al.*, 1981b; Sulzman *et al.*, 1977). Light intensity is also important, higher intensities producing higher regulated body temperatures. During constant light, the circadian rhythm of body temperature persists, but with decreased amplitude (Fuller *et al.*, 1979; Fuller and Sulzman, 1982; Sulzman *et al.*, 1979).

Ambient temperature exerts specific effects on the circadian rhythm of body temperature in the squirrel monkey. The 24-hr mean body temperature is positively correlated with the prevailing ambient temperature, the greatest increase in warm environments occurring during the periods of lights off (subjective night). The range of the temperature rhythm in both deep-body and brain temperature is maximal in the TNZ and becomes narrower in both cooler and warmer environments (Fuller *et al.*, 1981b; Baker and Fuller, 1982). As we have seen, the particular autonomic effector mechanism that regulates the body temperature at any given time depends on the prevailing ambient temperature (Fig. 1). Each effector response demonstrates rhythmicity, although the amplitude and phase relationships between individual effector responses may be complex (Fuller *et al.*, 1980).

It is now clear that the circadian timekeeping system of mammals is composed of multiple oscillators that are functionally related to light–dark cycles, feeding–fasting cycles, and perhaps sleep–waking cycles (Fuller, 1982). Analysis of the interrelationships between these oscillators is derived from experimental manipulation of the phase relationships between them.

One of the key oscillators or pacemakers of the circadian timekeeping system has been localized to the suprachiasmatic nucleus (SCN) of the hypothalamus. It may function to regulate sleep and wakefulness, drinking, and many rhythmic behaviors such as wheel-running in rodents (Richter, 1967). Fuller *et al.* (1981a) have reported that bilateral lesions of the SCN disrupted free-running drinking rhythms in the squirrel monkey but left the circadian rhythm of body temperature intact. These results provide strong evidence for the general notion of multiple oscillators in the circadian timekeeping system, but they also indicate that an oscillator located elsewhere than the SCN controls the circadian rhythm of deep-body temperature in the squirrel monkey.

2.8. Autonomic Thermoregulation during Exposure to Microwave Fields

Microwave energy coupled to living tissue generates heat. Therefore microwaves (electromagnetic radiation in the frequency range of 300 MHz to 300 GHz) must be considered as part of the thermal environment to which humans and animals may potentially be exposed. When the wavelength of the radiation is of the order of the dimensions of the body, microwaves are absorbed below the skin surface and thus provide a unique challenge to the thermoregulatory system. In other words, microwave exposure can be used as a tool with which to probe the relative input from subcutaneous and cutaneous thermosensitive structures in the control of individual effector responses of heat production and heat loss. Since 1977, the major emphasis of the research in the author's laboratory has concerned the impact of exposure to low-intensity microwave fields on the thermoregulatory capabilities of the squirrel monkey.

While a detailed account of results obtained within this specialized research program would be out of place in this chapter, a brief report of the more important findings illustrates the application of basic partitional calorimetry to the study of an unusual thermal stimulus. Figure 1 predicts that if a squirrel monkey were to be exposed to a microwave field, the particular response system that might be altered by absorbed radiant energy will be directly related to the ambient temperature at which the exposure occurs. Thus, at T_a below the LCT, microwave exposure should lower metabolic heat production; within the TNZ, microwave exposure should alter peripheral vasomotor tonus (conductance); and at T_a above the UCT, microwave exposure should increase evaporative heat loss, principally via an increase in thermoregulatory sweating.

To test such predictions, individual animals were chair-restrained inside a special nonreflecting chamber at sufficient distance from an antenna that the microwave field that irradiated them was very homogeneous. The animal's immediate vicinity was rigorously climate-controlled; standard measurements of body temperatures, oxygen consumption, sweat rate, etc., were made using special techniques adapted for use in microwave fields. Continuous (i.e., nonpulsed) microwaves at a frequency of 2450 MHZ (~13 cm wavelength) were presented for durations that ranged from 5 to 90 min in individual experiments.

Reliable reductions in the metabolic heat production of monkeys restrained in cool environments (15–25°C) were initiated by 10-min microwave exposures of the whole body at intensities (4–6 mW/cm^2) that are equivalent to about 15% of the resting metabolic rate of the animal (Adair and Adams, 1982). The magnitude of the reduction in heat production was linearly related to microwave intensity above this threshold level. Termination of the microwave field was always followed by a rapid rebound in heat production. When the microwave field was kept on for 90 min, the vigorous reduction of metabolic rate that occurred to microwave onset adapted slowly, ensuring continual precise regulation of the internal body temperature. These experiments demonstrated that cold-exposed endotherms will compensate readily for microwave-induced body heating by reducing endogenous heat production.

Vasomotor tonus in the squirrel monkey tail was indexed by changes in local skin temperature during exposure of the whole body to 13-cm microwaves (Adair and Adams, 1980b). At an ambient temperature (26°C) just below that at which the vessels of the tail normally vasodilate, criterion dilation was initiated by 5-min exposures at a microwave power density of 8 mW/cm^2. This intensity deposited energy equivalent to approximately 20% of the monkey's resting metabolic heat production, but produced no observable change in deep-body temperature. For each 1°C reduction in ambient temperature below 26°C, an increment of microwave intensity of 3–4 mW/cm^2 produced identical threshold responses. No vasodilation occurred during infrared exposures of equivalent intensity, indicating that noncutaneous thermosensitive structures may mediate microwave activation of thermoregulatory responses in the peripheral vasomotor system.

Thermoregulatory sweating from the foot of squirrel monkeys restrained in ambient temperatures just below the UCT was reliably initiated by microwave exposure of the whole body at a power density of 6–8 mW/cm^2 (Adair,

1981). Above this threshold, the magnitude of the change in sweat rate at a given ambient temperature was directly related to the intensity of the imposed microwave field.

All of the experimentally determined microwave thresholds for the alteration of autonomic thermoregulatory responses fall within a power density range of 4–8 mW/cm^2. This is also true for behavioral thermoregulatory responses; a microwave intensity of 6–8 mW/cm^2 will cause behaving animals to select a cooler environment for themselves (Adair and Adams, 1980a, 1983). The rate of energy absorption in all cases represents 15–20% of the resting metabolic heat production of the squirrel monkey (Stitt and Hardy, 1971; Lynch, 1976). The remarkable similarity of these thresholds implies that the same configuration of deep-body thermosensitive sites may be stimulated by absorbed microwaves to provide the neural impetus for response change. Experiments currently in progress in our laboratory have determined that the rise in local temperature within the PO/AH thermoregulatory center at a threshold power density of 6–8 mW/cm^2 is ~0.3°C, probably sufficient to trigger the changes in thermoregulatory effector responses we have observed.

3. BEHAVIORAL THERMOREGULATION

Voluntary behavioral action is an organism's first defense against exogenous thermal challenge. Endotherms and ectotherms (organisms lacking autonomic thermoregulatory capabilities) both use behavioral strategies whenever possible to counteract inhospitable alterations in the exchange of thermal energy between their bodies and the environment. Ordinarily, temperature sensors in the skin are activated when the ambient temperature changes, and information is relayed to the thermoregulatory centers where an error signal is generated that initiates a behavioral thermoregulatory response. Often characterized as anticipatory because it prevents perturbations of the deep-body temperature, thermoregulatory behavior is mobilized rapidly and in general has a higher gain than the autonomic effector responses that could be mobilized to do the same job.

3.1. Natural (Instinctive) Thermoregulatory Behaviors

All species exhibit instinctive behaviors that facilitate heat gain or heat loss, and the squirrel monkey is no exception. In cold environments they huddle with head tucked in and tail curled over the shoulders, a posture that minimizes the surface area exposed to the air. Piloerection increases the insulative value of the fur and often frank shivering is evident if the ambient temperature falls below 20°C. Huddling singly or in groups appears to be preferred to a substantial increase in activity (Hopf, 1978); the latter would have the potential to greatly increase heat loss as well as heat production. Over the long term, increased food intake will regulate body energy stores as greater amounts of energy are utilized for metabolic heat production. However, the most important behavior involves the seeking of external sources of heat (Jarosz and Dukelow, 1976).

This behavior is strongly motivated and lends itself easily to simplification in the laboratory setting, as we shall see.

In warm environments, squirrel monkeys spread their bodies and sprawl on shelves and perches or drape themselves over tree limbs with tail dangling in the air. These postures maximize the surface area exposed to the air and increase convective heat loss. They rub their faces and bodies against cool surfaces to increase heat loss by conduction. Water intake increases both to facilitate sweat secretion and to cool the body core if the water itself is cool. Urinating on the hands and feet and spreading urine over the body surface ("foot wash" or "hand wash") can aid evaporative cooling (Schmidt and Seitz, 1967). However, it is clear from Figs. 1 and 2 that these animals are extremely vulnerable to environmental temperatures above 38°C because of their limited capabilities to lose body heat. Once again, the most important behaviors involve seeking shade or a cooler environment, and the strong underlying motivation to these behaviors forms the basis for the learning of simple responses in the laboratory setting that can produce precise regulation of the body temperature.

3.2. *Operant Control of the Thermal Environment*

In the late 1950s, the operant conditioning paradigm was first applied to the measurement of thermoregulatory behavior; a lever was made available to an animal in a cold environment and each lever press was rewarded by a timed burst of radiant heat from an infrared lamp above the animal's cage. Sometimes the environment was warm and the reward was a puff of cool air or a few seconds' surcease from ongoing radiant heat. By this technique, heat- or cold-seeking behavior could be quantified in terms of response frequency, total number or magnitude of rewards (reinforcements), ambient temperature selected, and body temperatures achieved.

Although rats were the subjects of the earliest experiments, Carlisle (1966) soon tested the thermoregulatory behavior of several squirrel monkeys as well, and evaluated how heat intake varied with the conditions of reinforcement and, in one animal, how well the internal temperature was regulated during ongoing behavioral thermoregulation. He found that when the monkeys pressed a lever for infrared heat in a cold environment, they obtained fewer reinforcements as the ambient temperature was increased ($-5 \leq T_a \leq 10$°C) or as the duration or intensity of the reinforcement was increased. The response rate was adjusted so that a constant heat intake was maintained. Under these conditions, the temperature of the PO/AH varied only over a 1°C range. A subsequent study (Carlisle, 1969) demonstrated the ability of squirrel monkeys to thermoregulate behaviorally under conditions of a fixed ratio of responses to thermal reinforcements when both duration and intensity of the reinforcement were also varied. Already in these first studies, the superiority of monkeys over rats in the performance of this behavioral task was apparent. However, the fact that the monkeys were unrestrained in the test situation prohibited Carlisle from monitoring body temperatures in his animals routinely. From such measurements, he could have begun to understand the physiological mechanisms that initiate and control thermoregulatory behavior.

A somewhat different approach was adopted by Adair *et al.* (1970). Squirrel monkeys were restrained (Adair, 1969a) in a small test chamber that was heated and cooled by forced convection. A valve system allowed air from one of two thermostatically controlled sources to circulate through the chamber. The monkey learned to pull a chain to operate the valves, thereby selecting between two preset air temperatures, 10 and 50°C. Small differences in the air source temperatures offered were found to alter response frequency but not the final air temperature selected (Adair, 1970). Well-trained monkeys responded at a rate that produced a time-averaged air temperature of 35–36°C inside the test chamber. Since the animal was restrained, many body temperatures could be measured during the experiments. Rectal temperature, four representative skin temperatures [cf. Eq. (6)], and the air temperature selected by the animal were monitored routinely at 1-min intervals. In addition, the local temperature of the PO/AH or other brain structures was controlled and measured using a standard thermode–reentrant tube preparation (cf. Section 2.4.1). Figure 7 shows an animal in the behavioral test chamber during an experiment to probe the effects of PO/AH heating and cooling on normal thermoregulatory behavior.

3.3. Probing Changes in Thermoregulatory Behavior with Thermodes

Using the methods described above, Adair *et al.* (1970) altered the PO/AH temperature of squirrel monkeys within the range 35–43°C during ongoing thermoregulatory behavior. When the temperature displacements were relatively brief, the monkeys worked to change the temperature of the environment in the opposite direction by an amount proportional to the PO/AH temperature displacement. Typical results from this kind of experiment are depicted in the right half of Fig. 8. The figure shows the rectal temperature, mean skin temperature, and air temperature selected when the temperature of the hypothalamus (PO/AH) was raised and lowered every 10 min by different amounts relative to the neutral level. When the hypothalamic temperature was raised, the animal spent a high percentage of time in cool air, and skin and rectal temperatures fell. Conversely, when the hypothalamic temperature was lowered, the animal chose the warm air a high percentage of the time, and skin and rectal temperatures rose. The figure also shows that progressively smaller displacements of hypothalamic temperature around the neutral level produced body and air temperatures in the appropriate direction but of progressively smaller magnitude. Analysis shows that in this kind of experiment the air temperature chosen is a linear function of hypothalamic temperature and that a change of local PO/AH temperature as small as 0.2–0.3°C will result in a reliable change in thermoregulatory behavior.

The left half of Fig. 8 demonstrates that when a thermode implanted in the midbrain reticular formation was heated and cooled in a similar manner, holding the hypothalamic temperature as constant as possible, no correlated changes occurred in the animal's body temperature or in the air temperature selected (Adair and Stitt, 1971). Thus, although thermosensitive neurons have been identified in the midbrain, they appear to have no direct involvement with thermoregulatory behavior. Other brainstem sites that yielded similar negative

Fig. 7. View of restrained monkey in a behavioral test chamber during experiment to probe effects of hypothalamic warming and cooling on thermoregulatory behavior. The circulator above the animal's head is connected to a thermode implanted in the brainstem thermoregulatory center. The animal controls the ambient temperature by pulling the chain.

results were the lateral hypothalamus and zona inserta (Laudenslager, 1976) and the globus pallidus (E. R. Adair, unpublished results).

Thermal stimulation of the posterior hypothalamus, on the other hand, was found, in two of four monkeys tested, to alter thermoregulatory behavior nearly as effectively as comparable thermal stimulation of the PO/AH (Adair, 1974).

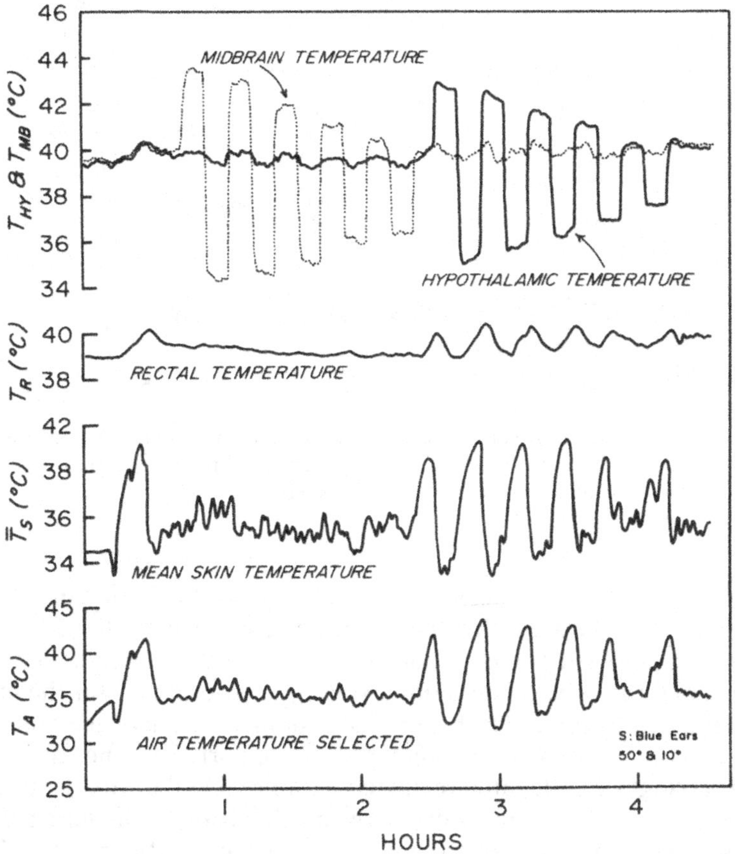

Fig. 8. Changes in rectal temperature, mean skin temperature, and the air temperature selected by a behaving animal when either midbrain or hypothalamic temperature is displaced at 10-min intervals relative to neutral level. [Reproduced from Adair and Stitt (1971), with permission of the publisher.]

When the temperatures of both brainstem areas were simultaneously displaced, the resulting behaviorally produced changes in body temperature in the steady state reflected additivity of action. However, the posterior hypothalamus may exert control over behavioral thermoregulation alone; no change in autonomic responses of heat production or heat loss could be detected when this area alone was thermally stimulated. With regard to this study, it must be pointed out that these findings have yet to be confirmed in any species by any other investigator.

Thermosensitive neurons have been located in the gut of several species, and these cells may act to initiate heat loss responses when the intra-abdominal temperature rises. By means of a flexible water-perfused thermode inserted in the colon, we heated and cooled the temperature of the abdominal cavity of the squirrel monkey during ongoing behavioral thermoregulation (Adair, 1971). Brief temperature changes of the deep viscera stimulated the monkeys to alter

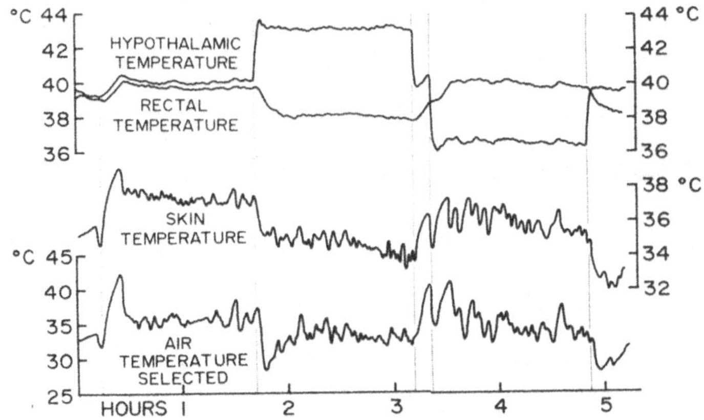

Fig. 9. Rectal temperature, mean skin temperature, and the air temperature selected by a behaving animal when the temperature of the medial preoptic/anterior hypothalamic area was heated or cooled for 90-min periods by means of implanted thermodes.

their skin (air) temperature in a manner comparable to stimulation of the PO/AH area, but only one-third as effectively. Prolonged warming (or cooling), superimposed on PO/AH cooling, modified behavioral thermoregulation by lowering (or raising) the skin temperature displacement necessary for the monkeys to achieve thermal balance. The behavioral response thus depended on whether the imposed colonic temperature opposed or supported the normal response to PO/AH cooling, i.e., elevated rectal temperature. On the other hand, PO/AH warming exerted such strong control over mechanisms for lowering the body temperature that superimposed colonic temperature displacement had no additional effect on the behavioral response. The best we can say is that the abdominal viscera of the squirrel monkey are slightly sensitive to temperature change, but that the significance of this sensitivity for normal thermoregulation in this species is unknown.

3.4. *Important Parameters of Central Thermal Stimulation*

3.4.1. *Duration of Temperature Change*

Nearly all of the experimental protocols described above involved brief (e.g., 10-min) warming or cooling of the PO/AH or other thermosensitive areas. If PO/AH thermal stimulation is prolonged (up to 2 hr), such that the animal attains a new steady state, a difference appears in the behavioral response to warming versus that to cooling (Adair *et al.*, 1970; Laudenslager, 1976). An example of such an experiment is illustrated in Fig. 9. The experiment contains three phases; (1) a 90-min baseline period for stabilization of thermoregulatory behavior, (2) a 90-min period of hypothalamic warming, and (3) following a brief interval, a 90-min period of hypothalamic cooling. Prolonged hypothalamic warming produces persistent changes in thermoregulatory behavior that are similar to those produced by shorter periods of warming, i.e., the warmer

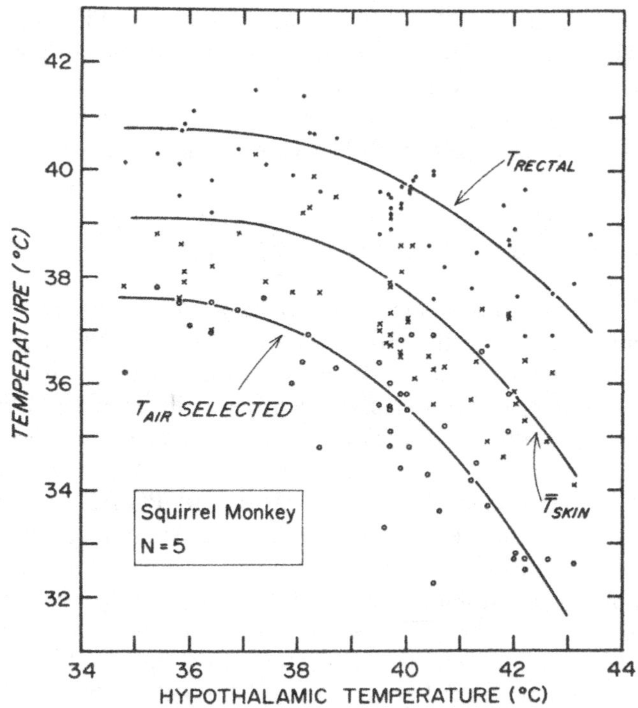

Fig. 10. Steady-state functions representing the rectal temperature (●), weighted mean skin temperature (×), and air temperature selected (○) when the medial preoptic/anterior hypothalamic temperature of five monkeys was displaced from neutral for 90-min periods. Each point is the mean temperature for the final 30 min of the period. [Reproduced from Adair (1971), with permission of the publisher.]

the hypothalamus, the cooler the environment selected. On the other hand, during prolonged hypothalamic cooling, the final stabilized behavior is little different from that under the baseline conditions. Figure 10 summarizes the thermal steady state achieved by five monkeys during experiments of this type. Each plotted point is a mean (for one monkey) of the final 30 min of the 90-min period during which the PO/AH was displaced. The variability is caused primarily by differences between animals, individual monkeys tending to reproduce their own responses. The curves are drawn by eye to outline the rectal temperature, the mean skin temperature, and the air temperature selected. Hypothalamic warming appears to be a far stronger stimulus than cooling for altering thermoregulatory behavior, at least when measured by the change in skin and deep-body temperatures required to oppose it. From the results of similar experiments, Laudenslager (1976) reached the same conclusion. He further deduced from behavioral and physiological data that the hypothalamic regulators of behavioral and autonomic thermoregulatory responses have similar operating characteristics.

During prolonged heating or cooling of the PO/AH, the local thermosensitivity of the area is retained. This important fact was demonstrated by superimposing small transient thermal displacements on a PO/AH temperature that was already clamped at a nonneutral level and measuring subsequent changes in thermoregulatory behavior. Whatever the clamped PO/AH temperature, changes in the selected ambient temperature were linear and inversely proportional to the magnitude of the PO/AH transient (Adair *et al.*, 1970; Adair, 1977).

3.4.2. Unilateral versus Bilateral PO/AH Stimulation

The advantages of implanting a bilateral multiple-thermode array in a specific brain locus, in terms of the efficient thermal control over that locus, have already been stated (Lynch *et al.*, 1980). Other experiments have probed the consequences of independent thermal control of the left and right PO/AH areas in both sheep and squirrel monkeys (Adair and Rawson, 1974). Behavioral thermoregulatory responses, together with skin and deep-body temperatures, were studied in two monkeys when either one or both PO/AH areas were heated and/or cooled. The response to unilateral heating or cooling was always augmented by simultaneous or successive stimulation of the contralateral area to the same temperature. Simultaneous heating of one side and cooling of the other could result in no change in thermoregulatory behavior if the two local PO/AH temperatures were chosen accurately. In similar fashion, a behavioral response initiated by unilateral heating (or cooling) could be abolished by subsequent contralateral cooling (or heating). A detailed analysis of the thermal conduction through the neural tissues indicated that the effect was not due to an overlap of thermal gradients from the bilateral thermodes at some intermediate thermosensitive site. These results imply that the vigor of a response to a change in PO/AH temperature depends on the volume of the neural substrate stimulated as well as on the local temperature imposed.

3.4.3. Rate of Change of Preoptic Temperature

During studies of the type we have been considering, the local temperature of the particular CNS site under study (e.g., PO/AH) was usually changed rapidly from one level to another. Such nonphysiological maneuvers may obscure subtleties in the operating characteristics of the site in question, specifically of the populations of temperature-sensitive elements that comprise part of the circuit to the central integrator for the control of body temperature. A study of changes in the thermoregulatory behavior of squirrel monkeys when the temperature of the PO/AH was increased very slowly (~2°C/hr) revealed that the ambient temperature selected by the animals often changed to a lower level in discrete steps (Adair and Rawson, 1973). When the PO/AH was warmed at higher rates, the phenomenon was not observed. A step change in the behavioral response occurred after each 0.8–1.0°C increase in the thermal stimulus, a result that might be expected if thresholds of discrete populations of thermosensitive afferent neurons had been exceeded. The occurrence of the step-response phenomenon evidently depends on a slow rate of change in one neural input to the CNS; thus, the basic step-function response could be masked in more typical stimulus–response relationships because several neural inputs vary concomitantly.

3.5. Interaction between Behavioral and Autonomic Thermoregulatory Responses

A fundamental question about thermoregulation that has been posed by many investigators may be stated as follows, "Since the purpose of both behavioral and autonomic thermoregulatory mechanisms is the same, i.e., the regulation of the body temperature, should not the elements of the central nervous system that control behavior logically be those that control the autonomic mechanisms?" Certain attempts to answer this question have involved the prediction of one type of mechanism from measurements of the other or, alternatively, simultaneous measurements of both behavioral and autonomic responses in the same experiments.

Stitt *et al.* (1971b) compared the control characteristics of autonomic and behavioral thermoregulation on the basis of two common parameters, mean skin temperature $\bar{T}_{sk}$ and PO/AH temperature T_{po}, using a squirrel monkey as the animal subject. A model of the "controller" for autonomic regulation was constructed from measurements of metabolic heat production and evaporative heat loss through sweating as each response varied with changes in $\bar{T}_{sk}$ and T_{po}. Then the $\bar{T}_{sk}$ (a linear function of T_a) selected by the animal during behavioral thermoregulation was determined for several T_{po}. When the behavioral data were plotted on the model of the "controller," they fell on a narrow band that was located within the thermoneutral zone. The implication of this result, confirmed in many other studies, was that the monkey has a preferred region within the thermoneutral zone that affords maximal thermal comfort and minimal physiological stress, and that it will create this microclimate behaviorally whenever possible. The manner in which behavioral compensation is made when T_{po} changes is in agreement with the idea that the controller for both behavioral and autonomic thermoregulatory responses is the same.

We know that in endotherms, autonomic mechanisms of heat production and heat loss are triggered by many of the same thermal stimuli that initiate thermoregulatory behavior. However, when endotherms have full control over the thermal characteristics of their environment, they use their behavioral capabilities almost exclusively. Man is no exception in this regard. If behavioral control were to become more and more difficult to execute, given a constant set of thermal stimuli, the autonomic responses would always be available to compensate for behavioral inadequacy and should appear at some critical point in time. We tested the validity of this idea by making the behavioral thermoregulatory response increasingly difficult (adding weights to the response chain), while monitoring changes in metabolic heat production in addition to measures of body temperature (Adair and Wright, 1976).

Figure 11 shows a representative experimental session for one of the animals, the individual tracings representing brain and body temperatures, metabolic heat production, and the air temperature selected by the animal. When the force required to pull the chain was increased in steps from 2.94 to 6.86

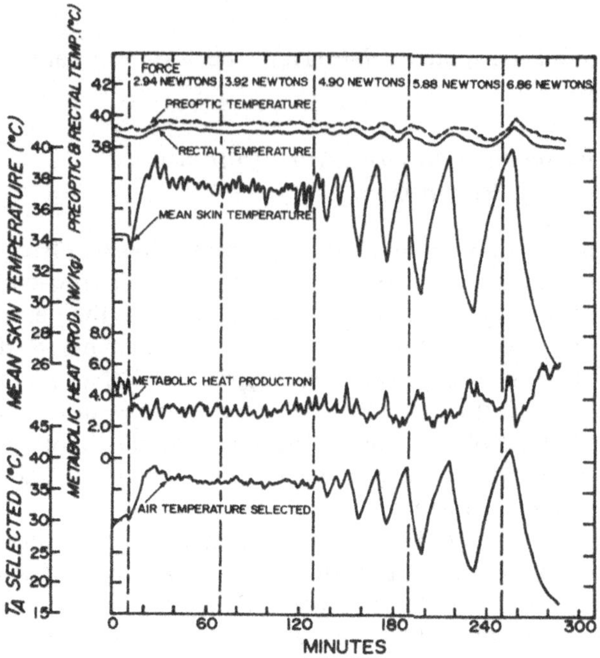

Fig. 11. Representative experiment showing changes in air temperature selected by one monkey (bottom tracing) when the force required to make a behavioral response was increased from 2.94 to 6.86 N. Other tracings depict medial preoptic/anterior hypothalamic, rectal, and mean skin temperatures and metabolic heat production. [Reproduced from Adair and Wright (1976), with permission of the publisher.]

N, the interresponse interval increased, resulting in wider air temperature swings within the test chamber. The average air temperature selected became progressively lower, producing a concomitant fall in skin temperature. However, internal body temperatures (rectal and PO/AH) remained nearly constant due to a compensatory increase in metabolic heat production. By allowing a cooler, rather than warmer, environment to result from a decreased response rate, the monkeys guarded against hyperthermia, which they are ill-equipped to handle autonomically. Thus, under conditions of restricted behavioral control over a thermal environment within which a given organism can survive by autonomic means, behavioral inadequacy is compensated by the mobilization of appropriate autonomic responses such that thermoregulation is achieved.

This experiment allowed us to observe autonomic thermoregulatory responses in a behaving animal by limiting the behavioral response, i.e., making it very difficult to perform. A second method was to set the environmental thermal stimuli at such levels that, even with maximal behavioral output, only limited control over the environment was available to the animal. Under these conditions also thermoregulation was accomplished by the mobilization of appropriate autonomic responses (Adair, 1976). The air sources supplying the behavioral test chamber were so set that a maximal response rate by the monkeys produced an average air temperature about 10°C below that normally preferred (35–36°C). A slight elevation of metabolic heat production accomplished thermoregulation. Preoptic warming and cooling under these conditions produced reductions and increases in rectal temperature, respectively, but these changes in body temperature were brought about by changes in metabolic heat production and vasomotor tonus, not by behavioral changes. The study demonstrated a fundamental fact of thermoregulation: even when the behavioral thermoregulation of an endotherm is inefficient, it is generally sustained with maximal vigor over long periods of time. This behavior is then supplemented by appropriate autonomic adjustments when necessary to achieve full and precise regulation of the body temperature.

ACKNOWLEDGMENTS. I wish to acknowledge the exciting and fruitful scientific interactions I have had over the past 15 years with the associates and colleagues whose work I have discussed in this chapter, particularly those who have joined me at the John B. Pierce Foundation in the probing of thermoregulatory mechanisms. I extend special thanks to James D. Hardy, who first suggested that I undertake study of behavioral thermoregulation in the squirrel monkey, and to Joseph C. Stevens for his continued support of my research effort. Lastly, I acknowledge with gratitude the technical expertise of Barbara W. Adams, whose understanding of squirrel monkeys clearly exceeds my own.

Research from my laboratory that is cited in this chapter was supported in part by USPHS Grants NS-11517 and ES-00354 and by Grant 77-3420 from the Air Force Office of Scientific Research.

REFERENCES

Adair, E. R., 1969a, An effective restraining chair for small primates, *Lab. Primate Newsl.* **8**:3–6.

Adair, E. R., 1969b, Method of protection of devices implanted in skulls of squirrel monkeys, *Lab. Primate Newsl.* **8**:22–23.

Adair, E. R., 1970, Control of thermoregulatory behavior by brief displacements of hypothalamic temperature, *Psychon. Sci.* **20**:11–13.

Adair, E. R., 1971, Displacements of rectal temperature modify behavioral thermoregulation, *Physiol. Behav.* **7**:21–26.

Adair, E. R., 1974, Hypothalamic control of thermoregulatory behavior: Preoptic–posterior hypothalamic interaction, in: *Recent Studies of Hypothalamic Function* (K. Lederis and K. E. Cooper, eds), Karger, Basel, pp. 341–358.

Adair, E. R., 1976, Autonomic thermoregulation in squirrel monkey when behavioral regulation is limited, *J. Appl. Physiol.* **40**:694–700.

Adair, E. R., 1977, Skin, preoptic, and core temperatures influence behavioral thermoregulation, *J. Appl. Physiol.: Respir. Environ. Exercise Physiol.* **42**:559–564.

Adair, E. R., 1981, Microwaves and thermoregulation, in: *Aeromedical Review, USAF Radiofrequency Radiation Bioeffects Research Program — A Review* (J. C. Mitchell, ed.), Brooks AFB, Texas, USAFSAM/AFSC.

Adair, E. R., and Adams, B. W., 1980a, Microwaves modify thermoregulatory behavior in squirrel monkey, *Bioelectromagnetics* **1**:1–20.

Adair, E. R., and Adams, B. W., 1980b, Microwaves induce peripheral vasodilation in squirrel monkey, *Science* **207**:1381–1383.

Adair, E. R., and Adams, B. W., 1982, Adjustments in metabolic heat production by squirrel monkeys exposed to microwaves, *J. Appl. Physiol.: Respir. Environ. Exercise Physiol.* **52**:1049–1058.

Adair, E. R., and Adams, B. W., 1983, Behavioral thermoregulation in the squirrel monkey: Adaptation processes during prolonged microwave exposure, *Behav. Neurosci.* **97**:49–61.

Adair, E. R., and Rawson, R. O., 1973, Step-wise changes in thermoregulatory responses to slowly changing thermal stimuli, *Pflügers Arch.* **339**:241–250.

Adair, E. R., and Rawson, R. O., 1974, Autonomic and behavioral temperature regulation: Unilateral *vs* bilateral preoptic thermal stimulation, *Pflügers Arch.* **352**:91–103.

Adair, E. R., and Stitt, J. T., 1971, Behavioral temperature regulation in the squirrel monkey: Effects of midbrain temperature displacements, *J. Physiol. (Paris)* **63**:191–194.

Adair, E. R., and Wright, B. A., 1976, Behavioral thermoregulation in the squirrel monkey when response effort is varied, *J. Comp. Physiol. Psychol.* **90**:179–184.

Adair, E. R., Casby, J. U., and Stolwijk, J. A. J., 1970. Behavioral temperature regulation in the squirrel monkey: Changes induced by shifts in hypothalamic temperature, *J. Comp. Physiol. Psychol.* **72**:17–27.

Baker, M. A., and Fuller, C. A., 1982, Differential regulation of brain and body temperature in the squirrel monkey, *Fed. Proc.* **41**:1695.

Bligh, J., and Johnson, K. G., 1973, Glossary of terms for thermal physiology, *J. Appl. Physiol.* **35**:941–961.

Byrd, L. D., 1980, Time-course effects of two barbiturates on cardiovascular activity and temperature in the squirrel monkey, *Life Sci.* **27**:935–942.

Byrd, L. D., 1982, Stimulant properties of dissociative anesthetics in the squirrel monkey, *Int. J. Primatol.* **3**:267.

Carlisle, H. J., 1966, Heat intake and hypothalamic temperature during behavioral temperature regulation, *J. Comp. Physiol. Psychol.* **61**:388–397.

Carlisle, H. J., 1969, Effect of fixed-ratio thermal reinforcement on thermoregulatory behavior, *Physiol. Behav.* **4**:23–28.

Chaffee, R. R. J., Allen, J. R., Brewer, M., Horvath, S. M., Mason, C., and Smith, R. E., 1966a, Cellular physiology of cold- and heat-exposed squirrel monkeys *(Saimiri sciurea)*, *J. Appl. Physiol.* **21**:151–157.

Chaffee, R. R. J., Horvath, S. M., Smith, R. E., and Welsh, R. S., 1966b, Cellular biochemistry and organ mass of cold- and heat-acclimated monkeys, *Fed. Proc.* **25**:1177–1184.

Chai, C. Y., and Lin, M. T., 1972, Effects of heating and cooling the spinal cord and medulla oblongata on thermoregulation in monkeys, *J. Physiol. (Lond.)* **225**:297–308.

Clark, S. M., and Lipton, J. M., 1981a, Effects of diazepam on body temperature of the aged squirrel monkey, *Brain Res. Bull.* **7**:5–9.

Clark, S. M., and Lipton, J. M., 1981b, Hypothermia produced in aged squirrel monkeys by central administration of Taurine, *Environ. Aging Res.* **7**:17–24.

Clark, S. M., Gean, J. T., and Lipton, J. M., 1980, Reduced febrile responses to peripheral and central administration of pyrogen in aged squirrel monkeys, *Neurobiol. Aging* **1**:175–180.

Crawshaw, L. I., and Stitt, J. T., 1975, Behavioural and autonomic induction of prostaglandin E_1 fever in squirrel monkeys, *J. Physiol. (Lond.)* **244**:197–206.

DuMond, F. V., 1968, The squirrel monkey in a seminatural environment, in: *The Squirrel Monkey* (L. A. Rosenblum and R. W. Cooper, eds.), Academic Press, New York, pp. 87–145.

Elizondo, R., 1977, Temperature regulation in primates, in: *International Review of Physiology, Environmental Physiology II*, Volume 15 (D. Robertshaw, ed.), University Park Press, Baltimore, pp. 71–118.

Emmers, R., and Akert, K., 1963, *A Stereotaxic Atlas of the Brain of the Squirrel Monkey (Saimiri sciureus)*, University of Wisconsin Press, Madison.

Fuller, C. A., 1982, Coupling between circadian temperature and feeding rhythms in the squirrel monkey, *Fed. Proc.* **41**:1695.

Fuller, C. A., and Sulzman, F. M., 1982, Circadian control of body temperature in primates, in: *Vertebrate Circadian Systems* (J. Aschoff, S. Daan, and G. Groos, eds.), Springer-Verlag, Berlin, pp. 224–236.

Fuller, C. A., Sulzman, F. M., and Moore-Ede, M. C., 1979, Circadian control of thermoregulation in the squirrel monkey *(Saimiri sciureus)*, *Am. J. Physiol.: Regul. Integr. Comp. Physiol.* **236**:R153–R161.

Fuller, C. A., Sulzman, F. M., and Moore-Ede, M. C. 1980, Circadian body temperature rhythms of primates in warm and cool environments, *Fed. Proc.* **39**:989.

Fuller, C. A., Lydic, R., Sulzman, F. M., Albers, H. E., Tepper, B., and Moore-Ede, M. C., 1981a, Circadian rhythm of body temperature persists after suprachiasmatic lesions in the squirrel monkey, *Am. J. Physiol.: Regul. Integr. Comp. Physiol.* **241**:R385–R391.

Fuller, C. A., Lydic, R., Sulzman, F. M., Albers, H. E., Tepper, B., and Moore-Ede, M. C., 1981b, Shift-work and the jet-lag syndrome: Conflicts between environmental and body time, in: *The 24-Hour Workday. A Symposium on Variations in Work–Sleep Schedules* (L. C. Johnson, D. I. Tepas, W. P. Colquhoun, and M. J. Colligan, eds.), NIOSH, Washington, D. C. (9DHHS No. 81-127).

Hammel, H. T., 1968, Regulation of internal body temperature, *Annu. Rev. Physiol.* **30**:641–710.

Hickey, T. E., and Kelly, W. A., 1972, Heatstroke in a colony of squirrel monkeys, *J. Am. Vet. Med. Assoc.* **161**:700.

Hopf, S., 1978, Huddling subgroups in captive squirrel monkeys and their changes in relation to ontogeny, *Biol. Behav.* **3**:147–162.

Jarosz, S. J., and Dukelow, W. R., 1976, Temperate season outdoor housing of *Saimiri sciureus* in the northern United States, *J. Med. Primatol.* **5**:176–185.

Kleiber, M., 1961, *The Fire of Life. An Introduction to Animal Energetics,* New York, Wiley.

Laudenslager, M. L., 1976, Proportional hypothalamic control of behavioral thermoregulation in the squirrel monkey, *Physiol. Behav.* **17**:383–390.

LeMaho, Y., Goffart, M., Rochas, A., Felbabel, H., and Chatonnet, J., 1981, Thermoregulation in the only nocturnal simian: The night monkey *(Aotus trivirgatus), Am. J. Physiol.: Regul. Integr. Comp. Physiol.* **240**:R156–R165.

Lipton, J. M., and Fossler, D. E., 1974, Fever produced in the squirrel monkey by intravenous and intracerebral endotoxin, *Am. J. Physiol.* **226**:1022–1027.

Lipton, J. M., and Kennedy, J. I., 1979, Central thermosensitivity during fever produced by intra-PO/AH and intravenous injections of pyrogen, *Brain Res. Bull.* **4**:23–34.

Lipton, J. M., and Trzcinka, G. P., 1976, Persistence of febrile response to pyrogens after PO/AH lesions in squirrel monkeys, *Am. J. Physiol.* **231**:1638–1648.

Lipton, J. M., Dinarello, C. A., and Kennedy, J. I., 1979, Fever produced in the squirrel monkey by human leucocytic pyrogen (40463), *Proc. Soc. Exp. Biol. Med.* **160**:426–428.

Liu, J. C., 1979, Tonic inhibition of thermoregulation in the decerebrate monkey *(Saimiri sciureus), Exp. Neurol.* **64**:632–648.

Lynch, W. C., 1976, Autonomic responses to changes in preoptic and ambient temperature in *Saimiri sciureus, Physiologist* **19**:279.

Lynch, W. C., and Adair, E. R., 1978, Ambient and hypothalamic temperatures alter vasomotor thresholds in squirrel monkey *(Saimiri sciureus),* in: *New Trends in Thermal Physiology* (Y. Houdas and J. D. Guieu, eds), Masson, Paris, pp. 130–132.

Lynch, W. C., Adair, E. R., and Adams, B. W., 1980, Vasomotor thresholds in the squirrel monkey: Effects of central and peripheral temperature, *J. Appl. Physiol.: Respir. Environ. Exercise Physiol.* **48**:89–96.

Machida, H., Perkins, E., and Hu, F., 1967, The skin of primates XXXV. The skin of the squirrel monkey *(Saimiri sciurea), Am. J. Phys. Anthropol.* **26**:45–54.

Malinow, M. R., and Wagner, R., 1966, Oxygen uptake in squirrel monkeys *(Saimiri sciurea), Lab. Anim. Care* **16**:105–108.

Nadel, E. R., and Stitt, J. T., 1970, Control of sweating in the squirrel monkey, *Physiologist* **13**:267.

Perkins, E. M., 1975, Phylogenetic significance of the skin of New World monkeys (Order Primates, Infraorder Platyrrhini), *Am. J. Phys. Anthropol.* **42**:395–424.

Richter, C. P., 1967, Sleep and activity: Their relation to the 24-hour clock, *Proc. Assoc. Res. Nerv. Ment. Dis.* **45**:8–27.

Schmidt, U., and Seitz, E., 1967, Waschen mit Harn zum Zweck der Thermoregulation bei Totenkopfaffen *(Saimiri sciureus), Anthropol. Anz.* **30**:162–165.

Smoake, J. A., Mulvey, P. F. Jr., Gerben, M., and Jones, L. G., 1977, Oxygen consumption and thyroid function in the squirrel monkey *(Saimiri sciureus), Lab. Anim. Sci.* **27**:655–659.

Stitt, J. T., and Hardy, J. D., 1971, Thermoregulation in the squirrel monkey *(Saimiri sciureus), J. Appl. Physiol.* **31**:48–54.

Stitt, J. T., Hardy, J. D., and Nadel, E. R., 1971a, The surface area of the squirrel monkey in relation to body weight, *J. Appl. Physiol.* **31**:140–141.

Stitt, J. T., Adair, E. R., Nadel, E. R., and Stolwijk, J. A. J., 1971b, The relation between behavior and physiology in the thermoregulatory response of the squirrel monkey, *J. Physiol. (Paris)* **63**:424–427.

Sulzman, F. M., Fuller, C. A., and Moore-Ede, M. C., 1977, Spontaneous internal desynchronization of circadian rhythms in the squirrel monkey, *Comp. Biochem. Physiol.* **58A**:63–67.

Sulzman, F. M., Fuller, C. A., and Moore-Ede, M. C., 1979, Tonic effects of light on the circadian system of the squirrel monkey, *J. Comp. Physiol.* **129**:43–50.

Thorington, R. W., Jr., 1968, Observations of squirrel monkeys in a Colombian forest, in: *The Squirrel Monkey* (L. A. Rosenblum and R. W. Cooper, eds.), Academic Press, New York, pp. 69–85.

Wasserstrum, N., and Herd, J. A., 1977, Elevation of arterial blood pressure in the squirrel monkey at 10°C, *Am. J. Physiol.: Heart Circ. Physiol.* **1**(5):H459–H463.

10

Sneezing Behavior in the Squirrel Monkey and Its Biological Significance

Gary G. Schwartz and Leonard A. Rosenblum

1. INTRODUCTION

When Ernst Mayr (1961) declared that there are "two biologies," a "How does it work?" and a "Why is it so?" biology, he stressed the division between two natural sciences: functional and evolutionary biology. Functional biologists study "proximate" causes—mostly physiological processes—and are not particularly concerned with the history or meaning of the genetic programs of the organisms that possess them. Evolutionary biologists, on the other hand, study "ultimate" causes—biological diversity and its meaning—and are much vexed by such questions as, "Why are there lions and tigers and things?" (Waddington, 1957). A familiar tale by the Brothers Grimm may illustrate the viewpoints of the two biologies. Little Red Riding Hood's observation, "Grandma, What big teeth you have!," exemplifies the view of a functional biologist; the Wolf's reply, "The better to eat you with," the point of an evolutionary biologist.

Uncovering the function of a structure is rarely as simple as the Wolf's pragmatism might suggest. Unlike the biologist who studies *how* something functions, the evolutionist must determine what that function *is*. And deciphering what something is "for" is a task steeped in philosophical difficulties, chief among them, the imprecision of everyday language:

> Just as a physicist might say that heating a gas causes it to expand, a biologist might say that heating a mammal causes it to sweat. But a biologist might also say that a mammal sweats in order to keep its temperature constant, while no physicist would

Gary G. Schwartz and Leonard A. Rosenblum • Primate Behavior Laboratory and Department of Psychiatry, Downstate Medical Center, Brooklyn, New York 11203.

> say that a gas expands in order to keep its temperature constant—even though that is exactly what happens (Hull, 1974, p. 102).

More serious than the problem of teleological language is the problem of underlying assumption: the goal of revealing something's function presupposes that it *has* one, and it is the salience of this assumption that distinguishes functional biology, where the focus is on mechanism, from disciplines as different as evolutionary biology and literary criticism, where the focus is on meaning (Mayr, 1961). Thus, in a commentary on Lewis Carroll's *The Hunting of the Snark*, Snarkophilus Snobbs (actually the American philosopher, F. C. S. Schiller), asks: "What is the meaning of the *Snark*? Or that we may not appear to beg the question let us first ask—how do we know that the *Snark* has a meaning?"

The answer, of course, is, "We don't." We assume that the *Snark* has a meaning as a necessary condition for discovering it. But there is no reason to assume that its meaning is obvious: credible interpretations of Carroll's ballad, once considered simply a "nonsense" poem, include satires on an expedition to the North Pole and on the quest of Hegelian philosophers for the Absolute (Gardner, 1981). Perhaps *The Hunting of the Snark* itself is an allegory of the search for meaning; that, in any event, is its borrowed significance here.*

The behavioral biology of the squirrel monkey is rich in meaning, or so we assume, and discovering that meaning depends upon posing "Why is it so?" questions. Here we probe one of the most conspicuous yet most overlooked of squirrel monkey behaviors: sneezing. Rather than a biologically trivial behavior, sneezing, we argue, may be a remarkable thermoregulatory one. (After all, who said that the meaning of sneezing would be obvious?)

2. *SNEEZING BEHAVIORS: A CLOSER LOOK*

> *Speak roughly to your little boy,*
> *And beat him when he sneezes:*
> *He only does it to annoy,*
> *Because he knows it teases.*
>
> ALICE'S ADVENTURES IN WONDERLAND

Sneezing behaviors have suffered the fate of "unglamorous familiarity" (Gould, 1975): frequently seen, but rarely studied. One reason for their neglect is semantic: sneezes in *Saimiri* look much like their namesakes in humans, and this identity in name has conferred an implied identity in function. But the functions of table legs and of tigers' legs are not the same (even though both may serve for support); their differences become evident on inspection. We turn, then, to a description of sneezing behaviors themselves.

* We have no wish to endorse the conventionalist view that the meaning of a thing may be determined by agreement. Our view—a realist one—is that the *Snark* has a definite meaning (i.e., the one Carroll intended), even though its explication has proven Snarkishly elusive. Luckily, probably few phenomena in nature are as bewildering as Carroll's deliberately cryptic verses.

Fig. 1. Nasal rub in a adult male squirrel monkey.

Sneeze. The sneeze consists of rapid side-to-side movements of the head coincident with an audible expulsion of air from the mouth and nares. Typically, the ears are drawn back close to the head, the mouth is closed, and the eyelids are shut as the head oscillates in a quick, lateral snap. Sneezes are observed primarily in monkeys that appear alert or agitated; they often punctuate the rapid pacing of an excited animal. The sneeze may occur without contact between the animal's nose and the substrate (aerosal sneeze) or with direct nasal contact against perch or wall (substrate sneeze).

Nasal Rub. In the nasal rub, the monkey presses its circumbuccal region in contact with the perch or other solid object and sweeps its nasal region from side to side. The nasal rub differs from the sneeze rub (below), in that the rubbing is not observed in immediate association with sneezing.

Sneeze Rub. The sneeze rub seems to be a rapidly coordinated version of the sneeze and nasal rub.

When the sneeze occurs directly on the substrate, the monkey typically crouches, elbows flexed, forearms approaching contact with the perch, hands directed outward (see Fig. 1). The neck appears outstretched and stiff. Several sneezes may occur in rapid succession, followed by a nasal rub. During a substrate sneeze, the monkey keeps its nasal region pressed firmly to the substrate, resulting in attenuation of the snaplike flexion of the head. Sneeze rubs may also begin as aerosol sneezes, with the monkey rubbing a neighboring surface immediately afterward.

Clonic Sneeze. The clonic sneeze denotes an extremely vigorous and protracted episode of rapid and repetitive sneezing. The latency between concatenated sneezes is extremely short, and the repeated sneezing may produce a characteristic whistling sound. During clonic sneezing the monkey's circumbuccal region usually remains pressed against the substrate. The clonic sneeze is often combined with a series of nasal rubs, and seems to be an exaggerated version of the sneeze rub.

Hand Sneeze. In the hand sneeze the monkey appears to elicit a sneeze manually by inserting one or two fingers into its external nares, tugging, and sneezing, frequently onto the outstretched palm. Following a hand sneeze, the animal may do any of the following: nasal rub, wipe the palm onto the substrate, or rub the tail with both hands. The hand sneeze to nasal rub sequence is depicted in Fig. 2. In the tail rub, the tail is held upright between the legs and grasped with both hands. The monkey rubs the tail in a vigorous hand-over-hand motion, starting from the middle and proceeding to the tip.

Herein we use "sneeze" and "sneezing" to refer to sneezes *per se* (i.e., exclusive of nasal rubbing) and "sneezing behaviors" to mean sneezing and/or nasal rubbing behaviors collectively.

3. ON FUNCTIONS AND EFFECTS

Previously we used the word "function" without troubling to define it, relying on the reader's sense of "purpose" or "goal." But the concept of purpose in biology—teleonomy—often is misunderstood, and so we must clarify what we mean by "function" and "adaptation."

"Adaptation," according to Williams (1966), refers to those cases in which the structure or behavior in question appears to be designed for a certain goal (its function) and has been shaped by natural selection for that purpose. Other fortuitous manifestations of that structure are "effects." Thus, if we say that the function of a hedgehog's spines is to repel predators, we do not merely mean that spines, do, in fact, repel predators. We mean that spines evolved *because* they repel predators, and that hedgehogs whose spines were more repulsive survived and left more descendants (Maynard Smith, 1982). The effects of a structure may be irrelevant to its function. Sundry things may stick to a hedgehog's spines—strawberries, say, and hedgehogs may eat them (Darwin, 1868). This is an effect; spines surely did not evolve for picking strawberries.

A structure may have more than one function. Typically, this results when some effect of one function is fortuitously suited to another (i.e., the structure is "preadapted"). Thus, in many mammals, secretions that function in thermoregulation also function in communication; the nasolacrimal secretions of rodents are a good example (see below) (Thiessen, 1977). In general, new struc-

Fig. 2. Hand sneeze to nasal rub sequence, approximately 1 sec between photographs.

tures do not evolve *de novo*, but, like the brass basin that became Don Quixote's helmet, old structures acquire new functions.*

4. SNEEZING BEHAVIORS: WHAT FUNCTION?

> *"If there's no meaning in it," said the King, "that saves a world of trouble, you know, as we needn't try to find any." "And yet I don't know," he went on, spreading out the verses on his knee . . . "I seem to see some meaning in them, after all."*
>
> THROUGH THE LOOKING GLASS

The sneezing behaviors of squirrel and other New World monkeys have generally been attributed to one of four principal causes: "display," "displacement," "infection," and "thermoregulation." We treat these alleged explanations as we would police suspects: first the lineup, then the interrogation.

Display. The display hypothesis proposes that sneezes are adaptations for precise visual and/or auditory communication (e.g., Moynihan, 1964, 1970). However, sneezes have nonsignaling uses as well (see below).

Displacement. The displacement hypothesis holds that sneezing behaviors serve as "comfort" patterns, and is based on the observation that sneezing behaviors often occur "out of context" or accompany agitated and presumably hostile behavior (Moynihan, 1976).

Infection. According to the infection hypothesis, sneezing behaviors are *prima facie* evidence of disease. These behaviors are reported to be especially conspicuous at low relative humidities (Levy and Artecona, 1964; Bantin, 1966).

Thermoregulation. This hypothesis proposes that in *Saimiri* and in other New World primates' sneezing behaviors function in thermoregulation. In short, sneezing behaviors are analogous as thermoregulatory responses to the face-washing behaviors of many rodents (Schwartz, 1981, 1985).

Display. According to the display hypothesis, sneezes have been pressed into service as social signals. This is the strategy of Douglas MacDougal:

> A bugler names Douglas MacDougal
>   Found ingenious ways to be frugal;
> He learned how to sneeze in various keys,
>   Thus saving the price of a bugle.

* Like many phenomena of biology, the concatenation of functions has parallels in economics, e.g., in product design (Ghiselin, 1978). Modern "antiperspirants" (the advertisers proclaim), "protect against wetness *and* odor." Or, in the present jargon, "antiperspirants" have acquired an antiodorant function to combat communicative effects acquired by thermoregulatory organs. Thus does commodity recapitulate phylogeny.

The idea is plausible: if human speech can evolve from grunts to grammar, sneezes may do likewise. Unfortunately, there is no evidence that sneezes have. Although sneezes presumably have been specialized as "social signals or releasers," Moynihan does not state what sneezes signal or release. Mostly they seem to "release" nothing at all—at least nothing readily observable: "Although sneezes would seem to be very noticeable (they certainly are conspicuous to human ears)," Moynihan wrote of *Callicebus*, "they seldom provoke overt reactions by other individuals of the same species" (1966, p. 95).

Displacement. One solution to this difficulty is to suggest that sneezes have nonsocial functions as well. Apart from "autochthonous" sneezes (sneezes to clear the nasal passages, for example, after feeding), Moynihan proposed that sneezing behaviors function as a sort of "displacement" activity. Thus his description of "muzzle-rubbing" in the cottontop tamarin (*Saguinus geoffroyi*):

> Like many Platyrrhini, these animals clean their faces, especially after eating messy food, by rubbing their "muzzles" along a branch. Rather surprisingly, individuals often perform identical or very similar movements . . . even when their faces are not dirty at the time. This is most likely to occur during obviously high-intensity disputes. (Moynihan, 1970, p. 20.)

Moynihan offered a similar explanation for the sneezes of night monkeys (*Aotus trivirgatus*), which were particularly evident during periods of physical activity and emotional arousal. Thus, "Both the wild and captive animals on Barro Colorado Island were heard to utter Sneeze-grunts," especially when "moving about very rapidly and energetically" (Moynihan, 1964, pp. 42–43). And, "Some of the captive Night Monkeys sneezed repeatedly during some high-intensity disputes . . . As the Night Monkeys that sneezed so frequently during disputes had not been feeding immediately beforehand, it is at least conceivable that some or all of their sneezing was a direct result or consequence, in one way or another, of their hostility" (p. 21).

The correlation of sneezing and activity was also noted for the golden lion tamarin (*Leontopithecus rosalia*) by McLanahan and Green (1977, p. 262): "The Snough sounds very similar to a sneeze. Its frequent occurrence at the end of feeding and locomotor bouts leads us to postulate that it may communicate completion of an activity."

Infection. The hypothesis that sneezing is a response to infection stems from a tacit analogy between sneezing in New World monkeys and sneezes as symptoms of disease in humans. The principal support for this hypothesis consists of reports of conspicuous sneezing behaviors associated with low relative humidities. As commonly interpreted, these reports suggest that respiratory infections may be more prevalent in drier air.

Despite its intuitive appeal, we suspect that much of the support for the infection hypothesis stems from semantic confusion. Let us examine the evidence.

The first report concerning the relationship between sneezing and humidity, in three marmoset species, is by Levy and Artecona (1964, p. 22): "While room temperature is not critical, we have maintained our marmoset rooms at about 80°F . . . On the other hand, humidity is critical . . . when humidity falls below 50%, upper respiratory infections are common." Later they add: "The animals are subject to few diseases. The marmoset develops typical cold symptoms, especially when the relative humidity falls below 50%" (p. 26).

We must distinguish observation and inference in the statement, "When humidity falls below 50%, upper respiratory infections are common." At low humidities, Levy and Artecona did not observe (or at any rate did not report) direct evidence of respiratory *infection*; they reported "typical cold *symptoms*," i.e., sneezes. Thus, it seems that Levy and Artecona observed sneezes, and by "the unconscious logic of common sense" (Huxley, 1929), they inferred infection was the cause.

The next report is by Bantin (1966). Referring to a squirrel monkey colony maintained at 75°F (±3°), he wrote: "Originally the relative humidity was 30–50% . . . and continual sneezing and sniffing was observed in the colony." However, when a humidifier was installed, "we were able to raise and maintain the relative humidity in excess of 55% and within a short time all signs of respiratory distress disappeared and have not recurred to date" (Bantin, 1966, p. 66). Bantin's remarks are important for two reasons: First, his observations of conspicuous sneezing at low humidities confirm the report of Levy and Artecona. Second, he seems to suggest that the changes in the frequency of sneezing were not immediate responses to the altered humidity ("sneezing was soon observed," and "within a short time" disappeared). Similarly, Deinhart *et al.* (1967) reported: "The sickness most commonly seen in the early days of maintaining marmosets in this laboratory was respiratory infection but this almost totally disappeared after the elevation of the humidity to 50% or greater" (p. 15). Unfortunately, whether these authors ascertained the existence of "respiratory infection" directly or simply inferred it from sneezing behaviors is not clear.

The potential problems of increased respiratory water loss induced by low humidities were considered in an influential paper by Thorington (1968). Although critical of an infectious etiology for sneezing, Thorington misconstrued Bantin (1966), claiming that the latter reported "troublesome respiratory *infections* at 75±3°F and 30–35% relative humidity" (Thorington, 1968, p. 83; cf. Bantin, 1966, p. 66; our emphasis). Like several authors before him, Thorington inadvertently substituted "infection" for "sneezing," thus conflating a supposed symptom with a supposed disease. With sneezes preconceived as symptoms, many failed to recognize that possibility that Moynihan did not; i.e., in New World monkeys, some sneezes may be behaviors *sui generis*.

If this analysis is correct, the facts of the infection/humidity affair appear to be these:

1. Sneezing behaviors seem to be more prevalent at low humidities; and these behaviors may be chronic responses to low vapor pressure.*

2. Apart from "autochthonous" sneezes [sneezes in response to nasal irritants (Moynihan, 1970)], there is no evidence that sneezes *per se* are symptoms of disease in New World monkeys.

* This statement rests on anecdotal evidence only. Long-term studies of sneezing behaviors under conditions of controlled humidities are required to ascertain the validity of this phenomenon.

Ironically, this semantic tale is complicated by a biological twist: if sneezing is a chronic response to low atmospheric moisture [due, perhaps, to irritation of the mucous membrane from prolonged desiccation; similar phenomena occur in humans (Andersen *et al.*, 1972; Thorington, 1968 p. 84)], then inflammation of the dry mucosa caused by frequent sneezing could well predispose it to infection. Thus, at low relative humidities, infection may not be the cause of sneezing, but the result!

Thermoregulation. The hypothesis that sneezing behaviors are thermoregulatory adaptations is based on an explicit analogy between sneezing in New World monkeys and autogrooming or "face washing" behaviors in many rodent taxa.

Consider the Mongolian gerbil (*Meriones unguiculatus*). In the face wash, the gerbil rubs its forepaws over its external nares, acquires nasal secretions from them, and rubs these vigorously over its face. Face washing in the gerbil occurs following increased ambient temperature, following increased activity, and in response to conditions of social stress and novelty. These conditions all produce an increase in core temperature. Not only does face washing reduce the gerbil's temperature, but chemicals contained in the nasal secretions serve in individual recognition and territorial marking [Thiessen (1977, 1983)]. Thus, in the gerbil, thermoregulatory secretions have acquired secondary communicative functions (Barnett, 1981).*

Sneezing behaviors in squirrel monkeys show many parallels to face washing in rodents (cf. Jolles *et al.*, 1979). Like the gerbil, squirrel monkeys (and many other platyrrhines; see above) sneeze during periods of activity and excitement. Moreover, sneezing behaviors display characteristics typical of olfactory marking (Eisenberg and Kleiman, 1972).

For example, employing the finding that secretions expelled from the nares in *Saimiri* fluoresce under ultraviolet light, Schwartz and Rosenblum (1980) exposed groups of socially living monkeys to novel objects. The quantity of fluorescent markings on these objects declined over a period of several days, but increased when monkeys were given visually identical objects previously exposed to conspecifics. Similarly, Hennessy *et al.* (1980) found that the frequency of sneezing was greatest during the breeding season, and was positively correlated with levels of circulating testosterone in males.

Allometry has also been invoked to bolster the thermoregulation hypothesis (Schwartz, 1985). Like most small mammals, squirrel monkeys are furred very densely, an adaptation to the problem of chronic heat loss. Yet in "opting" for insurance against chronic hypothermia, squirrel monkeys face acute hyperthermia during periods of activity or emotionality, when heat production may increase tenfold (Bartholomew, 1977). [Like most small primates, *Saimiri* sweats only minimally (Adair, this volume, Chapter 9).] In effect, squirrel monkeys are like Shakespeare's Katherina: "A little pot, and soon hot." The thermoregulation hypothesis holds that sneezes cool them.

* This phenomenon appears to be widespread among mammals. Hedgehogs (*Erinaceus europaeus*) (among others) annoint themselves with their own saliva, which may be used in thermoregulation and in social recognition (Brockie, 1976; cf. Eisenberg, 1981, p. 373).

5. THE HYPOTHESES ON TRIAL

Moynihan's twin hypotheses of "display" and "displacement" will not bear examination. Presumably, sneezing behaviors function either in communication or are displacement activities. But since Moynihan uses the term "displacement" to mean behaviors that occur "out of context" (Moynihan, 1955), this reduces to, "sneezes either function in communication, or they do not," which seems circular. Besides, Moynihan does not state in what context sneezes communicate in the first place.

The infection hypothesis fails less in accuracy than in adequacy: the alternate hypotheses all grant a (minor) role for sneezing in disease, but "disease" leaves the bulk of sneezing behaviors unexplained. Human sneezes may be signs of disease, as tremors may be of palsy. Yet as anyone who has ever quaked with fear, or, better still, with cold, knows, the relationship between such "signs" and the things they imply is imprecise and often misleading.

The thermoregulation hypothesis could harmonize the data underlying the competing hypotheses (the sights and sounds of sneezes are merely effects), but it faces a considerable difficulty: intuitively, it seems very unlikely. However implausible, this hypothesis boasts a major advantage over the competitors: it is readily falsifiable by experiment.

The thermoregulation hypothesis would be obviously falsified if the frequency of sneezing behaviors was not significantly elevated in the heat. Put simply, do increases in ambient temperature result in increased sneezing?

We subjected 16 socially living squirrel monkeys (Iquitos) to transient increases in their heat loads. A convection air source propelled either heated or unheated air into the monkeys' home pens at alternate 5-min intervals. The air temperature thus oscillated repeatedly, from approximately 23°C to nearly 33°C. Each trial lasted 30 min, and each monkey was tested twice. Air temperature in the pens was otherwise controlled at approximately 23°C, and the average relative humidity was 53%. [For experimental details, see Schwartz, (in press)].

The effect of the increases in air temperature on sneezing behaviors is shown in Fig. 3. The frequency of sneezing behaviors parallels the elevation in ambient temperature and was elevated markedly in the heat ($p < 0.005$, Friedman). Activity levels showed the reverse pattern and were significantly depressed in the heat.

Although these data are difficult to reconcile with the competing hypotheses, they are not, of themselves, unambiguous support for the thermoregulation hypothesis. The drawback stems from the use of *convection* heat: by raising the temperature of the air, convection heaters simultaneously reduce the relative humidity.*

* Relative humidity is the absolute vapor pressure of air (absolute humidity) divided by the saturation vapor pressure of air at that temperature. The vapor holding capacity of air increases with temperature; heating the air thus depresses the relative humidity (assuming that no additional moisture is introduced) (Mount, 1979).

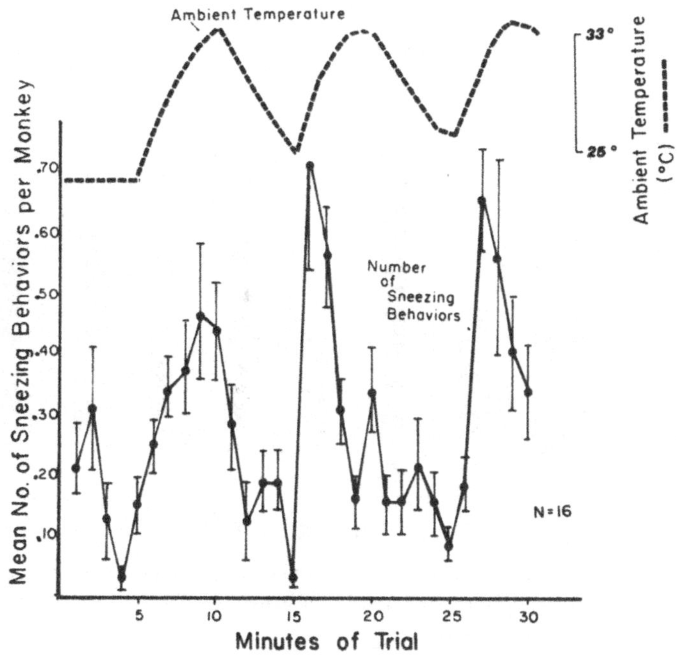

Fig. 3. Mean number of sneezing behaviors in 16 squirrel monkeys as a function of air temperature during 30-min trials. Bars are standard errors. [From Schwartz (1985).]

This difficulty is precluded with radiant heat. (Radiant heaters warm objects by direct electromagnetic radiation. Since they do not warm the air, radiant heaters leave the relative humidity unchanged.) The question we now address is, is radiant heat, uncontaminated by changes in humidity, sufficient to elicit increased sneezing behaviors?

We subjected eight squirrel monkeys to 20-min exposures to radiant heat. Heat was provided by three quartz heaters (Klinatron, 1400 W) mounted in the ceiling of their pen. The heaters were sufficient to raise surface temperatures at the pen's uppermost perch to approximately 32°C. Each monkey received 24 trials, 12 with the heaters operated and 12 without (controls). The mean relative humidity was 17.8 ± 4.5% immediately prior to testing and 17.7 ± 4.0% immediately thereafter. [Measurements were made with an electronic humidigraph (Humicheck II). For experimental details, see Schwartz (1985).]

The mean number of sneezing behaviors per subject under radiant and control conditions is shown in Fig. 4. As in convection heat, the frequency of sneezing behaviors increased dramatically in radiant heat (ANOVA, $p > 0.005$). Moreover, these behaviors showed no obvious habituation: there were no significant trials effects, even after repeated blocks of trials [$p < 0.005$ (Schwartz, 1985)].

We conclude from these data that heat alone is sufficient to elicit sneezing behaviors in squirrel monkeys. Thus, with respect to acute sneezing, "it *is* the heat; it's *not* the humidity."

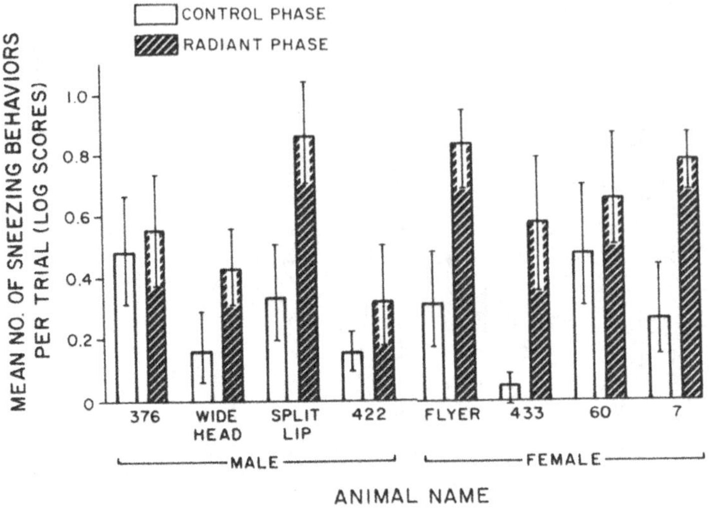

Fig. 4. Mean number of sneezing behaviors per monkey per 5 min in squirrel monkeys exposed to radiant heat. Bars are standard deviations. [From Schwartz (1985).]

These data are a far cry from demonstrating that sneezing behaviors are thermoregulatory. It remains to be shown that sneezing behaviors actually *regulate* temperature, not that *they* are regulated by *it*. Yet if we may introduce indirect evidence, it will be instructive to examine a behavior whose thermoregulatory significance is more immediately obvious: urine-washing.

The urine-wash is a highly stereotyped behavior in which a squirrel monkey spreads voided urine on its hands and ipsilateral feet. [Urine washing is exhibited by many ceboid and lorisoid primates (Robinson, 1980).] Schmidt and Seitz (1967) had suggested that the function of this behavior was thermoregulatory, but the idea has never been tested experimentally. Figure 5 depicts the frequency of urine-washing by these eight animals under the same conditions: control, and radiant heat. As Fig. 5 illustrates, urine-washing is clearly heat dependent, and shows a pattern strikingly similar to that observed for sneezing behaviors. Conversely, other spontaneous activities, e.g., locomotor behaviors, were significantly depressed by radiant heat.

With regard to sneezing, of course, this evidence is wholly circumstantial. Just as consorting with known criminals is no proof of criminality, the association of sneezing behaviors with suspected thermoregulatory ones does not sneezes thermoregulatory make. Yet the correlation between the two is suggestive. In fact, the major stumbling block now for the thermoregulation hypothesis pertains not to evolutionary, but to functional biology. In short, if sneezing behaviors are thermoregulatory, what is their *modus operandi*?

6. SNEEZING: FUNCTIONAL ASPECTS

There are finite ways in which sneezing behaviors could function in thermoregulation. Nasal secretions expelled from the nares and spread on the gla-

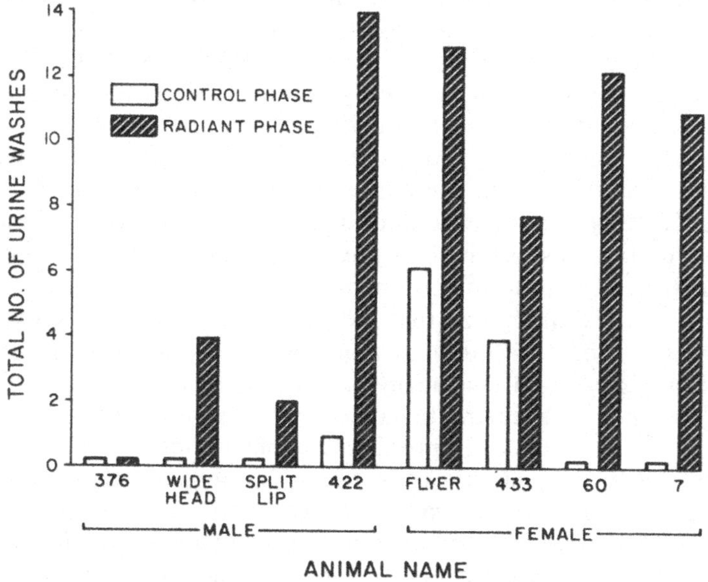

Fig. 5. Total number of urine washes in squirrel monkeys exposed to radiant heat. Totals are for 12 5-min trials in each condition.

brous thermolabile regions of the face could serve in evaporative cooling, as nasal and salivary secretions do in rodents. Yet there is at least one other intriguing possibility.

Recent research on many mammals indicates that the temperature of the brain is strongly influenced by the temperature of the nasal mucosa. This is because arterial blood en route to the brain first passes through the cavernous sinus, a cranial lake containing cool venous blood draining the nasal passages. Cooling of venous nasal blood results in cooling of brain-borne arterial blood [see Baker (1980, 1982) for reviews]. The ventral brain also loses heat to the cool cavernous sinus by direct conduction (Caputa *et al.*, 1976). This phenomenon—selective brain cooling—also occurs (in a more limited fashion) in humans.

In humans, selective brain cooling is achieved by a rerouting of cranial venous blood (veins in the face and brain have no values, so blood in them can flow in either direction). As Cabanac and Caputa (1979) have shown, the pattern of venous drainage of the brain and face depends upon a subject's thermal state. When subjects were cool, the rate of blood flow was low and directed from the brain to the face. However, when subjects were hyperthermic, the rate of blood flow was much greater and in the reverse direction: from the face, where the blood was cooled by evaporation, to the cavernous sinus. This hemodynamic detour enables humans to reduce cerebral temperature below that of the body core (Cabanac and Caputa, 1979).

But how is selective brain cooling relevant to sneezing? The answer is that a similar pattern of venous drainage occurs following a sneeze. During a sneeze, large increases in intrathoracic pressure literally "squeeze" blood out of the brain (Sharpey-Shafer, 1953). Blood effluent from the brain is forced into veins of the nose and face, where it could be cooled (Hamilton *et al.*, 1944). Since intrathoracic pressures remain elevated for many seconds following a sneeze, this facial blood would be prevented from draining into the thoracic veins directly. Instead, facial blood would take the path of least pressure, back to the cranial venous sinuses. Nasal rubbing could even act as venous "cuff" to expedite venous blood to the cavernous sinus, where it would cool the ventral brain by conduction. In short, in theory at least, squirrel monkeys could sneeze not to clear their heads, but to cool them.

In summary, we emphasize that the hypotheses sketched above are not truly competitors; they are different *types* of explanation (Mayr, 1961). The hypotheses of display and displacement are attempts to understand the immediate occurrence of sneezing behaviors by an analysis of immediate psychological events, i.e., they are proximate explanations. By comparison, the thermoregulation hypothesis has two senses. In the sense that it attempts to explain sneezing behaviors in terms of immediate (in this case, physiological) events, it, too, is a proximate explanation. Yet the thermoregulation hypothesis makes a broader, teleonomic claim: that squirrel monkeys sneeze *in order to* thermoregulate. It is in this sense an ultimate explanation (Mayr, 1961).

At different explanatory levels, these hypotheses all may be valid. Thus, in many mammals, psychogenic stimuli cause elevations in body temperature (Andrews, 1983; Baker *et al.*, 1976), which in turn trigger "displacement" grooming (Jolles *et al.*, 1979).

It is not our point to proselytize here on behalf of a pet hypothesis [We do that elsewhere (Schwartz, 1985)], but simply to emphasize that, like Carroll's "meaningless" verses, "meaningless" behaviors may be merely those whose significance we have yet to decipher.

7. EPILOGUE

> *"Tut, tut, child!," said the Duchess, "Everything's got a moral, if only you can find it."*
>
> ALICE'S ADVENTURES IN WONDERLAND

Of the many morals to be found in sneezes, we conclude with three.

The first maxim concerns the treacherousness of biological similarity. Sneezes in squirrel monkeys look much akin to sneezes in humans, yet (at least) some of the functions of monkey sneezes evidently are not. *Similarity in structure thus is no guarantee of similarity in function.* The case of eccrine glands is a good illustration. Eccrine glands are found throughout the skin of all primates thus far examined, yet only in humans and in patas monkeys, the most savannah-adapted primates, do these glands function (in any major way) in thermoregulatory sweating (Mahoney, 1980; Schwartz and Rosenblum, 1983).

Our second moral is a question, and for comparative psychologists, it is a particularly worrisome one. Behaviors in nonhuman primates are used widely as "models" (often in some unspecified sense) of behaviors in humans. But if the functions of sneezes differ among closely related taxa, and sneezes are reflexive behaviors, by what criteria do we recognize functional similarity among nonreflexive behaviors? Mere superficial similarity will not answer.

The final moral combines a couplet of encouragement and an item of advice.

Eberhard Curio (1973) neatly captured the inverse reasoning peculiar to research in teleonomy: "If one is to study adaptations, one must recognize them." This may sound a great deal like pulling oneself up by one's bootstraps, even for the most imaginative of scientists. But as Curio notes, there are easier ways to begin:

> As long as particular ecological factors have to be viewed only as potential selection pressures, their identification will always need much imagination and intuition. The search for selective forces is considerably facilitated if one investigates problems of adaptation for which it is compulsory for the organism to attain a solution. This is the case with climatic conditions. . . . (Curio, 1973, p. 1053).

Which is where we came in. But for the final word on the search for meaning, we turn to the advice of the Bellman, in *The Hunting of the Snark*:

> *For the Snark's a peculiar creature, that won't*
> *Be caught in a commonplace way.*
> *Do all that you know, and try all that you don't:*
> *Not a chance must be wasted to-day.*

ACKNOWLEDGMENTS. Preparation of this chapter was supported by Grant #MH15965 (USPHS) and by funds provided by the State University of New York.

REFERENCES

Andersen, I. B., Lindquist, G. R., and Proctor, D. F., 1972, Human nasal mucosal function under four controlled humidities, *Am. Rev. Respir. Dis.* **106**:438–449.

Andrews, R. V., 1983, Hyperthermic reactions of voles to novel and repetitive psychogenic stimuli, *Comp. Biochem. Physiol.* **75A**:589–592.

Baker, M. A., 1980, Anatomical and physiological adaptations of panting mammals to heat and exercise, in: *Environmental Physiology: Aging, Heat and Altitude* (S. Horvath and Y. Yousef, eds.), Elsevier/North-Holland, New York, pp. 121–146.

Baker, M. A., 1982, Brain cooling in endotherms in heat and exercise, *Ann. Rev. Physiol.* **44**:85–96.

Baker, M. A., Cronin, M. J., and Mountjoy, D. G., 1976, Variability of skin temperature in the waking monkey, *Am. J. Physiol.* **230**:449–455.

Bantin, G. C., 1966, Establishment of a squirrel monkey colony, *J. Inst. Anim. Tech.* **17**:66–73.

Barnett, S. A., 1981, *Modern Ethology: The Science of Animal Behavior*, Oxford University Press, New York.

Bartholomew, G. A., 1977, Body temperature and energy metabolism, in: *Animal Physiology: Principles and Adaptations* (M. S. Gordon, ed.), Macmillan, New York, pp. 364–449.

Brockie, R., 1976, Self-anointing by wild hedgehogs, *Erinaceus europaeus* in New Zealand, *Anim. Behav.* **24**:68–71.

Cabanac, M., and Caputa, M., 1979, Natural selective cooling of the human brain: Evidence of its occurrence and magnitude, *J. Physiol.* **286**:255–264.

Caputa, M., Kadziela, W., and Narebski, J., 1976, Significance of cranial circulation for the brain homeothermia in rabbits. II. The role of the cranial venous lakes in the defense against hyperthermia, *Acta Neurobiol. Exp.* **36**:625–638.

Curio, E., 1973, Towards a methodology of teleonomy, *Experientia* **29**:1045–1058.

Darwin, C., 1868, Hedgehogs, in: *Hardwicke's Science Gossip: An Illustrated Medium of Interchange and Gossip for Students and Lovers of Nature* p. 280. [Reprinted in P. H. Barett, ed., 1977, *The Collected Papers of Charles Darwin*, University of Chicago Press, Chicago, p. 137.]

Deinhart, J. B., Devine, J., Passovoy, M., Pohlman, R., and Deinhart, F., 1967, Marmosets as laboratory animals, *Lab. Anim. Care* **17**:11–29.

Eisenberg, J. F., 1981, *The Mammalian Radiations*, University of Chicago Press, Chicago.

Eisenberg, J. F., and Kleiman, D. G., 1972, Olfactory communication in mammals, *Annu. Rev. Ecol. Syst.* **3**:1–32.

Gardner, M., 1981, The annotated Snark, in: *Lewis Carroll's The Hunting of the Snark* (J. Tanis and J. Dooley, eds.), William Kaufman, Los Altos, California, pp. 13–80.

Ghiselin, M. T., 1978, The economy of the body, *Am. Economic Rev.* **68**:233–237.

Gould, S. J., 1975, Allometry in primates, with an emphasis on scaling and the evolution of the brain, in: *Contributions to Primatology*, Volume 5, pp. 244–292.

Hamilton, W. F., Woodbury, R. A., and Harper, H. T., Jr., 1944, Arterial, cerebrospinal and venous pressures in man during cough and strain, *Am. J. Physiol.* **141**:42–50.

Hennessy, M. B., Mendoza, S. P., Coe, C. L., Lowe, E. L., and Levine, S., 1980, Androgen-related behavior in the squirrel monkey: An issue that is nothing to sneeze at, *Behav. Neural Biol.* **30**:103–108.

Hull, D., 1974, *Philosophy of Biological Science*, Prentice-Hall, Englewood Cliffs, New Jersey.

Huxley, T. H., 1929, On the method of Zadig, in: *Essays of Thomas H. Huxley* (F. Barry, ed.), Macmillan, New York, pp. 160–180.

Jolles, J., Rompa-Barendregt, J., and Gispen, W. H., 1979, Novelty and grooming behavior in the rat, *Behav. Neural Biol.* **25**:563–572.

Levy, B. M., and Artecona, J., 1964, The marmoset as an experimental animal in biological research: Care and maintenance, *Lab. Anim. Care* **14**:20–27.

Mahoney, S. A., 1980, Cost of locomotion and heat balance during rest and running from 0 to 55°C in a patas monkey, *J. Appl. Physiol. Respir. Environ. Exercise Physiol.* **49**:789–800.

Maynard Smith, J., 1982, Overview—Unsolved evolutionary problems, in: *Genome Evolution* (G. A. Dover and R. B. Flavell, eds. Academic Press, New York, pp. 375–382.

Mayr, E., 1961, Cause and effect in biology, *Science* **134**:1501–1506. [Reprinted in E. Mayr, 1976, *Evolution and the Diversity of Life*, Belknap Press, Cambridge, Massachusetts, pp. 359–371.]

McLanahan, E. B., and Green, K. M., 1977, The vocal repertoire and analysis of the contexts of vocalizations in *Leontopithecus rosalia*, in: *The Biology and Conservation of the Callitrichidae* (D. G. Kleiman, ed.), Smithsonian Institution Press, Washington, D. C., pp. 251–269.

Mount, L. E., 1979, *Adaptation to Thermal Environment*, Edward Arnold, London.

Moynihan, M., 1955, Types of hostile display, *Auk* **72**:247–259.

Moynihan, M., 1964, Some behavior patterns of Platyrrhine monkeys I. The night monkey (*Aotus trivirgatus*), *Smithson. Misc. Coll.* **146**:1–84.

Moynihan, M., 1966, Communication in the titi monkey, *Callicebus*, *J. Zool.* (*Lond.*) **150**:77–127.

Moynihan, M., 1970, Some behavior patterns of Platyrrhine monkeys. II. *Saguinus geoffroyi* and some other tamarins, *Smithson. Contrib. Zool.* **28**:1–7.

Moynihan, M., 1976, *The New World Monkeys*, Princeton University Press, Princeton, New Jersey.

Robinson, J. G., 1980, Correlates of urine washing in the wedge-capped capuchin *Cebus nigrivittatus*, in: *Vertebrate Ecology in the Northern Neotropics*, (J. F. Eisenberg, ed.), Smithsonian Institution Press, Washington, D. C., pp. 137–143.

Schmidt, U., and Seitz, E., 1967, Washen mit Harn zum Zweck der Thermoregulation bei Totenkopfaffen (*Saimiri sciureus*), *Anthropol. Anz.* **30**:162–165.

Schwartz, G. G., 1981, Thermally-induced sneezing in the squirrel monkey: An hypothesis for the origins of platyrrhine sneezing, *Am. J. Primatol.* **3**:325–326 (Abstract).

Schwartz, G. G., 1985, Thermal Influences on the Behavior of Squirrel Monkeys (*Saimiri sciureus*): Does *Saimiri* Sneeze to Thermoregulate?, Ph.D. Dissertation, Downstate Medical Center of the State University of New York (unpublished).

Schwartz, G. G., and Rosenblum, L. A., 1980, Novelty, arousal, and nasal marking in the squirrel monkey, *Behav. Neural Biol.* **28:**116–122.

Schwartz, G. G., and Rosenblum, L. A., 1983, Allometric influences on primate mothers and infants, in: *Symbiosis in Parent–Offspring Interactions* (L. A. Rosenblum and H. Moltz, eds.), Plenum Press, New York, pp. 215–248.

Sharpey-Schafer, E. P., 1953, Effects of coughing on intrathoracic pressure, arterial pressure and peripheral blood flow, *J. Physiol.* **122:**351–357.

Thiessen, D. D., 1977, Thermoenergetics and the evolution of pheromone communication, in: *Progress in Psychobiology and Physiological Psychology*, Volume III (J. M. Sprague and A. N. Epstein, eds.), Academic Press, New York, pp. 91–191.

Thiessen, D. D., 1983, The thermoenergetics of communication and social interaction among Mongolian gerbils, in: *Symbiosis in Parent–Offspring Interactions* (L. A. Rosenblum and H. Moltz, eds.), Plenum Press, New York, pp. 113–144.

Thorington, R. W., Jr., 1968, Observations of squirrel monkeys in a Colombian forest, in: *The Squirrel Monkey*, (L. A. Rosenblum and R. W. Cooper, eds.), Academic Press, New York, pp. 69–85.

Waddington, C. H., 1957, *The Strategy of the Genes*, Allen & Unwin, London.

Williams, G. C., 1966, *Adaptation and Natural Selection: A Critique of Some Current Evolutionary Thought*, Princeton University Press, Princeton, New Jersey.

11

Visual System of the Squirrel Monkey

Gerald H. Jacobs

1. INTRODUCTION

A few minutes' casual observation of the arboreal acrobatics of a group of squirrel monkeys (*Saimiri sciureus*) will serve to convince one that, as for other primates, the visual system provides these animals an extraordinarily important access to the environment. This intuition is confirmed through formal studies, which show that (1) a large portion of the neocortex in the squirrel monkey receives some visual inputs, and (2) squirrel monkeys can make keen discriminations among many features of their visual environments. This chapter reviews the literature on vision and the visual system of the squirrel monkey. To accomplish this, I first provide a sketch of the organization of the visual system, and then examine several functional features of vision that have been subjected to experimental scrutiny. Particular attention is paid to those cases where vision can be related to the anatomy and physiology of the visual system.

2. ORGANIZATION OF THE VISUAL SYSTEM

Although much has been learned about the organization of the squirrel monkey visual system, perhaps more than about any other platyrrhine species, the picture remains incomplete. The following summarizes the current state of knowledge.

2.1. The Eye

2.1.1. Gross Features

Considering its external appearance in the intact animal, the eye of the squirrel monkey is surprisingly large, with an axial length of 14.5 mm (Cowey

Gerald H. Jacobs • Department of Psychology, University of California, Santa Barbara, Santa Barbara, California 93106.

and Ellis, 1967). For comparison, the equivalent figure for the macaque eye is 20.0 mm. A schematic eye has not yet been derived for the squirrel monkey, but Rolls and Cowey (1970) measured the radius of curvature of the cornea as 4.75 mm and computed the posterior nodal distance as 10.24 mm. Based on that measurement, the magnification factor of the eye is such that 1° of visual angle corresponds to 0.180 mm of retinal extent. The nominal refractive state of the squirrel monkey eye has not yet been specifically studied. There are reports that, as in many other species, retinoscopic refraction suggests the presence of a slight hyperopia (Cowey, 1964; Kadoya *et al.*, 1971a), but in refracting three squirrel monkeys, Merigan (1976) found two to be slightly hyperopic, while the third was slightly myopic. In the absence of more extensive measurements, it is probably prudent to suppose that squirrel monkeys are on average emmetropic, with the possibility for individual variations around this value similar to those demonstrated for other primate species.

Primate eyes usually contain two spectrally selective prereceptor filters: the lens and the macula. Spectrophotometric measurements of excised lenses done in my laboratory show lens absorbance in the squirrel monkey to be similar to that of other primates. Specifically, over the spectral range from 420 to 620 nm, the lens has an optical density that does not differ by more than 0.15 log unit from those values typically reported for human lenses (e.g., Wyszecki and Stiles, 1967). The presence of a macular pigment in the squirrel monkey retina has been noted (Cowey, 1964; Wolin and Massopust, 1967), but so far not measured.

2.1.2. Retina

The squirrel monkey retina is generally similar to those found in other diurnal primates. A striking feature of this retina is a large, clearly developed fovea. Indeed, the foveola (that portion of the fovea from which ganglion cells are completely absent) is larger in the squirrel monkey than in the macaque: it has a diameter of 550 μm versus 300 μm for the macaque monkey (Stone and Johnston, 1981). The outer nuclear layer of the central retina contains 4–5 rows of nuclei, while at the same location the inner nuclear layer typically has four rows of nuclei (Wolin and Massopust, 1970). Stone and Johnston (1981) compute that the total area of the squirrel monkey retina is 345 mm^2.

The cones of the squirrel monkey retina are typical in appearance and densely packed in the central retina. At that location, there are about 100,000 cones/mm^2 (Rolls and Cowey, 1970). Viewed in terms of angular separation, these same authors compute central intercone distances to be about 1.3 min of arc. The decline in cone packing toward the peripheral retina is substantial, as can be seen in Fig. 1. Rods are also abundantly present in this retina. They seem to have the same sort of heterogeneous retinal distribution as is found in other primate retinas, but so far no documentation of their number or spatial distribution is available.

The spectral absorbance characteristics of photopigments in the squirrel monkey retina have been recently measured by microspectrophotometry (MSP) (Jacobs *et al.*, 1981, 1982; Bowmaker *et al.*, 1983). These measurements reveal

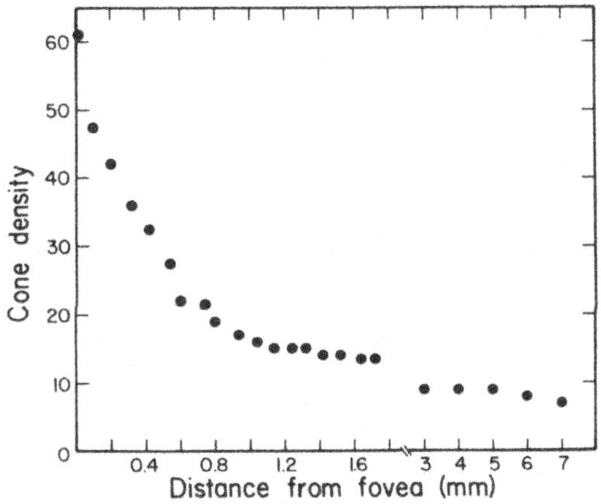

Fig. 1. Cone density in the squirrel monkey retina. Plotted are average values for three eyes in which the numbers of cones in areas of 600 μm^2 were counted along the vertical and horizontal meridians. [Data from Rolls and Cowey (1970).]

that the rods contain a typical rhodopsin having a peak absorbance λ_{max} at about 500 nm. This value corresponds closely to previous estimates derived from the electroretinogram (Jacobs *et al.*, 1963) and from behavioral tests (Jacobs, 1972). Whereas the rod pigment is like that of other primate rods, MSP measurements of the cone photopigments show a highly unusual arrangement. The essential finding is that individual squirrel monkeys have either two or three classes of cone photopigments. All individuals appear to have a small number of cones containing a pigment absorbing maximally in the short wavelengths ($\lambda_{max} \approx$ 433 nm). An additional three classes of cone pigments with absorbance peaks at about 535, 549, and 564 nm have also been found. Individual animals have either any one or two (in various combinations) of the latter three types. The substantial individual variability in cone pigment classes obviously has powerful implications for vision. That issue is considered in later sections dealing with studies of visual capacity in the squirrel monkey.

The sizes and spatial distributions of ganglion cells in the squirrel monkey retina have been charted in two investigations (Rolls and Cowey, 1970; Stone and Johnston, 1981). These features bear qualitative similarities to patterns detected in other primate retinas. Among the commonalities are the facts that the size range and the mean soma size of ganglion cells is smaller at the fovea than in the peripheral retina, and that ganglion cells are larger and less densely packed in the temporal retina than in the nasal retina. Figure 2 summarizes the results of ganglion cell counts made from retinal flat mounts (Stone and Johnston, 1981). So measured, cell density declines from peak values of greater than 6000/mm^2 near the fovea to less than 500/mm^2 in the peripheral retina. The isodensity lines shown in Fig. 2 are somewhat elongated in the horizontal plane, suggesting that the squirrel monkey retina has something resembling a visual streak. Doty *et al.* (1964) have estimated that the squirrel monkey retina contains a total of about one million ganglion cells. That number is surprisingly large; it approaches the figure usually given for the human, a species have a much larger optic nerve.

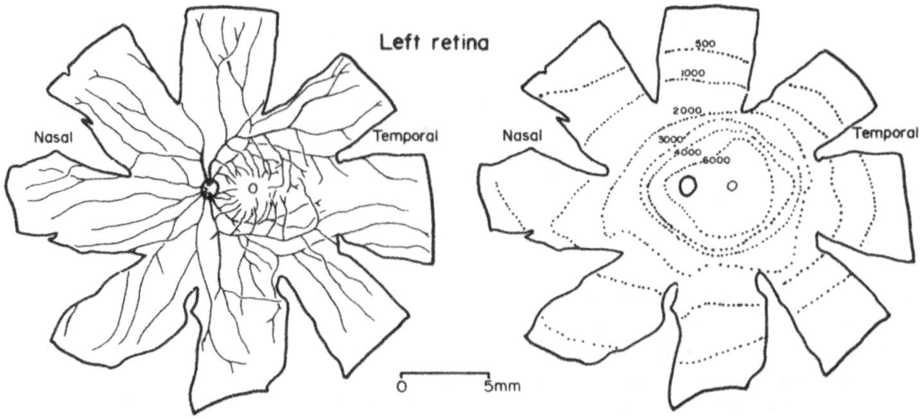

Fig. 2. Flat-mounted sections from the squirrel monkey retina. The drawing to the left shows the distribution of retinal blood vessels and the locations of the fovea and optic disc. To the right, ganglion cell densities (counts/mm^2) are shown on the same projection. [From Stone and Johnston (1981).]

2.1.3. *Optic Tract Projections*

The projections and terminations of optic tract fibers in the squirrel monkey have been studied using silver impregnation of degenerating fibers and autoradiographic methods. Based on degeneration patterns, it is estimated that 50–60% of all ocular fibers cross at the chiasm (Tigges and Tigges, 1969). The principal termination of these fibers is the lateral geniculate nucleus (LGN), a structure specifically considered in the next section. Fibers originating in both eyes emerge from the apex of the LGN to terminate in the pregeniculate nucleus (Campos-Ortega and Glees, 1967) Many optic tract fibers bypass these structures to join the brachium of the superior colliculus, and a majority of these enter the superior colliculus. This input is mostly from the contralateral eye and the principal termination is the stratum opticum layer of the superior colliculus. There is some additional input to the pretectum. A smaller number of fibers leave the brachium of the superior colliculus to bypass the medial geniculate nucleus and terminate in the nuclei of the accessory optic tract. This portion of the projection is mostly contralateral. Although examinations of degenerating fibers produced no evidence for a retinohypothalamic projection in the squirrel monkey, autoradiography does indicate the presence of such a pathway (Tigges and O'Steen, 1974). Its termination is the suprachiasmatic nucleus.

2.2. *Subcortical Visual Centers*

2.2.1. *Lateral Geniculate Nucleus*

The LGN, the principal relay nucleus in the mammalian visual system, is typically a laminated structure with its separate layers segregating inputs from the two eyes. The variation in the number of laminae from species to species

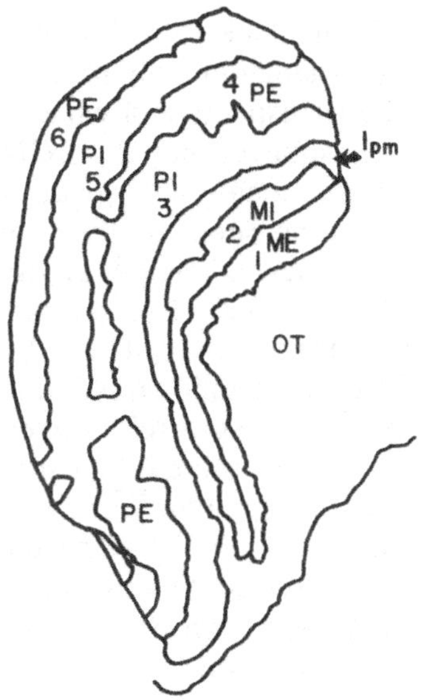

Fig. 3. Drawing made from a frontal section through the LGN of the squirrel monkey. The laminae are identified according to conventional terminology (layers numbered 1–6), and according to the suggestion of Kaas *et al.* (1978). In their scheme, layers 3–6 represent leaflets from two basic laminae, PE and PI. Abbreviations: OT, optic tract; ME, external magnocellullar layer; MI, internal magnocellular layer; Ipm, interlaminar zone; PI, internal parvocellular layer; PE, external parvocellular layer.

and the functional significance of the layering are enduring problems. The prototypical LGN pattern seem in the Old World primates has six layers—two ventral magnocellular layers and four dorsal parvocellular layers. By contrast, in normally stained material, the squirrel monkey LGN shows two magnocellular layers and a large undifferentiated parvocellular mass. Starting in the mid-1960s, a number of investigators examined transneuronal degeneration patterns to determine if this parvocellular mass might contain a "concealed" lamination (Doty *et al.*, 1966; Campos-Ortega and Glees, 1967; Jacobs, 1969; Tigges and Tigges, 1969; Wong-Riley, 1971). Each of these experimenters found degeneration patterns in the LGN suggesting the presence of six layers, four in the parvocellular mass, such that, numbered from the base dorsally, layers 1, 4, and 6 received inputs from the contralateral eye, while layers 2, 3, and 5 received inputs from the ipsilateral eye.

Despite this concordance of opinion, the lamination pattern observed was variable from location to location within the nucleus and, perhaps, from animal to animal. A more thoroughgoing analysis of LGN lamination has been provided by Kaas *et al.* (1978). They suggest that the basic primate pattern consists of two basal layers, two magnocellular layers, and two parvocellular layers, with the latter divided into variable numbers of leaflets. They show specifically that the four parvocellular laminae identified in the degeneration experiments done on squirrel monkeys indicate the presence of leaflets from the basic parvocellular laminae. That arrangement is illustrated in Fig. 3.

Although in some other species there are indications that the laminae of LGN contain some functional segregations of inputs in addition to the segregation of ocular origin, there is as yet no such evidence for the squirrel monkey. Doty *et al.* (1964) discovered that cells in the magnocellular portion of the LGN respond with shorter latencies than cells located in the parvocellular portion of the nucleus. A likely reason is that the more peripherally located (hence, larger) ganglion cells project more heavily to the magnocellular layers of the LGN. Irrespective of what other significance may be attributed to the layering of the squirrel monkey LGN, the various layers are in receptotopic register, so that a location in space is represented by an approximate column of cells running perpendicularly through the nucleus (Hollander, 1974).

In view of the popularity of the LGN as a site for conducting electrophysiological analyses of visual coding, there is surprisingly little such work on the squirrel monkey LGN. In addition to a series of investigations on the effects of mesencephalic stimulation on the transmission of information through the LGN of the squirrel monkey (Doty *et al.*, 1973; Wilson *et al.*, 1973; Bartlett *et al.*, 1973), there is an early study of the spectral response properties of LGN cells (Jacobs, 1964), an examination of the effects of adaptation on LGN responses (Jacobs, 1965), and a description of some receptive field properties of LGN cells (Jacobs and Yolton, 1970). Perhaps part of the reason for this relative lack of attention is that the squirrel monkey has usually been considered to provide a much less hardy preparation for electrophysiology than other common laboratory subjects—in particular, than the cat or the macaque monkey. Doty and his colleagues have documented some of the concerns to be faced in using squirrel monkeys in electrophysiological experiments. Anyone contemplating such a venture should consult the articles authored by this group listed in the reference section.

The total output from the squirrel monkey LGN terminates in the ipsilateral visual cortex. It was originally claimed that this LGN projection is to V-2 as well as to V-1 (Wong-Riley, 1976), but subsequent examination revealed that all geniculocortical fibers terminate in V-1; specifically in layers IV-A and -C and in layer III of that structure (Hendrickson *et al.*, 1978).* Recently, projections from the interlaminar zones of the LGN (one such zone is shown in Fig. 3) have been shown to terminate in layer I of striate cortex (Weber *et al.*, 1983). Nonretinal afferents to the squirrel monkey LGN appear similarly straightforward. There is a significant projection from the ipsilateral striate cortex to all layers of the IGN (Spatz *et al.*, 1970; Hollander, 1974; Ogren and Hendrickson, 1976; Wong-Riley, 1976). This projection originates in the deep layers of the cortex (layers V and VI). The fact that electrical stimulation delivered to a wide variety of sites in the mesencephalon can influence synaptic transmission through the LGN in both facilitatory and inhibitory directions indicates that there must also be inputs from this region to the LGN (Wilson *et al.*, 1973; Bartlett and Doty, 1974a). The pathway(s) involved in these effects remain for the most part obscure, but Weber *et al.* (1983) have found that there is an important input from the superficial layers of the superior colliculus to the interlaminar zones of the LGN.

* In the literature the identification of component areas in the visual cortex has been made on both cytoarchitectonic (Areas 17, 18, etc.) and functional (Visual Areas 1, 2, etc.) grounds. There is a general correspondence between these two classification systems, reasonably exact in some cases (Visual Area 1 = Area 17), but less exact for others, and often highly variable from species to species. For consistency, in this chapter the component regions of the squirrel monkey visual cortex are identified as Visual Area 1 (V-1), etc.

2.2.2. Superior Colliculus

It was noted that the squirrel monkey superior colliculus receives a robust input from the retina. As judged by measurements of conduction velocities, it appears that at least two separate classes of fibers are included in this input (Doty *et al.*, 1964). The receptotopic map of the superior colliculus is similar to those found in cat and macaque monkey. The horizontal meridian of this map runs anterior/posterior. High elevations are located on the medial side of the structure and low elevations on the lateral surface. On this same map, the vertical meridian runs from medial to lateral across the rostral portion of the superior colliculus (Kadoya *et al.*, 1971a). There is also a topographically organized input to the superior colliculus from the ipsilateral striate cortex (Kadoya *et al.*, 1971b; Spatz *et al.*, 1970).

Kadoya *et al.* (1971a,c) studied the response properties of single units in the superior colliculus and the pretectum of the squirrel monkey. Among an assortment of cell types, they found substantial numbers of units in the upper layers of this structure that responded vigorously to long-wavelength lights, but not at all to other monochromatic lights or to achromatic light. Units responding in this fashion are obvious candidates to have a role in the production of color vision. To date such units have not been seen at any subcortical locations in any other mammal, and, specifically, the presence of units in the macaque superior colliculus that might transmit information useful for color vision has been flatly denied (Marrocco and Li, 1977). If the observations of Kadoya and colleagues are correct, they indicate a functional uniqueness for the squirrel monkey superior colliculus that bears further investigation.

2.2.3. Pulvinar

In the squirrel monkey, as in other primates, the pulvinar (made up of lateral, medial, and inferior divisions) comprises a large portion of the posterior thalamus. There is a direct input from the superior colliculus to the pulvinar, particularly to its inferior portion (Mathers, 1971). There is also widespread corticofugal input to the pulvinar in the squirrel monkey. Fibers originating in layers V and VI of the striate cortex are retinotopically mapped onto the lateral and inferior portions of the pulvinar (Spatz *et al.*, 1970; Hollander, 1974; Ogren and Hendrickson, 1976). Mathers (1972) placed large lesions in the nonstriate visual cortex, in the posterior temporal cortex, and in the posterior parietal cortex. All of these areas were found to provide inputs to the pulvinar, with the regions more closely associated with visual cortex projecting specifically to the inferior portion of the pulvinar. In turn, the primate pulvinar is generally found to provide inputs to those portions of the cortex that have visuotopic representations, and to other classical visual association areas such as the inferotemporal cortex (Allman, 1977). However, with the exception of demonstrations that both the lateral and inferior portions of the pulvinar project to striate cortex (Ogren and Hendrickson, 1976) and that there is a pulvinar projection to V-2 (Livingstone and Hubel, 1982), these pathways appear to remain unspecified in the squirrel monkey.

Both the size of the structure and its rich innervation from other portions of the visual system suggest that the pulvinar must have an important role in the processing of visual information. What that role might be in the squirrel monkey, indeed in any of the primates, remains obscure. One study has been done in which responses were recorded from pulvinar cells in the squirrel monkey (Mathers and Rapisardi, 1973). A majority of the cells found to respond to light were in the inferior pulvinar, as would be predicted from the arrangement of inputs just noted. Some of these had relatively simple receptive fields, suggesting principal input from the retinocollicular leg of the system, while others had more complex response properties, implying inputs from cortical locations.

2.3. Visual Cortex

The squirrel monkey brain has a relatively unfissured cortex. The striate cortex (V-1), the target for LGN afferents, covers the dorsolateral suface of the occipital lobe and extends over the medial bank of the lobe into the region around the calcarine fissure. According to Cowey (1964), the extent of the striate cortex can be discerned in fresh tissue by the presence of an overlaying fine capillary network. The total area of V-1 has been computed as 722 mm^2, a value roughly half that of V-1 in various macaque species (Cowey, 1964).

In Old World primates there is a clear spatial segregation of inputs from the two eyes into layer IV of the striate cortex. This patchwork segregation provides the anatomical substrate for the orderly arrangement of striate cortex into ocular dominance columns—discrete collections of cells arranged in columns running from the pial surface toward the white matter in which all cells receive the same relative degree of influence from the two eyes. It has recently been shown that whereas this same segregation of inputs occurs in at least one New World monkey, the spider monkey, it is lacking in several other platyrrhine species. The squirrel monkey shows no such segregation of LGN afferents, and therefore presumably no ocular dominance columns (Hendrickson *et al.*, 1978; Humphrey and Hendrickson, 1983). The functional implications of this difference have not been established, but it at least suggests that the first locations where binocular interactions occur in the visual system differ in the squirrel monkey from that typical of Old World monkeys.

As in other species, the contralateral half visual field of the squirrel monkey is receptotopically mapped onto V-1 in each hemisphere. Figure 4 illustrates this receptotopic mapping onto the dorsolateral surface of the occipital lobe as determined through an analysis of visually evoked potentials (Cowey, 1964). The foveal projection (the region of the intersection of the vertical and horizontal meridians in Fig. 4) occupies the same relative position in the squirrel monkey that it does in other monkeys. More peripheral locations in the visual field are represented in the tissue along the medial bank of the lobe. The dashed lines in Fig. 4 indicate the approximate positions of the radii in the visual field. It is apparent that the tissue dedicated to analysis of the central visual field is greatly expanded relative to the representation of more peripheral locations. This relationship has obvious implications for the analysis of spatial information from various parts of the visual field, a topic to be considered later.

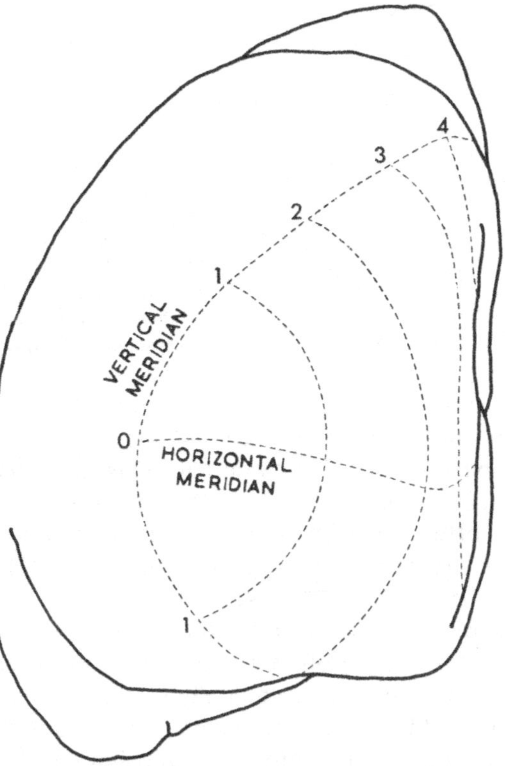

Fig. 4. Receptotopic organization of V-1 of the squirrel monkey. The left hemisphere is shown. The projection was determined through an analysis of visually evoked cortical potentials (Cowey, 1964). The numbers and dashed lines indicate radii in the visual field. Anterior is upward in this drawing.

Multiple cortical projection areas are characteristic of the sensory cortices in mammals. Information about the number of such visual projection areas and their interrelationships has expanded greatly in recent years, due principally to the advent of new anatomical tracing methods. A combination of anatomical and physiological evidence indicates that there are at least four such separate projection areas in the squirrel monkey (Tigges *et al.*, 1981). The central field representations of these four areas (here designated V-1, V-2, V-3, and the middle temporal area MT) are shown schematically in Fig. 5. As indicated there, the central field representations of areas V-2 and V-3 lie in narrow strips located successively anterior to V-1, while MT is a small region at the top of the superior temporal sulcus.

Although the receptotopic map for V-1 has been described in detail, similar specification is not yet available for the other three projection areas. The boundaries of V-2 in the squirrel monkey can be visualized through the use of several different anatomical markers (Tootell *et al.*, 1983). However, the absence of

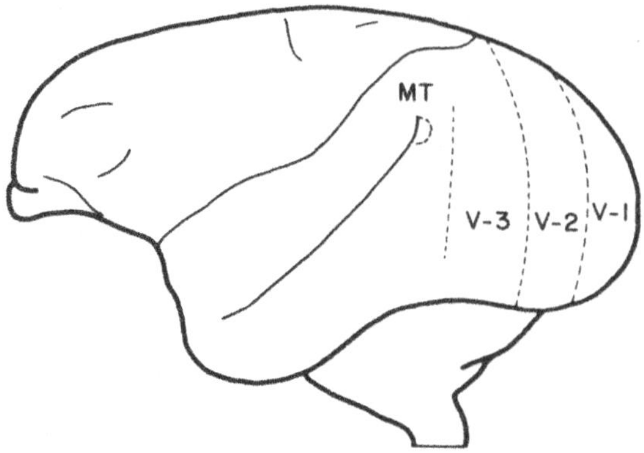

Fig. 5. Central visual field representations of V-1, V-2, V-3, and MT in the squirrel monkey. MT is the small area indicated by dashes at the dorsal end of the superior temporal sulcus.

receptotopic maps for the extrastriate areas means that the specification of these regions should be considered provisional. This is perhaps particularly true for the region designated here as V-3. In the *Aotus* monkey, a platyrrhine primate whose visual cortex has been more extensively mapped, there are five receptotopically distinct regions adjoining V-2 (Allman, 1977). A similar parcellation may exist in the squirrel monkey.

Two general principles of the organization of the visual cortex are, first, that various projection regions are often reciprocally connected, and, second, these connections are organized topographically so that representations of the same parts of the visual field are interconnected. For the squirrel monkey this latter arrangement has been verified both electrophysiologically (Cowey, 1964; Sandell and Schiller, 1982) and anatomically (Wong-Riley, 1974, 1978; Tigges *et al.*, 1981). Tigges and colleagues (1981) also studied the interconnections among the visual areas in the squirrel monkey by injecting horseradish peroxidase into each of the four projection regions and then examining its retrograde transport to cortical targets in the ipsilateral hemisphere and, via the corpus callosum, to contralateral targets. The connections so established are summarized in Fig. 6.

The data of Fig. 6 illustrate clearly that the flow of information in the squirrel monkey visual system is not unidirectional. Rather, at each point information converges from more peripheral structures and from more central locations. That these reciprocal pathways function to provide some sort of feedback control seems obvious, but actual evidence for this is scarce. Sandell and Schiller (1982) showed that cooling cells located in V-2 (hence inactivating them) changed the responses to visual stimuli of about one-third of all cells recorded from V-1. Most of these became less responsive in the absence of V-2 input. From this result they suggest that the normal role of this particular reciprocal pathway is to facilitate the flow of information from V-1 to subcortical locations, although they also found that the responses of some V-1 units were qualitatively changed, indicating that this feedback also may alter the nature of information transmitted to subsequent cortical locations.

Pathways between the visual cortex and the three principal diencephalic visual centers (LGN, superior colliculus, and pulvinar) were noted in previous

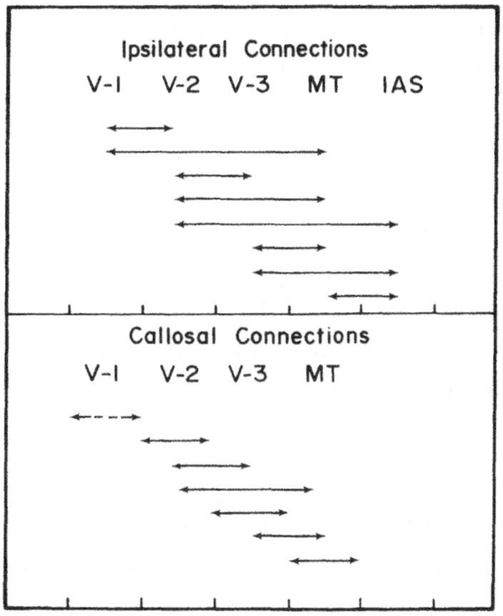

Fig. 6. Connections between visual cortical areas in the squirrel monkey. The ipsilateral connections are indicated at the top, those passing contralaterally via the corpus callosum are shown at the bottom. Among the latter, those between V-1 in the two hemispheres are described as weak. Arrows in both directions indicate that the connections are reciprocal. The connections shown were established by examining the retrograde transport of horseradish peroxidase (Tigges *et al.*, 1981). IAS, Inferior arcurate sulcus (location of the frontal eye fields).

sections. In addition, the squirrel monkey visual cortex has been found to receive afferents from a broad range of other subcortical sites. Tigges *et al.* (1982) have shown that structures in the limbic system, the extrapyramidal system, and the nonspecific thalamic system, as well as a variety of structures in the lower brainstem all supply inputs to the visual cortex. Further, with minor exceptions, these structures project to all four of the visual cortical locations identified above. Unlike the LGN afferents, inputs from these structures come from both ipsilateral and contralateral locations and they appear to show no orderly topographic arrangement. Tigges *et al.* (1982) estimate that the total number of fibers from these subcortical structures may equal the number of fibers in the projection from the LGN to V-1. The functions of this massive, heterogeneous input are presently only subjects for vague speculations.

Although the anatomy of the visual cortex in the squirrel monkey is at least broadly understood, the same cannot be said of the physiology of this region. Thus far, only one study involving an analysis of the response properties of visual cortex units has been reported. Bartlett and Doty (1974b) recorded from single units in the foveal representation of striate cortex in unanesthetized, immobilized squirrel monkeys. They found that a large number of units there responded to spatially nonspecific stimuli (e.g., flashes from a strobe light). Many of the units responded to moving stimuli, and about 60% of all units showed some directional selectivity. They concluded that the receptive field categories classically applied to cat cortical units ("simple cells" and the like) were not useful to summarize the responses elicited from squirrel monkey cortical units. Perhaps the most novel finding of this investigation was that in about 40% of all units the ongoing discharge rate was uniquely related to the luminance of a steadily viewed light. These units (termed "luxotonic") showed either monotonic increases or decreases in firing rate as a function of the luminance of a steady light.

3. VISUAL CAPACITIES

Although clear inferences about vision may be drawn from studies of the electrophysiology and anatomy of the visual system, it is still obviously necessary to have direct measurements of visual behavior in order to determine how close these inferences come to matching reality. In this section the literature on behavioral studies of vision in the squirrel monkey is reviewed.

3.1. Some Practical Issues

Squirrel monkeys have been used as subjects in a wide range of behavioral discrimination experiments, and consequently copious details about animal handling, deprivation variables, stimulus and response contingencies, and the like are already available in the literature. Few unique problems arise in using squirrel monkeys as subjects for visual discrimination experiments. In our laboratory, among a dozen or so different mammalian species that have been tested on visual discriminations over the years, both primate and nonprimate, squirrel monkeys are neither the easiest nor the most difficult to train to perform visual discriminations. We have found that test–trial spacing is frequently an important factor in obtaining a good yield of visual discrimination data from these monkeys. Generally, the more rapidly stimulus sequences follow one another the better. This is probably because most squirrel monkeys are excitable, and rapid trial sequencing helps to inhibit the occurrence of all sorts of extraneous behavior.

Once trained, squirrel monkeys produce very reliable psychophysical data. For instance, in one experiment involving the measurement of visual thresholds (Jacobs, 1983a), it was found that these threshold determinations are highly repeatable (an average test/retest correlation value of 0.90 for 25 animals). The actual within-animal variations in threshold in this situation were on average between 01. and 0.2 log unit. This value is close to those obtained from human observers tested in analogous situations. An extreme example of how reliably visual thresholds can be measured in squirrel monkeys was seen in this same situation when one monkey was tested on two occasions with an interval of slightly more than 8 years between the two tests. In this case the two thresholds differed by less than 0.1 log unit! Clearly, squirrel monkeys can be apt subjects for behavioral studies of vision.

A potentially important practical issue for anyone using squirrel monkeys is that there are now known to be significant interanimal variations in visual sensitivity and color vision. This fact was alluded to in an earlier section, and the behavioral evidence to support it will be spelled out below. Although this within-species variation obviously has great significance for studies of vision and visual mechanisms in squirrel monkeys, it is important to emphasize that it also may be significant in any experimental situation where one attempts to

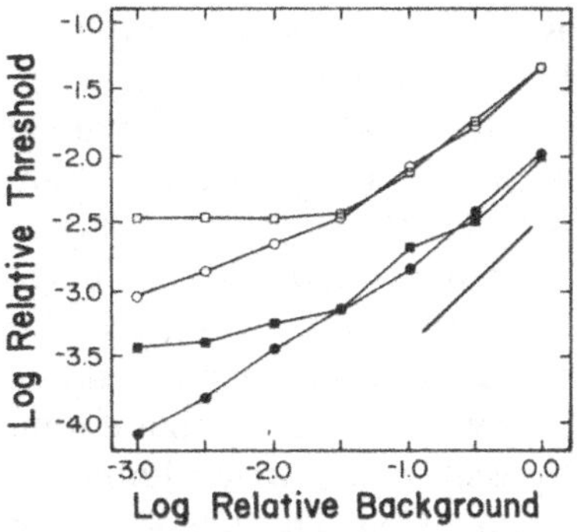

Fig. 7. Mean increment-threshold values for three squirrel monkeys (solid symbols) and one human subject (open symbols), all obtained in the same behavioral test situation. Background light: achromatic (4400°K). Test lights for the monkeys were (■) 610 nm and (●) 500 nm; for the human subjects they were (□) 640 nm and (○) 500 nm. The functions for the two species are arbitrarily positioned on the vertical axis. The solid reference line to the right has a slope of + 1.0. [Data from Jacobs (1973).]

specify, standardize, or control the visual effectiveness of light for squirrel monkey subjects. For example, suppose ambient light level was a controlled variable in some nonvision experiment, perhaps one involving an endocrinological measure. Since some individual squirrel monkeys are as much as 1 log unit less sensitive to long-wavelength lights than others (Jacobs, 1983a), it could well be the case that although all animals were exposed to the same light sources, the effective light level (the illuminance) varied greatly from animal to animal, thus in effect placing some animals in dimmer environments than others. How large an illuminance variation is introduced will depend on the spectral energy distribution of the lights used and on the vision of the particular monkeys selected as subjects. It is at any rate an issue that should be kept in mind when designing experiments involving squirrel monkeys.

3.2. Visual Sensitivity

The minimal amount of light necessary to produce a discriminative visual response in the squirrel monkey is similar to that required by the human. Two highly trained, dark-adapted squirrel monkeys could reliably detect a 3-sec flash of achromatic light having a mean luminance of 9.3×10^{-6} cd/m^2 (Jacobs, 1973). This value is not significantly different from that for three human subjects equivalently tested.

That the presence of rods and cones in the squirrel monkey retina leads to the usual separation of vision into two functionally coupled systems (scotopic and photopic vision) has been demonstrated on several occasions. Figure 7 shows increment-threshold functions obtained from three squirrel monkeys (Jacobs, 1973). In that experiment thresholds were obtained for two test wavelengths, 500 and 610 nm, presented on a series of achromatic adapting backgrounds covering a luminance range of 3 log units. On the brighter backgrounds these monkeys were equally sensitive to the two test wavelengths. However, when the luminance of the adapting light was set below about 0.05 cd/m^2, the threshold for the 610-nm light showed no further decline, while that for the 500-nm light continued to decline monotonically. This change indicates the presence of a Purkinje shift, with photopic vision operational above that point and a shift toward scotopic vision below that point. Shown for comparison in

Fig. 7 are results obtained from a human subject tested in the same way. The photopic/scotopic transition point is the same for the two species. Note also that the slopes of the increment-threshold functions are the same for the two species; both have values of about 1.0, and thus obey Weber's law. The presence of a Purkinje shift in the squirrel monkey has also been demonstrated in experiments involving the discrimination of flickering lights (Jacobs, 1963).

In the human, as the luminance of an adaptation light is increased above that required for cone threshold, the threshold for rod vision begins to increase rapidly. The result is that if luminance is increased still further it soon becomes impossible to provide luminance increments sufficiently large to be detected by rod vision. This phenomenon, termed "rod saturation," effectively removes rod-based signals from the visual system under these luminance conditions. Rod saturation has also been shown to occur in the squirrel monkey visual system. In this case rods begin to saturate when the luminance of an adapting light reaches about 30 cd/m^2 (Jacobs, 1973). Consequently, rod vision in the squirrel monkey, from absolute threshold to the point of saturation, spans a luminance range of 7 log units.*

In the course of measuring increment-threshold spectral sensitivity in squirrel monkeys it was discovered that although individual monkeys showed no systematic variations in sensitivity to short and middle wavelengths (<600 nm), there were alarmingly large variations in sensitivity to longer wavelengths (Jacobs, 1972). Since these variations appeared significant, an investigation was made to determine their extent and nature. The results are summarized in Fig. 8, which shows the differences in increment threshold for 540- and 640-nm lights for 41 squirrel monkeys,† The large range of threshold differences illustrated there arises from the fact that although there were only modest variations in 540-nm thresholds, there were large variations in thresholds determined with the 640-nm test light. As a consequence, there is a span of about 1.5 log units in threshold differences among this sample of squirrel monkeys. Additional measurements made using various adaptation paradigms show that these variations cannot be attributed to individual variations in the presence of an ocular filter (Jacobs, 1983a).

These individual variations in sensitivity make it impossible to specify *the* spectral luminosity function for squirrel monkeys. Complete increment-threshold spectral luminosity functions have been measured on five monkeys (Jacobs, 1972). Included is an indication of the range of individual variation across the visible spectrum. Average spectral luminosity functions measured in a flicker discrimination test have also been reported for this species (Jacobs, 1963).

* Throughout this discussion the luminance units given are based on the human spectral luminous efficiency function. Given the within-species variations in visual sensitivity among squirrel monkeys, these values will be more or less correct for individual animals, depending on the details of the test situation and the capacities of the particular individuals involved. The latter were not known at the time these experiments were run, but the small between-subjects variation in these tests suggests that, whatever the capacity variation, it was not very significant.

† All of the squirrel monkeys on whom these sensitivity measurements were made were of Peruvian origin. A small number of monkeys of Colombian origin have also been tested. They, too, show within-species variations in visual sensitivity, but it is not known if the nature and extent of their variation match those described here.

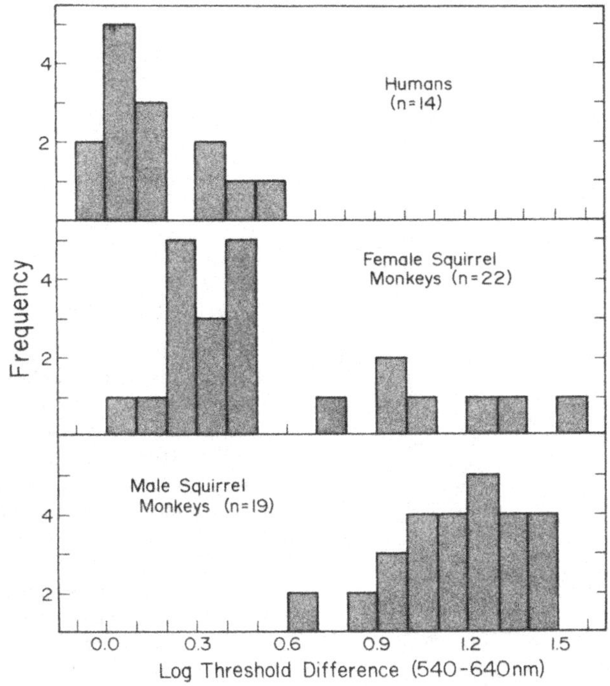

Fig. 8. Summary of increment-threshold measurements made on squirrel monkeys in a behavioral test situation (Jacobs, 1983b). Plotted are the differences in threshold for 540- and 640-nm lights. These stimuli were presented as flashes of light added to an achromatic adapting field. The top panel shows results obtained from a group of normal human trichromats who were tested in the same situation as the monkeys.

The results in Fig. 8 also indicate that the sensitivity variation among squirrel monkeys has a clear gender-related component. Specifically, taken as a group, the male monkeys were very significantly less sensitive to the 640-nn light than were the female monkeys. However, this sex difference is statistical, not absolute, since a minority of the females (seven of 22) also showed low sensitivity to the 640-nm light. The result is that these sensitivity variations are bimodally distributed: one mode is made up of results from females and the other from males plus a minority proportion of females. These sensitivity differences among squirrel monkeys suggest the likelihood of some within-species variations in color vision, a subject considered next.

3.3. *Color Vision*

Three studies published prior to 1975 all concluded that the squirrel monkey has color vision, and that it differs from the color vision of the normal human trichromat. Two of these studies presented evidence that squirrel monkeys were trichromats shifted rather strongly in the direction of protanomaly (Jacobs, 1963; De Valois and Morgan, 1974). The third investigator believed squirrel monkeys had dichromatic color vision, probably protanopia (Miles,

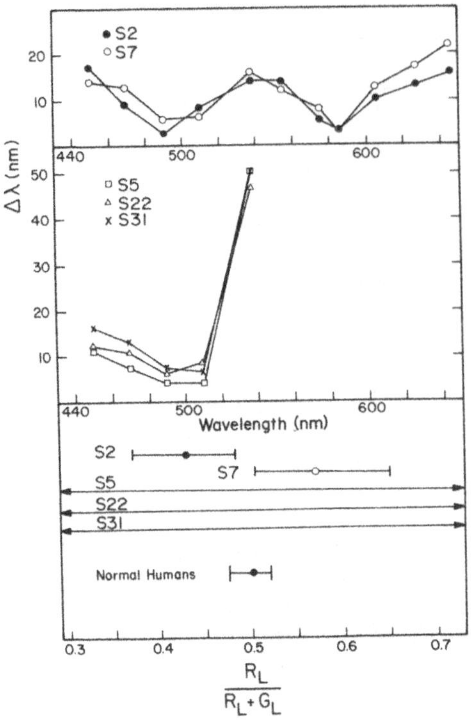

Fig. 9. Results from behavioral tests of color vision for five squirrel monkeys (Jacobs, 1983b). The top two panels show wavelength discrimination functions for each animal. The bottom panel summarizes Rayleigh match determinations in which monkeys were tested for their abilities to discriminate various mixtures of red (625 nm) and green (536 nm) lights from a fixed yellow light (585 nm). The average Rayleigh match for 41 normal human trichromats is indicated by the solid circle, and the total range of these matches by the horizontal line.

1958). All of these conclusions were in accord with the then generally held view that platyrrhine monkeys have color vision similar to that of human protan observers.

On the basis of the within-species variations in visual sensitivity described above, I have recently carried out a fresh investigation of squirrel monkey color vision. The summary conclusion is that there are indeed significant individual variations in color vision among members of this species. The evidence for this is summarized in Fig. 9. The top two panels of Fig. 9 show the three different types of wavelength discrimination functions given by individual squirrel monkeys. Monkeys whose results are plotted in the top panel show relatively good wavelength discrimination across the spectrum, with points of best discrimination at about 500 and 575 nm. Note that although the functions in the top panel are qualitatively similar for values up to about 600 nm, at longer wavelengths they diverge, so that S7 has poorer wavelength discrimination in the long wavelengths than S2. The monkeys whose results are plotted in the middle panel of Fig. 9 have reasonable discrimination around 500 nm, but are quite unable to discriminate among long spectral wavelengths, the result being a U-shaped function with a single minimum.

Figure 9 also gives the results of the second color test: the Rayleigh match. In this case animals were required to discriminate various mixtures of red and green light from a fixed yellow light. The horizontal lines indicate the range of red/green mixtures they were unable to so discriminate. As shown, some animals (S5, S22, S31) could not discriminate any red/green mixtures from yellow. Others (S2, S7) failed over relatively restricted ranges, but these ranges differed. Relative to the discrimination failures of normal human observers tested in the same situation, monkey S7 failed the test when there was more red light in the mixture, while S2 failed when the mixture contained more green light.

The results shown in Fig. 9 make it clear that there are very striking color vision variations among squirrel monkeys. For convenience it is possible to summarize the variations in color vision among squirrel monkeys according to the color vision categories used for man. If this is done, animals who fail the Rayleigh discrimination when the mixture contains an excess of red light, and whose wavelength discrimination is like that of S7, would be considered protanomalous trichromats. Animals whose wavelength discriminations are like those of S2, and who fail the mixture problem when excessive green light is present, are deuteranomalous trichromats. Squirrel monkeys unable to perform the Rayleigh discrimination at all, and whose wavelength discriminations are U-shaped, are dichromatic. However, additional color matching tests indicate that these dichromats can also be subdivided into different classes. If, again, the human classification scheme is used, these dichromatic squirrel monkeys are either like deuteranopes *or* protanopes *or* they have a third type of dichromacy that does not have a parallel in the human scheme. In sum, these experiments indicate that there are at least five different color vision phenotypes among squirrel monkeys.

As noted in the previous discussion of the retina, there are now some direct measurements of the cone photopigments in the squirrel monkey (Jacobs *et al.*, 1982; Bowmaker *et al.*, 1983). These measurements were made in animals whose color vision had been previously established, and so it was possible to determine if the color vision variations are directly tied to variations in cone pigment complement. They are. The MSP measurements reveal that in addition to a short-wavelength pigment that all animals appear to have in small proportions, there are three other classes of pigments absorbing maximally in the middle wavelengths with peaks at about 535, 549, and 564 nm. The dichromatic animals each had one of these three pigment classes, which varied from monkey to monkey in accord with the type of dichromacy. Animals classified as trichromatic had two of these pigment classes: those described above as deuteranomalous had the 549- and 565-nm pigments, while those classed as protanomalous in the behavioral tests had 535- and 549-nm pigments. Thus the dramatic variations in color vision among squirrel monkeys are directly traceable to the pigments found in individual retinas.

The individual variations in visual sensitivity and color vision among squirrel monkeys summarized here are substantial. So far not enough animals have been examined to yield confidence about the statistics of the variation, or, indeed, to know if the full extent of the variations have been seen. Perhaps even more challenging is the question of what implications, if any, these variations have for visually guided behavior in natural situations.

3.4. Spatial Vision

The visual acuity of the squirrel monkey has been determined on several occasions. The most extensive investigation was that of Cowey and Ellis (1967), who tested 12 squirrel monkeys to find the finest grating pattern these animals could discriminate from an equiluminant blank field. The resulting acuity thresholds, specified in minutes of arc, ranged from 0.61 to 0.90, with a mean value of 0.74 min. A like number of rhesus monkeys were tested on the same discrimination. Although the results from the two genera showed extensive overlap, the rhesus monkeys on average had significantly higher visual acuity (a mean of 0.65 min). Two other investigations of grating acuity in the squirrel monkey (Woodburne, 1965; Ordy and Samorajski, 1968) yielded values in reasonable agreement with those of Cowey and Ellis.

Given the large difference in eye and brain size between macaque and squirrel monkeys, the closeness of their acuity values is perhaps surprising. In comparing retinal and cortical anatomy to visual acuity in the two genera, Rolls and Cowey (1970) noted the following relationships. First, squirrel monkeys have a slightly lower density of foveal cones per degree of visual angle than macaques, in correspondence with their slightly lower foveal acuity. Second, from the fovea toward the periphery, cone density falls off rather sharply in the squirrel monkey (see Fig. 1) and, in particular, this falloff is faster than that found in the macaque retina. This was seen by Rolls and Cowey to correlate with the fact that peripheral acuity is relatively poorer in the squirrel monkey than in the macaque (Cowey and Ellis, 1969). Finally, the cortical magnification factor (that is, the size of the cortical region dedicated to each part of the retina, note Fig. 4) is lower for both foveal and peripheral regions in the squirrel monkey in comparison to the same measure in the macaque. With all these relative disadvantages, the squirrel monkey still has excellent foveal acuity—nearly as good as that of Old World monkeys, who are, in turn, only slightly poorer at these tasks than man.

Classical measurements of visual acuity, like those just described for the squirrel monkey, only index a subject's ability to discriminate fine detail in targets of fixed (usually high) contrast. In recent years it has become clear that a more useful description of spatial resolution can be obtained from tests that measure sensitivity to targets of various sizes over a range of contrast settings. The simplest such measure is the contrast sensitivity function, which plots the luminance contrast required for a criterion level discrimination at each of a range of different spatial frequencies. Merigan (1976) measured contrast sensitivity functions for three squirrel monkeys. The results are shown in Fig. 10. Each of the subjects showed highest sensitivity to targets having spatial frequencies of 2–5 cycles/deg of visual angle, and all showed the decline in sensitivity to larger targets characteristic of many viewing conditions and which presumably reflects the operation of inhibitory mechanisms in the visual system. The high-frequency cutoff location was somewhat variable—for two monkeys it was located at about 35 cycles/deg and for the other, it was somewhat lower, the difference perhaps due to the fact that the latter subject was slightly myopic. For comparison, human subjects tested in the same situation gave

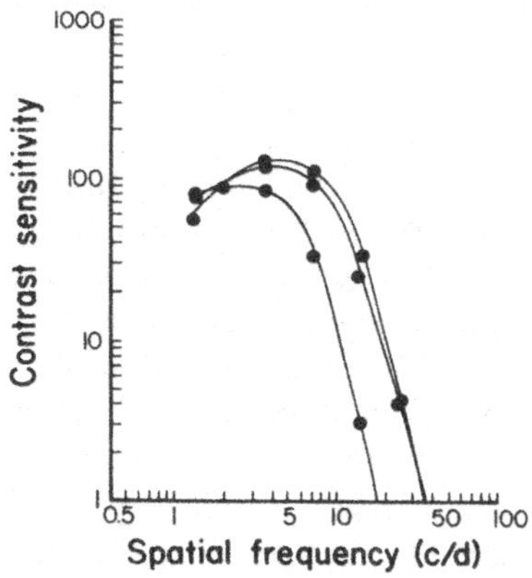

Fig. 10. Spatial contrast sensitivity functions for three squirrel monkeys. [Data from Merigan (1976).]

similar results, both in the location of their peak sensitivities and in the location of the high-frequency cutoff.

For other species, particularly the cat and macaque monkey, considerable progress has been made toward determining how nervous system mechanisms might be related to contrast sensitivity functions like those shown in Fig. 10. Regretably, that is not yet possible for the squirrel monkey, because there is a near absence of electrophysiological investigations on the basis of this behavior. In a brief abstract, Campbell *et al.* (1969) reported measurements of the spatial selectivity of single cells in the squirrel monkey LGN. For each of a number of cells they determined some aspects of the cell's responsivity to various spatial frequencies. From these results they suggest that the acuity of the squirrel monkey might turn out to be about three times that of the cat. This underestimates substantially the actual difference between the two species (squirrel monkey acuity is about 4–7 times that of the cat). This underestimate may reflect the small sample of LGN units tested in the squirrel monkey. At any rate, additional studies of this sort would be useful to help understand how the squirrel monkey's sensitivity to spatial stimuli results from operations of the visual nervous system.

3.5. Binocular Vision

Although binocular viewing was permitted in all of the experiments on visual capacities so far reviewed, none of these capacities probably depended directly on the use of conjoint information from the two eyes. However, the frontally placed eyes, and the well-coordinated movements of the two eyes that can be easily observed in squirrel monkeys, suggest that binocularity is a keystone of squirrel monkey vision. There is almost no scientific evidence on the issue. Schmidt (1968) required squirrel monkeys to discriminate between two targets presented at separate distances in front of the animal. Manipulation of distance yielded a measure of depth discrimination. The threshold for depth discrimination was about the same in squirrel monkeys and humans. Schmidt also tested squirrel monkeys for their abilities to jump to various targets. The fact that in such tasks the monkey typically engaged in right–left head movements prior to the jump suggested to Schmidt that visual orientation in this monkey depends heavily on motion parallax.

A particularly engaging visual behavior in squirrel monkeys, which may or may not be related to binocular vision, is head cocking. Head cocking is the rotation of the head about the body axis while holding some object in fixed regard. It occurs very frequently in infant squirrel monkeys, particularly during the time they are passengers on the mothers back. Menzel (1980) carried out a survey of head cocking in 40 simian and prosimian species. He found that this behavior occurs at particularly high frequency in several cebid and callitrichid species. In a longitudinal study of two infant squirrel monkeys, head-cocking frequency rose rapidly during early life to a peak at about 5–7 weeks of age, and it then declined just as sharply, as the infants achieved greater independence of movement.

In seeking an explanation for head cocking, Menzel (1980) suggests that the only obvious anatomical difference between species that do and do not head cock is the presence of ocular dominance columns in V-1. As noted earlier, the squirrel monkey lacks such columns. Why the absence of ocular dominance columns should be correlated with head cocking is not known, and even more puzzling is why, if they are related, the behavior should have such a strong ontogenetic component. Menzel suggests also that head cocking may provide a means for providing the animal with depth information by substituting horizontal disparities for vertical disparities. Although a plausible reason, this again fails to explain why there is a developmental component. Like many other aspects of vision and the visual system in the squirrel monkey, this behavior will require further analysis.

ACKNOWLEDGMENTS. I thank Roger Tootell and Jay Neitz for helpful comments, and the National Eye Institute (Grant EY-02052) for supporting my research on the squirrel monkey.

REFERENCES

Allman, J., 1977, Evolution of the visual system in the early primates, in: *Progress in Psychobiology and Physiological Psychology*, Volume 7 (J. M. Sprague and A. N. Epstein, eds.), Academic Press, New York, pp. 1–53.

Bartlett, J. R., and Doty, R. W., 1974a, Influence of mesencephalic stimulation on unit activity in striate cortex of squirrel monkeys, *J. Neurophysiol* **37**:642–652.

Bartlett, J. R., and Doty, R. W., 1974b, Response of units in striate cortex of squirrel monkeys to visual and electrical stimuli, *J. Neurophysiol.* **37**:621–641.

Bartlett, J. R., Pecci-Saavedra, J., and Wilson, P. D., 1973, Mesencephalic control of lateral geniculate nucleus in primates III. Modification with states of alertness, *Exp. Brain. Res.* **18**:214–224.

Bowmaker, J. K., Mollon, J. D., and Jacobs, G. H., 1983, Microspectrophotometric results for Old and New World primates, in: *Colour Vision: Physiology and Psychophysics* (J. D. Mollon and L. T. Sharpe, eds.), Academic Press, London, pp. 57–68.

Campbell, F. W., Cooper, G. F., Robson, J. G., and Sachs, M. B., 1969, The spatial selectivity of visual cells of the cat and the squirrel monkey, *J. Physiol* **204**:120–121P.

Campos-Ortega, J. A., and Glees, P., 1967, The subcortical distribution of optic fibers in *Saimiri sciureus* (squirrel monkey, *J. Comp. Neurol.* **131**:131–142.

Cowey, A., 1964, Projection of the retina onto striate and prestriate cortex in the squirrel monkey, *Saimiri sciureus*, *J. Neurophysiol.* **27**:366–393.

Cowey, A., and Ellis, C. M., 1967, Visual acuity of rhesus and squirrel monkeys *J. Comp. Physiol. Psychol.* **64**:80–84.

Cowey, A., and Ellis, C. M., 1969, The cortical representation of the retina in the squirrel and rhesus monkeys and its relation to visual acuity, *Exp. Neurol.* **24**:374–385.

De Valois, R. L., and Morgan, H. C., 1974, Psychophysical studies of monkey vision. II. Squirrel monkey wavelength and saturation discrimination, *Vis. Res.* **14**:69–73.

Doty, R. W., Kimura, D. S., and Mogenson, G. J., 1964, Photically and electrically elicited responses in the central visual system of the squirrel monkey, *Exp. Neurol.* **10**:19–51.

Doty, R. W., Glickstein, M., and Calvin, W. H., 1966, Lamination of the lateral geniculate nucleus in the squirrel monkey, *Saimiri sciureus*, *J. Comp. Neurol.* **127**:335–340.

Doty, R. W., Wilson, P. D., Bartlett, J. R., and Pecci-Saavedra, J., 1973, Mesencephalic control of lateral geniculate nucleus in primates. I. Electrophysiology, *Exp. Brain. Res.* **18**:189–203.

Hendrickson, A. E., Wilson, J. R., and Ogren, M. P., 1978, The neuroanatomical organization of pathways between the dorsal lateral geniculate nucleus and visual cortex in Old World and New World primates, *J. Comp. Neurol.* **1822**:123–136.

Hollander, H., 1974, Projections from the striate cortex to the diencephalon in the squirrel monkey (*Saimiri sciureus*). A light microscopic radioautographic study following intracortical injection of H3 leucine, *J. Comp. Neurol.* **155**:425–440.

Humphrey, A. L., and Hendrickson, A. E., 1983, Background and stimulus-induced patterns of high metabolic activity in the visual cortex (Area 17) of the squirrel and macaque monkey, *J. Neurosci.* **3**:345–358.

Jacobs, G. H., 1963, Spectral sensitivity and color vision of the squirrel monkey, *J. Comp. Physiol. Psychol.* **56**:616–621.

Jacobs, G. H., 1964, Single cells in squirrel monkey lateral geniculate nucleus with broad spectral sensitivity, *Vis. Res.* **4**:221–232.

Jacobs, G. H., 1965, Effects of adaptation on the lateral geniculate response to light increment and decrement, *J. Opt. Soc. Am.* **55**:1535–1540.

Jacobs, G. H., 1969, Transneuronal changes in the lateral geniculate nucleus of the squirrel monkey, *Saimiri sciureus*, *J. Comp. Neurol.* **135**:81–84.

Jacobs, G. H., 1972, Increment-threshold spectral sensitivity in the squirrel monkey, *J. Comp. Physiol. Psychol.* **79**:425–431.

Jacobs, G. H., 1973, Visual sensitivity in the squirrel monkey, *Am. J. Phys. Anthropol.* **38**:371–376.

Jacobs, G. H., 1983a, Within-species variations in visual capacity among squirrel monkeys (*Saimiri sciureus*): Sensitivity differences, *Vis. Res.* **23**:239–248.

Jacobs, G. H., 1983b, Variations in color vision among nonhuman primates, in: *Colour Vision: Physiology and Psychophysics* (J. D. Mollon and L. T. Sharpe, eds.), Academic Press, London, pp. 39–50.

Jacobs, G. H., and Yolton, R. L., 1970, Center–surround balance in receptive fields of cells in the lateral geniculate nucleus, *Vis. Res.* **10**:1127–1144.

Jacobs, G. H., Jones, A. E., and De Valois, R. L., 1963, Electroretinogram of the squirrel monkey, *J. Comp. Physiol. Psychol.* **56**:405–409.

Jacobs, G. H., Bowmaker, J. K., and Mollon, J. D., 1981, Behavioural and microspectrophotometric measurements of colour vision in monkeys, *Nature* **292**:541–543.

Jacobs, G. H., Bowmaker, J. K., and Mollon, J. D., 1982, Colour vision variations in monkeys: Behavioural and microspectrophotometric measurements on the same individuals, *Doc. Ophthalmol. Proc. Ser.* **33**:269–280.

Kaas, J. H., Huerta, M. F., Weber, J. T., and Harting, J. K., 1978, Patterns of retinal terminations and laminar organization of the lateral geniculate nucleus of primates, *J. Comp. Neurol.* **182**:517–554.

Kadoya, S., Wolin, L. R., and Massopust, L. C., Jr., 1971a, Photically evoked unit activity in the tectum opticum of the squirrel monkey, *J. Comp. Neurol.* **142**:495–508.

Kadoya, S., Wolin, L. R., and Massopust, L. C., Jr., 1971b, Striate cortex–superior colliculus projection in squirrel monkey, *Exp. Neurol.* **32**:98–110.

Kadoya, S., Wolin, L. R., and Massopust, L. C., Jr., 1971c, Collicular unit responses to monochromatic stimulation in squirrel monkey, *Brain Res.* **32**:251–254.

Livingstone, M. S., and Hubel, D. H., 1982, Thalamic inputs to cytochrome oxidase-rich regions in monkey visual cortex, *Proc. Natl. Acad. Sci. USA* **79**:6098–6101.

Marrocco, R. T., and Li, R. H., 1977, Monkey superior colliculus: Properties of single cells and their afferent inputs, *J. Neurophysiol.* **40**:844–860.

Mathers, L. H., 1971, Tectal projection to the posterior thalamus of the squirrel monkey, *Brain Res* **35**:295–298.

Mathers, L. H., 1972, The synaptic organization of the cortical projection to the pulvinar of the squirrel monkey, *J. Comp. Neurol.* **146**:43–60.

Mathers, L. H., and Rapisardi, S. C., 1973, Visual and somatosensory receptive fields of neurons in the squirrel monkey pulvinar, *Brain Res.* **64**:65–83.

Menzel C. R., 1980, Head-cocking and visual perception in primates, *Anim. Behav.* **28**:151–159.

Merigan, W. H., 1976, The contrast sensitivity of the squirrel monkey (*Saimiri sciureus*), *Vis. Res.* **16**:151–159.

Miles, R. C., 1958, Color vision in the squirrel monkey, *J. Comp. Physiol. Psychol.* **51**:328–331.

Ogren, M., and Hendrickson, A., 1976, Pathways between striate cortex and subcortical regions in *Macaca mulatta* and *Saimiri sciureus*: Evidence for a reciprocal pulvinar connection, *Exp. Neurol.* **53**:780–800.

Ordy, J., and Samorajski, T., 1968, Visual acuity and ERG-CFF in relation to morphologic organization of the retina among diurnal and nocturnal primates, *Vis. Res.* **8**:1205–1225.

Rolls, E. T., and Cowey, A., 1970, Topography of the retina and striate cortex and its relationship to visual acuity in rhesus monkeys and squirrel monkeys, *Exp. Brain. Res.* **10**:298–310.

Sandell, J. H., and Schiller, P. H., 1982, Effect of cooling area 18 on striate cortex, *J. Neurophysiol.* **48**:38–48.

Schmidt, U., 1968, Untersuchungen zur visuellen raumorientierung bei totenkopfaffen (*Saimiri sciureus L.*), *Z. Vgl. Physiol.* **60**:176–208.

Spatz, W. B., Tigges, J., and Tigges, M., 1970, Subcortical projections, cortical associations, and some intrinsic interlaminar connections of the striate cortex in the squirrel monkey (*Saimiri*), *J. Comp. Neurol.* **140**:155–174.

Stone, J., and Johnston, E., 1981, The topography of primate retina: A study of the human, bushbaby, and New- and Old-World monkeys, *J. Comp. Neurol.* **196**:205–223.

Tigges, J., and O'Steen, W. K., 1974, Termination of retinofugal fibers in squirrel monkey: A reinvestigation using autoradiographic methods, *Brain Res.* **19**:489–495.

Tigges, J., and Tigges, M., 1969, The accessory optic system and other optic fibers of the squirrel monkey, *Folia Primatol.* **10**:245–262.

Tigges, J., Tigges, M., Anschel, S., Cross, S. A., Letbetter, W. D., and McBride, R. L., 1981, Areal and laminar distribution of neurons interconnecting the central visual cortical areas 17, 18, 19 and MT in squirrel monkey (*Saimiri*), *J. Comp. Neurol.* **202**:539–560.

Tigges, J., Tigges, M., Cross, N. A., McBride, R. L., Letbetter, W. D., and Anschel, S., 1982, Subcortical structures projecting to visual cortical areas in squirrel monkey, *J. Comp. Neurol.* **209**:29–40.

Tootell, R. B. H., Silverman, M. S., DeValois, R. L., and Jacobs, G. H., 1983, Functional organization of the second cortical visual area (V2) in the primate, *Science* **220**:737–739.

Weber, J. T., Huerta, M. F., Kaas, J. H., and Harting, J. K., 1983, The projections of the lateral geniculate nucleus of the squirrel monkey: Studies of the interlaminar zones and the S layers, *J. Comp. Neurol.* **213**:135–145.

Wilson, P. D., Pecci-Saavedra, J., and Doty, R. W., 1973, Mesencephalic control of lateral geniculate nucleus in primates. II. Effective loci, *Exp. Brain Res.* **18**:204–213.

Wolin, L. R., and Massopust, L. C., Jr., 1967, Characteristics of the ocular fundus in primates, *J. Anat.* **101**:693–699.

Wolin, L. R., and Massopust, L. C., Jr., 1970, Morphology of the primate retina, in: *The Primate Brain*, Volume I (C. R. Noback and W. Montagna, eds.), Appleton-Century-Crofts, New York, pp. 1–27.

Wong-Riley, M., 1971, Terminal degeneration and glial reactions in the lateral geniculate nucleus of the squirrel monkey after eye removal, *J. Comp. Neurol.* **144:**61–92.

Wong-Riley, M., 1974, Demonstration of geniculocortical and callosal projection neurons in the squirrel monkey by means of retrograde axonal transport of horseradish peroxidase, *Brain Res.* **79:**267–272.

Wong-Riley, M., 1976, Projections from the dorsal lateral geniculate nucleus to prestriate cortex in the squirrel monkey as demonstrated by retrograde transport of horseradish peroxidase, *Brain Res.* **109:**595–600.

Wong-Riley, M., 1978, Reciprocal connections between striate and pestriate cortex in the squirrel monkey as demonstrated by combined peroxidase histochemistry and autoradiography, *Brain. Res.* **147:**159–164.

Woodburne, L. S., 1965, Visual acuity of "*Saimiri sciureus*," *Psychonom. Sci.* **147:**307–308.

Wyszecki, G., and Stiles, W. S., 1967, *Color Science*, Wiley, New York.

12

Use of Squirrel Monkeys in Cardiovascular Research

Harriet L. Strickland and Thomas B. Clarkson

1. INTRODUCTION

Squirrel monkeys (*Saimiri sciureus*) have been the most widely used of the New World monkeys in cardiovascular research. The majority of the cardiovascular research done with squirrel monkeys has concerned atherosclerosis both as it occurs naturally in these animals and with regard to the effect of diet on the progression of atherosclerotic lesions. Of the several subspecies of squirrel monkeys, Brazilian-type squirrel monkeys have been used almost exclusively in the research on atherosclerosis. To a somewhat lesser extent, squirrel monkeys have been used to study behavioral influences on hypertension, myocardial necrosis, and cholelithiasis as a complication of unsaturated fat diets. In this review, we summarize naturally occurring atherosclerosis, the effects of dietary manipulations on atherosclerosis progression, the effects of physiological and psychological manipulations on diet-induced atherosclerosis, the complications of atherosclerosis, stress-induced myocardial disease, behaviorally induced hypertension, and, finally, some of the diseases that complicate the use of squirrel monkeys in cardiovascular research.

2. NATURALLY OCCURRING ATHEROSCLEROSIS

2.1. Early Reports

The first report of naturally occurring atherosclerosis in the squirrel monkey was made by Rewell (1954), who described atherosclerotic plaques in the thoracic and abdominal aorta of a female known to be over 20 years old. The

---

Harriet L. Strickland • Laboratory of Behavioral Medicine, Southwest Foundation for Biomedical Research, San Antonio, Texas 78284. *Thomas B. Clarkson* • Department of Comparative Medicine and Arteriosclerosis Research Center, Bowman Gray School of Medicine, Wake Forest University, Winston-Salem, North Carolina 27103.

Table I
Prevalence of Aortic Atherosclerosis and Total Serum Cholesterol Concentrations in Squirrel Monkeys Trapped near Leticia, Colombia[a]

	Juveniles[b]		Adults[b]		Total serum cholesterol concentration[c] (mg/dl)
	Male (n = 20)	Female (n = 20)	Male (n = 90)	Female (n = 90)	
Normal	95	100	58	37	97 ± 2
Atherosclerotic	5	0	42	63	114 ± 3

[a] Adapted from Middleton *et al.* (1967b) and Clarkson *et al.* (1976).
[b] Expressed as percent of animals normal or affected.
[c] Expressed as mean ± SEM.

results of a study by Middleton *et al.* (1964) revealed that out of 74 Brazilian squirrel monkeys maintained under laboratory conditions and fed a monkey chow diet for 3–12 months, 85% had aortic fatty streaking, 7% had raised plaques in the abdominal aorta, and 30% had coronary artery atherosclerosis affecting the small intramyocardial arteries.

2.2. Leticia, Colombia Study 1967

In order to eliminate the effects of confinement and diet on the extent of atherosclerosis developed under laboratory conditions, Middleton *et al.* (1967b) studied 220 free-living squirrel monkeys in Leticia, Colombia (Table I). The animals studied were 180 adults and 40 juveniles, with cases evenly distributed between the sexes. Fifty-three percent of the adults had aortic fatty streaking, while 10% had coronary artery atherosclerosis affecting the small intramyocardial arteries. Aortic fatty streaking was essentially absent in juveniles but was prevalent in adults, occurring more frequently in adult females than in adult males. Adult females also had a higher mean extent of aortic fatty streaking, suggesting a trend for a sex difference. Occasionally, raised plaques were found in adults, usually located in the abdominal aorta. Monkeys with fatty streaks or other atherosclerotic lesions had significantly higher total serum cholesterol (TSC) and β^+ pre-β lipoprotein cholesterol concentrations than did monkeys without lesions.

2.3. Squirrel Monkeys Compared to Other New World Monkeys

When squirrel monkeys were compared with five other species of New World monkeys (black spider, *Ateles*; capuchin, *Cebus apella*; ringtail, *Cebus albifrons*; white-lipped marmoset, *Saguinus nigricollis*; and woolly, *Lagothrix lagothrix*), they ranked third highest in TSC concentration (behind the black spider and woolly monkeys), yet ranked highest in prevalence of naturally occurring atherosclerosis in the thoracic and abdominal aorta (Lofland *et al.*, 1967) (Fig. 1).

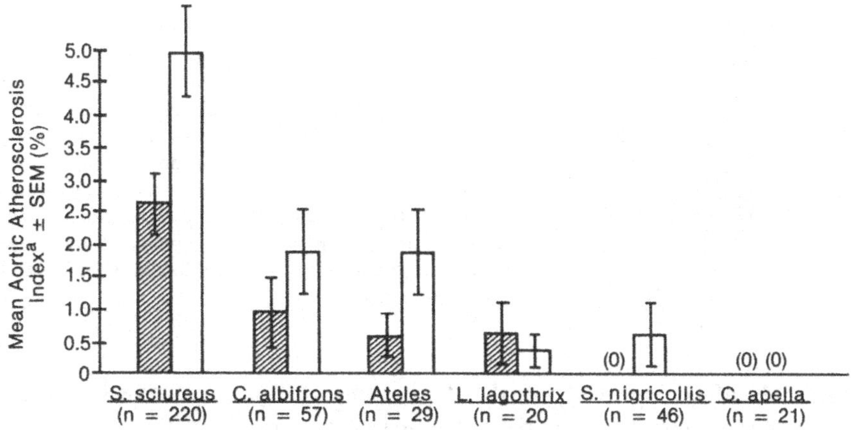

Fig. 1. Extent of aortic atherosclerosis in six species of New World monkeys, including the squirrel monkey (*S. sciureus*, $n = 220$), the ringtail monkey (*C. albifrons*; $n = 57$), the black spider monkey (*Ateles*; $n = 29$), the woolly monkey (*L. lagothrix*; $n = 20$), the white-lipped marmoset (*S. nigricollis*; $n = 46$), and the capuchin monkey (*C. apella*; $n = 21$). Values are expressed as percentage of aortic intimal surface area affected by plaques or fatty streaks. ▨, Thoracic aorta atherosclerosis index; □, abnormal aorta atherosclerosis index. [Adapted from Lofland *et al.* (1967).]

2.4. Histological Characteristics

McCombs *et al.* (1969) have described carefully the histological characteristics of naturally occurring atherosclerosis in squirrel monkeys housed under laboratory conditions and fed a semipurified control diet for 1 year. These investigators felt that the ultrastructure of aortic intimal lesions in the squirrel monkey had numerous similarities to those in human beings and other primates. Major morphological differences were found between intimal lesions in the thoracic and abdominal aorta. Intimal lesions in the thoracic aorta were 3–5 cell layers thick and contained densely packed lipid-laden cells resembling blood monocytes. Smooth muscle cells underlying the lesions also contained lipid. Intimal lesions in the abdominal aorta were 1–2 cell layers thick and contained no lipid-laden cells resembling monocytes. Rather, the lesions were composed of smooth muscle cells containing lipid, as well as abundant extracellular lipid.

3. DIETARY EFFECTS ON ATHEROSCLEROSIS

3.1. Summary of Lipid Changes

The serum lipid and lipoprotein characteristics of squirrel monkeys are more similar to man's than are those of many other nonhuman primates. Of importance in planning nutritional studies is the observation that squirrel monkeys undergo significant changes in plasma lipids with the stress of relocation.

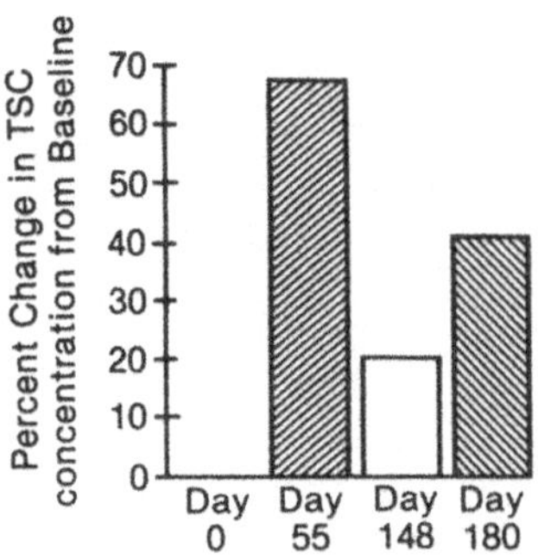

Fig. 2. Changes in total serum cholesterol (TSC) concentration in squirrel monkeys under the stress of importation from South America and following acclimation to living in outdoor group cages and subsequent relocation to indoor laboratory facilities. Day 0, Baseline value (116 mg/dl) at time of capture; ▨, change at time of importation; □, change following period of acclimation; ▧, change at time of relocation. [Adapted from St. Clair *et al.* (1967).]

Immediately after capture from the wild, squirrel monkeys have lower serum cholesterol concentrations than do human beings; however, the distribution of the lipoproteins at that time is similar to that of man. After capture from the wild, squirrel monkeys have a mean TSC level of approximately 100 mg/dl, with the very low density lipoprotein (VLDL) plus low density lipoprotein (LDL) fraction accounting for around 65–75% and the high density lipoprotein (HDL) fraction around 25–35% of serum cholesterol (Rudel and Lofland, 1976; St. Clair *et al.*, 1967). Under the stress of importation and acclimation to laboratory conditions, the squirrel monkey has an increased serum cholesterol concentration (up to 175–200 mg/dl) such that it approximates the levels measured in human beings; however, this change in TSC is entirely due to an increase in the LDL function from 65% to 80% of the TSC (St. Clair *et al.*, 1967) (Fig. 2).

The squirrel monkey can be considered to be intermediate in its responsiveness to dietary cholesterol, with macaques being very sensitive and human beings relatively refractile to serum lipid changes in response to consumption of a cholesterol-containing diet. Furthermore, the redistribution of lipoprotein cholesterol classes during the hypercholesterolemic response to dietary cholesterol follows a different pattern in squirrel monkeys than in macaques: macaques have increased levels of LDL cholesterol and decreased levels of HDL cholesterol, while squirrel monkeys have increased levels of VLDL and intermediate density lipoprotein (IDL) cholesterol, and slightly but relatively insignificantly increased levels of LDL and HDL cholesterol (Rudel and Lofland, 1976).

3.2. Composition of Diet

For reasons not clear, the type of diet to which fat and cholesterol are added plays a role in determining the degree of hypercholesterolemic response of squirrel monkeys and most macaques: when fat and cholesterol are fed in a monkey chow-based diet or other natural products diet, only a small elevation in serum cholesterol is observed. However, when the same amount of fat and cholesterol is added to a semipurified diet, the hypercholesterolemic response of monkeys consuming the diet is approximately twice as great as that seen with the natural products base. In a study by Clarkson *et al.* (1976), squirrel monkeys with baseline TSC levels around 150 mg/dl were fed one of three different diets, all containing 1 mg cholesterol/kcal. TSC levels in monkeys eating the natural products diet plus cholesterol reached approximately 290 mg/dl at 32 months, and then plateaued at around 230 mg/dl. Monkeys fed the semipurified diet plus cholesterol had average TSC levels of about 500 mg/dl, and those fed the purified diet plus cholesterol had average TSC levels of about 350 mg/dl.

Even larger amounts of cholesterol (10 mg cholesterol/kcal) fed in a natural products diet by Malinow *et al.* (1966) yielded similarly small increases in TSC levels in squirrel monkeys. One of five different diets was fed for 10 months to squirrel monkeys with baseline TSC levels approximately equal to 200 mg/dl. During the last 5 months of the test period, control monkeys eating plain monkey chow had TSC levels below 180 mg/dl, while monkeys consuming a monkey chow-based diet with added cholesterol, butter, or margarine had TSC levels of approximately 200 mg/dl. Monkeys eating a monkey chow-based diet with added cholesterol plus butter or cholesterol plus margarine had TSC levels around 270 mg/dl.

A range of hypercholesterolemic responses in squirrel monkeys fed a semipurified diet containing stepwise increases in the amount of cholesterol was reported by Clarkson *et al.* (1976). Feeding a semipurified diet containing 0.3 mg cholesterol/kcal to squirrel monkeys with baseline TSC levels of about 170 mg/dl raised TSC levels to about 250 mg/dl (range of human normal). Feeding them a semipurified diet containing 0.75 mg cholesterol/kcal and 1.0 mg cholesterol/kcal raised TSC levels to about 400 and 500 mg/dl, respectively (range of human hyperlipidemia). Feeding a semipurified diet containing 2.0 mg cholesterol/kcal resulted in a transitory rise in TSC above that seen with lower amounts of dietary cholesterol, but the TSC levels then returned to the same mean levels as for monkeys fed 1.0 mg cholesterol/kcal.

A cholesterol–protein interaction was suggested as a factor in atherogenesis in squirrel monkeys in a study by Middleton *et al.* (1967a). Compared to squirrel monkeys fed a diet low in protein (9% by weight) plus cholesterol (0.5% by weight), squirrel monkeys fed a diet containing a high protein level (25% by weight) plus cholesterol (0.5% by weight) had higher TSC levels and more extensive aortic atherosclerosis (higher percentages of intimal surface area of the distal portions of aortas stained positively with Sudan IV), and more prevalent and extensive coronary atherosclerosis (higher numbers of animals with coronary artery atherosclerosis, and higher percentages of the total number of coronary arteries in each animal with plaques). Feeding the high-protein and low-protein diets without cholesterol had no effect on TSC levels nor on the degree of aortic or coronary atherosclerosis.

3.3. *Type of Dietary Fat*

The response of the squirrel monkey to consumption of cholesterol-containing diets has been shown to be influenced by type of dietary fat. Like other New World monkeys, the squirrel monkey responds independently to dietary fat and cholesterol. In a study by Portman and Andrus (1965), the squirrel monkey responded similarly to the woolly and cebus monkeys when fed diets containing saturated or unsaturated fats with or without cholesterol, although

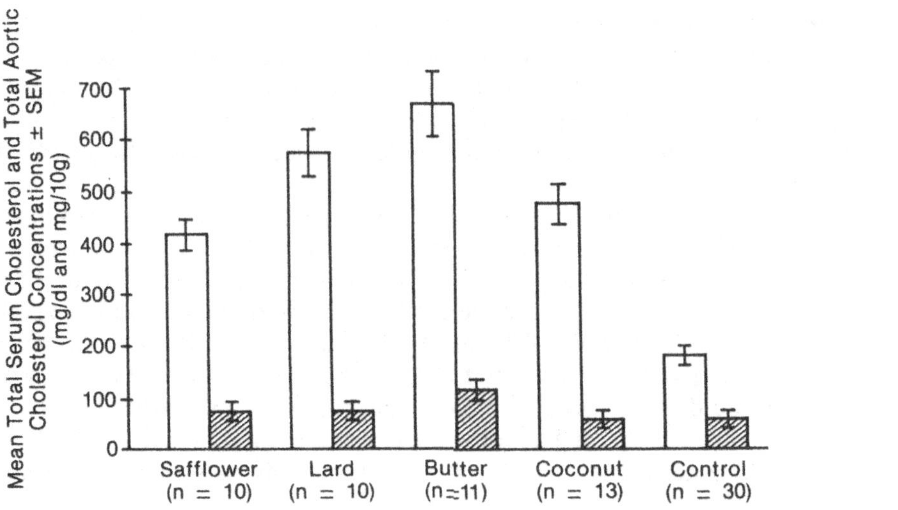

Fig. 3. Total serum cholesterol (TSC) and total aorta cholesterol (TAC) concentrations of squirrel monkeys fed diets containing cholesterol and various fats. □, Mean total serum cholesterol concentration ± SEM (mg/dl); ▨, mean total aortic cholesterol concentration ± SEM (mg/10g wet weight). [Adapted from Lofland *et al.* (1970) and Clarkson *et al.*, 1976).]

the squirrel monkey developed the most extensive aortic atherosclerosis (highest percentage of intimal surface area with sudanophilic lesions). Squirrel monkeys fed diets containing corn or coconut oil as 45% of dietary calories had lower TSC levels than did squirrel monkeys fed diets containing either of the oils plus 1.0 mg cholesterol/kcal. The extent of aortic atherosclerosis was found to be less in squirrel monkeys fed diets containing corn oil as 45% of dietary calories with or without the addition of cholesterol than in those fed diets containing coconut oil as 45% of dietary calories with or without the addition of cholesterol.

In a study by Lofland *et al.* (1970) in which the serum and aorta cholesterol concentrations and the extent of coronary artery atherosclerosis were compared for squirrel monkeys fed diets containing 1.0 mg cholesterol/kcal plus 45% of dietary calories as safflower or coconut oil, lard, or butter, it was found that safflower oil induced the least rise in TSC level and that safflower oil was the least atherogenic, except for coconut oil (Fig. 3). Butter produced the highest level of TSC and the most extensive coronary artery atherosclerosis. The squirrel monkey also has been found to respond independently to dietary fats and cholesterol even at low levels of supplementation (22% of dietary calories as coconut oil and 0.3 mg cholesterol/kcal), developing modest yet significant increases in TSC, similar to the rises in TSC seen in human beings fed 22% of dietary calories as coconut oil (Robison *et al.*, 1971).

Malinow *et al.* (1966) studied the atherosclerotic lesions in the aorta and coronary arteries of squirrel monkeys fed a monkey chow-based diet containing 10 mg cholesterol/kcal, butter, or margarine as 30% of dietary calories or combinations thereof. The amount of aortic sudanophilia was increased by all di-

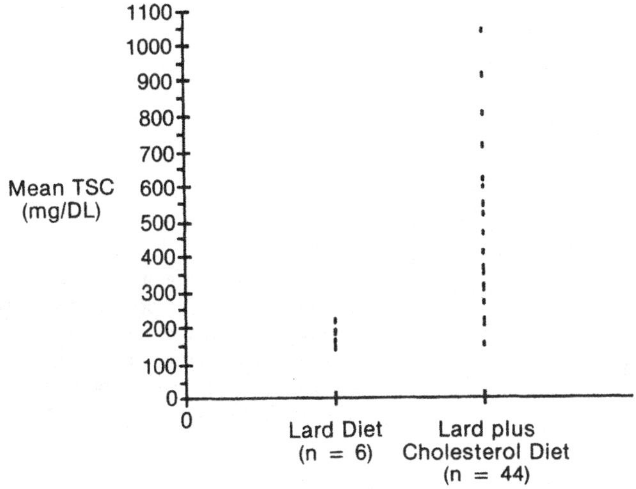

Fig. 4. Mean total serum cholesterol (TSC) concentrations for squirrel monkeys fed a lard-containing diet without added cholesterol for 24 months (n = 6) and for squirrel monkeys fed the same diet plus 5.0 g cholesterol/kg for 30 months (n = 44). [Adapted from Lofland *et al.* (1970) and Clarkson *et al.* (1971).]

etary lipid additions, especially, though not only, when cholesterol was also added to the diet. The amount of sudanophilia and fibrosis in the coronary arteries was increased only when cholesterol was added to the diet, with or without fat. The most extensive sundanophilia in both the coronary arteries and the aorta was seen in animals eating diets containing cholesterol plus butter. The investigators saw a protective effect against the development of sudanophilic lesions in the coronary arteries when margarine was added to cholesterol in the diet, though the addition of margarine did not prevent hypercholesterolemia.

3.4. Genetic Influences on Diet-Induced Atherosclerosis

Individuality in response to consumption of cholesterol-containing diets has been studied extensively in the squirrel monkey. Although marked differences in plasma cholesterol concentrations are not apparent in monkeys fed cholesterol-free diets, some monkeys (hyperresponders) fed cholesterol-containing diets develop considerable hypercholesterolemia, with TSC levels in the range of 700–1000 mg/dl (Fig. 4). Other monkeys (hyporesponders) fed the same diet maintain TSC concentrations around 170–250 mg/dl (similar to those of controls), while many other individuals are intermediate in their responsiveness (Lofland *et al.*, 1970; Clarkson *et al.*, 1971). In studies by Clarkson *et al.* (1971, 1976), hyporesponders fed cholesterol-containing diets for long periods of time were found to have minimal arterial lesions, while hyperresponders fed the same diets had markedly exacerbated atherosclerosis.

Approximately 65% of the variability in the level of TSC in cholesterol-fed squirrel monkeys is attributable to genetic factors, a figure similar to that

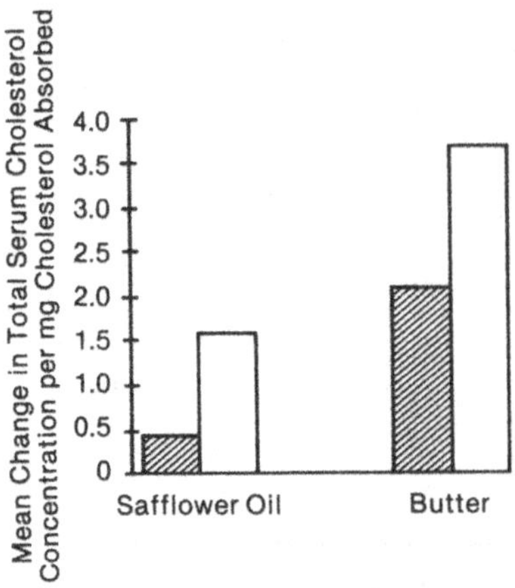

Fig. 5. Mean change in total serum cholesterol (TSC) concentration per mg of dietary cholesterol absorbed by hyporesponsive (▨, $n = 2$) and hyperresponsive (□, $n = 2$) squirrel monkeys fed diets containing safflower oil plus cholesterol or butter plus cholesterol. [Adapted from Jones *et al.* (1975).]

describing the human situation. This strong heritability of variation of TSC level in response to dietary cholesterol had made possible the development of colonies of hyper- and hyporesponsive squirrel monkeys through selective breeding programs (Clarkson *et al.*, 1971). By studying such colonies, Lofland *et al.* (1972) demonstrated that hyporesponders fed a cholesterol-containing diet have a much greater and more rapid increase in fecal excretion of bile salts than do hyperresponders. The investigators speculated that the mechanism of control of the level of plasma cholesterol is related to the rate of conversion of cholesterol to bile salts.

Jones *et al.*, (1975) analyzed further the mechanism for control of plasma cholesterol levels in hyporesponders by monitoring the absorption of cholesterol and the excretion of bile acids and cholesterol in hyper- and hyporesponders fed safflower oil (unsaturated fat) or butter (saturated fat) as 40% of dietary calories In hyporesponders fed unsaturated fat, almost no increase in the TSC level occurred; however, when the monkeys were fed saturated fat, an early increase in TSC of nearly 100% developed and was found to be due to increased absorption of ingested cholesterol. This early rise in TSC level was followed by a decrease and plateauing to concentrations near baseline due to an increase in bile acid excretion. It was noted that the excretion of cholesterol via enhanced catabolism to bile acids occurred too late to prevent the initial rise in TSC seen in hyporesponders. Hyperresponders fed unsaturated fat developed 100% increases in TSC levels due to increased absorption of ingested cholesterol and decreased bile acid excretion, and when they were fed saturated fat, their TSC levels climbed even higher. In both cases of dietary manipulation of hyperresponders, increased amounts of bile acid excretion eventually allowed for a plateauing of TSC level, but unlike that in hyporesponders, the TSC in hyperresponders plateaued at a much higher level. Hyperresponders absorbed a higher mean percentage of ingested cholesterol than did hyporesponders, and hyperresponders were more sensitive to a given amount of absorbed cholesterol than were hyporesponders (Fig. 5).

4. ADDITIONAL EFFECTS ON DIET-INDUCED ATHEROSCLEROSIS

The extent to which dietary cholesterol and fat affect levels of TSC and atherosclerosis in squirrel monkeys is influenced by many nondietary factors, including age, sex, presence of a disease state such as hypothyroidism, hypertension, or diabetes mellitus, and the presence or absence of psychic stress.

4.1. Effects of Age and Sex

Age has been shown to have a significant effect on the degree of hypercholesterolemia and the extent of atherosclerosis in squirrel monkeys consuming cholesterol-containing diets. In studies by Clarkson *et al.* (1976), young squirrel monkeys under 1 year of age developed less hypercholesterolemia than older monkeys fed the same cholesterol-containing diet. Although the prevalence of aortic atherosclerosis was similar in both young and old squirrel monkeys fed cholesterol, the extent of atherosclerosis was much greater in the adults (based on percentage of aortic intimal surface area affected with atherosclerotic lesions or on aortic cholesterol concentration). Middleton, *et al.*, (1967a) found that young squirrel monkeys fed a diet containing 0.5% cholesterol by weight for 12 months developed an increase in TSC levels during the first 6 months and then a decline in levels during the last 6 months, while older monkeys had a steady rise in TSC levels over the entire 12-month dietary period. Older monkeys developed more severe coronary artery atherosclerosis when fed the cholesterol-containing diet than did younger monkeys; however, no significant differences were found in the severity of aortic atherosclerosis between the two age groups.

Estrus in female squirrel monkeys during their breeding season (March through June) significantly affects cholesterol metabolism, but has little or no effect on the extent of atherosclerosis. TSC levels in cholesterol-fed anestrous females range from about 375 to 560 mg/dl. While these same females are cycling, their TSC levels drop to about 315 mg/dl in March and 220 mg/dl in June (Clarkson *et al.*, 1976) (Fig. 6).

4.2. Effects of Hypothyroidism, Insulin Deficiency, and Renally Induced Hypertension

Lehner *et al.* (1971) have studied the effects of experimentally induced hypothyroidism (^{131}I injection method), insulin deficiency (alloxan monohydrate injection method), and hypertension (Grollman renal manipulation method) on diet-induced atherosclerosis in squirrel monkeys. A natural products diet containing 1.0 mg cholesterol/kcal was fed to test and control monkeys for a period of 3 years, and all monkeys were monitored for changes in levels of TSC, LDL cholesterol, systolic blood pressure, and glucose tolerance. When necropsy findings were correlated with *antemortem* physiological parameters, the levels of systolic blood pressure, TSC, and LDL cholesterol were significantly and positively correlated with coronary and aortic atherosclerosis. The

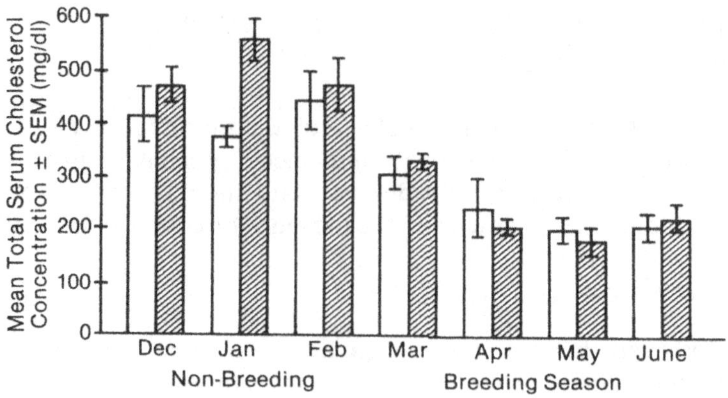

Fig. 6. Mean total serum cholesterol (TSC) concentrations of female squirrel monkeys fed a diet containing 1 mg cholesterol/kcal plus safflower oil or butter during nonbreeding and breeding seasons. □, Safflower oil diet ($n = 5$); ▨, butter diet ($n = 7$). [Adapted from Clarkson *et al.* (1976).]

hypothyroid monkeys and the insulin-deficient monkeys had significantly higher levels of TSC and LDL cholesterol than did control monkeys, though these differences were not seen between hypertensive monkeys and controls. All three groups of test monkeys had more extensive coronary and aortic atherosclerosis than did the control group, the lesions being most extensive and severe in the insulin-deficient monkeys (Figs. 7 and 8).

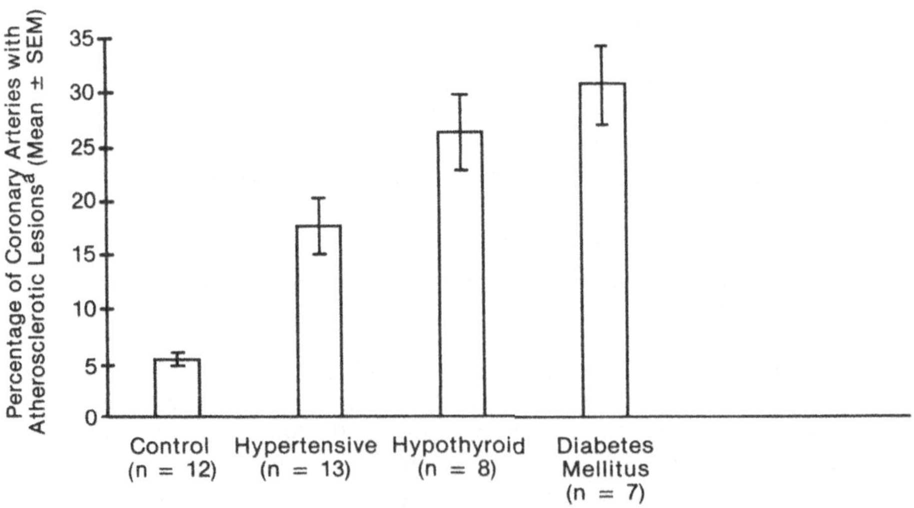

Fig. 7. Coronary artery atherosclerosis in control ($n \approx 12$), hypertensive ($n = 13$), hypothyroid ($n = 8$), and diabetic ($n = 7$) squirrel monkeys fed a diet containing 1 mg cholesterol/kcal for 3 years. Values are expressed as total number of coronary arteries observed with intimal lesions containing fat times 100, divided by the total number of arteries observed in 15 sections of heart. [Adapted from Lehner *et al.* (1971) and Clarkson *et al.* (1976).]

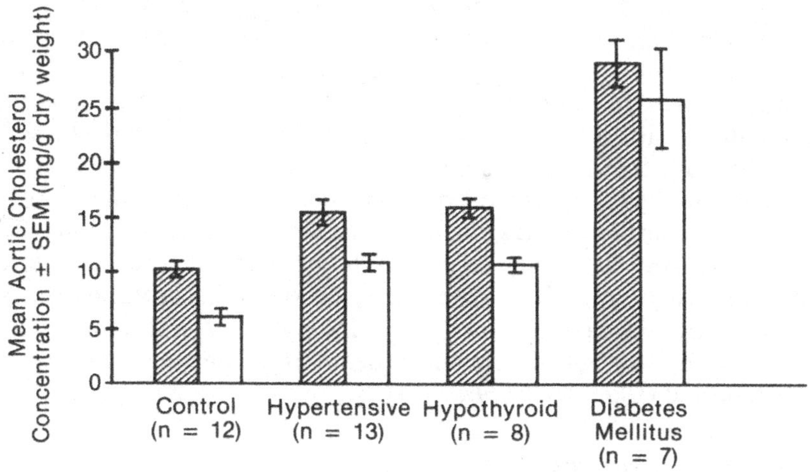

Fig. 8. Thoracic (▨) and abdominal (□) aorta cholesterol concentrations of control (n = 12), hypertensive (n = 13), hypothyroid (n = 8), and diabetic (n = 7) squirrel monkeys fed a diet containing 1 mg cholesterol/kcal for 3 years. [Adapted from Lehner *et al.* (1971) and Clarkson *et al.* (1976).]

4.3. Effects of Psychic Stress

Psychic stress has been found to have an effect on the response of squirrel monkeys to consumption of a cholesterol-containing diet. In a 2-year study by Lang (1967), 18 squirrel monkeys (fed a diet containing 0.5–1.0% cholesterol and 8–14% fat by weight) were divided into three groups. Group I monkeys were placed in a Skinner box and subjected to additional psychological stress by the Sidman avoidance procedure for 1 hr/day for 5 days/week. Group II monkeys were placed in a Skinner box for the same amount of time but were not stressed further, and group III monkeys were left alone in their cages, serving as controls. Despite differences in protocol, all test monkeys (groups I and II) appeared to have been subjected to similar amounts of stress. Both test groups had higher mean TSC levels following periods of stress than did control monkeys, and had more marked coronary artery atherosclerosis at the end of the study than did controls. Group I monkeys also had significantly higher levels of urinary 17-ketosteroids (cortisol metabolite) than did controls.

5. SUMMARY OF LESIONS AND COMPLICATIONS OF ATHEROSCLEROSIS

5.1. Summary of Lesions by Location

The most frequently occurring atherosclerotic lesion in the squirrel monkey and the predominant lesion of the aorta of squirrel monkeys is fatty streaking. In studies by Middleton *et al.* (1964) on naturally occurring atherosclerosis in the squirrel monkey, fatty streaks were located principally at the aortic arch and iliac bifurcation of the aorta, with small, raised plaques near the iliac bifurcation. When lesions in squirrel monkeys were enhanced with dietary lipids, Malinow *et al.* (1966) found that the incidence of intimal sudanophilia and fibrosis was greatest in the thoracic aorta, followed by the abdominal aorta, branches of the aortic arch, and the common iliac arteries. In squirrel monkeys selectively bred for their hyperresponsiveness to cholesterol-containing diets, fatty streaks as well as plaques have been found extending from the aortic arch into the common carotid arteries and involving the carotid bifurcations (Clarkson *et al.*, 1971). Atherosclerotic lesions in the intracranial cerebral arteries of the squirrel monkey have led to vascular occlusion and cerebral infarction in at least one reported case (Bullock and Moossy, 1972). Mesenteric artery plaques with secondary bowel ischemia or death from intestinal infarction have also been reported (Clarkson *et al.*, 1971).

5.2. Congestive Heart Failure

Clarkson *et al.* (1971) found clinically significant foam cell lesions in the aortic valves of hypercholesterolemic squirrel monkeys selectively bred for their hyperresponsiveness to cholesterol-containing diets. Death due to congestive heart failure secondary to the development of aortic valve lesions was reported. Affected valve leaflets were thickened by the accumulation of foam cells, cholesterol clefts, and inflammatory cells.

5.3. Myocardial Infarction

Unlike the human situation, coronary artery atherosclerosis in the squirrel monkey primarily affects the small and medium-sized intramyocardial arteries with no pathological consequences. However, among hyperresponsive monkeys coronary artery lesions are more extensive, often calcified, and more frequently involve the proximal main branches (Clarkson *et al.*, 1971).

5.4. Aneurysms

More reported cases of extracranial aneurysm have occurred in the squirrel monkey than in any other species of nonhuman primate. Case reports have described a fusiform aneurysm of the common carotid artery in one squirrel monkey (Malinow *et al.*, 1966), multiple saccular aortic aneurysms in two squirrel monkeys (Malinow *et al.*, 1966; Lehner *et al.*, 1971), and dissecting aortic aneurysms in three squirrel monkeys (Middleton *et al.*, 1964; Boorman *et al.*, 1976; Kessler *et al.*, 1979).

In a retrospective study by Strickland and Bond (1983) of aneurysms documented at necropsy in a large colony of squirrel monkeys, 11 animals out of the 730 reviewed were found to have aneurysms. The majority of aneurysms were dissecting, although some saccular and fusiform types were also observed. Dissecting aneurysms occurred exclusively in the aorta, and saccular aneurysms were found only in the common carotid artery. The development of dissecting aneurysms was not limited to animals with diet-induced hypercholesterolemia; however, saccular and fusiform aneurysms occurred only in animals consuming atherogenic diets. Similarities were seen between the gross anatomical pathology of aneurysms in the squirrel monkey and those observed in human beings. Although the rate of incidence for all aortic aneurysms in the colony squirrel monkeys at necropsy was only about half that reported for human beings at autopsy, the incidence rate for dissecting aortic aneurysm was threefold greater in the squirrel monkey than in human beings.

5.5. *Regression of Diet-Induced Atherosclerosis*

There is only a limited number of studies to date of the regression of diet-induced atherosclerotic lesions in squirrel monkeys. Maruffo and Portman (1968) examined the reversibility of coronary artery atherosclerosis in squirrel monkeys by first inducing lesions during a 3-month period of feeding a butter-rich diet and then withdrawing cholesterol from the diet of the test group for the next 3–5 months. Lesions in the distal branches of coronary arteries did not progress in the test group as they did in control monkeys fed the atherogenic diet for all 8 weeks. Although morphological changes were observed in the lesions of the test group, the investigators could not definitely interpret them as being representative of lesion regression. At necropsy the lesions in the distal branches of the coronary arteries in the test group were characterized by intimal accumulations of mucopolysaccharide along with fragmentation of elastic laminae, proliferation of collagen, and calcification. Two possible interpretations were proposed: either the lesions were lipid-rich during induction and then regressed by loss of lipid, or the lesions represented an early stage of development of a type of lesion similar to the gelatinous lesions of human beings. The second interpretation would suggest that an intermittent rather than a continuous lipid effect on the vessel wall may be a mechanism in human atherogenesis.

Hayes *et al.* (1972) have suggested that early diet-induced lesions of squirrel monkeys will regress only if they are not complicated by aortitis. Their findings also indicate that the degree of diet-induced atherosclerosis in squirrel monkeys is increased when aortitis is present.

6. *STRESS-INDUCED CARDIOMYOPATHY*

Several studies have been conducted on the use of the squirrel monkey as a model for stress-induced cardiomyopathy. Functional cardiac changes and myocardial necrosis have been associated with enhanced sympathetic activity in monkeys subjected to the stress of shock avoidance testing. Corley *et al.* (1975) found that the cardiac response of squirrel monkeys to stress varies depending on whether the stress occurs in an avoidance or a yoked situation. Monkeys in a yoked situation were found to have a parasympathetic-like response with bradycardia and ventricular arrest. Further studies compared the effects of stress of shock avoidance testing on propranolol-treated monkeys with the effects on untreated vagotomized and nonvagotomized animals (Table

Table II
Myocardial Lesions in Squirrel Monkeys Subjected to the Stress of Shock Avoidance Testing[a]

Monkeys by condition	Myofibrillar degeneration[b]	Myocardial necrosis[b]
Untreated vagotomized (n = 6)	2.0	2.5
Untreated, intact vagus (n = 22)	2.5	1.0
Propranolol-treated, intact vagus (n = 6)	1.3	0.5
Nonstressed controls (n = 10)	0.4	0.1

[a] Adapted from Corley *et al.* (1980).
[b] Lesions graded under light microscopy and rated on a scale from 0 to 4 (Corley *et al.*, 1977).

II). There was a higher frequency of occurrence of myofibrillar degeneration, necrosis, and leukocytic infiltration, similar to catecholamine-induced pathology, in untreated vagotomized and nonvagotomized test animals than in propranolol-treated test animals or in unstressed controls (Corley *et al.*, 1980).

Extensive myocardial necrosis, distributed in a pattern inconsistent with sudden occlusion of a coronary artery, has been seen in squirrel monkeys at the Bowman Gray School of Medicine. It has been found most often in squirrel monkey infants that have died while undergoing the stress of weaning at about 4 months of age. The necrosis appears to be of metabolic origin related to stress and high cortisol levels (Prichard and Bullock, 1965).

7. BEHAVIORALLY INDUCED HYPERTENSION

Herd *et al.* (1969) have studied arterial hypertension in squirrel monkeys trained under operant-conditioning schedules. In the training sessions, monkeys were required to press a key in order to turn off a light and prevent electrical shock. Significant elevations in mean arterial blood pressure were produced in four of the six squirrel monkeys studied over a 4- to 11-month period. Eventually the elevations in blood pressure were present before, during, and after each test session, even if noxious stimuli were not delivered. These results suggested that operant-conditioning schedules can induce marked persistent elevations in systemic mean arterial blood pressure.

In another study, Benson *et al.* (1969) modified the operant-conditioning procedure in order to eliminate any effects of key pressing *per se* on cardiovascular function. In the first phase of the experiment, each monkey was required to raise its mean arterial blood pressure in order to avoid electrical shock. Blood pressures in the test animals rose to hypertensive levels during each training session. In the second phase of the study, each monkey was required to decrease its mean arterial blood pressure in order to avoid electrical shock, and blood pressures in the test animals returned toward control values.

8. DISEASES COMPLICATING THE USE OF SQUIRREL MONKEYS IN CARDIOVASCULAR RESEARCH

Many diseases have the potential for complicating the use of nonhuman primates in cardiovascular research; however, when using squirrel monkeys, certain diseases are of particular concern because of their high rate of incidence in the species, their relationship to the consumption of diets containing fat and cholesterol, or their effect on the myocardium, serum cholesterol level, and progression or regression of atherosclerotic lesions. These diseases include cholelithiasis, glomerulonephritis, aortitis, and trypanosomiasis.

8.1. Cholelithiasis

Cholelithiasis resulting from consumption of diets containing fat and cholesterol occurs frequently in the squirrel monkey. Osuga and Portman (1971) were able to induce formation of cholesterol gallstones in squirrel monkeys by feeding them a semipurified diet containing corn oil as 15% of dietary calories and no cholesterol. No gallstones were found in monkeys fed commercial monkey chow. The highest incidence of cholelithiasis (83%) was in monkeys fed a semipurified diet containing 1 mg cholesterol/kcal and butter as 45% of dietary calories. The gallstones first appeared in the monkeys 3 months after the start of the atherogenic diet and increased in size, number, and total mass during the 9-month period of feeding.

Other studies have shown an inverse relationship between TSC concentration and the frequency of stone formation in the squirrel monkey, with polyunsaturated fats exacerbating the formation of gallstones (Lofland, 1975; Lofland *et al.*, 1974; Lofland and Clarkson, 1975). Monkeys fed a diet containing 1 mg cholesterol/kcal and safflower oil as 45% of dietary calories had a lower mean TSC concentration, a higher incidence of cholesterol gallstones, and a higher mean stone weight than did animals fed diets containing the same amount of cholesterol with butter or lard or coconut oil as 45% of dietary calories.

The susceptability of squirrel monkeys to cholelithiasis is age dependent. In a study by Clarkson *et al.* (1976), the extent and prevalence of cholelithiasis in squirrel monkeys fed semipurified diet containing lard and added cholesterol were greater in adults (over 5 years old) than in juveniles (about 2 years old).

Portman *et al.* (1980) found no clear differences in incidence of cholelithiasis between male, female, castrated male, and ovariectomized female Brazilian-type squirrel monkeys, but did find that male Bolivian-type squirrel monkeys were resistant to cholesterol gallstone formation. The Bolivian-type squirrel monkeys also maintained lower TSC levels than did Brazilian-type squirrel monkeys when fed identical diets containing butter and added cholesterol.

8.2. Glomerulonephritis

Old squirrel monkeys and squirrel monkeys that have consumed cholesterol-containing diets for long periods of time develop a nephrotic syndrome relatively frequently (Clarkson *et al.*, 1976). Affected monkeys may have hypercholesterolemia, hypoalbuminemia, as well as sudanophilic globules in the urine. Even in the absence of clinical signs of renal disease, histological changes consistent with glomerulonephritis are often found in the kidneys. Lesions appear to be more extensive in animals consuming cholesterol-containing diets than in animals consuming commercial monkey chow; however, it is not known whether the renal changes are induced by dietary cholesterol or occur naturally and are merely aggravated by cholesterol. Lesions include changes in the basement membrane and mesangium with or without cell proliferation. Foam cells may be present in the glomeruli of monkeys fed cholesterol-containing diets (Clarkson *et al.*, 1976).

8.3. Aortitis

In a study by Hayes *et al.* (1972), 44% of wild-caught squirrel monkeys were found to have an aortitis, which enhanced the progression and prevented the regression of diet-induced atherosclerosis. The most prevalent lesion was a periaortitis in the ascending aorta and aortic arch characterized by cellular infiltrates in the adventitia. The degree of diet-induced atherosclerosis was increased in monkeys with aortitis, and in lesion regression studies, the regression of diet-induced atherosclerosis was absent in monkeys with aortitis, yet was significant in monkeys without aortitis. Because aortitis enhanced the atherogenic process in squirrel monkeys consuming cholesterol-containing diets, it represents a potential bias when this species is used in atherosclerosis research. It is not known whether the prevalence of aortitis in wild populations of squirrel monkeys contributes to the relatively high incidence of naturally occurring atherosclerosis in this species.

8.4. Trypanosomiasis

The squirrel monkey is reportedly among the most favored hosts of *Trypanosoma cruzi*. Although some references state that *T. cruzi* is nonpathogenic in its natural primate host, knowledge of the course of trypanosomiasis in the squirrel monkey is limited, and severely infected monkeys probably are symptomatic. One Bolivian-type squirrel monkey was reported to have numerous leishmanic forms in the heart muscle at necropsy (Rosenblum and Cooper, 1968). Malinow *et al.* (1966) found *T. cruzi*-like protozoa in the peripheral blood of 29 out of 120 recently imported squirrel monkeys. In infected human beings, trypanosomiasis passes from an acute septicemic phase to a chronic phase with a very low parasitemia, and parasitic pseudocysts are rarely found in the body tissues of chronically infected human patients (Koberle, 1974).

Based on the extent of cardiac pathology known to be caused by *T. cruzi* infection in human beings, it might be assumed that severe infections in squirrel monkeys could complicate the use of these animals in cardiovascular research. In human beings, trypanosomiasis frequently produces an acute myocarditis with localized inflammation and ganglion cell destruction leading to the development of a chronic cardiomyopathy. This chronic form of the disease,

Chagas cardiomyopathy, is characterized by varying degrees of cardiomegaly, myocardial necrosis, apical aneurysm, and change in cardiac function. Atrioventricular conduction disturbances, arrhythmias, and thromboembolism, cardiac failure, and sudden death are not uncommon in human patients (Koberle, 1974).

9. CONCLUSION

The squirrel monkey has been a valuable model for the study of human cardiovascular disease. Several similarities exist between atherosclerosis in squirrel monkeys and in human beings, which has made the squirrel monkey especially useful to researchers. Squirrel monkeys develop atherosclerosis under natural conditions or when fed cholesterol-containing diets, and the only major difference between atherosclerosis in the squirrel monkey and in human beings is the location of coronary lesions (in the squirrel monkey, coronary atherosclerosis is found primarily in the small intramyocardial arteries rather than in the proximal main coronary arteries). The TSC concentration in squirrel monkeys housed under laboratory conditions and the distribution of serum lipoproteins in squirrel monkeys in the wild are similar to those found in human beings. Furthermore, the genetic heritability of TSC level variation in squirrel monkeys is similar to that of human beings.

Squirrel monkeys also seem to be good models for the study of other cardiovascular diseases, including stress-related cardiomyopathy, behaviorally induced hypertension, and dissecting aneurysm. Several diseases can complicate the use of the squirrel monkey in cardiovascular research; however, affected monkeys may serve as valuable models for further study of some of those diseases, as has been the case with cholelithiasis.

REFERENCES

Benson, H., Herd, J. A., Morse, W. H., and Kelleher, R. T., 1969, Behavioral induction of arterial hypertension and its reversal, *Am. J. Physiol.* **217**:30–34.

Boorman, G. A., Silverman, S., and Anderson, J. H., 1976, Spontaneous dissecting aortic aneurysm in a squirrel monkey (*Saimiri sciureus*), *Lab. Anim. Sci.* **26**:942–947.

Bullock, B. C., and Moossy, J., 1972, Cerebral infarction with artherocierotic occlusion of the middle cerebral artery in a squirrel monkey, (*Saimiri sciureus*), fed an atherogenic diet, *J. Neuropathol. Exp. Neurol.* **31**:177.

Clarkson, T. B., Lofland, H. B., Bullock, B. C., and Goodman, H. O., 1971, Genetic control of plasma cholesterol, *Arch. Pathol.* **92**:37–45.

Clarkson, T. B., Lehner, N. D. M., Bullock, B. C., Lofland, H. B., and Wagner, W. D., 1976, Atherosclerosis in New World monkeys, in: *Primates in Medicine*, Volume 9 (J. P. Strong, ed.), Karger, Basel, pp. 90–144.

Corley, K. C., Mauck, H. P., and Shiel, F. O'M., 1975, Cardiac responses associated with "yolked-chair" shock avoidance in squirrel monkeys, *Psychophysiology* **12**:439–444.

Corley, K. C., Shiel, F. O'M., Mauck, H. P., Clark, L. S., and Barber, J. H., 1977, Myocardial degeneration and cardiac arrest in squirrel monkey: Physiological and psychological correlates, *Psychophysiology* **14**:322–328.

Corley, K. C., Shiel, F. O'M., and Mauck, H. P., 1980, Stress-induced cardiomyopathy in squirrel monkey, in: *The Use of Nonhuman Primates in Cardiovascular Diseases* (S. S. Kalter, ed.), University of Texas Press, Austin, pp. 357–360.

Hayes, K. D., Westmoreland, N. P., and Faherty, T. P., 1972, An aortitis in the squirrel monkey and its effect on atherosclerosis, *Exp. Mol. Pathol.* **17**:334–347.

Herd, J. A., Morse, W. H., Kelleher, R. T., and Jones, L. G., 1969 Arterial hypertension in the squirrel monkey during behavioral experiments, *Am. J. Physiol.* **217**:24–29.

Jones, D. C., Lofland, H. B., Clarkson, T. B., and St. Clair, R. W., 1975, Plasma cholesterol concentrations in squirrel monkeys as influenced by diet and phenotype, *J. Food Sci.* **40**:2–7.

Kessler, M. J., Kupper, J. L., Brown, R. K., and O'Neill, T. P., 1979, A prolonged hematopoietic disturbance associated with aortic aneurysm and splenomegaly in a squirrel monkey (*Saimiri sciureus*), *J. Med. Primatol.* **8**:377–381.

Koberle, F., 1974, Pathogenesis of Chagas' disease, in: *Trypanosomiasis and Leishmaniasis, with Special Reference to Chagas' Disease* (Ciba Foundation Symposium on Trypanosomiasis and Leishmaniasis, Caracas, 1973), Associated Scientific Publishers, Amsterdam, pp. 137–158.

Lang, C. M., 1967, Effects of psychic stress on atherosclerosis in the squirrel monkey (*Saimiri sciureus*), *Proc. Soc. Exp. Biol. Med.* **126**:30–34.

Lehner, N. D. M., Clarkson, T. B., and Lofland, H. B., 1971, The effect of insulin deficiency, hypothyroidism, and hypertension on atherosclerosis in the squirrel monkey, *Exp. Mol. Pathol.* **15**:230–244.

Lofland, H. B., 1975, Animal model: Cholelithiasis in Brazilian squirrel monkeys, *Am. J. Pathol.* **79**:619–622.

Lofland, H. B., and Clarkson, T. B., 1975, Cholesterol-lowering agents and cholelithiasis, *Artery* **1**:226–227.

Lofland, H. B., St. Clair, R. W., MacNintch, J. E., and Prichard, R. W., 1967, Atherosclerosis in New World primates, *Arch. Pathol.* **83**:211–214.

Lofland, H. B., Clarkson, T. B., and Bullock, B. C., 1970, Whole body sterol metabolism in squirrel monkeys (*Saimiri sciureus*), *Exp. Mol. Pathol.* **13**:1–11.

Lofland, H. B., Clarkson, T. B., St. Clair, R. W., and Lehner, N. D. M., 1972, Studies on the regulation of plasma cholesterol levels in squirrel monkeys of two genotypes, *J. Lipid Res.* **13**:39–47.

Lofland, H. B., Clarkson, T. B., and Melchior, G. W., 1974, Cholesterol gallstones in nonhuman primates, in: *Atherosclerosis III* (G. Schettler and A. Weizel, eds.), Springer-Verlag, Berlin, pp. 392–395.

Malinow, M. R., Maruffo, C. A., and Perley, A. M., 1966, Experimental atherosclerosis in squirrel monkeys (*Saimiri sciureus*), *J. Pathol. Bacteriol.* **92**:491–510.

Maruffo, C. A., and Portman, O. W., 1968, Nutritional control of coronary artery atherosclerosis in the squirrel monkey, *J. Atherosclerosis Res.* **8**:237–247.

McCombs, H. L., Zook, B. C., and McGandy, R. B., 1969, Fine structure of spontaneous atherosclerosis of the aorta in the squirrel monkey, *Am. J. Pathol.* **55**:235–252.

Middleton, C. C., Clarkson, T. B., Lofland, H. B., and Prichard, R. W., 1964, Atherosclerosis in the squirrel monkey, *Arch. Pathol.* **78**:16–23.

Middleton, C. C., Clarkson, T. B., Lofland, H. B., and Prichard, R. W., 1967a, Diet and atherosclerosis in squirrel monkeys, *Arch. Pathol.* **83**:145–153.

Middleton, C. C., Rosal, J., Clarkson, T. B., Newman, W. P., and McGill, H. C., 1967b, Arterial lesions in squirrel monkeys, *Arch Pathol.* **83**:352–358.

Osuga, T., and Portman, O. W., 1971, Experimental formation of gallstones in the squirrel monkey, *Proc. Soc. Exp. Biol. Med.* **136**:722–726.

Portman, O. W., and Andrus, S. B., 1964, Comparative evaluation of three species of New World monkeys for studies of dietary factors, tissue lipids, and atherogenesis, *J. Nutr.* **87**:429–438.

Portman, O. W., Alexander, M., Tanaka, N., and Osuga, T., 1980, Relationships between cholesterol gallstones, biliary function, and plasma lipoproteins in squirrel monkeys, *J. Lab Clin. Med.* **96**:90–101.

Prichard, R. W., and Bullock, B. C., 1965, Spontaneous myocardial necrosis in a squirrel monkey, *Fed. Proc.* **24**:997.

Rewell, R. E., 1954, Uterine fibromyomas an bilateral ovarian granulosa cell tumors in a senile squirrel monkey, *Saimiri sciurea*, *J. Pathol. Bacteriol.* **68:**291–293.

Robison, R. L., Hayes, K. C., McCombs, H. L., and Faherty, T. P., 1971, Effect of dietary fat and cholesterol on circulating lipids and aortic ultrastructure of squirrel monkeys, *Exp. Mol. Pathol.* **15:**281–303.

Rosenblum, L. E., and Cooper, R. W., 1968, *The Squirrel Monkey*, Academic Press, New York, pp. 44–45.

Rudel, L. L., and Lofland, H. B., 1976, Circulating lipoproteins in nonhuman primates, in: *Primates in Medicine*, Volume 9 (J. P. Strong, ed.), Karger, Basel, pp. 224–266.

St. Clair, R. W., MacNintch, J. E., Middleton, C. C., Clarkson, T. B., and Lofland, H. B., 1967, Changes in serum cholesterol levels of squirrel monkeys during importation and acclimation, *Lab. Invest.* **16:**828–832.

Strickland, H. L., and Bond, M. G., 1983, Aneurysms in a large colony of squirrel monkeys (*Saimiri sciureus*), *Lab. Anim. Sci.* **33:**589–592.

13

Behavioral Pharmacology of the Squirrel Monkey

James E. Barrett

1. INTRODUCTION: PROCEDURES IN BEHAVIORAL PHARMACOLOGY

Experiments on the effects of drugs on the behavior of the squirrel monkey have increased considerably in the 15 years since Hanson's (1968) report on the use of this species in pharmacological experiments. More widespread usage of the squirrel monkey in both behavioral and pharmacological studies has generated a sizable amount of information. The primary emphasis of this chapter will be on summarizing characteristic behavior and drug effects under operant conditioning procedures. However, the effects of drugs on the social behavior of the squirrel monkey, as well as species comparisons of the behavioral effects of drugs, will also be treated briefly. Detailed procedures for handling squirrel monkeys and for administering drugs will not be covered, since earlier treatments of this material (Hanson, 1968; Kelleher *et al.*, 1963) remain valid.

1.1. Behavioral Pharmacology

The field of behavioral pharmacology has traditionally focused on the ways in which drug effects on behavior are related to the manner in which that behavior is controlled by its environment. An integral part of behavioral pharmacology research has also been the study of the way drugs control behavior, e.g., when they are self-administered and/or abused, as well as when they serve as discriminative stimuli to set the occasion for a particular response.

Historically, many individuals who conducted early research in behavioral pharmacology were also interested in behavior controlled by its consequences—operant conditioning—and sought to use operant techniques as a

James E. Barrett • Department of Psychiatry, Uniformed Services University of the Health Sciences, School of Medicine, Bethesda, Maryland 20814.

means of evaluating the effects of drugs (Dews, 1955; Pickens, 1977; Sidman, 1959). Almost from the outset, it became apparent that certain features or characteristics of the way behavior was controlled by its consequences could play an important role in determining the behavioral effects of drugs. The subtle but complex and significant interplay among behavior, the environment, and drug effects has produced a substantial body of information with certain fundamental principles that are of widespread importance in understanding drug action. The squirrel monkey has been an intimate and integral part of this progress.

1.2. Schedule-Controlled Behavior

Behavioral pharmacology has relied extensively on behavior maintained under schedules of reinforcement. Briefly, a schedule of reinforcement refers to the specific manner in which environmental events are related to behavior. The use of schedules of reinforcement in behavioral pharmacology has been reviewed recently (Dews and DeWeese, 1977; McKearney and Barrett, 1978) and will not be treated in detail here. Some of the advantages of schedule-controlled behavior over certain types of unconditioned behavior or behavior not maintained explicitly by its consequences (e.g., spontaneous locomotor activity) are that it: (1) is orderly and reproducible within and across species; (2) is stable over long periods of time; (3) is sensitive to a variety of variables, including drug administration, and is generally recoverable after such interventions; and (4) can be manipulated in almost infinite ways by parametric alterations in the schedule or by changes in type of schedule. These procedures permit the determination of an entire dose-effect curve in a single subject. Perhaps most significant, however, is that by modifying the manner in which consequent events control behavior, principles of drug action can be revealed that would otherwise be unknown.

In most experiments using the squirrel monkey, the subject is placed in a Plexiglas restraint chair (Fig. 1) that is equipped with a lever, a means for delivering a suitable reinforcer such as liquid food or food pellets, and visual stimuli to provide discriminative control over behavior. In experiments using noxious stimuli, such as electric shock, the chair is also provided with brass electrodes that rest on a shaved portion of the tail. This type of chair, adapted from a version described initially by Hake and Azrin (1963), provides minimal restraint only at the waist.

1.3. Schedules Using Behavior Maintained by Food Presentation or Drug Injection

Performances of squirrel monkeys under various schedules of reinforcement are generally comparable to those of other laboratory animals. For example, under a schedule where, in the presence of one visual stimulus (e.g., red lights), the first response after 3 min produces food (a *fixed-interval* schedule) and, under a different visual stimulus (e.g., white lights), every 30th re-

Fig. 1. Squirrel monkey seated in a typical restraint chair used in behavioral pharmacology experiments.

sponse produces food (a *fixed-ratio* schedule), performances of squirrel monkeys, rabbits, and pigeons can be quite similar (Fig. 2).

In addition to the similarities in schedule performances across species, schedules can also impose a degree of uniformity across many different types of maintaining events. For example, performances of squirrel monkeys can be quite similar, whether maintained by food, drug administration, or other consequent events, as long as the schedules under which those events are delivered are also similar (Fig. 3). This ability to develop similar performances with different events is experimentally crucial for studies that compare drug effects on behaviors controlled by different environmental consequences. For example, if one is interested in assessing whether food- or drug-maintained behavior is differentially sensitive to punishment procedures, it is essential that the behaviors prior to the introduction of punishment be as comparable as possible in rate or frequency of occurrence, execution, and temporal distribution. Inequalities in these features are known to contribute directly to the sensitivity of behavior to other variables.

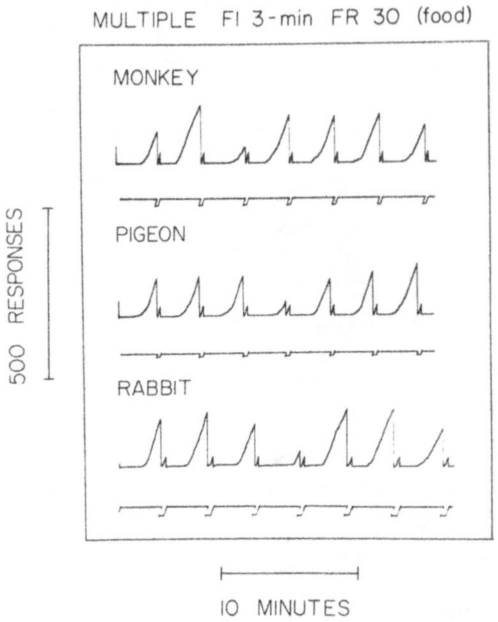

Fig. 2. Performances of a squirrel monkey, pigeon, and rabbit under a multiple fixed-interval (3-min), fixed-ratio (30-response) schedule of food presentation. The pen beneath each record was displaced during the fixed-ratio component; the response pen was reset after each food presentation. The pigeon pecked a Plexiglas key, whereas the monkey pressed and rabbit lifted a lever. All animals were reduced to 85% of their free-feeding body weights. The reinforcer was 4 sec of access to mixed grain for the pigeon, a 1.0-ml dipper of 0.25% saccharin solution for the rabbit, and a 300-mg banana-flavored pellet for the monkey. Note that patterns of responding were similar across the different species.

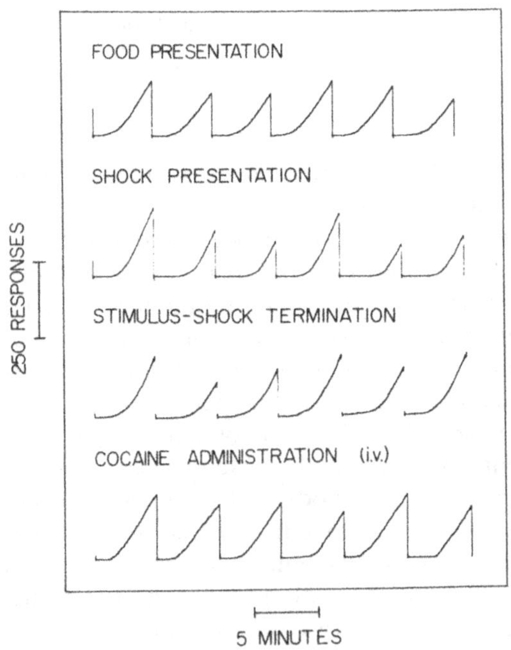

Fig. 3. Comparable performances of squirrel monkeys responding under 5-min fixed-interval schedules with different maintaining events. Food presentation consisted of the delivery of a 300-mg banana-flavored pellet; shock presentation was a 200-msec, 9-mA electric shock delivered to the shaved portion of the monkey's tail. Under the stimulus-shock termination schedule, 9-mA shocks were scheduled to occur beginning 1 sec after the 5-min interval had elapsed; a response after the 5-min interval terminated the prevailing stimuli and shock schedule and produced a 1-min timeout period. Cocaine hydrochloride was injected via an indwelling intravenous catheter connected to a motor-driven syringe; a response after the 5-min fixed interval produced a 50 μg/kg infusion of cocaine. The pens reset to baseline at reinforcement. [From Barrett and Katz (1981).]

1.4. Schedules Using Noxious Stimuli

A wide variety of procedures has been developed with the squirrel monkey for controlling behavior under various schedules of noxious stimuli. Significantly, it has been possible to develop performances maintained by noxious stimuli that are similar to those maintained by other events, such as food presentation or drug administration. These procedures have been especially useful in comparing drug effects on behaviors maintained by noxious events with those maintained by events such as food or drug administration (see Section 2.3.). Early experiments in behavioral pharmacology focused on behavior controlled by noxious events because of the role noxious environments theoretically played in the development of psychiatric disorders for which pharmacological management was sought. The procedures summarized in the next section represent the major types of techniques that have been used with the squirrel monkey both in preclinical psychopharmacology (e.g., drug screening and evaluation) and in studies of the behavioral determinants of drug action. The use of schedules using noxious stimuli in behavioral pharmacology was recently reviewed by Morse *et al.* (1977).

1.4.1. Punishment

Punishment refers to a reduction in responding due to the response-dependent presentation of some event. With squirrel monkeys, electric shock presentation typically has been used as a punisher, although other consequences, such as pressurized air (Spealman, 1979) and histamine injections (Goldberg, 1980), have also been studied. Generally, responding is developed initially under a schedule of food presentation and the event to be used as a punisher is added as an additional consequence of responding. One type of punishment schedule that has been studied with the squirrel monkey is illustrated in Fig. 4. Responding was first established under a 3-min fixed-interval schedule of food presentation in the presence of two different visual stimuli (top record). When shock was scheduled to follow every 30th response during alternate components, responding was reduced during that component but was not altered during the other (lower record). Procedures that develop both punished and unpunished responding in the same animal are useful because they allow one to determine whether drugs exert selective effects on punished and unpunished behavior. Punished performances of this type, frequently characterized as "conflict" procedures, have been used extensively in behavioral pharmacology and in the evaluation of new compounds, particularly those with antianxiety properties (Sepinwall and Cook, 1978).

1.4.2. Shock Postponement and Termination

Whereas punishment schedules *reduce* responding by shock presentation, other schedules *increase* responding by the postponement or termination of shock. Under shock-postponement schedules (sometimes called continuous avoidance), shocks are scheduled to occur every x seconds; a response post-

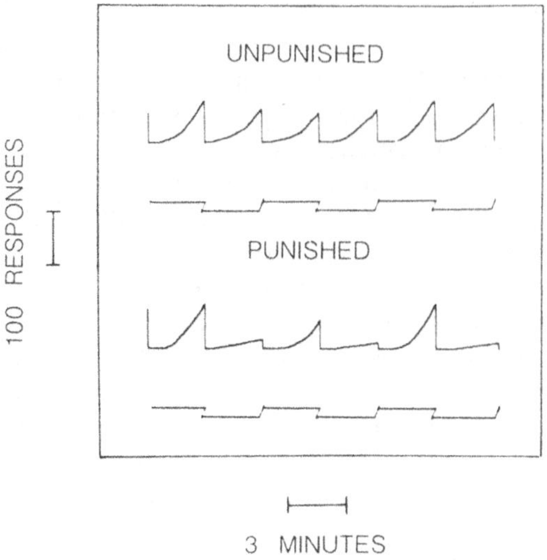

Fig. 4. Cumulative response records showing suppression of responding by punishment in the squirrel monkey. In the top record responding was maintained under a multiple 3-min fixed-interval schedule; the first response after 3 min produced food in the presence of each of two different visual stimuli. When every 30th response during one component produced a 7-mA electric shock, responding during that component was suppressed (lower record).

pones the next scheduled shock for *y* seconds. Squirrel monkeys generally develop steady rates of responding under this schedule and receive few shocks once stable performances have been established.

A number of studies have also been conducted with squirrel monkeys under a procedure in which responding terminates a stimulus in the presence of which shocks occur (Azrin *et al.*, 1962; Morse and Kelleher, 1966). As with other schedules, the specific contingencies under which the stimulus and correlated shocks are terminated determine the rate and temporal pattern of responding. Morse and Kelleher (1966), for example, maintained responding under fixed-ratio and fixed-interval stimulus-shock termination schedules. Under the fixed-ratio schedule, responding terminated a stimulus in the presence of which shocks would occur if a specified time elapsed without the completion of the ratio requirement. Completion of the ratio produced a distinctive stimulus period during which shocks were not delivered. Under the fixed-interval schedule shocks were delivered at various times after the interval elapsed, and the first response after the interval terminated the prevailing stimulus and produced a shock-free period. Under these schedules, performances were comparable to those obtained when other events maintain responding (see Fig. 3 for an example of fixed-interval performances under this schedule).

1.4.3. Shock Presentation

Although shock presentation can suppress responding, as under punishment procedures, there are also conditions under which the presentation of shock will increase and maintain responding of the squirrel monkey. In many cases, these performances can be remarkably similar to those maintained by food delivery, stimulus-shock termination or drug administration (Fig. 3).

Performances maintained by response-produced shock have been studied extensively since they were first reported by Kelleher and Morse (1968a), and have been reviewed in several recent articles (Barrett and Katz, 1981; McKearney and Barrett, 1978; Morse and Kelleher, 1977; Morse *et al.*, 1977). Although detailed treatment of this intriguing finding is beyond the scope of this chapter, a few essential points will be made. First, the development and maintenance of responding under schedules of response-produced shock depend on prior experience; performances of the type shown in Fig. 3 do not occur without a suitable behavioral history. Because most reinforcers require a suitable history or specified set of conditions for them to be effective (e.g., food deprivation), the maintenance of responding by shock is not unique in this respect. Second, environmental stimuli can exert multiple effects on behavior, depending on how they are scheduled. Electric shock, as well as other types of consequences, can simultaneously both *maintain* and *suppress* responding in the same monkey and can therefore function both as a reinforcer or punisher. These findings, based on the results of research conducted with the squirrel monkey, have been important contributions to a more complete and accurate understanding of factors determining the effects consequent events can have on behavior. Additionally, the maintenance of responding under schedules of response-produced shock has permitted other direct questions concerning factors that determine the effects of drugs on behavior.

2. *DETERMINANTS OF THE BEHAVIORAL EFFECTS OF DRUGS*

Research in behavioral pharmacology has focused largely on an analysis of the manner in which the behavioral effects of drugs are altered or determined by variables that also affect behavior. In early studies that blended the procedures of operant conditioning with those of pharmacology, the first principle to emerge was that of the contribution of the schedule of reinforcement. Initial studies by Dews (1955) with pigeons indicated that different schedule-controlled performances were differentially sensitive to the effects of pentobarbital. For example, pentobarbital *increased* responding under fixed-ratio schedules at doses that *decreased* responding of the same animals under fixed-interval schedules. This differential sensitivity of behavior to the effects of pentobarbital depending on the schedule of reinforcement strongly suggested that the schedule of reinforcement could play an important role in determining the behavioral effects of drugs.

2.1. Response Rate

Schedules of reinforcement produce different rates and patterns of responding which, it appears, can directly affect changes in behavior produced by many drugs. A second experiment by Dews (1958), also with pigeons, reported that the effects of methamphetamine were related to the rate at which responding occurred. Relatively high rates of responding were decreased by doses of methamphetamine that increased the responding that occurred at lower rates. This relationship between response rate and drug effects appears to be

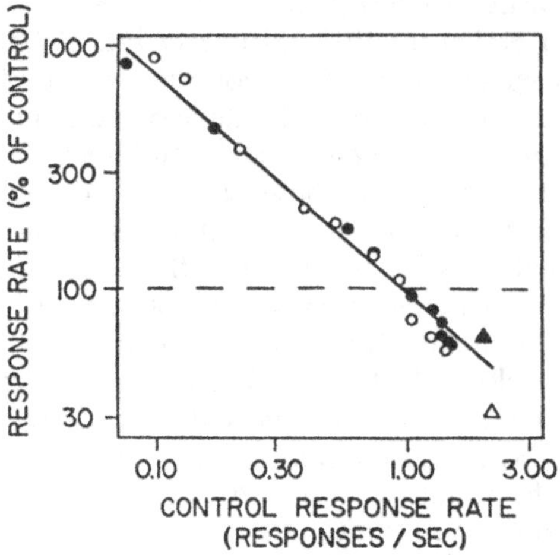

Fig. 5. Rate-dependent effects of *d*-amphetamine in the squirrel monkey. Abscissa: average rate of responding in successive minutes of a 10-min fixed-interval schedule (circles) and under a 30-response fixed-ratio schedule (triangles) of stimulus-shock termination. Ordinate: percentage change in rate of responding after 0.3 mg/kg *d*-amphetamine, intramuscularly. Open and filled symbols represent data taken from two different sessions. [From Kelleher and Morse (1968b).]

a general characteristic of the effects of many psychomotor stimulants as well as other drugs in many species, including the squirrel monkey (Fig. 5), and has been confirmed repeatedly across many conditions (Dews and Wenger, 1977; Robbins, 1982). The rate-dependent effect of drugs, though certainly not pervasive and all-determining, has been an important guiding principle in behavioral pharmacology. The possibility that a drug's behavioral effects may be related to the rate at which that behavior normally occurs must always be considered, particularly when comparing drug effects across experimental conditions that engender different response rates.

2.2. Behavioral and Pharmacological History

The schedule-controlled rate of responding is a variable contributing to the behavioral effects of drugs that is evident in current behavior. However, recently it has been shown that conditions that have occurred in the past can play a significant role in determining drug effects even though those influences are not apparent in ongoing behavior. For example, the effects of *d*-amphetamine on punished behavior of squirrel monkeys can depend on whether those monkeys were trained previously under a shock-postponement or avoidance schedule. Without prior avoidance training, *d*-amphetamine will decrease punished responding (left panel, Fig. 6). However, if administered to the same subjects *after* exposure to a shock-postponement schedule, *d*-amphetamine produces large increases in punished responding (right panel, Fig. 6). These different effects of *d*-amphetamine on punished behavior of the squirrel monkey prior to and following exposure to an avoidance schedule occurred even though there were no discernible differences in the rate or temporal patterning of punished behavior at the different experimental phases (Barrett, 1977; Bacotti and McKearney, 1979). Similar effects demonstrating the importance of prior experience have also been obtained in the squirrel monkey with morphine (Barrett and Stanley, 1983).

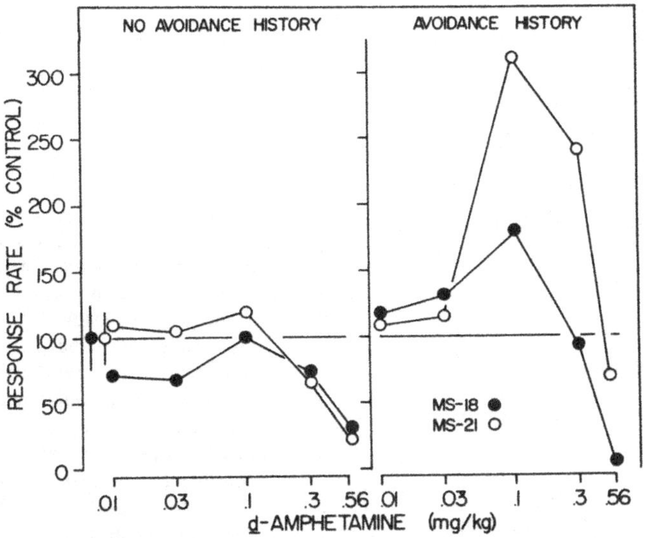

Fig. 6. Modification of the effects of *d*-amphetamine on punished responding by behavioral history. The left portion of the figure shows the effects of *d*-amphetamine on punished responding prior to training under a shock-postponement or avoidance schedule. Responding was maintained under a 5-min fixed-interval schedule of food delivery and was suppressed by a 7-mA shock that occurred following every 30th response within each interval (punishment). Note that prior to a history of avoidance, *d*-amphetamine did not affect or decreased punished responding. Both monkeys were then trained under an avoidance schedule and, after approximately 2 weeks, were returned to the punishment condition. The effects of *d*-amphetamine were then redetermined under the punishment schedule where responding was again suppressed. In marked contrast to the effects obtained prior to training, *d*-amphetamine produced substantial increases in punished responding. [Adapted from Barrett (1977).]

Experience with other drugs can also be important. Experiments with squirrel monkeys have shown that the effects of pentobarbital can be modified if monkeys have previously been administered morphine (Glowa and Barrett, 1983). Thus, behavioral and pharmacological experience can contribute substantially to determining the specific ways in which a drug will affect behavior.

2.3. Nature of the Maintaining Event

In the introduction to this chapter procedures were described for controlling behavior of squirrel monkeys by different types of consequent events such as food or shock presentation. A number of compounds produce effects that depend on whether responding is maintained by food or shock. For example, chlordiazepoxide, widely used in the treatment of anxiety, markedly increases food-maintained responding at doses that virtually eliminate responding maintained by shock presentation (Fig. 7). These differential effects occurred even though control performances maintained by food or shock were remarkably comparable. Other sedative-hypnotic drugs, such as pentobarbital and ethanol, as well as other antianxiety drugs, produce effects similar to those

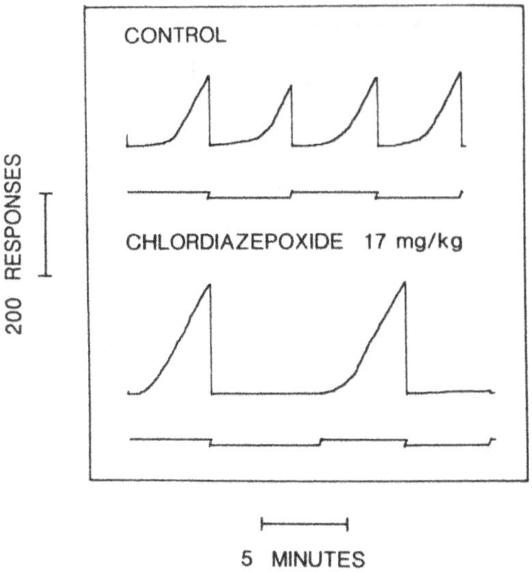

Fig. 7. Selective effects of chlordiazepoxide on performances maintained either by food or shock presentation in the squirrel monkey. Control performances show comparable rates and patterns under the food or shock-presentation schedules (the line beneath the cumulative record was displaced during the shock-presentation component). Chlordiazepoxide increased responding maintained by food but nearly eliminated responding under the shock-presentation schedule.

of chlordiazepoxide under these conditions. Although the type of maintaining event is an important factor with some compounds, it appears less important with others. For example, many psychomotor stimulants (e.g., amphetamine and cocaine) increase performances of squirrel monkeys maintained by both food- and shock-presentation schedules, as well as by electrical stimulation of the brain (Spencer and Revzin, 1976), whereas the antipsychotic drugs (e.g., chlorpromazine, haloperidol) generally decrease responding regardless of the type of consequent event [see review by Barrett and Katz (1981) and summary later in this chapter].

Thus, the behavioral effects of drugs are related in significant ways to the specific characteristics of ongoing behavior (e.g., rate of responding), the organism's previous behavioral and pharmacological experience, and to the precise ways in which behavior is controlled by consequent events. Clearly, behavioral factors can play an important role in determining the manner in which a drug affects behavior.

2.4. Environmental Context

The behavioral effects of drugs can also be related to the entire context in which behavior occurs. Conditions or consequences occurring simultaneously or sequentially can directly affect drug action. For example, when squirrel monkeys responded under a procedure in which a schedule of punishment alternated with a second component in which responding had no consequences (extinction), *d*-amphetamine did not increase punished behavior. However, when a shock-postponement schedule replaced the component previously correlated with extinction, *d*-amphetamine produced large increases in punished responding even though punished performances did not differ (Fig. 8). Thus, drug effects under one condition can depend on factors occurring under a dif-

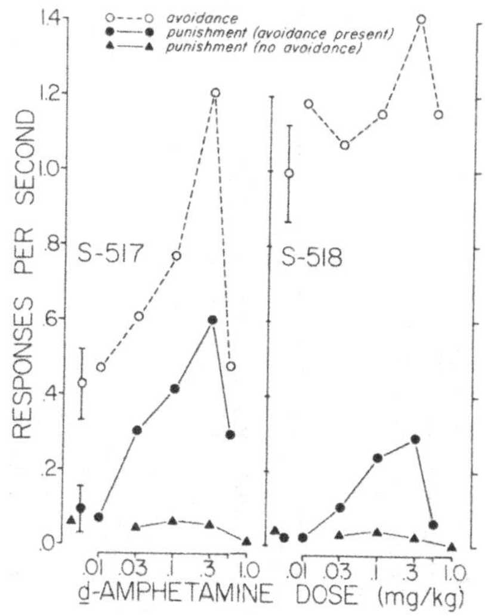

Fig. 8. Context-dependent drug effects. Effects of *d*-amphetamine on punished responding of two squirrel monkeys under a multiple schedule when the schedule of punishment either alternated with an extinction component in which responding had no consequences (triangles) or when the extinction component was replaced by an avoidance schedule (filled circles). *d*-Amphetamine increased punished responding only when the avoidance schedule was in effect in the alternate component. Thus, effects of *d*-amphetamine on punished behavior were influenced by behavior occurring under the alternate schedule component. Note that punished rates under the two conditions did not change. [From McKearney and Barrett (1975).]

ferent condition and therefore on the total environmental context in which behavior occurs. The finding that *d*-amphetamine will continue to increase punished responding even after the shock-postponement schedule is removed (cf. Fig. 6) suggests that conditions experienced under one context may continue to be influential in determining drug effects even when the context or environmental situation subsequently changes.

3. BEHAVIORAL EFFECTS OF DRUGS IN THE SQUIRREL MONKEY

The preceding sections have reviewed briefly those behaviors used most frequently to study drug effects and have outlined some of the factors that determine those effects in the squirrel monkey. It is clear from the material presented thus far that it is exceedingly difficult to characterize *the* effect a drug will have on behavior. Drugs do not have singular effects, but depend on many factors. The discovery and characterization of the diverse effects that a drug will have and the variables that determine those effects has been the hallmark of research in behavioral pharmacology. Despite these difficulties, this section will provide a broad overview of the "typical" effects of drugs from diverse pharmacological classes on the schedule-controlled behavior of the squirrel monkey. Table I summarizes the effects of a wide variety of drugs on responding maintained under fixed-interval and fixed-ratio schedules of food presentation or shock delivery (either response-produced shock or stimulus-shock termination schedules). Table II provides similar information for punished responding and for responding maintained under shock-postponement or avoidance schedules.

Table I
Comparison of Drug Effects on Performances of Squirrel Monkeys Maintained under Schedules of Food or Electric Shock Delivery[a]

Drug (dose in mg/kg)	Fixed interval		Fixed ratio		Reference
	Food	Shock	Food	Shock	
Psychomotor stimulants					
d-Amphetamine (0.01–3.0)	↑	↑	↓	↓ ↑[b]	Barrett *et al.* (1977), Barrett and Katz (1981), Branch (1979), Byrd (1982), Dalrymple and Stretch (1971), Katz and Barrett (1978a), Kelleher and Morse (1964), McKearney (1974, 1980)
l-Amphetamine (0.01–3.0)	↑	↑	—	—	Katz (1982)
Cocaine (0.1–3.0)	↑	↑	↓	↓	Barrett (1976), Gonzalez and Goldberg (1977), Spealman *et al.* (1977)
Methylphenidate (1.0–4.0)	—	↑	—	↓	Stretch and Skinner (1969)
Caffeine (1.0–17.0)	↑	↑	↓	—	Davis *et al.* (1973), Valentine and Spealman (1983)
Sedative-hypnotics					
Chlordiazepoxide (1.0–30.0)	↑	↓	↓	↓	Barrett (1976), Barrett *et al.* (1977), Katz and Barrett (1978b), Spealman *et al.* (1983)
Pentobarbital (1.0–17.0)	↑	↓	↓	↓	Barrett (1976), Katz and Barrett (1978b)
Ethanol (1000–3500)[c]	↑	↓	↓	↓	Barrett (1976), Katz and Barrett (1978a,b)
Serotonin antagonists					
Cyproheptadine (0.1–3.0)	↑	↓	—	—	Barrett (1983a)
Cinanserin (1.0–17.0)	↑	↓	—	—	Barrett (1983a)
Methysergide (0.01–3.0)	↑	↑	—	—	Barrett (1983a)
Antipsychotics					
Chlorpromazine (0.001–0.3)	↓	↓	↓	↓	McKearney (1974), Barrett (1982, 1983b)

Haloperidol (0.001–0.03)	↓	↓	↓	↓	Barrett (1982, 1983b), Spealman *et al.* (1983)
Molindone (0.001–0.1)	↓	—	↓	—	Barrett (1982, 1983b)
Thiothixene (0.001–0.1)	↓	—	↓	—	Barrett (1982, 1983b)
Thioridazine (0.1–10.0)	↓	↓	↓	↓	Spealman *et al.* (1983)
Clozapine (0.01–3.0)	↑	↑	↓	↓	Bacotti (1981), Barrett (1982, 1983b), McKearney (1980), Spealman and Katz (1980)
Opiates and opiate antagonists					
Morphine (0.01–3.0)	↑	↑ ↓[d]	↓	↑	Barrett and Katz (unpublished), Byrd (1976, 1982), Goldberg *et al.* (1976, 1981), McKearney (1974, 1975, 1980), Witkin *et al.* (1983)
Meperidine (0.3–10.0)	↓	↓	—	—	Witkin *et al.* (1983)
Methadone (0.1–1.7)	↓	↑	—	—	McKearney (1975)
Nalorphine (1.0–17.0)	↓	↑	↓	—	Goldberg *et al.* (1976), McKearney (1975)
Naloxone (0.3–17.0)	↓	↑ ↓[d]	↓	↓	Byrd (1976), Goldberg *et al.* (1976), McKearney (1975), Witkin *et al.* (1983)
Naltrexone (0.1–10.0)	↓	—	↓	—	Goldberg *et al.* (1981)
Miscellaneous					
Atropine (0.05–0.32)	↓	↑	—	—	Chait and Balster (1979), McKearney (1981)
Nicotine (0.01–5.0)	↑	↑	↓	↓	Davis *et al.* (1973), Spealman *et al.* (1981)
Phencyclidine (0.01–0.60)	↑	↑	↓	—	Byrd (1982), Chait and Balster (1978)
Physostigmine (0.175–0.56)	↓	↓	—	—	Vaillant (1967), Chait and Balster (1979)

[a] Data collected usually under 3- or 5-min fixed-interval and 30-response fixed-ratio schedules. Fixed-interval shock schedules were typically response-produced shock, except in a few instances where responding was maintained under stimulus-shock termination schedules. All fixed-ratio shock schedules were termination schedules. Arrows refer to whether individual doses increased (↑) or decreased (↓) responding. A dash indicates that no data appear to be available.

[b] Certain studies have reported decreases with *d*-amphetamine in responding under fixed-ratio stimulus-shock termination schedules, whereas other studies have shown increases. Variables related to these effects are presently unclear.

[c] Given orally.

[d] Morphine and naloxone have been shown to increase or decrease responding under fixed-interval schedules of response-produced shock. At the present time it is not clear what factors contribute to these differences.

Table II
Comparison of Drug Effects in Squirrel Monkeys under Punishment and Shock-Postponement Schedules[a]

Drug (dose in mg/kg)	Punishment	Shock-postponement	Reference
Psychomotor stimulants			
d-Amphetamine (0.01–1.0)	↓	↑	Davis *et al.* (1973), Hanson *et al.* (1967), Holtzman (1979), McKearney and Barrett (1975), Spealman (1979)
Cocaine (0.1–3.0)	↓	—	Spealman (1979)
Caffeine (1.0–30)	↑	↑	Davis *et al.* (1973), Valentine and Spealman (1983)
Sedative hypnotics			
Chlordiazepoxide (1.0–30)	↑	↓	Cook and Catania (1964), Hanson *et al.* (1967), Miczek (1973), Sepinwall *et al.* (1978), Spealman (1979)
Ethanol (1000–4000)[b]	↑	↓	Glowa and Barrett (1976)
Pentobarbital (1.0–17.0)	↑	↓	Hanson *et al.* (1966, 1967, 1970), McKearney (1976), Spealman (1979)
Meprobamate (30–180)[b]	↑	↓	Hanson *et al.* (1967)
Serotonin antagonists			
Cyproheptadine (0.1–3.0)	↑	↑	Barrett (1983a)
Cinanserin (1.0–17.0)	—	↑	Barrett (1983a)
Methysergide (0.1–3.0)	—	↑	Barrett (1983a)

Antipsychotics			
Chlorpromazine (0.001–0.3)	↓	↓	Bacotti and McKearney (1979), Barrett (1982), Hanson *et al.* (1966, 1967, 1970), Holtzman (1979), Spealman and Katz (1980)
Haloperidol (0.001–0.03)	—	↓	Barrett (1982), Hanson *et al.* (1966, 1970)
Molindone (0.001–0.1)	—	↓	Barrett (1982)
Thiothixene (0.001–0.1)	—	↓	Barrett (1982)
Clozapine (0.01–3.0)	↑	↑	Barrett (1982), Spealman *et al.* (1983), Spealman and Katz (1980)
Opiates and opiate antagonists			
Morphine (0.01–3.0)	—	↓	Barrett and Stanley (1983), Dworkin and Branch (1982), Holtzman (1976), Houser and Cash (1975)
Cyclazocine (0.01–1.0)	—	↑	Holtzman (1976)
Nalorphine (0.01–100)	—	↑	Holtzman (1976)
Naloxone (0.1–10.0)	—	↑	Stretch and Sanchez-Ramos (1979)
Miscellaneous			
Scopolamine (0.02–0.5)	↓	↓	Hanson *et al.* (1967), Miczek (1973)
Nicotine (0.03–5.0)	—	↑	Davis *et al.* (1973)

[a] Punishment data generally collected under conditions where punished responding was maintained by food and suppressed by electric shock. Shock-postponement data usually taken from continuous avoidance schedules. Dose ranges are given in parentheses and are IM unless noted. Arrows refer to whether any drug dose increased or decreased responding. A dash indicates that data do not appear to be available.

[b] Given orally.

3.1. Psychomotor Stimulants

Compounds such as cocaine, the amphetamines, and caffeine generally increase responding of squirrel monkeys under fixed-interval schedules of reinforcement regardless of the type of maintaining event. Although these drugs increase responding under fixed-interval schedules, performances under fixed-ratio schedules of food delivery are almost always decreased. At the present time, both increases and decreases have been reported with *d*-amphetamine under fixed-ratio stimulus-shock termination schedules, a finding that appears to be related to certain parameter values of the schedule, such as the time during which the ratio must be completed before shock occurs (McKearney, 1980) or the duration of timeout (J. E. Barrett, J. L. Katz, and J. R. Glowa, unpublished).

Under shock-postponement or avoidance schedules psychomotor stimulants generally produce increases in responding, whereas these drugs typically decrease responding under punishment schedules (Table II). Caffeine, a methylxanthine that is pharmacologically unrelated to either cocaine or the amphetamines, is an exception to the effects of the stimulant drugs on punished responding in that caffeine increases this behavior.

Although amphetamine generally does not increase operant responding such as lever pressing that has been punished by electric shock, increases in punished consummatory behavior with *d*-amphetamine have been reported (Miczek, 1973). Under conditions where intake by squirrel monkeys of an 8% dextrose solution was suppressed by delivering shock through the drinking tube, *d*-amphetamine increased drinking. Thus, the effects of amphetamine on punished behavior may depend upon the particular response under investigation or the type of reinforcer used to maintain responding. Similar conclusions were drawn from a study in which the effects of atropine, administered directly into the ventromedial area of the hypothalamus of squirrel monkeys, were studied under various punishment procedures (Miczek and Grossman, 1972). Although lever pressing suppressed by electric shock delivery was increased with atropine, this drug did not increase the intake of a dextrose solution that was also suppressed by shock.

3.2. Antianxiety Drugs, Sedative-Hypnotics, and Serotonin Antagonists

These compounds, which include antianxiety agents such as chlordiazepoxide, produce different effects on behavior under fixed-interval schedules, depending on whether responding is maintained by food or by noxious stimuli (Table I). Performances maintained under fixed-interval food-presentation schedules are increased, whereas comparable performances maintained by shock are only decreased. As indicated earlier, this effect is often extremely selective (Fig. 7). Compounds that are classified as serotonin antagonists also produce effects that are comparable to the sedative-hypnotic drugs under fixed-interval food- and shock-presentation schedules.

Although these drugs increase responding under fixed-interval schedules of food delivery, generally performances under fixed-ratio schedules, regardless of the type of event, are only decreased. This effect may depend on response rate, however, with lower rates of fixed-ratio responding (e.g., 2–3 responses per second) more likely to show increases with sedative-hypnotic drugs than higher rates (5.0 responses per second and above).

One of the striking features of the sedative-hypnotic compounds is that these agents produce marked increases in punished behavior; this feature also applies to the serotonin antagonists (Table II). The sedative-hypnotics, however, decrease behavior under shock-postponement schedules, while the serotonin antagonists appear to increase performances under these schedules.

3.3 Antipsychotic Drugs

The classical antipsychotic compounds (e.g., chlorpromazine and haloperidol) produce decreases in responding under most schedules regardless of the type of maintaining event and whether responding is controlled by punishment or shock-postponement schedules (Tables I and II). The one exception appears to be clozapine, an atypical antipsychotic drug, which increases performances under fixed-interval schedules of food and shock presentation and also increases responding under punishment and shock-postponement schedules. The effects of clozapine on responding maintained under stimulus-shock termination schedules may depend upon the monkey's history with shock. Clozapine only decreased responding under fixed-interval termination schedules in squirrel monkeys without a history of response-produced shock (Spealman *et al.*, 1983). However, clozapine increased responding under a similar stimulus-shock termination schedule in squirrel monkeys with previous experience under schedules of response-produced shock (Bacotti, 1981).

Because the clinical potency of many of the drugs with antipsychotic activity correlates with their potency in decreasing responding under continuous shock-postponement schedules, this procedure has been used extensively with squirrel monkeys in preclinical drug evaluation. A number of studies have analyzed the ability of various anticholingergic drugs (e.g., scopolamine and benztropine) to antagonize the decreases in avoidance behavior produced by the antipsychotic compounds (Hanson *et al.*, 1970; Leibman *et al.*, 1981). The ability of the anticholinergic drugs to reverse the effects of certain antipsychotic drugs is also correlated with the tendency of the antipsychotics to produce acute dyskinesias, which are predictive of extrapyramidal liability. Several compounds have been studied under these conditions and, although not summarized in the table because only single doses were used, these experiments add considerably to our knowledge of the basic behavioral pharmacology of the squirrel monkey (Liebman and Neale, 1980; Neale *et al.*, 1981).

3.4. Opiates and Opiate Antagonists

Morphine and other opiate substances, as well as opiate antagonists, produce dose-dependent decreases in food-maintained behavior of the squirrel monkey under both fixed-ratio and fixed-interval schedules (Table I). Under some conditions, these compunds can increase performances maintained by shock presentation or stimulus-shock termination, although this finding is not consistent.

Surprisingly, there appear to be no published accounts of the effects of these compounds on punished behavior of the squirrel monkey, although unpublished studies with morphine in our laboratory across a range of doses indicate that responding is only decreased. Under shock-postponement schedules morphine generally decreases responding (Table II). In contrast, opiates with antagonist activity, such as naloxone, cyclazocine, and nalorphine, increase responding under shock-postponement schedules.

Because of their profound analgesic activity, opiates have also been studied in squirrel monkeys under several procedures purported to measure pain threshold or the detection of pain (e.g., shock titration or shock discrimination schedules). Under the titration procedure, shock levels are incremented periodically if no responses occur; however, responses decrease shock intensity. Most subjects maintain the shock at a fairly constant (threshold) level. A large number of drugs with known analgesic activity have been shown to increase the level at which shock is maintained. These include morphine (0.3–3.0 mg/kg), methadone (0.3–3.0 mg/kg), pentazocine (1.0–10 mg/kg), and cyclazocine (0.01–1.0 mg/kg) (Dykstra, 1979a; Dykstra and McMillian, 1977). Naloxone antagonized the effects of morphine, pentazocine, and cyclazocine, with higher naloxone doses required to reverse the effects of pentazocine and cyclazocine. Unlike other clinically effective analgesics, nalorphine (1.0–30.0 mg/kg) decreased median shock levels under the titration procedure (Dykstra, 1980a) and meperidine, anileridine, and alphaprodine (all studied at doses of 0.3–10 mg/kg) did not increase shock intensity levels (Dykstra and Leander, 1978). It is not clear what factors may be responsible for the differences in these effects obtained with squirrel monkeys and clinical efficacy in humans. Factors controlling behavior under shock titration schedules (e.g., the level and rate at which shock intensity increments) are very complex. These factors may directly or indirectly affect response rate and may play a role in determining the specific drug effects obtained under these procedures (Smith and McKearney, 1977).

A related approach to understanding the possible role of opiates in the detection of pain has involved the study of drug effects on the discrimination of electric shock. In brief, under this procedure squirrel monkeys are trained to press one response lever when shock is present and to press a different lever when shock is absent. Psychophysical procedures are used to establish the lowest shock intensity at which discrimination can be maintained. Under this procedure, morphine, methadone, cyclazocine, pentazocine, and nalorphine, all at doses similar to those studies under the shock titration procedure, decreased the percentage of correct responses in the presence of shock, indicating an elevation of the threshold of sensitivity. These same drugs had little effect on correct responding in the absence of shock (Dykstra, 1979b, 1980b, 1981). Other compounds, such as diazepam, chlorpromazine, ketamine, and scopolamine, were without such effects. As under the shock titration procedure, meperidine's effects also differed from other narcotic analgesics under the shock discrimination procedure; rather than decreasing the percentage of correct responses, meperidine had little effect. However, when the metabolic conversion of meperidine to normeperidine was inhibited by SKF-525A, effects comparable to those of morphine were obtained (Dykstra, 1980b). This finding suggests that the metabolism of meperidine in the squirrel monkey may be somewhat different or may have different behavioral effects than that in other species and may account for the atypical effects of this drug.

3.5. Miscellaneous Compounds

Various other compounds that have been studied in the squirrel monkey are listed in Tables I and II. The effects of other drugs (not listed) that have been studied include compounds from a wide range of classes. Several cannabinoids have been reported to decrease responding of squirrel monkeys under various schedules of food presentation after both intraperitoneal and intraventricular administration (Carney *et al.*, 1979); Δ^9–tetrahydrocannabinol has also been shown to decrease various performances maintained by food and, with chronic administration, to show diminished effects (tolerance) (Branch *et al.*, 1980; Gablicka *et al.*, 1980). Dimethylheptyl tetrahydrocannabinol (0.03–1.0 mg/kg IM), however, produces slight increases in food-maintained responding under fixed-ratio, fixed-interval, and shock-avoidance schedules (Davis *et al.*, 1973). The behavioral effects of these compounds are poorly understood as a general rule.

Various histamine H_1 antagonists (e.g., benztropine, 0.03–1.7 mg/kg; promethazine, 0.3–10.0 mg/kg; and diphenhydramine, 0.3–17 mg/kg) can produce marked increases in responding of squirrel monkeys under fixed-interval schedules of either food delivery or stimulus-shock termination (McKearney, 1982a). Dopamine uptake inhibitors such as bupropion (0.3–5.6 mg/kg), mazindol (0.01–0.3 mg/kg), and nomifensine (0.1–1.0 mg/kg) that are used clinically as antidepressants increased responding under a fixed-interval schedule of food presentation at doses that produced little effect under a fixed-ratio schedule that also maintained responding with food (McKearney, 1982b). These effects are similar to those shown with psychomotor stimulants (Table I). In other studies with fixed-interval schedules of food presentation or stimulus-shock termination, antidepressant drugs such as amitriptyline (0.1–17 mg/kg), imipramine (0.3–17 mg/kg), chlorimipramine (0.3–17 mg/kg), nortriptyline (1–17 mg/kg), and desmethylimipramine (1–17 mg/kg) produced different effects in individual squirrel monkeys that correlated with effects obtained with the anticholinergic drugs atropine (0.03–1.7 mg/kg) and scopolamine (0.003–0.3 mg/kg) (McKearney, 1981, 1982c). Atropine and scopolamine increased responding only in monkeys that showed an increase with antidepressant drugs. This differential sensitivity to the antidepressant and antimuscarinic cholinergic drugs suggests that some of the effects of antidepressants in squirrel monkeys may be mediated by an antimuscarinic action.

Research with the squirrel monkey has generated a fairly extensive and extremely reliable set of behavioral pharmacological data. Although gaps certainly exist (e.g., systematic studies of drugs with anticholinergic properties), basic information is available that characterizes the effects of a wide variety of drugs in this species under a wide range of conditions employing schedule-controlled behavior.

4. DISCRIMINATIVE STIMULUS PROPERTIES OF DRUGS

Recent research has focused on the ability of drugs to function as discriminative stimuli that control behavior occurring during the time the drug is active. As such, drugs function as antecedent stimuli rather than consequent events. Usually, one response (such as pressing a lever positioned on the left) is reinforced after the administration of a drug such as morphine, whereas after a saline injection another response (e.g., pressing the right lever) produces food or avoids shock. After a period of training, most subjects respond appropriately (i.e., discriminate), depending on whether the drug or saline has been administered. Subsequently, other compounds may be substituted to provide information on whether they produce responses appropriate to the training drug or to saline. It is believed that drug discrimination procedures may be useful methods for preclinical drug evaluation and that these studies provide a means for making qualitative comparisons between drugs (Schuster and Balster, 1977). Thus, if a squirrel monkey has been trained to discriminate morphine from saline, drugs that produce behavior similar to that occurring after the administration of morphine would presumably produce the same subjective effects, be effective analgesics, and also, would likely be abused.

Several experiments have examined the discriminative effects of drugs in the squirrel monkey. Although a complete review and analysis of this literature are beyond the scope of this chapter, the results of a number of studies are summarized in Table III. Clearly, the predominant focus has been with opioid compounds. Work with the discriminative properties of these agents has shown that it is possible to obtain different effects with opiate compounds that are active at different opiate receptor subtypes, which may mediate different activities of the opiates (Teal and Holtzman, 1980). This remarkable degree of behavioral sensitivity to drugs with relatively specific pharmacological activity may help characterize opiates with maximal therapeutic utility and minimal side affects or abuse liability.

5. DRUGS AS CONSEQUENT EVENTS

As mentioned previously, a large number of drugs can function effectively as reinforcers for the squirrel monkey when arranged as consequent events under schedules of reinforcement. Under most conditions drugs have been delivered through chronically implanted intravenous catheters. However, several studies with squirrel monkeys have also shown that responding can be maintained by intramuscular injections of cocaine or morphine [Goldberg *et al.* (1976); see reviews by Barrett and Katz (1981) and Spealman and Goldberg

(1978)]. Studies in which behavior is maintained by drugs are directly relevant to the study of drug abuse and permit an experimental analysis of abuse potential or liability of various compounds and of factors that influence drug-taking behavior. A number of different compounds have been shown to maintain responding of squirrel monkeys, such as morphine, *d*-amphetamine, cocaine, nicotine, and phencyclidine. In studies that have compared food-maintained responding with that maintained by drug administration (e.g., Goldberg, 1973) performances have often been quite comparable (see Fig. 3).

A relatively recent development in behavioral pharmacology has been the study of responding under second-order schedules of reinforcement. Under these schedules, responding produces scheduled presentations of a brief stimulus that is eventually paired with the terminal maintaining event. Second-order schedules have been used extensively in experiments where responding is maintained by drug administration. Unlike studies where a drug is injected upon completion of each schedule component, second-order schedules typically arrange for the drug to be injected only at the end of the session. This procedure prevents the direct action of the maintaining drug on behavior that precedes its presentation. Further, because the drug is not administered until after the session, it is possible to study drug pretreatments on drug-maintained behavior without concern for drug interactions until after the session has been completed (e.g., Katz, 1980). Figure 9 shows responding maintained under second-order schedules by postsession food or cocaine.

In addition to demonstrating that drugs can serve as reinforcing events, recent studies have also shown that drug injections can suppress or punish behavior. For example, Goldberg (1980) has shown that response-dependent injections of histamine can suppress behavior of squirrel monkeys maintained by food (see Fig. 10) and that this behavior can be increased by histamine antagonists and by chlordiazepoxide. Other studies have shown that nicotine injections can either maintain *or* suppress behavior, depending on the manner in which those injections are scheduled (Goldberg and Spealman, 1982, 1983). Experiments of this type show that many similarities exist between drugs and other consequent events that exert powerful control over behavior. As mentioned previously when discussing performances maintained by electric shock, salient events such as drugs do not possess singular behavioral properties but are capable of functioning in many different ways. Experimentation with the squirrel monkey has permitted the clarification of this important principle and opened new areas of research.

6. *DRUG EFFECTS ON SOCIAL BEHAVIOR*

Although primate social behavior has been studied extensively, relatively few experiments have examined the effects of drugs on various social behavior patterns. In the squirrel monkey, the majority of these studies have focused on aggressive behavior. For example, Miczek (1978) found the Δ^9-tetrahydrocannabinol (0.25–2.0 mg/kg) decreased the intense attack and threat behavior normally shown by resident male squirrel monkeys when confronted with an

Table III
Cross-Drug Generalization in Squirrel Monkeys[a]

Training drug	Generalization	No generalization	Reference
Ethanol (1600)	Pentobarbital (7.5)	Morphine (5.0–20)	York and Bush (1982)
Cyclazocine (0.1)	Ketocyclazocine (0.1) Butorphanol (1.0) Oxilorphan (3.0) Levallorphan (3.0) Ethylketocyclazocine (0.01) SKF 10,047[b] (0.3)	Nalorphine (0.1–100) Morphine (0.1–5.6) Nalbuphine (1.0–30) Pentazocine (0.1–3.0) Scopolamine (0.01–0.3) *d*-Amphetamine (0.03–1) Mescaline (1–30) Pentobarbital (1–5.6) Apomorphine (0.03–0.3)	Schaefer and Holtzman (1978), Teal and Holtzman (1980), Holtzman *et al.* (1977)
Morphine (3.0)	Fentanyl (0.003) Oxymorphone (0.1) Levorphanol (0.3) Methadone (1.0) Meperidine (17.5) Profadol (3.0)	Butorphanol (0.003–3.0) Nalbuphine (1.0–30) Cyclazocine (0.01–0.3) Pentazocine (0.1–3.0) Dextrorphan (0.3–3) Naloxone (0.1–10) *d*-Amphetamine (0.03–1) Pentobarbital (1–5.6) Ethylketocyclazocine (0.01–0.1) SKF 10,047[b] (0.1–3.0) Apomorphine (0.03–0.3)	Schaefer and Holtzman (1977, 1981), Teal and Holtzman (1980)

Pentazocine (3.0)	Morphine (3.0)	Ketocyclazocine (0.03–1)	White and Holtzman (1982)
	Levorphanol (0.3)	Ethylketocyclazocine (0.01–0.3)	
	Levomethorphan (3.0)	Nalbuphine (1–30)	
	Cyclazocine (0.3)		
	Phencyclidine (0.3)		
	Dextrorphan (5.6)		
Phencyclidine (0.25)	SKF 10,047[b] (2.0)	Ethylketocyclazocine (0.062–0.25)	Holtzman (1982)
	d-Cyclazocine (2.0)	Ketocyclazocine (0.125–0.5)	
	Dextrorphan (4.0)	Levorphanol (0.125–1.0)	
	Normetazocine (16.0)	Levallorphan (1.0–4.0)	
		Pentazocine (1.0–4.0)	
		Naltrexone (1.0–16)	
Phencyclidine (0.16)	Etoxadrol (0.32)	(−)-*N*-Allylnormetazocine	Brady *et al.* (1981, 1982)
	Dexoxadrol (1.92)		
	(±)-*N*-Allylnormetazocine		
	(+)-*N*-Allylnormetazocine		

[a] All doses (shown in parentheses) are mg/kg and administered intramuscularly except the drugs in the York and Bush (1982) study, which were given intragastrically. Most studies used either shock-avoidance procedures or food-maintained responding to establish discriminative performances.

[b] SKF 10,047-*N*-allylnormetazocine.

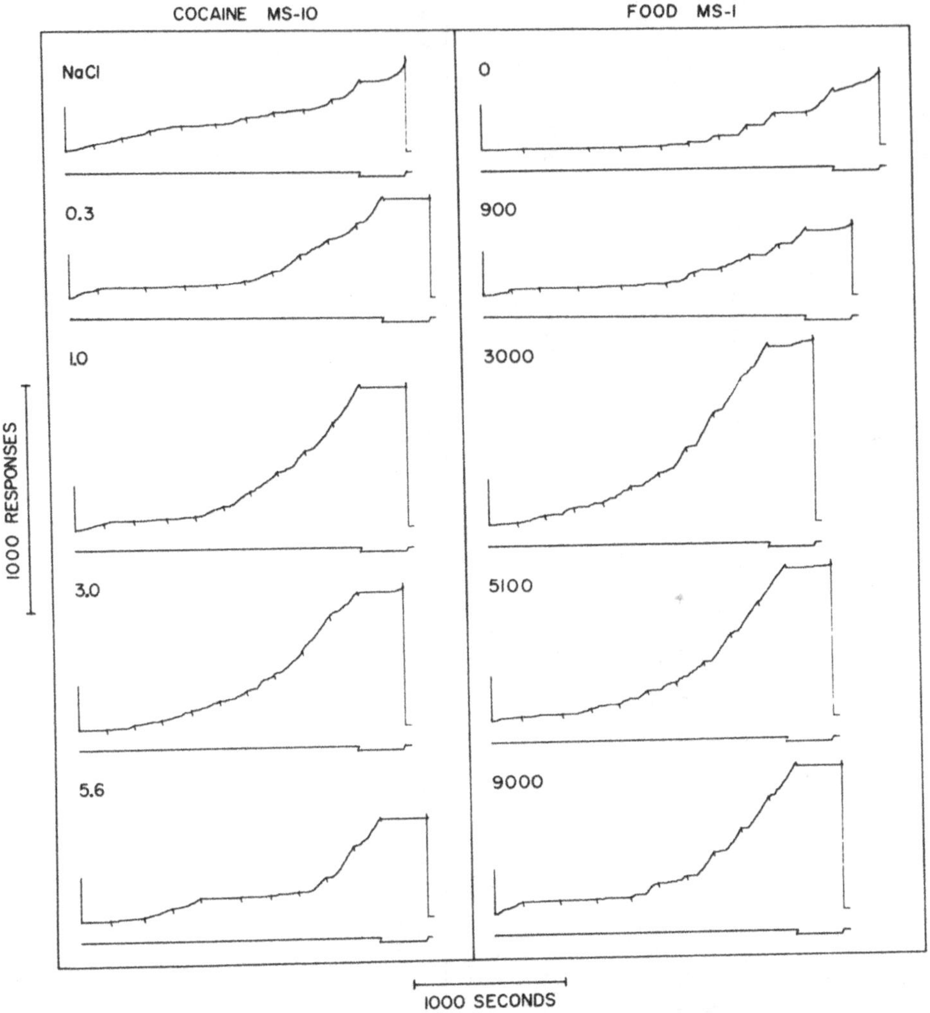

Fig. 9. Performances of squirrel monkeys maintained under second-order schedules of intramuscular cocaine hydrochloride injection or food delivery. Numbers refer to cocaine dose (in mg) or food amount (in mg). The first response after 3 min produced a 1–sec brief stimulus change and the tenth brief stimulus change was followed either by the intramuscular cocaine injection (MS-10) or by the delivery of food (MS-1). Diagonal slashes denote brief stimulus presentations. The simultaneous displacement of the response and event pens denotes a 5–min period during which food was delivered or cocaine was injected. The records are from an entire session. [From Katz (1980).]

intruder. In a different study of aggressive behavior, scopolamine (0.005–0.05 mg/kg) produced different effects on aggressive responses, depending on the environment and social rank of the animal (Plotnik *et al.*, 1975). When the squirrel monkeys were tested in a group cage, aggressive responses decreased with scopolamine. However, in a shuttle-box situation, where groups of four monkeys competed for a banana-sugar mixture, scopolamine markedly in-

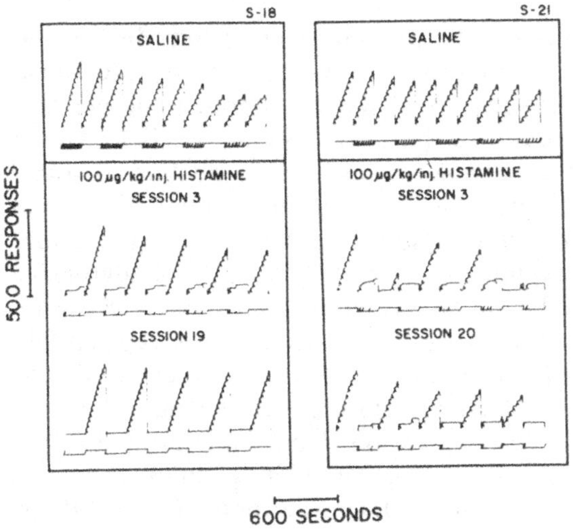

Fig. 10. Suppression of responding by histamine injections. The top record shows the last session with saline injections and the middle and bottom records show the effects of delivering histamine following each 30th lever-pressing response during one component. Diagonal strokes on the record denote food presentations; during punishment components the event line beneath each record was displaced. Histamine injections are indicated by the mark on the event line during punishment components. [From Goldberg (1980).]

creased the aggressive behavior of the normally dominant animal; aggressive responses in less dominant monkeys were decreased by scopolamine. Physostigmine (6.25–25.0 mg/kg, SC) also has been shown to produce greater increases in aggression in dominant than in submissive male squirrel monkeys (Thomas *et al.*, 1978).

Miczek and Gold (1983) have also reported on the importance of social rank as a determinant of the effects of amphetamine. In their study, increasing doses of amphetamine caused dominant animals to reverse their characteristic pattern from that of initiating agonistic behavior to actually being the recipient. Thus, it would appear that, as with the effects of drugs on schedule-controlled behavior, the effects of certain drugs on social and aggressive behavior depend on the context in which the behavior is studied and on the characteristics of the behavior under nondrug conditions (see also Poignant and Avril, 1978).

More recent studies with squirrel monkeys have examined the effects of drugs on both agonistic and affiliative behavior. In one study, *d*-amphetamine (0.5 mg/kg) decreased threat and attack displays shown by males to an intruder monkey and also decreased the affiliative response of an adult male toward a familiar adult male group member (Miczek *et al.*, 1981). In this same experiment, *d*-amphetamine (0.5 mg/kg) also decreased affiliative responses by juvenile squirrel monkeys toward their mothers when alone but did not alter this behavior when another adult male was present. Morphine (2.0 mg/kg), in contrast, increased time spent near the mother both in the absence and presence of an unfamiliar male, whereas chlordiazepoxide did not affect behaviors under either of these conditions. The findings with morphine and chlordiazepoxide, indicating that neither of these drugs reduced affiliative behavior in the presence of an unfamiliar male, presumably an index of "distress," are interesting because of the widespread belief that these compounds attenuate behavior under stressful conditions.

Additional experiments by Miczek and Yoshimura (1982) and Miczek and Gold (1983) have shown that both *d*-amphetamine and cocaine reduce social interactive and agonistic behavior of squirrel monkeys. Although haloperidol, chlorpromazine, or physostigmine reversed the stereotypic motor behavior of amphetamine and cocaine, they were unable to reverse the decline in social behavior. Thus, it appears that the mechanisms underlying social behavior in the squirrel monkey differ from those underlying stereotypic motor behavior.

A few experiments have also examined the effects of drugs on aggressive behavior of squirrel monkeys elicited by noxious stimuli. Generally, these studies have delivered intermittent shock to the tail of a restrained monkey (Hutchinson and Emley, 1972). Immediately after shock is delivered, these monkeys will bite a rubber hose that is available; hose-attacks are recorded automatically and used as a measure of shock-elicited aggression. Under this procedure, *d*-amphetamine (0.06–2.0 mg/kg) increased shock-elicited biting; chlorpromazine (0.06–2.0 mg/kg), chlordiazepoxide (1–24 mg/kg), and morphine (0.06–2.0 mg/kg), however, generally only decreased this behavior. Phencyclidine (0.1–1.0 mg/kg) has also been shown to increase attack behavior elicited by shock, although this effect was quite variable both within and across individual monkeys and dose replications (Emley and Hutchinson, 1983).

Biting has also been studied in squirrel monkeys when induced by schedules that maintain lever pressing by food or shock delivery (DeWeese, 1977). In this study, *d*-amphetamine increased lever pressing under both schedules, but decreased biting whether induced by food or shock.

Pharmacological studies of social and aggressive behaviors, though relatively small in number at the present time, complement research conducted with schedule-controlled behavior and promise to contribute to a more complete understanding of both behavioral and neurochemical factors involved in the effects of drugs. On a broad level, it appears that general principles have emerged from both areas of research on the importance of nonpharmacological factors in drug action.

7. COMPARISON OF BEHAVIORAL EFFECTS OF DRUGS IN THE SQUIRREL MONKEY WITH OTHER SPECIES

Very few studies have directly compared the effects of drugs on the behavior of squirrel monkeys with those of other species. Cross-species comparisons of drug effects pose several formidable problems in equating behavioral measures and performances. These include the nature of the response, suitability of the reinforcer, general comparability of behavior, as well as differences in the way different species metabolize drugs. This section will only review studies that directly compared drug effects in squirrel monkeys with those of another species. A more general review of species differences in drug effects can be found in Dews (1976).

The effects of morphine, naloxone, and nalorphine were compared in the squirrel monkeys and pigeons responding under a multiple fixed-interval (10-min), fixed-ratio (30-response) schedule of food reinforcement (Goldberg *et al.*,

1976). Morphine typically produced dose-related decreases in responding in both monkeys and pigeons; monkeys were approximately ten times more sensitive to the rate-decreasing effects of morphine than pigeons. Naloxone decreased responding in monkeys under both schedules, whereas in the pigeon, naloxone increased responding under the fixed-interval schedule at a dose (10.0 mg/kg) that reduced responding of monkeys to 50% of control levels. Nalorphine produced increases in responding of pigeons under both schedules, but only decreased responding of squirrel monkeys (0.001–30 mg/kg). Naloxone reversed the rate-suppressant effects of morphine in both species, with a tenfold greater dose of naloxone required to reverse suppression in pigeons than in monkeys. This same experiment showed that *d*-amphetamine produced similar effects in both species, but that a dose of *d*-amphetamine approximately ten times higher was required in the pigeon to produce comparable effects in the monkey.

In a subsequent study, also with squirrel monkeys and pigeons, Goldberg *et al.* (1981) found naltrexone to be 3–10 times more potent than naloxone as an antagonist of morphine in both species, with each of these compounds being nearly ten times more potent in the pigeon when given alone. After chronic treatment with high doses of either naltrexone or naloxone, squirrel monkeys showed enhanced sensitivity to these compounds, whereas pigeons did not.

Although differences in the effects of opiate agonists and antagonists exist between squirrel monkeys and pigeons, norcocaine and several norcocaine derivatives produce similar effects on schedule-controlled behavior of these species (Spealman *et al.*, 1979). Most compounds were ten times more potent in the squirrel monkey.

Several studies have suggested there are species differences in the discriminative effects of opioid compounds. For example, some mixed action opiate agonist/antagonists that produce effects similar to those of morphine in the rat (e.g., pentazocine) do not do so in the squirrel monkey; i.e., squirrel monkeys trained to discriminate morphine from saline will not generalize to pentazocine, whereas rats will (Table III). Another apparent species difference comes from studies where in squirrel monkeys the discriminative effects of cyclazocine are blocked by the opiate antagonist naloxone; in rats, however, this is not so, suggesting that in the rat cyclazocine may have a relatively greater nonopiate component. It is presently unclear whether these differences reflect basic species differences in receptor subpopulations or whether they are related to details of the procedure, such as the training dose of morphine (Herling and Woods, 1981; Holtzman *et al.*, 1977).

The effects of clozapine (0.3–3.0 mg/kg), chlordiazepoxide (10.0–56.0 mg/kg), and chlorpromazine (0.01–0.3 mg/kg in monkeys and 0.1–3.0 mg/kg in mice) were studied on punished responding of squirrel monkeys and mice (Spealman and Katz, 1980). Clozapine and chlordiazepoxide increased punished responding in mice and monkeys, whereas chlorpromazine only decreased responding of both species. Clozapine also increased unpunished responding in mice, and chlordiazepoxide increased unpunished responding of both mice and monkeys. The range of effective doses was generally similar in both species except for chlorpromazine, which was ten times more potent in the squirrel monkey than in the mouse.

A recent study examined the effects of several antipsychotic drugs (chlorpromazine, haloperidol, molindone, promazine, thiothixene, and clozapine) on performances of squirrel monkeys and pigeons maintained under a multiple fixed-interval (3-min), fixed-ratio (30-response) schedule of food delivery (Barrett, 1983b). Doses are given in Table I. Generally, most antipsychotic drugs, excluding clozapine, were between 5 and 160 times more potent in monkeys than in pigeons. Additionally, in pigeons, most drugs except clozapine produced decreases in fixed-interval performances at doses lower than those required to reduce responding under the fixed-ratio schedule. In monkeys, the effects of these drugs were more similar on the different schedule-controlled behaviors. In contrast to the other antipsychotic drugs, clozapine was equipotent in the monkey and pigeon and produced increases in responding under the fixed-interval schedules at intermediate doses.

Thus, it would appear that with few exceptions (e.g., naloxone and nalorphine as well as some discriminative stimulus properties), direct comparisons of the effects of most drugs studied thus far in squirrel monkeys and other species are qualitatively quite similar. The main difference appears to be that squirrel monkeys are much more sensitive to the effects of opiate and opiate antagonists, as well as to antipsychotic compounds and amphetamine.

8. SUMMARY AND CONCLUSIONS

This chapter has reviewed the basic behavioral procedures used with the squirrel monkey and has summarized the behavioral effects of drugs in this species. On the basis of the studies summarized here, it is apparent that the squirrel monkey has played a vital role in shaping the characteristics and principles in the field of behavioral pharmacology. The data summarized in this chapter are remarkably orderly and reproducible across many laboratories, thereby providing ample testimony to the validity of the data and justification for the continued use of the squirrel monkey in behavioral pharmacology.

ACKNOWLEDGMENTS. I would like to thank Myra J. Zimmerman for expert help in the preparation of this chapter and L. S. Brady and J. M. Witkin for comments on an earlier version. Preparation of this chapter was supported in part by PHS Grant DA-02873.

REFERENCES

Azrin, N. H., Holz, W. C., and Hake, D. F., 1962, Intermittent reinforcement by removal of a conditioned aversive stimulus, *Science* **136**:781–782.

Bacotti, A. V., 1981, Clozapine effects on responding maintained under shock presentation and shock termination schedules, *Pharmacol. Biochem. Behav.* **15**:415–418.

Bacotti, A. V., and McKearney, J. W., 1979, Prior and ongoing experience as determinants of the effects of *d*-amphetamine and chlorpromazine on punished behavior, *J. Pharmacol. Exp. Ther.* **211**:80–85.

Barrett, J .E., 1976, Effects of alcohol, chlordiazepoxide, cocaine and pentobarbital on responding maintained under fixed-interval schedules of food or shock presentation, *J. Pharmacol. Exp. Ther.* **196:**605–615.

Barrett, J. E., 1977, Behavioral history as a determinant of the effects of *d*-amphetamine on punished behavior, *Science* **198:**67–69.

Barrett, J. E., 1982, Antipsychotic drug effects on the behavior of squirrel monkeys differentially controlled by noxious stimuli, *Psychopharmacology* **77:**1–6.

Barrett, J. E., 1983a, Effects of serotonergic drugs on operant behavior of squirrel monkeys controlled by different schedules of noxious stimuli, *Fed. Proc.* **42:**621.

Barrett, J. E., 1983b, Comparison of the effects of antipsychotic drugs on the schedule-controlled behavior of squirrel monkeys and pigeons, *Neuropharmacology* **22:**519–524.

Barrett, J. E., and Katz, J. L., 1981, Drug effects on behaviors maintained by different events, in: *Advances in Behavioral Pharmacology*, Volume III (T. Thompson, P. B. Dews, and W. A. McKim, eds.), Academic Press, New York, pp. 119–168.

Barrett, J. E., and Stanley, J. A., 1983, Prior behavioral experience can alter the effects of morphine, *Psychopharmacology* **81:**107–110.

Barrett, J. E., Dworkin, S. I., and Zuccarelli, R. R., 1977, Effects of *d*-amphetamine, chlordiazepoxide and promazine on responding of squirrel monkeys maintained under fixed-interval schedules of food presentation and stimulus-shock termination, *Pharmacol. Biochem. Behav.* **7:**529–535.

Brady, K. T., Balster, R. L., and May, E. L., 1981, Stereoisomers of *N*-allylnormetazocine: Phencyclidine-like behavioral effects in squirrel monkeys and rats, *Science* **215:**178–180.

Brady, K. T., Woolverton, W. L., and Balster, R. L., 1982, Discriminative stimulus and reinforcing properties of etoxadrol and dexoxadrol in monkeys *J. Pharmacol. Exp. Ther* **220:**56–62.

Branch, M. N., 1979, Consequent events as determinants of drug effects on schedule-controlled behavior: Modification of effects of cocaine and *d*-amphetamine following chronic amphetamine administration, *J. Pharmacol. Exp. Ther.* **210:**354–360.

Branch, M. N., Dearing, M. E., and Lee, D. M., 1980, Acute and chronic effects of Δ^9-tetrahydrocannabinol on complex behavior of squirrel monkeys, *Psychopharmacology* **72:**247–256.

Byrd, L. D., 1976, Effects of morphine alone and in combination with naloxone or *d*-amphetamine on shock-maintained behavior in the squirrel monkey, *Psychopharmacology* **49:**225–234.

Byrd, L. D., 1982, Comparison of the behavioral effects of phencyclidine, ketamine, *d*-amphetamine and morphine in the squirrel monkey, *J. Pharmacol. Exp. Ther.* **220:**139–144.

Carney, J. M., Balster, R. L., Martin, B. R., and Harris, L. S., 1979, Effects of systemic and intraventricular administration of cannabinoids on schedule-controlled responding in the squirrel monkey, *J. Pharmacol. Exp. Ther.* **210:**399–404.

Chait, L. D., and Balster, R. L., 1978, The effects of acute and chronic phencyclidine on schedule-controlled behavior in the squirrel monkey, *J. Pharmacol. Exp. Ther.* **204:**77–87.

Chait, L. D., and Balster, R. L., 1979, Effects of phencyclidine, atropine and physostigmine, alone and in combination, on variable-interval performance in the squirrel monkey, *Pharmacol. Biochem. Behav.* **11:**37–42.

Cook, L., and Catania, A. C., 1964, Effects of drugs on avoidance and escape behavior, *Fed. Proc.* **23:**818–835.

Dalrymple, S. D., and Stretch, R., 1971, Effects of amphetamine and chlorpromazine on second-order escape behavior in squirrel monkeys, *Psychopharmacologia* **21:**268–282.

Davis, T. R. A., Kensler, C. J., and Dews, P. B., 1973, Comparison of behavioral effects of nicotine, *d*-amphetamine, caffeine and dimethylheptyl tetrahydrocannabinol in squirrel monkeys, *Psychopharmacologia* **32:**51–65.

DeWeese, J., 1977, Schedule-induced biting under fixed-interval schedules of food or electric-shock presentation, *J. Exp. Anal. Behav.* **27:**419–431.

Dews, P. B., 1955, Studies on behavior. I. Differential sensitivity to pentobarbital of pecking performance in pigeons depending on the schedule of reward, *J. Pharmacol. Exp. Ther.* **113:**393–401.

Dews, P. B., 1958, Studies on behavior, IV. Stimulant actions of methamphetamine, *J. Pharmacol. Exp. Ther.* **122:**137–147.

Dews, P. B., 1976, Interspecies differences in drug effects: Behavioral, in: *Psychotherapeutic Drugs* (E. Usdin and I. S. Forrest, eds.), Marcel Dekker, New York, pp. 175–224.

Dews, P. B., and DeWeese, J., 1977, Schedules of reinforcement, in: *Hankbook of Psychopharmacology*, Volume 7 (L. L. Iversen, S. D. Iversen, and S. H. Snyder, eds.), Plenum Press, New York, pp. 107–150.

Dews, P. B., and Wenger, G. R., 1977, Rate-dependency of the behavioral effects of amphetamine, in: *Advances in Behavioral Pharmacology*, Volume 1 (T. Thompson and P. B. Dews, eds.), Academic Press, New York, pp. 167–227.

Dworkin, S. I., and Branch, M. N., 1982, Behavioral effects of morphine and naloxone following chronic morphine administration, *Psychopharmacology* **77**:322–326.

Dykstra, L. A., 1979a, Effects of morphine, pentazocine and cyclazocine alone and in combination with naloxone on electric shock titration in the squirrel monkey, *J. Pharmacol. Exp. Ther.* **211**:722–732.

Dykstra, L. A., 1979b, Effects of morphine, diazepam and chlorpromazine on discrimination of electric shock, *J. Pharmacol. Exp. Ther.* **209**:297–303.

Dykstra, L. A., 1980a, Nalorphine's effect under several schedules of electric shock titration, *Psychopharmacology* **70**:69–72.

Dykstra, L. A., 1980b, Discrimination of electric shock: Effects of some opioid and nonopioid drugs, *J. Pharmacol. Exp. Ther.* **213**:234–240.

Dykstra, L. A., 1981, Effects of morphine, levonantradol, and *N*-methyllevonantradol on shock intensity discrimination, *J. Clin. Pharmacol.* **21**:341S-347S.

Dykstra, L. A., and Leander, J. D., 1978, Electric shock titration: Effects of meperidine, anileridine and alphaprodine, *Pharmacol. Biochem. Behav.* **8**:387–389.

Dykstra, L. A., and McMillan, D. E., 1977, Electric shock titration: Effects of morphine, methadone, pentazocine, nalorphine, naloxone, diazepam and amphetamine, *J. Pharmacol. Exp. Ther.* **202**:660–669.

Emley, G. S., and Hutchinson, R. R., 1983, Effects of phencyclidine on aggressive behavior in squirrel monkeys, *Pharmacol. Biochem. Behav.* **18**:163–166.

Gablicka, G., Lee, D. M., and Branch, M. N., 1980, Schedule-dependent tolerance to behavioral effects of Δ^9-tetrahydrocannabinol when reinforcement frequencies are matched, *Pharmacol. Biochem. Behav.* **12**:85–91.

Glowa, J. R., and Barrett, J. E., 1976, Effects of alcohol on punished and unpunished responding of squirrel monkeys, *Pharmacol. Biochem. Behav.* **4**:169–173.

Glowa, J. R., and Barrett, J. E., 1983, Drug history modifies the behavioral effects of pentobarbital, *Science* **220**:333–335.

Goldberg, S. R., 1973, Comparable behavior maintained under fixed-ratio and second-order schedules of food presentation, cocaine injection or *d*-amphetamine injection in the squirrel monkey, *J. Pharmacol. Exp. Ther.* **186**:18–30.

Goldberg, S. R., 1980, Histamine as a punisher in squirrel monkeys: Effects of pentobarbital, chlordiazepoxide and H_1-receptor antagonists on behavior and cardiovascular responses, *J. Pharmacol. Exp. Ther.* **214**:726–736.

Goldberg, S. R., and Spealman, R. D., 1982, Maintenance and suppression of behavior by intravenous nicotine injections in squirrel monkeys, *Fed. Proc.* **41**:216–220.

Goldberg, S. R., and Spealman, R. D., 1983, Suppression of behavior by intravenous injections of nicotine or by electric shocks in squirrel monkeys: Effects of chlordiazepoxide and mecamylamine, *J. Pharmacol. Exp. Ther.* **224**:334–340.

Goldberg, S. R., Morse, W. H., and Goldberg, D. M., 1976, Some behavioral effects of morphine, naloxone and nalorphine in the squirrel monkey and the pigeon, *J. Pharmacol. Exp. Ther.* **196**:625–636.

Goldberg, S. R., Morse, W. H., and Goldberg, D. M., 1981, Acute and chronic effects of naltrexone and naloxone on schedule-controlled behavior of squirrel monkeys and pigeons, *J. Pharmacol. Exp. Ther.* **216**:500–509.

Gonzalez, F. A., and Goldberg, S. R., 1977, Effects of cocaine and *d*-amphetamine on behavior maintained under various schedules of food presentation in squirrel monkeys, *J. Pharmacol. Exp. Ther.* **201**:33–43.

Hake, D. F., and Azrin, N. H., 1963, An apparatus for delivering pain shock to monkeys, *J. Exp. Anal. Behav.* **6**:297–298.

Hanson, H. M., 1968, Use of the squirrel monkey in pharmacology, in: *The Squirrel Monkey* (L.A. Rosenblum and R. W. Cooper, eds.), Academic Press, New York, pp. 365–392.

Hanson, H. M., Witoslawski, J. J., Campbell, E. H., and Itkin, A. G., 1966, Estimation of relative antiavoidance activity of depressant drugs in squirrel monkeys, *Arch Int. Pharmacol. Ther.* **161**:7–16.

Hanson, H. M., Witoslawski, J. J., and Campbell, E. H., 1967, Drug effects in squirrel monkeys trained on a multiple schedule with a punishment contingency, *J. Exp. Anal. Behav.* **10**:565–569.

Hanson, H. M., Stone, C. A., and Witoslawski, J. J., 1970, Antagonism of the antiavoidance effects of various agents by anticholinergic drugs, *J. Pharmacol. Exp. Ther.* **173**:117–124.

Herling, S., and Woods, J. H., 1981, Discriminative stimulus effects of narcotics: Evidence for multiple receptor-mediated actions, *Life Sci.* **28**:1571–1584.

Holtzman, S. G., 1976, Effects of morphine and narcotic antagonists on avoidance behavior of the squirrel monkey, *J. Pharmacol. Exp. Ther.* **196**:145–155.

Holtzman, S. G., 1979, Failure of naloxone to modify the effects of chlorpromazine and *d*-amphetamine on avoidance behavior in the squirrel monkey, *Psychopharmacology* **63**:207–209.

Holtzman, S. G., 1982, Phencyclidine-like discriminative stimulus properties of opioids in the squirrel monkey, *Psychopharmacology* **77**:295–300.

Holtzman, S. G., Shannon, H. E., and Schaefer, G. J., 1977, Discriminative properties of narcotic antagonists, in: *Discriminative Stimulus Properties of Drugs* (H. Lal, ed.), Plenum Press, New York, pp. 47–72.

Houser, V. P., and Cash, R. J., 1975, The effects of chronic morphine administration upon a Sidman avoidance schedule that utilized response-dependent shock, *Psychopharmacologia* **41**:255–262.

Hutchinson, R. R., and Emley, G. S., 1972, Schedule-independent factors contributing to schedule-induced phenomena, in: *Schedule-Effects: Drugs, Drinking, and Aggression* (R. M. Gilbert and J. E. Keehn, eds.), University of Toronto Press, Toronto, pp. 174–202.

Katz, J. L., 1980, Second-order schedules of intramuscular cocaine injection in the squirrel monkey: Comparisons with food presentation and effects of *d*-amphetamine and promazine, *J. Pharmacol. Exp. Ther.* **212**:405–411.

Katz, J. L., 1982, Rate-dependent effects of *d*- and *l*-amphetamine on schedule-controlled responding in pigeons and squirrel monkeys, *Neuropharmacology* **21**:235–242.

Katz, J. L., and Barrett, J. E., 1978a, Effects of *d*-amphetamine and ethanol on responding of squirrel monkeys maintained under fixed-ratio schedules of food presentation and stimulus-shock termination, *Pharmacol. Biochem. Behav.* **8**:35–39.

Katz, J. L., and Barrett, J. E., 1978b, Ethanol, pentobarbital and chlordiazepoxide effects in squirrel monkeys responding under fixed-ratio food and stimulus-shock termination, *Psychopharmacology* **56**:153–155.

Kelleher, R. T., and Morse, W. H., 1964, Escape behavior and punished behavior, *Fed. Proc.* **23**:808–817.

Kelleher, R. T., and Morse, W. H., 1968a, Schedules using noxious stimuli. III. Responding maintained with response-produced electric shocks, *J. Exp. Anal. Behav.* **11**:819–838.

Kelleher, R. T., and Morse, W. H., 1968b, Determinants of the specificity of the behavioral effects of drugs, *Ergeb. Physiol. Chem. Exp. Pharmakol.* **60**:1–56.

Kelleher, R. T., Gill, C. A., Riddle, W. C., and Cook, L., 1963, On the use of the squirrel monkey in behavioral and pharmacological experiments, *J. Exp. Anal. Behav.* **6**:249–262.

Liebman, J. M., and Neale, R., 1980, Neuroleptic-induced acute dyskinesias in squirrel monkeys: Correlation with propensity to cause extrapyramidal side effects, *Psychopharmacology* **68**:25–29.

Liebman, J. M., Neale, R., Noreika, L., and Braunwalder, A., 1981, Differential reversal of various dopamine antagonists by anticholinergics in Sidman avoidance: Possible relationship to adrenergic blockade, *Psychopharmacology* **75**:248–253.

McKearney, J. W., 1974, Effects of *d*-amphetamine, morphine and chlorpromazine on responding under fixed-interval schedules of food presentation or electric shock presentation, *J. Pharmacol. Exp. Ther.* **190**:141–153.

McKearney, J. W., 1975, Effects of morphine, methadone, nalorphine and naloxone on responding under fixed-interval schedules in the squirrel monkey, *Fed. Proc.* **34**:766.

McKearney, J. W., 1976, Punishment of responding under schedules of stimulus-shock termination: Effects of *d*-amphetamine and pentobarbital, *J. Exp. Anal. Behav.* **26**:281–287.

McKearney, J. W., 1980, Fixed ratio schedules of food presentation and stimulus shock termination: Effects of *d*-amphetamine, morphine, and clozapine, *Psychopharmacology* **70:**35–39.

McKearney, J. W., 1981, Individual differences in effects of tricyclic antidepressants and anticholinergic drugs on operant behaior in the squirrel monkey, *Life Sci.* **28:**2049–2054.

McKearney, J. W., 1982a, Stimulant actions of histamine H_1 antagonists on operant behavior in the squirrel monkey, *Psychopharmacology* **77:**156–158.

McKearney, J. W., 1982b, Effects of dopamine uptake inhibitors on schedule-controlled behavior in the squirrel monkey, *Psychopharmacology* **78:**377–379.

McKearney, J. W., 1982c, Effects of tricyclic antidepressant and anticholinergic drugs on fixed-interval responding in the squirrel monkey, *J. Pharmacol. Exp. Ther.* **222:**215–219.

McKearney, J. W., and Barrett, J. E., 1975, Punished behavior: Increases in responding after *d*-amphetamine, *Psychopharmacologia* **41:**23–26.

McKearney, J. W., and Barrett, J. E., 1978, Schedule-controlled behavior and the effects of drugs, in: *Contemporary Research in Behavioral Pharmacology* (D. E. Blackman and D. J. Sanger, eds.) Plenum Press, New York, pp. 1–68.

Miczek, K. A., 1973, Effects of scopolamine, amphetamine and chlordiazepoxide on punishment, *Psychopharmacologia* **28:**373–389.

Miczek, K. A., 1978, Δ^9-Tetrahydrocannabinol: Antiaggressive effects in mice, rats, and squirrel monkeys, *Science* **199:**1459–1461.

Miczek, K. A., and Gold, L. H., 1983, Ethological analysis of amphetamine action on social behavior in squirrel monkeys (*Saimiri sciureus*), in: *Ethopharmacology: Primate Models of Neuropsychiatric Disorders* (K. A. Miczek, ed.), Alan R. Liss, New York, pp. 137–155.

Miczek, K. A., and Gold, L. H., 1983, *d*-Amphetamine in squirrel monkeys of different social status: Effects on social and agonistic behavior, locomotion and Stereotypies, Psychopharmacology **81:**183–190.

Miczek, K. A., and Grossman, S. P., 1972, Punished and unpunished operant behavior after atropine administration to the VMH of squirrel monkeys, *J. Comp. Physiol. Psychol.* **81:**318–330.

Miczek, K. A., and Yoshimura, H., 1982, Disruption of primate social behavior by *d*-amphetamine and cocaine: Differential antagonism by antipsychotics, *Psychopharmacology* **76:**163–171.

Miczek, K. A., Woolley, J., Schlisserman, S., and Yoshimura, H., 1981, Analysis of amphetamine effects on agonistic and affiliative behavior in squirrel monkeys (*Saimiri sciureus*), *Pharmacol. Biochem. Behav.* **14:**103–107.

Morse, W. H., and Kelleher, R. T., 1966, Schedules using noxious stimuli. I: Multiple fixed-ratio and fixed-interval termination of schedule complexes, *J. Exp. Anal. Behav.* **9:**267–290.

Morse, W. H., and Kelleher, R. T., 1977, Determinants of reinforcement and punishment, in: *Handbook of Operant Behavior* (W. K. Honig and J. E. R. Staddon, eds.), Prentice-Hall, Englewood Cliffs, New Jersey, pp. 174–200.

Morse, W. H., McKearney, J. W., and Kelleher, R. T., 1977, Control of behavior by noxious stimuli, in: *Handbook of Psychopharmacology*, Volume 7 (L. L. Iversen, S. D. Iversen, and S. H. Snyder, eds.), Plenum Press, New York, pp. 151–180.

Neale, R., Fallon, S., Gerhardt, S., and Liebman, J. M., 1981, Acute dyskinesias in monkeys elicited by halopemide, mexilamine and the "antidyskinetic" drugs, oxiperomide and tiapride, *Psychopharmacology* **75:**254–257.

Pickens, R., 1977, Behavioral pharmacology: A brief history, in: *Advances in Behavioral Pharmacology*, Volume 1 (T. Thompson and P. B. Dews, eds.), Academic Press, New York, pp. 230–257.

Plotnik R., Mollenauer, S., Gore, W., and Popov, A., 1975, Comparing the effects of scopolamine on operant and aggressive responses in squirrel monkeys, *Pharmacol. Biochem. Behav.* **3:**739–748.

Poignant, J. C., and Avril, A., 1978, Pharmacological studies on drugs acting on the social behaviour of the squirrel monkey: Effects of amineptine, piribedil, *d*-amphetamine, and amitriptyline, *Arz. Forsch./Drug Res.* **28:**267–271.

Robbins, T. W., 1982, Behavioral determinants of drug action: Rate dependency revisited, in: *Theory of Psychopharmacology* (S. J. Cooper, ed.), Academic Press, New York, pp. 1–63.

Schaefer, G. J., and Holtzman, S. G., 1977, Discriminative effects of morphine in the squirrel monkey, *J. Pharmacol. Exp. Ther.* **201:**67–75.

Schaefer, G. J., and Holtzman, S. G., 1978, Discriminative effects of cyclazocine in the squirrel monkey, *J. Pharmacol. Exp. Ther.* **205**:291–301.

Schaefer, G. J., and Holtzman, S. G., 1981, Morphine-like stimulus effects in the monkey: Opioids with antagonist properties, *Pharmacol. Biochem. Behav.* **14**:241–245.

Schuster, C. R., and Balster, R. L., 1977, The discriminative stimulus properties of drugs, in: *Advances in Behavioral Pharmacology*, Volume 1 (T. Thompson and P. B. Dews, eds.), Academic Press, New York, pp. 85–138.

Sepinwall, J. and Cook, L., 1978, Behavioral pharmacology of antianxiety drugs, in: *Handbook of Psychopharmacology*, Volume 13 (L. L. Iversen, S. D. Iversen, and S. H. Snyder, eds.), Plenum Press, New York, pp. 345–393.

Sepinwall, J., Grodsky, F. S., and Cook, L., 1978, Conflict behavior in the squirrel monkey: Effects of chlordiazepoxide, diazepam and *N*-desmethyldiazepam, *J. Pharmacol. Exp. Ther.* **204**:88–102.

Sidman, M., 1959, Behavioral pharmacology, *Psychopharmacologia* **1**:1–19.

Smith, J. B., and McKearney, J. W., 1977, Effects of morphine, methadone, nalorphine and naloxone on responding under schedules of electric shock titration, *J. Pharmacol. Exp. Ther.* **200**:508–515.

Spealman, R. D., 1979, Comparison of drug effects on responding punished by pressurized air or electric shock delivery in squirrel monkeys: Pentobarbital, chlordiazepoxide, *d*-amphetamine and cocaine, *J. Pharmacol. Exp. Ther.* **209**:309–315.

Spealman, R. D., and Goldberg, S. R., 1978, Drug self-administration by laboratory animals: Control by schedules of reinforcement, *Annu. Rev. Pharmacol. Toxicol.* **18**:313–339.

Spealman, R. D., and Katz, J. L., 1980, Some effects of clozapine on punished responding by mice and squirrel monkeys, *J. Pharmacol. Exp. Ther.* **212**:435–440.

Spealman, R. D., Goldberg, S. R., Kelleher, R. T., Goldberg, D. M., and Charlton, J. P., 1977, Some effects of cocaine and two cocaine analogs on schedule-controlled behavior of squirrel monkeys, *J. Pharmacol. Exp. Ther.* **202**:500–509.

Spealman, R. D., Goldberg, S. R., Kelleher, R. T., Morse, W. H., Goldberg, D. M., Hakansson, C. G., Nieforth, K. A., and Lazer, E. S., 1979, Effects of norcocaine and some norcocaine derivatives on schedule-controlled behavior of pigeons and squirrel monkeys, *J. Pharmacol. Exp. Ther.* **210**:196–205.

Spealman, R. D., Goldberg, S. R., and Gardner, M. L., 1981, Behavioral effects of nicotine: Schedule-controlled responding by squirrel monkeys, *J. Pharmacol. Exp. Ther.* **216**:484–491.

Spealman, R. D., Kelleher, R. T., Goldberg, S. R., DeWeese, J., and Goldberg, D. M., 1983, Behavioral effects of clozapine: Comparison with thioridazine, chlorpromazine, haloperidol and chlordiazepoxide in squirrel monkeys, *J. Pharmacol. Exp. Ther.* **224**:127–134.

Spencer, J., and Revzin, A., 1976, Amphetamine, chlorpromazine and clonidine effects on self-stimulation in caudate or hypothalamus of the squirrel monkey, *Pharmacol. Biochem. Behav.* **5**:149–156.

Stretch, R., and Sanchez-Ramos, J. R., 1979, Effects of morphine, naloxone, and diallylnormorphine upon behaviour maintained by electric-shock postponement in squirrel monkeys, *Can. J. Physiol. Pharmacol.* **57**:98–105.

Stretch, R., and Skinner, N., 1969, Fixed-ratio and fixed-interval control of responding, using noxious stimuli: Effects of methylphenidate and reserpine, *Psychopharmacologia* **16**:89–104.

Teal, J. J., and Holtzman, S. G., 1980, Discriminative stimulus effects of prototype opiate receptor agonists in monkeys, *Eur. J. Pharmacol.* **68**:1–10.

Thomas, G. S., Caccamise, D. J., and Clark, D. L., 1978, Aggression increase and water competition decrease in squirrel monkeys given physostigmine injections, *Pharmacol. Biochem. Behav.* **8**:633–639.

Vaillant, G. E., 1967, A comparison of antagonists of physostigmine-induced suppression of behavior, *J. Pharmacol. Exp. Ther.* **157**:636–648.

Valentine, J. O., and Spealman, R. D., 1983, Effects of caffeine and chlordiazepoxide on schedule-controlled responding of squirrel monkeys, *Fed. Proc.* **42**:1158.

White, J. M., and Holtzman, S. G., 1982, Properties of pentazocine as a discriminative stimulus in the squirrel monkey, *J. Pharmacol. Exp. Ther.* **223**:396–401.

Witkin, J. M., Leander, J. D., and Dykstra, L. A., 1983, Modification of behavioral effects of morphine, meperidine and normeperidine by naloxone and by morphine tolerance, *J. Pharmacol. Exp. Ther.* **225**:275–283.

York, J. L., and Bush, R., 1982, Studies on the discriminative stimulus properties of ethanol in squirrel monkeys, *Psychopharmacology* **77**:212–216.

14

Nutrition and Metabolism of the Squirrel Monkey

Lynne M. Ausman, Daniel L. Gallina, and Robert J. Nicolosi

1. DIET

1.1. Natural Habitat Diets

Feeding and foraging occupy a large percentage of time for the squirrel monkey. Animals have been reported to consume a variety of natural foods, including nuts, fruits, seeds, leaves, and insects. These data have been collected from visual observation of materials that enter the animal's mouth, from studying the stomach contents of primates that have been collected, from identification of damage done to surrounding fauna and flora, and from studying the contents of fecal material left from the animals (Brodie and Marshall, 1963a,b; Fooden, 1964; Thorington, 1968, 1970). With these techniques, however, it is difficult to quantitate the nutrient intakes over long periods of time and it is impossible to make estimates of the mean calorie, protein, vitamin, and mineral intakes of these animals.

1.2. Diets Fed to Captive Animals

Several types of diets may be offered to animals in a captive situation, whether the animals are free-ranging or kept in relatively close confinement

Lynne M. Ausman • School of Nutrition, Tufts University, Medford, Massachusetts 02155, and Department of Nutrition, Harvard School of Public Health, Boston, Massachusetts 02115. *Daniel L. Gallina* • Department of Nutrition, Harvard School of Public Health, Boston, Massachusetts 02115; *Present Address:* Division of Diabetes, Department of Medicine, University of Massachusetts Medical School, Worcester, Massachusetts 01605. *Robert J. Nicolosi* • Harvard Medical School, New England Regional Primate Research Center, Nutrition Division, Southborough, Massachusetts 01772.

(Harris, 1970; Lang, 1970; Navia, 1970; Knapka and Morin, 1979; Newberne and Hayes, 1979). These include natural, semipurified, and synthetic or chemically defined diets.

1.2.1. Natural Diets

Natural diets may be comprised of relatively unprocessed foodstuffs, such as potatoes, cereal, grains, fruits, vegetables, meat, and eggs on the one hand, and protein meal (meat, legume, or fish), processed grains and cereals, and oils on the other. These natural diets may be formulated using a specified type and amount of ingredients (closed formula) or may simply be designed to provide the nutrients in the guaranteed minimum or maximum analysis utilizing the least cost mixture (open formula) (Knapka and Morin, 1979; Newberne and Hayes, 1979). Commercially prepared monkey biscuits from at least four companies (Ralston Purina Company, Agway Inc., Allied Mills, and Hills) fall into this latter category and are intended to provide sufficient nutrients for most primates. The biscuits generally provide 24–26% protein (28% of calories), 5–6% fat (about 13% of calories), 2.5–3.5% fiber (w/w), and several other dietary components. Biscuits containing 15–18% (w/w) protein are also available, but these are used primarily for the larger Old World primates, which appear to have lower protein requirements (Kerr, 1972; Ausman and Gallina, 1979; Ausman *et al.*, 1979; Nicolosi and Hunt, 1979). There appears to be little reported evidence, however, that these lower protein diets would also not be sufficient for growth, maintenance, and reproduction of the squirrel monkey.

The fat content of these biscuits, set at 12–13% of calories, may also not be optimal for the New World squirrel monkey. The markedly increased weight gain of squirrel monkeys fed with a high-fat, semipurified diet (Ausman *et al*, 1981) appears to indicate that the fat content of the diet may be a substantial factor regulating the body weight of the animal, as is true for rodents.

Although the commercially prepared dry biscuits have a good acceptance for most nonhuman primates as well as squirrel monkeys, the size and texture for the biscuits may represent a problem for small, weak, or aged animals or those with poor dentition. Receptive to the above needs, the food technologists and animal scientists are presently working on a formula intended primarily for squirrel monkeys that will have a slightly higher fat and sucrose content and be of a smaller size and a different texture (Damon Shelton, Ralston Purina, personal communication).

Although the commercially prepared diets are formulated to be nutritionally complete, they are frequently supplemented by a variety of natural foods to provide a margin of safety for some vitamins as well as to increase palatability. Examples of these supplements are hard-cooked eggs, spinach, oranges, apples, bananas, applesauce, a marshmallow with a dose of vitamins, or a semipurified diet cube impregnated with a concentrated vitamin mixture. Supplementation with folate has been shown to decrease the incidence of stillbirths (Rasmussen *et al.*, 1980) and with ascorbate, oranges, peppers, or other foods (Lang, 1968; Hayes *et al.*, 1972) to avoid the development of scurvy in an animal that is quite susceptible (see Section 2.7). The advantages of using these natural products include having a diet that is palatable, readily acceptable, available, and easy to feed, involves no preparation, and is cost effective. In addition, these natural product diets may be associated with better dental health (Corruccini and Beecher, 1982). However, since the exact composition of these diets is not known, they are unsuitable for many nutrition experiments, in which the concentration of both nutritive and nonnutritive properties must be controlled.

1.2.2. Semipurified Diets

Semipurified diets are those formulated from refined ingredients. Briefly, lactalbumin, casein, soy protein, or gluten serve as protein sources; sucrose, dextrin, corn starch, or glucose are common dietary carbohydrates; and several animal (lard, tallow) and vegetable (corn, coconut, soybean, cottonseed, safflower) sources provide lipids. The vitamins and minerals are traditionally provided as mixtures of chemically pure ingredients. Fiber may be the nondigestible carbohydrate portion of seeds, grains, wood, and vegetable pulp. As in the commercial biscuits, a variety of other ingredients, such as agar, emulsifying agents, stabilizing, coloring, or flavoring agents, preservatives, antioxidants, and antibiotics, may also be incorporated into these diets. The semipurified diets offer the advantage of allowing relative control over the composition of the diet when nutritional variables must be known and specified. The reader is referred to several references for semipurified diets utilized successfully for squirrel monkeys to maintain body weight and support growth and/or reproduction (Portman, 1970; Manocha and Long, 1977; National Research Council, 1978; Ausman and Gallina, 1979; Newberne and Hayes, 1979; Russo *et al.*, 1980).

Based on personal experience and a reevaluation of the nutrient requirements of these animals, the skeleton diet given in Table I can be recommended for adult squirrel monkeys. As alluded to earlier, the texture of the diet is an important variable for the primate, which prefers to manipulate the diet while feeding. Semipurified powdered-type diets popular for use with rodents do not find wide acceptance, and in addition are easily scattered about and lost through the cage floor. Most investigators use a method of formulation that produces a solid product, whether through incorporation of a gelling agent (agar or gelatin), baking, or pelleting. Liquid formulations, of course, are utilized for the infant nonhuman primates, and in certain experimental and therapeutic situations in all animals (see Sections 1.2.3–1.2.5).

1.2.3. Chemically Defined Diets

These synthetic diets are formulated from chemically pure ingredients. One advantage of chemically defined diets is that the composition of the food is known with certainty and may be manipulated at will; only amino acids, triglycerides, essential fatty acids, sugars, vitamins, and minerals in addition to other ingredients are used. These diets, in liquid form, may also be useful for therapeutic situations in which ill animals may be fed by gavage allowing

Table I
Semipurified Agar Cake Diet for Squirrel Monkeys[a]

Ingredients	g/100 g	kcal
Protein[b]	16.08	64.32
Corn oil[c]	9.0	81
Dextrin	39.7	158.8
Sucrose	20.0	80
Mineral mix[d]	4.5	0
Vitamin mix[e]	0.45	1.22
Choline chloride	0.27	0
Fiber (cellulose, bran, and/or other)	10	0
Total	100.00	385.34

[a] Mixing directions: Combine all ingredients but corn oil in suitable size mixer and mix well. Add oil and continue to mix. A warm 1.25% agar solution (800 ml/kg dry diet) is added and the resulting mixture is blended well before being poured into pans to harden. The diet is stored in the refrigerator (34–38°F).

[b] Assume 78% crude protein (12.5% nitrogen × 6.25) and that protein quality is 100%. If vegetable proteins such as soy protein are utilized, the amount in the diet should be increased in inverse proportion to potency (Ausman *et al.*, 1979).

[c] Other fats may be used as desired.

[d] Ausman–Hayes mineral mix contains (in g/kg): calcium carbonate, 290.5; calcium phosphate (dibasic), 72.6; potassium phosphate (dibasic), 327.9; sodium chloride, 162.4; magnesium sulfate, 98.7; magnesium oxide, 32.0; ferric citrate, 13.3; manganese sulfate, 1.2; zinc chloride, 0.915; cupric sulfate, 0.290; potassium iodide, 0.077; chromium acetate, 0.044; sodium fluoride, 0.023; sodium selenite, 0.004. Available as premix (Bio-Mix 1321), BioServ, Frenchtown, New Jersey.

[e] Newberne–Hayes vitamin mix contains (in g/kg): dextrin, 686.58; ascorbic acid beadlets, 120; inositol, 100; taurine, 50; *dl*-alpha-tocopheryl acetate (500 IU/g), 20; vitamin A acetate (500,000 IU/g), 5; vitamin D_3 (200,000 IU/g), 1.25; menadione, 0.1; thiamine HCl, 0.80; riboflavin, 1.6; niacinamide, 8.00; calcium pantothenate, 5.00; pyridoxine HCl, 0.80; folic acid, 0.80; biotin, 0.04; cyanocobalamin, 0.03. [Newberne and Hayes (1979); available from Bio-Serv, Frenchtown, New Jersey, # V2500.]

for fluid and nutrient support in the face of anorexia, diarrhea, malnutrition, or other illness. One product intended for humans postsurgically (Vivonex, Norwich-Eaton Pharmaceuticals) has, in our hands, been useful for this pou-pose for short periods of time. Intravenous hyperalimentation in noncatheterized animals is difficult at best and is not possible for long periods of time.

1.2.4. Total Parenteral Nutrition

There are yet no data demonstrating tha total parenteral nutrition has been accomplished in these animals for long periods of time. This laboratory has used a technique of indwelling arterial and venous catheters (Ausman and Gallina, 1978) to provide hydration during an experiment designed to assess the metabolic response during 4–7 days of complete fasting (D. L. Gallina and L. M. Ausman, unpublished data). Approximately 100 ml/day of 0.45% saline plus potassium adjusted to pH 7.4 with bicarbonate was infused into the animal without untoward effects. Utilizing a similar technique, Finn *et al.*, (1982) infused a hypocaloric solution containing amino acids, glucose, and electrolytes into adult squirrel monkeys. Although in both cases the animals lost substantial amounts of weight, the length of the experimental periods testifies to the utility of the infusions.

1.2.5. Liquid Formulas

Liquid diets may be the method of choice in certain experimental situations with adult animals in which, for example, accurate control of dietary intake is essential, where the liquid diet is the vehicle for alcohol or other compound, or where the liquid diet is appropriate to a reinforcement schedule (Herndon *et al.*, 1958; Ellison and Riddle, 1961; Lofland *et al.*, 1972; Jones *et al.*, 1975). In addition, diets may consist of solid food plus nutrients such as sugars or alcohol offered in the water bottle. One example of this are the experiments of Kaplan *et al.* (1982), who studied the effect of oral ethanol intake on blood alcohol levels in the adult squirrel monkey using several different schedules of alcohol administration. Liquid diets may also be utilized for the nutritional support of ill and/or anoretic animals. Liquid diets as outlined in this review, or even those prepared from blending equal parts of the agar cake diet and water, may be fed by gavage where appropriate to stimulate appetite and provide some nutrient intake.

Newborn and young squirrel monkeys must be fed liquid diets. A variety of commercial liquid formulas (Similac with Iron, Ross Laboratories; SMA, Wyeth Laboratories) have been utilized as a dam's milk substitute for short periods after birth. Most of the commercial formulas based on cow's milk contain about 9–10% of the dietary calories as protein. Those formulated with soy protein contain somewhat more at 12%. As discussed in Section 2.3, these formulas may not contain adequate protein to meet the needs for maximal growth in the 2–4 weeks following birth. Before the development of semipurified liquid formulas for the neonatal period, infant monkeys were routinely kept on these commercial formulas for periods of 6–8 weeks before being switched to liquid diets intended for older animals or to solid food. The reader is cautioned about the adequacy of these commercial diets for long-term usage. In one experiment focusing on iron metabolism, infant squirrel monkeys were fed pasteurized milk and a commercial liquid formula with and without added iron (Amine *et al.*, 1973). However, by 12–16 weeks of age, animals fed solely the commercial formula had developed a yellow hair coat without the stripes of black that are charcteristically found (Ausman, 1973). This same change in hair coat could be reproduced in adult animals fed the commercial preparation solidified with an agar–water solution. A shaved area on the back, approximately 4×5 cm^2, grew in as hair that was a brilliant yellow. Although the cause of the change in pigmentation is still unknown, this observation reemphasizes that diets intended for humans may not always be adequate for squirrel monkeys.

The desire to manipulate the diet of the infant squirrel monkey from birth or soon after birth brought a number of dietary trials utilizing semipurified liquid diets in formulas intended for the very young (Ausman and Gallina, 1979). Rules for the formulation of human infant diets served as a guidline in the formulation of these diets (ESPGCN, 1977). Because of the immature adrenal and kidney, it is important to keep the salt concentration of the formula low but adequate. Excess protein should be avoided to prevent the amino acid toxicity that may occur before the amino acid catabolic enzymes in the brain and liver have matured. Pertaining to this issue, Wen and colleagues (1973) utilized an infant squirrel monkey model to investigate the possible detrimental effects of oral monosodium glutamate (MSG) delivered in a commercial liquid formula; they found no significant effect of the MSG on growth rate, electroretinograms, electroencephalograms, blood pressure, bone development, and tissue histology.

For reasons of digestibility, the carbohydrate of choice during the period of infancy is lactose. However, to keep the osmolarity of the diet within reasonable bounds, the lactose content of the diet is kept low and the fat content at levels ranging from 40 to 60% of calories, a number that appears to hold for dam's milk of several species (Altman and Dittmer, 1971).

Animals of older ages, i.e., greater than 6–8 weeks of age, can easily tolerate higher salt concentrations, more concentrated diet, higher osmolarity, and nonlactose carbohydrates.

Table II details the complete recipes for formulating liquid diets intended for the first 1–2 months of life (DL-16) and thereafter (L-13). Spray-dried lactalbumin, fine-ground casein, isolated soy protein, soy protein concentrate, and gluten have served as protein sources (although used in different amounts) in these diets. Carbohydrates incorporated have ranged from lactose in infancy to sucrose, dextrin, corn starch, glucose, and fructose after a few weeks of age. Fat sources incorporated into diets of animals of all ages have included corn oil, safflower oil, cottonseed oil, coconut oil, lard, tallow, and butter. The solid oils are first melted and then incorporated into the liquid diets in a blender. The emulsifying agents help prevent the total aggregation of the oil when the bottle is introduced into the refrigerator. Rewarming and shaking are usually sufficient to resuspend these diets.

The physical characteristics as well as nutrient qualities of the dietary ingredients are important in determining their usage for liquid diets. The ingredients described in Table II are examples of products that are especially suitable for liquid diets. The importance of a long-lasting emulsion and/or suspension varies in direct proportion to the length of time the bottle is left unattended in the animal's cage. Furthermore, unless a good suspension is obtained, it is difficult to aliquot the diet uniformly into individual bottles. It should be obvious that errors here may jeopardize the entire experiment.

2. *REQUIREMENTS FOR SPECIFIC NUTRIENTS*

2.1. *Introduction*

The word requirement stresses an accuracy that in most cases does not exist with respect to the squirrel monkey. In a few cases investigators have determined the needs for a particular nutrient. In the majority of cases, however, the needs of these animals were assumed to be met by diets in which

Table II
Liquid Diets for Newborn and Young Squirrel Monkeys

Ingredient	DL-16[a] (g/liter)	L-13[b] (g/liter)
Lactalbumin, spray-dried[c]	35.44	41.27
Lactose[d]	57.26	0
Sucrose	0	50.00
Dextrin[e]	0	82.58
Corn oil[f]	36.19	33.33
Water-soluble mineral mix[g]	4.79	9.80
Vitamin mix[h]	0.8	1.14
Choline	0.48	0.688
Tween 80[i]	0.67	0.96
Suspending agent[j]	0.67	0.96
Fiber (cellulose, bran, pectin, or other)[k]	0	9.80
Total	136.30	230.53

[a] These diets are fed to neonates soon after they are removed from their dams and have been fed Similac for 1–2 days. These semipurified liquid diets were approximately 0.7 kcal/ml, contained lactose, and served to wean the animals to semipurified diets (mixing directions are similar to those given in footnote *b*).

[b] One liter of liquid diet was prepared by blending about 700 ml water with the oil, Tween 80, and suspending agent in a blender for 2 min. The remaining ingredients (protein, carbohydrate, salt and vitamin mixtures, choline, and fiber) were then added and the mixture blended for an additional 2 min. The contents of the blender was then poured into a beaker, and an additional 150 ml of water was used to clean the blender and scraper; this was also poured into the beaker. The entire mixture was left to defoam in the refrigerator overnight. The following morning, the mixture was quantitatively poured into a dry, tared 1-liter volumetric flask. The flask was brought up to volume and the weight of the filled flask was recorded. This procedure allows the calculation of the amount of water added to produce a recipe of 1.0 kcal/ml. If less precision is desired in the future, the calculated amount of water may simply be added. Diet may be aliquoted into containers of convenient sizes and frozen. In practice, we have made this diet in batches of 8 liters at a time for highly standardized nitrogen balance experiments.

[c] Spray-dried lactalbumin, Teklad, Madison, Wisconsin. The following could also be used, although due consideration should be given to the potency of the protein before formulation of the diet: fine-ground casein (Bio-Serv, Frenchtown, New Jersey); Promosoy 100 (Central Soya, Gibson City, Illinois); Supro 710 (Ralston Purina, St. Louis, Missouri).

[d] *d*-Lactose hydrate, Grand Island Biological Co., Grand Island, New York.

[e] Bio-Serv, Frenchtown, New Jersey.

[f] Mazola, Best Foods, CPC International, Englewood Cliffs, New Jersey.

[g] Ausman water-soluble mineral mix for liquid diets contains (in g/kg): calcium phosphate tribasic $Ca_3(PO_4)_2$, 356.51; potassium carbonate, 260.19; sodium chloride, 162.37; magnesium sulfate (anhydrous), 143.89; potassium phosphate, 67.71; ferrous sulfate, 6.74; manganese sulfate ($MnSO_4 \cdot H_2O$), 1.22; zinc chloride, 0.91; copper sulfate ($CuSO_4 \cdot 5H_2O$), 0.29; potassium iodide, 0.077; chromium acetate $Cr(C_2H_3O_2)_3 \cdot H_2O$, 0.048; sodium fluoride, 0.023; and sodium selenite, 0.004.

[h] Identical to Newberne–Hayes vitamin mixture. Ascorbic acid (USP-FCC) Type 5, Hoffman LaRoche, is always utilized for liquid diets.

[i] Polyoxyethylene monosorbate, Grand Island Biological Co., Grand Island, New York.

[j] Suspending agent K, Bio-Serv, Frenchtown, New Jersey.

[k] Nonnutritive fiber, Bio-Serv, Frenchtown, New Jersey. Other fiber may be substituted or added as desired.

Table III
Protein and Calorie Requirements of Squirrel Monkeys Fed Casein-Based Diets[a]

Body weight (g)	Growth rate (g/day)	Calorie intake		Protein requirements for maintenance + growth		
		kcal/day	kcal/kg per day	g/day	g/kg per day	percent kcal
150	3.4	72	480	2.66	17.7	14.78
200	2.5	90	450	2.27	11.3	10.09
300	1.8	124	413	2.19	7.3	7.06
400	1.3	137	342	2.25	5.6	6.57
500	0.6	150	300	2.17	4.3	5.79

[a] Derived from Ausman *et al.* (1979). Based on data from 42 male and female monkeys.

animals were clinically asymptomatic. Experiments looking at the effect of the nutrient on life span, ability to reproduce, resistance to infection, and so forth are largely missing, in part due to their great expense. This dilemma has faced both the National Academy when making recommendations for nutrient intakes (National Research Council, 1978) as well as other authors who have reviewed this area (Portman, 1970; Kerr, 1972; Nicolosi and Hunt, 1979).

2.2. Calorie Requirements

Calorie requirements of infant to adult squirrel monkeys for weight maintenance, growth, and reproduction have not been determined and it is doubtful, except under the most highly standardized conditions, whether this would be possible. The observation that restriction of *ad libitum* food intake by 15% does not decrease weight gain in 300- to 500-g monkeys (L. M. Ausman, unpublished data) is an example of the difficulties involved in assessing calorie requirements.

Instead, information on calorie needs must be derived from reports of *ad libitum* intakes.

Both Beischer (1968) and Finn *et al.* (1982) report that a diet comprised of 25–30 grams per day of commercial monkey biscuits or Purina Chow (100–130 kcal) will maintain an adult squirrel monkey. Since their male animals are reported to weigh 800–1100 g, maximum calorie intakes calculated per kg body weight probably varied from 120 to 160 kcal/kg per day. In contrast, data from animals fed semipurified diets indicate significantly greater intakes.

The data of De la Iglesia *et al.* (1967) indicated a mean caloric intake of about 200–250 kcal/kg per day for healthy, growing, young male squirrel monkeys (initial body weight of about 695 g). Data from this laboratory (Table III) (Ausman *et al.*, 1972, 1979, 1981) summarize experience with semipurified liquid and solid diets for neonates through obese adults. Intakes approached 500 kcal/kg per day during infancy and decrease to levels of 238 $\pm$ 50 kcal/kg per day for lean adults (mean body weight of 788 g) and 155 $\pm$ 24 kcal/kg per day for obese animals (mean body weight of 1411 g).

It is well known that expression of food intake data on a per kg basis instead of metabolic body size may lead to highly variable results, especially in comparing small to large animals, or those with different body composition (Brody, 1945; Ausman *et al.*, 1972, 1981). In addition, animal activity, markedly influenced by cage size, socialization, and other factors, will affect the calorie needs of the animal. It is difficult, therefore, to judge whether differences in reported calorie consumptions are due to the diet *per se* or to differences in some of the other variables discussed.

2.3. Protein Requirements

The protein requirements of these animals for maintenance and growth are given in Table III. These data were derived from slope-ratio growth assays in which young squirrel monkeys were fed diets of varying protein concentrations (Ausman *et al.*, 1979). The protein requirements of these young animals tend to be higher than those of humans or other Old World monkeys such as *Macaca fascicularis* or *Macaca mulatta* (Kerr, 1972; Ausman *et al.*, 1979; Ausman and Gallina, 1979). In the early postnatal period (birth to 2 weeks) the minimum protein requirements may be as high as 15% of the calories, but decrease to 10% of calories (levels found in the commercial formulas) by 200 g of body weight. It is interesting to note that protein comprises about 16% of dietary calories from squirrel monkey dam's milk (Buss and Cooper, 1972).

Although juveniles and adults could be maintained on diets containing a minimum of 6–7% protein (as percent of calories), doubling this amount would allow for individual variance, stress, illness, pregnancy, or other mediating factors. Thus, it has been our practice to maintain nonpregnant animals on diets containing approximately 13% protein (as percent of calories). Indeed, Manocha and colleagues have shown that 8% protein during gestation was insufficient to produce normal infants, as judged by an increased abortion rate, decreased birth weight, and histochemical abnormalities (Manocha and Long, 1977; Manocha and Sharma, 1978; Sharma and Manocha, 1978). Since dietary fat concentrations frequently vary, protein and other nutrient concentrations are often expressed as percent of dietary calories using the conventional factors of four, four, and nine for the caloric equivalent of protein, carbohydrate, and fat, respectively. It should be remembered that the nitrogen (and thus crude protein) content of these protein sources varies considerably according to type of protein and lot number. If protein content is critical, a Kjeldahl analysis for total nitrogen must be carried out on every lot. The factor of 6.25 to convert from nitrogen to protein is convenient to use, again to enable discussion in terms of crude protein or total nitrogen without introducing different variables into each diet.

The potencies of lactalbumin and casein as measured in the infant squirrel monkey do not appear to be statistically different (Ausman *et al.*, 1979) although in *Cebus albifrons*, another New World primate, casein has a potency approximately 65–75% that of lactalbumin (Ausman *et al.*, 1983). The potency of soy protein, as measured in either species, is approximately 50% of lactalbumin. Therefore, if this protein source is selected, a methionine or cystine supplement must be utilized, or, in the case of older animals, additional unsupplemental protein could be used. De la Iglesia *et al.* (1967) maintained 12 young squirrel monkeys on soy protein diets containing 6.1, 9, 12.5, and 25% of the calories as protein and included a methionine supplement. Although original small group numbers and deaths during the experimental period precluded a strong statistical evaluation, it was apparent from body weight, serum protein, and albumin concentrations, hematological analyses, and other parameters that the diets containing both 6.1 and 9% were inadequate. It was impossible to deduce whether the performances of the 12.5 and 25% protein levels were significantly different. Taken together, these data appear to confirm that additional soy protein is needed (as compared to lactalbumin) to maintain the adult squirrel monkey.

2.4. *Fat Requirements*

Little is known about the essential fatty acid requirements of the squirrel monkey. One report suggests that linolenic acid is a requirement for the *Cebus* sp. (Fiennes *et al.*, 1973). In the absence of other information, it appears that the requirements for essential fatty acids, as least as measured by serum 20:3/20:4 fatty acid ratios (this laboratory, unpublished data), can be satisfied with a diet in which the linoleic acid (supplied by coconut oil) is only 0.4% of the dietary calories.

As detailed earlier, squirrel monkeys adapt well to a variety of dietary fat sources. However, the vitamin E and perhaps other dietary requirements may vary, depending on fat source. In one experiment in *Cebus albifrons* and *Macaca fascicularis* in which safflower oil was utilized as the sole fat source, the dietary vitamin E was increased from 50 to 100 mg/kg nutritive diet to correct marginal serum tocopherol levels that appeared after 1 year of feeding (Ausman and Hayes, 1974).

2.5. *Carbohydrate Requirements*

There is no specified requirement for carbohydrate in the diet of these animals, although the natural diet contains generous amounts from leaves, nuts, grains, and fruits. In practice in the laboratory, several simple to complex carbohydrate sources have been used in semipurified diets. Several of these have been used in research experiments to investigate factors that modulate diet-induced atheroma or influence circulating lipids (Hayes, 1979; Malinow, 1979). Aside from a consideration of experimental and physiological requirements, factors such as palatability and prior performance are important in the selection of diets for nonhuman primates (Ganchrow and Fisher, 1968; Lang, 1970). This is especially true for the case of carbohydrates, as demonstrated by the results of one 2-year trial in which dextrin was the sole carbohydrate, providing about 70% of the dietary calories. Body weights of both male and female animals after 1.5 years were substantially decreased from body weights of those fed diets containing both sucrose and dextrin. In addition, several animals developed clinical evidence reminiscent of an allergic conjuctivitis. Introduction of sucrose into the diet was associated with amelioration of the clinical symptoms, increased food intake, and catchup growth.

Table IV
Mineral Allowances for New World Monkeys

Element	Content in Ausman–Hayes mineral mix	Content in diet[a]	Recommended daily allowance[b]
	(g/kg)	(g/kg diet)	(g/kg diet)
Ca	133.2	6.660	6–8
K	147.2	7.360	2.4–10.9
P	71.40	3.570	3–4
Mg	29.06	1.453	1–1.5
S	13.11	0.656	—
Na	63.89	3.194	2.2–4.4
Cl	98.98	4.949	2.7–6.2
	(mg/kg)	(mg/kg diet)	(mg/kg diet)
Fe	2220.0	111	100
I	59.1	2.95	2
Mn	396.9	19.84	20
Zn	438.8	21.94	20
Cu	73.8	3.69	2
Cr	10.05	0.502	0.5
Se	1.96	0.098	0.1
F	10.5	0.525	0.5

[a] When incorporated into the diet at 5 g/100 g.
[b] Derived from Nicolosi and Hunt (1979).

2.6. Mineral Requirements

With no data to the contrary, the mineral requirements of this species were assumed to be similar to those of man and Old World monkeys, whose requirements are better defined. Several different mineral mixes, many of which were formulated before the importance of all the trace elements were known, have been used with these New World primates. Deficiencies of the trace elements have usually been avoided due to contribution of the other dietary ingredients and galvanized cages.

The mineral requirements of nohuman primates have been summarized for all species (Kerr, 1972; Nicolosi and Hunt, 1979; National Research Council, 1978). Deficiencies of zinc (Barney *et al.*, 1967; Macapinlac *et al.*, 1967), chromium (Davidson and Blackwell, 1968; Martin *et al.*, 1972), and selenium (Muth *et al.*, 1971) have specifically been produced in squirrel monkeys; therefore, amounts recommended for these minerals rely in part on these reports. Considering what is currently known about mineral requirements for squirrel and other monkeys, a mineral mix suitable for New World monkeys was formulated and is detailed in the footnotes to Table I. This is a modification of the Hegsted IV mineral mix used for so many years. The elemental contents in this mineral mix and the resulting diet as well as the recommended requirements for the nonhuman primate are summarized in Table IV.

Table V
Vitamin Allowances for New World Monkeys

Vitamin	Vitamin content	Recommended allowance[a]
	(mg/kg diet)	
Thiamin hydrochloride	4.0	4–6
Riboflavin	8.0	8–10
Pyridoxine hydrochloride	4.0	4–6
Calcium pantothenate	25.0	11[b]
Niacinamide	40.0	40–50
Folic acid	4.0	4–6
Biotin	0.2	0.2
Cyanocobalamin	0.15	0.004[c]
Ascorbic acid	612.5	500–600
Menadione	5.0	0.800[d]
	(IU/kg diet)	
dl-Alpha-tocopheryl acetate (500 IU/g)	50.0	20–100
Vitamin A acetate (500,000 IU/g)	12,500	12,500
Vitamin D_3 (200,000 IU/g)	1,250	1,250

[a] Taken from Nicolosi and Hunt (1979).
[b] Calculated using a U.S. intake of 7 mg/kg, a 2500 kcal/day intake, and a diet density of 4 kcal/g.
[c] Calculated using the human RDA of 3 μg/day, a 2500 kcal intake, and a diet density of 4 kcal/g.
[d] Calculated using the usual intake of 500 μg/day (U.S. RDA), a 2500 kcal/day intake, and a diet density of 4 kcal/g.

2.7. Vitamin Requirements

Vitamin requirements for nonhuman primates are reviewed elsewhere (Kerr, 1972; National Research Council, 1978; Nicolosi and Hunt, 1979). Since few studies have been done detailing vitamin needs of the squirrel monkey, recommended allowances are based on amounts in natural diets and amounts present in semipurified diets that appear to promote normal growth and development. A vitamin mixture developed for use in both adult and infant squirrel monkey diets has been detailed in Table I. This mixture contains the following modifications of the vitamin mix formerly used in this laboratory (Ausman and Gallina, 1979): (1) the folic acid content has been quadrupled, (2) taurine (a putative growth factor for the New World monkey) has been added (Hayes *et al.*, 1980), (3) inositol was incorporated directly into the vitamin mixture, and (4) coated ascorbic acid beadlets have been used so that the mixture can be of use for both solid and liquid diets. It should be emphasized that vitamin D_3 is obligatory for this species (Hunt *et al.*, 1967; Lehner *et al.*, 1967; Hunt *et al.*, 1969).

The vitamin content of the diet (vitamins formulated at 0.5% of the nutritive diet) and the recommended allowances are summarized in Table V. In addition, as discussed in Section 1.2.1, folic acid and ascorbic acid seem to be particularly important, since spontaneous abortion and hematological abnormalities attributable to deficiencies of the former (Rasmussen *et al.*, 1979, 1980; Rosenblum, 1968) and outright scurvy attributable to a deficiency of the latter (L. M. Ausman, unpublished observations; Lehner *et al.*, 1968; Blackwell *et al.*, 1974; Demaray *et al.*, 1978; Kessler, 1980) are frequently found when supplements are insufficient. Scurvy is also seen in squirrel monkeys fed outdated, improperly stored, and/or water-leached commercial monkey biscuits (Demaray *et al.*, 1978). The ease with which this animal may be depleted of ascorbate makes it a good model for studying the effect of vitamin C nutriture on the handling of toxic compounds such DDT (Chadwick *et al.*, 1971).

As discussed in Section 1.2.1, dietary supplements are frequently used to provide extra ascorbic acid and folic acid as well as other vitamins. Vitamin D_3 may also be directly supplemented or can be provided via the use of special fluorescent lights or sunlight. Choline, due to its hygroscopic nature, is routinely added separately to the diet at levels of about 3 g/kg diet. This is done chiefly as a safety factor since there are yet no data to indicate the actual requirement of this species. Furthermore, this lipotropic factor may become more important during experimental situations involving methionine and/or protein deficiency.

2.8. *Fiber*

It is impossible to set a fiber requirement for this species of monkey. Undoubtedly, the feral diet contains generous amounts of celluloses, hemicelluloses, pectins, gums, and mucilages, as judged by their intakes of leaves, seeds, grains, fruits, and nuts. The influence of fiber type on lipid and sterol metabolism is the most active arena of investigation in these animals. Cellulose has been used in past experiments in this laboratory, primarily to add consistency to the stools of animals fed semipurified diets, although it lacks many of the desirable metabolic effects evidenced by other forms of fiber. Since several types of fiber undoubtedly exist in natural diets, there is a definite need to establish recommendations for fiber usage in the squirrel monkey fed semipurified diets.

3. *GROWTH AND BODY COMPOSITION AS AFFECTED BY DIET*

3.1. *Infants*

The growth of the neonatal and infant squirrel monkey has been reported by several investigators (Long and Cooper, 1968; Ausman *et al.*, 1970, 1976, 1979, 1981; Portman, 1970; Hinkle and Session, 1972; Kaplan, 1974, 1977, 1979; Kaack *et al.*, 1979; Manocha, 1979; Russo *et al.*, 1980).

These references appear to indicate that there are real differences in the birth weights and growth rates of the various subspecies of squirrel monkeys. This is not unexpected, since adult weights of the subspecies also vary (Long and Cooper, 1968; Middleton and Rosal, 1972; Delort *et al.*, 1976). It is difficult to make comparisons between many of these growth curves because estimates of variance are omitted, numbers of animals are small, or the subspecies is not

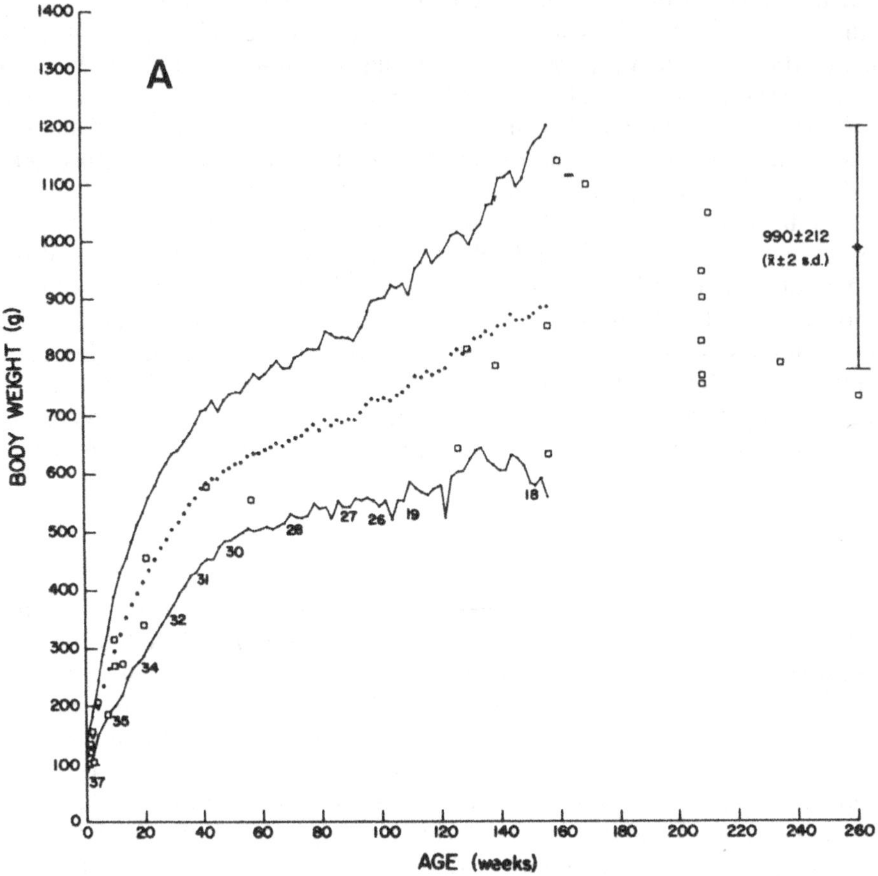

Fig. 1. Mean longitudinal body weights (···) of (A) male and (B) female squirrel monkeys hand-reared from birth and fed semipurified "control" diets. The solid lines enclose the 95% confidence interval for these data. Small numbers indicate sample size at each age. For explanation of small squares, see Russo *et al.* (1980). The mean body weight (mean ± 2 SD) of adult monkeys from the breeding colony was arbitrarily plotted at 5 years of age for comparison. [Reprinted from Russo *et al.* (1980), with permission.]

specified. There does appear to be a remarkable similarity among the various growth curves of the Colombian species (Leticia); this has occurred despite the differences in dam or surrogate rearing as well as the differences in dietary regimens utilized in each study.

Human infant formulas do not contain as much protein as squirrel monkey milk (Buss and Cooper, 1972). The use of human formulas could affect the growth rate of the rapidly growing or premature squirrel monkey infant, who could be expected to have requirements of 15–16% protein and not 10% as is found in the commercial human infant formulas. In particular, this would appear to have its impact, however marginal, in the first month of life.

Survival in the 2–4 weeks after birth appears to be compatable with a liquid diet that does not contain excess salt, protein, or osmolarity, and is

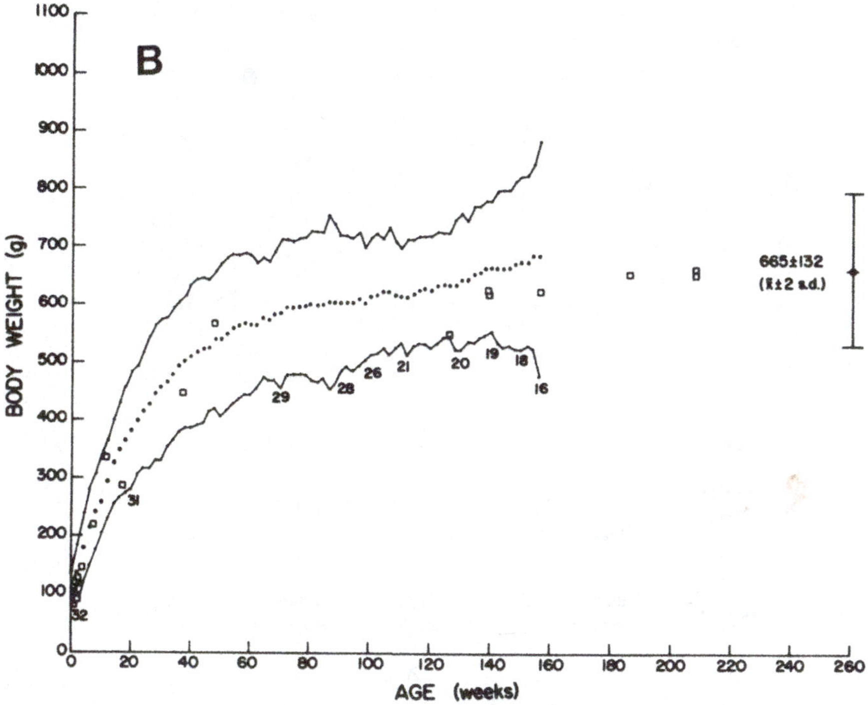

Fig. 1. *(continued)*

sufficiently dilute (0.7 kcal/ml) to adequately hydrate animals who do not drink water *ad libitum*. It should come as no surprise that diets low in fat and high in lactose (high osmolarity) are associated with a poor neonatal survival (this laboratory, unpublished data). Once the animals are a few weeks old, those consuming diets of 1 kcal/ml demonstrate a greater weight gain than those fed the traditional 0/.7 kcal.ml (this laboratory, unpublished data), although it is unclear whether or not this is primarily fat. Nevertheless, continuation of the dilute diets for longer than 12 weeks should be discouraged, since introduction of foodstuffs adequate for adults usually begins by this age.

Postnatal growth rates of the infant monkeys appear to be correlated with birth weight. Whether or not this is of nutritional or genetic origin is not known. Evidence from one laboratory (Manocha and Long, 1977) indicates that inadequate (low-protein) diets during pregnancy cause low-birth-weight infants, although it remains to be shown if the gestational history will affect postnatal growth rate if an "adequate" diet is provided.

3.2. Adults

Fewer growth curves exist depicting long-term growth of the squirrel monkey (Long and Cooper, 1968; Russo *et al.*, 1980; Ausman *et al.*, 1981); these are shown for monkeys fed semipurified diets in Fig. 1A and 1B. It appears

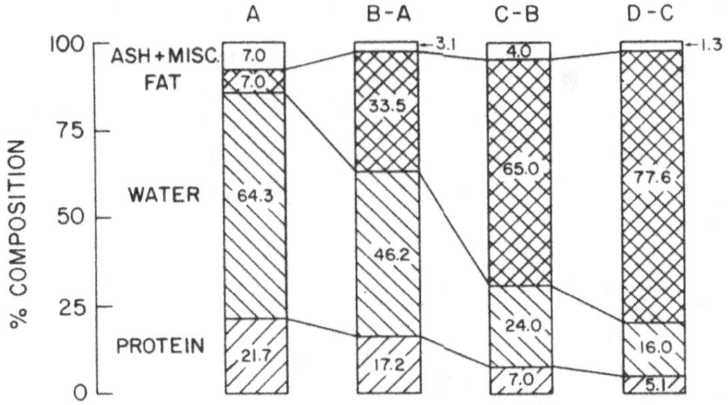

GROUP		A	B	C	D
BODY WEIGHT (g)	mean	818	1097	1380	1668
	range	733-950	1050-1140	1370-1390	1562-1822
CARCASS WEIGHT (g)	mean	700	927	1185	1422
	range	625-821	880-971	1150-1220	1320-1527
N		10	3	2	3

Fig. 2. Carcass analysis of 18 male squirrel monkeys calculated to show the composition of carcasses of (A) ten lean control monkeys (body weights 733–950 g) and composition of incremental tissue for each of (B) fatted control and (C, D) obese groups. [Reprinted from Ausman *et al.* (1981), with permission.]

that by 3 years of age, the animals have attained the body weights of their dams and sires: 665 ± 122 and 990 ± 212 g (mean ± 2 SD) The body weights of the adult animals are reminiscent of adult weights for these Colombian animals (Long and Cooper, 1968; Delor *et al.*, 1976; Raymond *et al.*, 1976). Feral weights of these animals normally fluctuate 100–200 g with the season and are highest during times of lush vegetation, rainfall, and, obviously, food (DuMond, 1968; Coe and Rosenblum, 1978). Animals fed commercial biscuits appear to approximate the weights of these feral animals. As discussed previously, these commercially prepared monkey biscuits may not be optimal for these animals. Furthermore, body weights derived from animals that naturally endure periods of low food availability may not represent the genetic potential of this species. On the other hand, it was of concern when a high proportion of these hand-reared squirrel monkeys developed a postpubertal obesity (Ausman *et al.*, 1981). Careful scrutiny of the growth curves of these obese animals for their first few months of life versus those of dam-reared or those fed commercial milk preparations suplemented with biscuits indicated no obvious differences in the rates of weight gain. The obesity is attributed, at least in part, to an interaction of postpubertal hormonal changes with a lack of exercise and excessive consumption of the highly palatable high-fat and moderate-sucrose diet. The maximum body weight of 14 males classified as obese was 1527 ± 246 g (mean ± SD) and of 18 females was 1032 ± 229 g (mean ± SD). Figure 2 demonstrates clearly that the majority of the excess added weight was fat, with one male having 39.3% fat. As shown previously, the adult feral animal, those maintained on commercial monkey biscuits, and juvenile squirrel monkeys fed semipurified diet are normally quite lean, with body fat averaging 6–8% in nonfatted males and females (Russo *et al.*, 1980; Ausman *et al.*, 1981).

The effect of dietary nutrients on both the rate and nature of chemical maturation of the infant squirrel monkey carcass has been determined for the "normal" situation and could serve as a standard should the effect of nutritional intervention on carcass composition be studied (Russo *et al.*, 1980). These animals appear to show the normal decrease in body water and associated increase in body nitrogen that occurs in most species studied within weeks to months of birth. In the squirrel monkey, greater than three-quarters of the normal maturation appears to have taken place by 4-11 weeks of age, and the carcass composition of animals 20 weeks of age was not distinguishable from that of an adult. Furthermore, it appeared from biochemical and hematological indicies that these parameters had reached adult values by 6–8 weeks of age (Ausman *et al.*, 1976). These developmental milestones could be important in determining the protocol for experiments dealing with infant or young animals.

4. *NUTRITIONAL AND METABOLIC INTERACTIONS*

4.1. *Introduction*

The study of nutritional requirements *per se* may not hold much excitement, but is extremely important in the maintenance of a healthy animal and preservation of a sound experimental design. Increasingly, the animals are being used in several avenues of research, among them pharmacology, behavior, reproduction, and nutrition. Due to some innate peculiarities of the squirrel monkey as well as its ease of handling in a research situation, the animal may serve as a model for human nutritional and metabolic disorders. The following sections briefly review some aspects of nutrition and metabolism in this species.

4.2. *Carbohydrate Metabolism: Glucose Tolerance*

Two main observations regarding carbohydrate metabolism in this species are prevalent in the literature.

First, two laboratories found that 50% or more of the adult squirrel monkey population presented an impaired response to an oral glucose load or to tolbutamide stimulation (Lang, 1966; Davidson *et al.*, 1967; Wilson *et al.*, 1971). Chromium supplementation was reported to partially ameliorate this impaired response (Davidson and Blackwell, 1968). In addition, the response to IV glucose tolerance in several laboratories indicated that even the "normal" responses (Lehner *et al.*, 1971; Wilson *et al.*, 1971; Ausman, 1973) were less than expected for metabolic body size (Gallina and Ausman, 1979).

Second, plasma insulin concentrations were reported to be undetectable (Mann and Crofford, 1970), or presented a poor response to the beta cell stim-

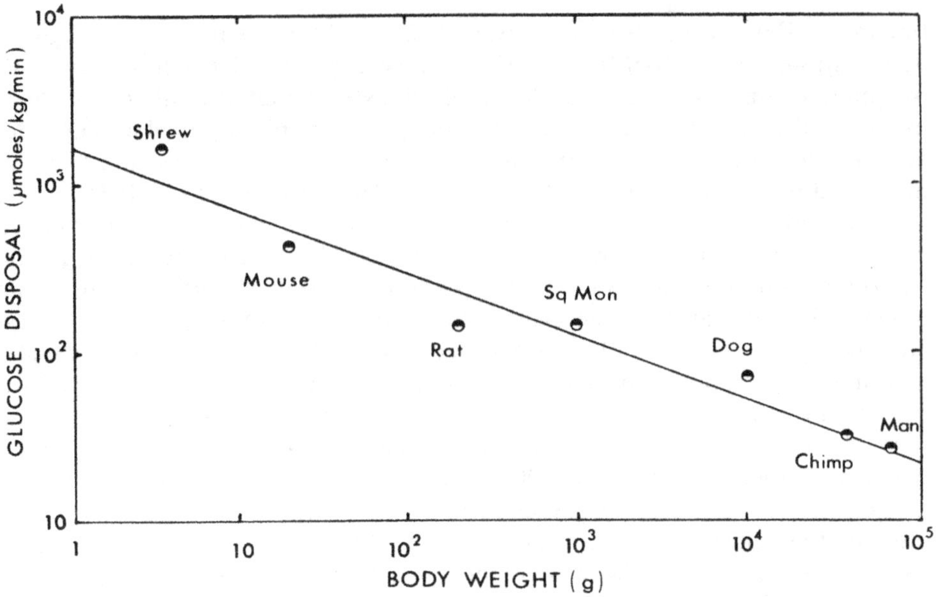

Fig. 3. Glucose disposal rates (oxygen consumption divided by six) are plotted versus body weight according to Coulson *et al.* (1977). Data from the squirrel monkey (Malinow and Wagner, 1966; Smoake *et al.*, 1977) and chimpanzee (Brody, 1945) were calculated in the same manner and are included for comparison. The regression equation $Y = -0.378X + 3.234$ was calculated using log values for both X and Y. [Reprinted from Gallina and Ausman (1979), with permission.]

ulation (Wilson *et al.*, 1971). These and more recent findings concerning the first observations have been reviewed in detail (Ausman and Gallina, 1978; Gallina and Ausman, 1979), and can be summarized as follows. Several environmental variables, such as the length of fasting, access to water, type of anesthesia, and acclimatization to the experimental conditions, have profound impact on the blood glucose response to a glucose load. When a group of adult and juvenile animals was studied in unrestrained and acclimatized conditions through permanently implanted indwelling arterial and venous catheters (Ausman and Gallina, 1978), the response to the oral dose fell well within the values accepted as a "normal" response in virtually all monkeys. On the basis of the blood glucose and free fatty acid concentrations as well as the response of several polypeptide hormones, glucose given at 2 g/kg for the oral test appeared to approximate the physiological load for this species (Gallina and Ausman, 1979).

The response to an IV load (0.5 g/kg) showed a clearance rate (K value) of about 7–8% min, a value that would be expected for animals of this body size and metabolic rate when compared on a scale extending from shrews to man (Fig. 3). The low rates of glucose disposal found in the initial studies may be attributable to several procedural and environmental factors. Since the average age of the monkeys studied in the other laboratories was probably greater than those tested in this laboratory, and since glucose intolerance in known to increase with age, it is possible that some of the impaired response to glucose loading may have been due to age. The one animal shown to have impaired glucose tolerance utilizing our catheterized model was an old, breeder male (Gallina and Ausman, 1979).

Insulin concentrations in these animals, particularly after pancreatic stimulation, are generally accepted as low, even though in one study the crude pancreatic concentration of insulin was found to be similar to that of the rhesus, an Old World species (Wilson *et al.*, 1971). In comparison, the rhesus shows a 5- to 10-fold increase in insulin after glucose loading (Wilson *et al.*, 1971), which is similar to the human response. Since the adipose tissue depot in the squirrel monkey is very low (6–8%), especially before the age of 4 years (Russo *et al.*, 1980), it was postulated that the small increase in insulin concentration following the administration of glucose or arginine could correspond to a relatively high efficiency of a modest insulin release such as seen in thin humans. This explanation appeared to be supported by both a high glucose disposal and an antilypolitic effect in the absence of major surprises in the behavior of the "anti-insulin" hormones such as glucagon or growth hormone. However, it is well recognized that the squirrel monkey polypeptide hormones may cross-react poorly with antibodies intended for human assays or those for "higher" primates (Tashjian *et al.*, 1966; Wang *et al.*, 1968). Insulin is probably not an exception. With many of the common antibodies prepared to pork or beef insulin it is impossible to get consistent values on dilution of squirrel monkey plasma. For this reason we are currently preparing a new assay utilizing purified squirrel monkey insulin with a pork tracer and antibody. At the same time, assays for glucagon (Gallina and Ausman, 1979), growth hormone (Ausman and Gallina, 1978; Gallina and Ausman, 1979), and prolactin (C. L. Coe, personal communication) may conveniently be done with assays intended to measure human circulating levels, since values are reproducible regardless of the sample dilution assayed.

4.3. Protein Metabolism and PCM

It has been shown that the young squirrel monkey has a low efficiency for protein utilization, which does not improve by dietary supplementation with sulfur amino acids nor by extending the feeding schedule to 24 hr daily (Ausman *et al.*, 1979). Indeed, the requirement of the juvenile to adult animal of at least 4 g protein/kg per day as found in this laboratory (see Table III) and the work of Finn *et al.* (1982) support the conclusion that these animals must consume greater amounts of protein to maintain body weight than expected as compared to Old World monkeys and man (Ausman and Gallina, 1979; Ausman *et al.*, 1979). The cause of this inefficiency is not known, although it can be speculated that two main mechanisms might be responsible. First, as seen in the obligatory requirement of the cat for increased levels of dietary arginine (Rogers and Morris, 1980), the squirrel monkey could have an absolute or facultative higher requirement for one of the amino acids. On the other hand, the inefficient protein utilization might represent a defect in the metabolic adaptation of the liver degradative enzymes to low nitrogen levels, as seen in at least one species

(Rogers and Morris, 1980). Although one experiment showed that the addition of sulfur amino acids to casein failed to improve the utilization of that protein (Ausman *et al.*, 1979), to our knowledge, no data have been gathered on either of these hypotheses.

Several studies designed to characterize the animal with respect to its response to fasting, feeding, and loading tests have failed to show major differences as compared to man. Based on metabolic rate (Malinow and Wagner, 1966; Smoake *et al.*, 1977), the postabsorptive period and metabolic response in the squirrel monkey are predictably 3–4 times faster than in man. The response to glucose and to amino acid stimulation, however, as well as to total fasting provides hormone and metabolite patterns similar to those found in man, with perhaps the exception of a lower insulin/glucagon ratio. Even this possibility may be revised when accurate insulin assays for the squirrel monkey become available. In relative terms, therefore, we should classify the squirrel monkey as a hypoanabolic species when compared with Old World nonhuman primates or man.

Aside from the characterization of protein requirements and "normal" protein metabolism, these animals have also been utilized experimentally to approach a very difficult issue: protein–calorie malnutrition (PCM). Since it is not possible to individualize the single causal factors involved in the pathogenesis of human PCM, several monkey species, the squirrel monkey included, have been utilized for this purpose.

Manocha and colleagues have carried out a series of studies in the juvenile, adult, and pregnant squirrel monkey. Animals were either fed a low-protein (2% for juvenile and adult and 8% for pregnant) are high-protein (25%) diet. Evidence of protein deficiency after a few weeks is easily detectable using histological and histochemical techniques on various aspects of the nervous system (Manocha and Olkowski, 1973; Manocha and Sharma, 1978; Sharma and Manocha, 1978). The same changes can be found in neonates of dams fed 8% protein during pregnancy.

Our laboratory utilizes two species of New World monkeys, *Cebus albifrons* (Samonds and Hegsted, 1978) and *Saimiri sciureus* (Ausman, 1973; Ausman *et al.*, 1977) to access the impact of protein, calorie, or a combined restriction upon the parameters associated with either growth and development of the infant or metabolic homeostasis of the adult. One key feature of this animal model is that the animals maintain their calorie intakes in the face of moderate to severe protein restriction. Clinical symptoms observed in both infant and adult protein-deficient animals include the classical signs of edema, discoloration and loss of hair, and desquamation. Results from both species showed that induced weight maintenance (prevention of growth) could be achieved in infant animals through either protein or calorie restriction; plasma protein and albumin levels were decreased only in the former case. A third group fed both a protein- and calorie-deficient diet appeared no different in body weight change or biochemical parameters than the protein-deficient group alone. Transferrin levels in all three groups appeared to correlate with the instantaneous rate of growth rather than the type of nutrient available (D. L. Gallina *et al.*, unpublished data).

The aforementioned studies focused on the effects of protein deficiency utilizing high biological value proteins, such as lactalbumin or casein. However, protein quality has also been varied. When weight maintenance was induced in very young animals using soy isolate as the sole protein source, a striking anemia developed as a consequence of an early blockade in the erythroid maturation, the cause of which is still unknown (Ausman *et al.*, 1977). Findings of this type opened new avenues for the understanding of the metabolic adaptation of humans to PCM, as well as to the possible long-term effects of some protein sources when used improperly.

Another component has been studied within the context of the response of these animals to protein restriction. Infant squirrel monkeys have shown a spontaneous glycosuria and impaired response to oral glucose load when diets contained inadequate amounts of protein to maintain weight (Ausman *et al.*, 1972). This finding was confirmed in adult animals studied in unrestrained conditions, although the response to the intravenous load was apparently not different from the controls (this laboratory, unpublished data). This finding suggests that some aspect of the GI tract might be involved in the impaired response to oral loading.

4.4. Lipid and Lipoprotein Metabolism

The search for appropriate experimental animals to model human disease processes has stimulated the utilization of nonhuman primates. The development of a monkey species that is both responsive to dietary manipulation and susceptible to human diseases such as atherosclerosis has led to the investigation of the squirrel monkey, *Saimiri sciureus*.

4.4.1. Diet-Induced Lipemias

Squirrel monkeys, like humans, are quite responsive to dietary fat-induced alterations of plasma cholesterol. For example, the consumption of saturated fats such as coconut oil (Nicolosi *et al.*, 1977a,b) and butterfat (Portman *et al.*, 1976, 1977) by squirrel monkeys results in higher plasma cholesterol levels than those consuming the polyunsaturated fats corn oil or safflower oil. The hypercholoesterolemic effect of dietary cholesterol is also influenced by the level of dietary cholesterol and the polyunsaturated/saturated fat ratio (P/S) of the dietary fat consumed along with the dietary cholesterol. Thus, it has been demonstrated that a semipurified diet containing 45% of the calories as lard and 0.05 mg cholesterol per kcal fed to squirrel monkeys was less hypercholesterolemic than a diet containing the same level of fat but with 1 mg cholesterol per kcal (Rudel and Lofland, 1976), an observation also supported by other investigations (Portman *et al.*, 1976; Srinivasan *et al.*, 1976).

The importance of the degree of interaction of dietary fat and cholesterol is underscored by the observation in squirrel monkeys that dietary cholesterol consumed with saturated fat is more hypercholesterolemic than when added to a polyunsaturated fat-containing diet (Lichtenstein *et al.*, 1980).

Table VI
Effect of Dietary Fat Saturation on Plasma and Lipoprotein Cholesterol Levels in Squirrel Monkeys[a]

Diet	Corn oil (mg/dl)	Coconut oil (mg/dl)
Plasma cholesterol	170 ± 49[a]	273 ± 44
VLDL + LDL cholesterol[b]	79 ± 34	140 ± 36
HDL cholesterol[c]	91 ± 18	133 ± 23

[a] Values represent mean ± SD for four monkeys in each dietary group.
[b] Obtained as difference between plasma and HDL cholesterol.
[c] Obtained after heparin–Mn^{2+} precipitation.

Another dietary variable that can influence plasma cholesterol levels in squirrel monkeys is the interaction between the type of dietary carbohydrate and degree of dietary fat saturation. Thus, one study reported and increase in plasma cholesterol levels in squirrel monkeys as a result of the synergistic effects of dietary sucrose and the saturated fat coconut oil compared to monkeys fed a diet containing safflower oil as the dietary fat and glucose or dextrin as the dietary carbohydrate (Westmoreland and Hayes, 1974; Hayes, 1979).

4.4.2. Plasma Lipoprotein Profiles and Cholesterol Distribution

In normocholesterolemic squirrel monkeys fed commercial monkey biscuits or a semipurified diet containing polyunsaturated fat, high-density lipoproteins (HDL) represent the predominant circulating lipoprotein fraction (Rudel and Lofland, 1976; Illingworth, 1975). However, our earlier studies (Nicolosi *et al.*, 1977b) and more recent data (Table VI) indicate that when squirrel monkeys are fed a saturated fat such as coconut oil, the cholesterol concentrations of the lower density lipoproteins, i.e., very low-density lipoprotein (VLDL) and low-density lipoprotein (LDL), increase to often equal or surpass HDL cholesterol levels. Thus, the degree of dietary fat saturation in squirrel monkeys can influence the partition of plasma cholesterol between VLDL + LDL and HDL.

4.4.3. HDL Subclass Distribution

Although information pertaining to the distribution of HDL subclasses, i.e., HDL_2 and HDL_3, in monkeys is sparse, it is apparent that New World monkeys such as the squirrel monkey have relatively more HDL_2 than HDL_3, in contrast to Old World monkeys such as the rhesus (Morello and Nicolosi, 1981; Nelson *et al.*, 1980). The impact of this observation takes on more significance in light of the inverse correlation of HDL_2 and coronary heart disease in humans (Anderson *et al.*, 1978) and the greater HDL_2 levels in atherosclerosis-susceptible monkey species such as the squirrel monkey.

The significance of HDL_2/HDL_3 ratios as predictors of disease risk has been further investigated in Brazilian and Bolivian squirrel monkeys, in which the former, with the higher HDL_2 levels, are more susceptible than the latter to gallstone formation (Portman *et al.*, 1980b). Thus, while in humans, increased HDL_2 levels are generally correlated with reduced risk of atherosclerosis, an increase in this HDL subclass in squirrel monkeys is associated with not only increased risk of atherosclerosis but gallstone formation as well. The discrepancy remains to be resolved.

Although polyunsaturated fat feeding has been demonstrated to reduce HDL_2 levels in humans (Shepherd *et al.*, 1978), similar studies have not been carried out in monkeys. However, increasing levels of dietary cholesterol have been shown to reverse the HDL_2/HDL_3 ratio in squirrel monkeys (Rudel and Lofland, 1976).

4.4.4. Lipoprotein Apoproteins

The apoproteins of squirrel monkey lipoproteins have been reported (Illingworth, 1975), but only apoprotein B (apo B) and the apo C complex were tentatively identified. In unpublished studies from our laboratory, squirrel monkey lipoproteins were isolated by density gradient ultracentrifugation, and ten fractions were collected and then subjected to 10% sodium dodecyl sulfate polyacrylamide gel electrophoresis (SDS–PAGE) and gave the following apoprotein distribution as a function of diet (Fig. 4, Table VII): fraction 1 = VLDL, $d < 1.008$; fractions 2,3 = LDL, $d = 1.017–1.065$; fractions 4–8 = HDL, $d = 1.068–1.170$; bottom fraction 9–10, $d > 1.170$.

The decrease in lipoprotein apoproteins with polyunsaturated fat feeding was in line with the lipoprotein cholesterol reductions (Table VI) under the same dietary treatment.

4.4.5. Cholesterol and Lipoprotein Metabolism

The squirrel monkey has been used extensively in studies of cholesterol metabolism. *In vitro* investigations of cholesterol synthesis of various squirrel monkey tissues in monkeys on low-cholesterol diets revealed that the liver accounted for 86% of the cholesterol synthesized and the gastrointestinal (GI) tract 11%. Feeding a high-cholesterol diet resulted in nearly complete inhibition of hepatic cholesterol synthesis, while the GI tract was only reduced 50%, implying substantial intestinal cholesterol synthesis during hypercholesterolemia (Dietschy and Wilson, 1968, 1970). This difference in feedback inhibition of liver and intestinal lipogenesis was also found in squirrel monkeys fed saturated fat (Corey *et al.*, 1974). In related studies, hyporesponding squirrel monkeys to high-cholesterol diets were able to maintain low plasma cholesterol levels compared to hyperresponders by increasing cholesterol catabolism (Lofland *et al.*, 1972). In addition, cholesterol turnover studies in squirrel monkeys (Corey *et al.*, 1976) demonstrated an enhanced clearance of circulating cholesterol from the body in monkeys fed safflower oil and coconut oil. The implication that diet-induced hypercholesterolemia may be associated with a ca-

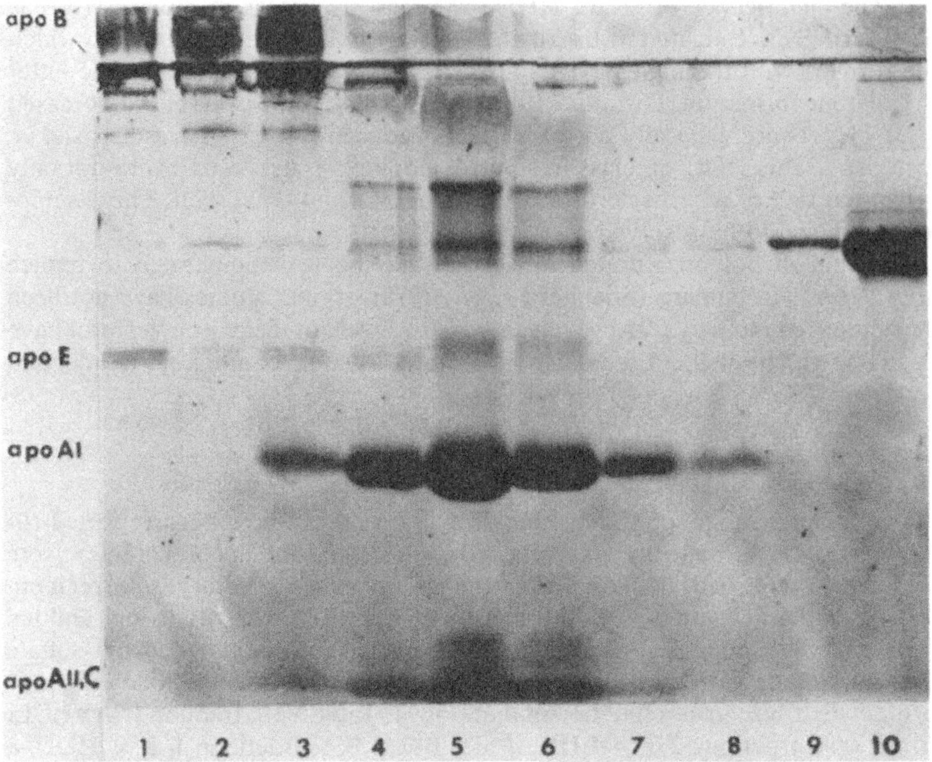

Fig. 4. Apoprotein composition of squirrel monkey lipoproteins. Lane 1 = VLDL, $d < 1.008$; lanes 2, 3 = LDL, $d = 1.017$–1.065; lanes 4–8 = HDL, $d = 1.068$–1.170; lanes 9–10 = bottom fraction, $d > 1.170$.

tabolic defect is supported by several studies of lipoprotein metabolism in squirrel monkeys. Studies of VLDL metabolism using the detergent Triton WR1339 (Illingworth *et al.*, 1975; Portman *et al.*, 1977; Nicolosi *et al.*, 1977a) implied reduced clearance of VLDL in hypercholesterolemic monkeys as a funtion of dietary cholesterol and fat saturation. Studies of LDL apoprotein turnover (Portman *et al.*, 1976) demonstrated a reduced fractional catabolic rate in monkeys fed the saturated butterfat. The fractional catabolic rate of HDL apoproteins was significantly reduced in squirrel monkeys fed the saturated fat coconut oil (Nicolosi *et al.*, 1982). In related studies of HDL turnover, it was found that HDL-free cholesterol is the preferred substrate for biliary cholesterol and bile acids (Portman *et al.*, 1980a).

Dietary fat-induced hypercholesterolemia can also influence the rate and composition of hepatic nascent lipoprotein synthesis and secretion. Thus, livers from squirrel monkeys fed coconut oil secreted more triglyceride and cholesterol-rich VLDL than those fed corn oil (Nicolosi and Hayes, 1980). Obviously, diet-induced hypercholesterolemia can be the result of alterations in both catabolic and/or synthetic processes.

Table VII
Concentration of Lipoprotein Apoproteins in Squirrel Monkeys Fed Corn Oil and Coconut Oil[a]

	apo B (mg/dl)	apo E[b] (mg/dl)	apo A-I[b] (mg/dl)	apo A-II, C[b] (mg/dl)
LDL				
Corn oil	60 ± 6	0.8 ± 0.1	ND	1.5 ± 0.3
Coconut oil	77 ± 8	2.0 ± 0.5	ND	1.6 ± 0.2
HDL				
Corn oil	ND	ND	75.2 ± 10.1	20.1 ± 2.0
Coconut oil	ND	4.1 ± 0.9	91.1 ± 7.8	26.2 ± 1.0

[a] Values represent mean ± SE obtained by densitometric scanning of 3.5% SDS–PAGE gels containing albumin standards. ND, Not detected; sample size of VLDL too small to measure apoprotein concentrations.
[b] Obtained by densitometric scanning of 10% SDS–PAGE gels containing albumin standards. Not corrected for differences in chromogenicity between apoproteins.

One of the major enzymes associated with lipoprotein metabolism that has been studied in squirrel monkeys is the lecithin:cholesterol acyltransferase (LCAT) enzyme (Lichtenstein *et al.*, 1980). In monkeys fed the coconut oil diets, the percent of cholesterol esterified was lower than monkeys fed the corn oil diets. Addition of 0.1% cholesterol to the diet further reduced cholesterol esterification by the LCAT enzyme. Thus, the squrrel monkey is very responsive to various dietary manipulations. This small, relatively inexpensive monkey species should be a useful model for investigating the mechanism(s) associated with diet-induced alterations of lipid and lipoprotein metabolism as they relate to the development of hypercholesterolemia and atherosclerosis.

ACKNOWLEDGMENTS. Supported in part by the Armand G. Erpf Fund, School of Nutrition, Tufts University, Medford, Massachetts; Public Health Service Research grants HL-10098 and HD-07032, Department of Nutrition, Harvard School of Public Health, Boston, Massachusetts; and NIH grants HL23792 and RR00168 from the Division of Research Resources, NERPRC, Southborough, Massachusetts.

REFERENCES

Altman, P. L., and Dittmer, D. S., eds., 1971, *Blood and Other Body Fluids*, Federation of American Societies for Experimental Biology, Bethesda Maryland.

Amine, E. K., Ausman, L. M., and Hegsted, D. M., 1973, Iron in absorption and occult blood loss infant squirrel monkeys fed cow's milk and commercially prepared formulae, in: Nutrition and Development of the Infant Squirrel Monkey, Ph.D. Dissertation by L. M. Ausman, Harvard School of Public Health, Boston, Massachusetts.

Anderson, D. W., Nicholes, A. V., Pan, S. S., and Lindgren, F. T., 1978, High density lipoprotein distribution: Resolution determination of three major components in a normal population, *Atherosclerosis* **29:**161–179.

Ausman, L. M., 1973, Nutrition and Development of the Infant Squirrel Monkey (*Saimiri sciureus*), Ph.D. Dissertation, Harvard School of Public Health, Boston, Massachusetts.

Ausman, L. M., and Gallina, D. L., 1978, Response to glucose loading of the lean squirrel monkey in unrestrained conditions, *Am. J. Physiol.* **234**:R20–R24.

Ausman, L. M., and Gallina, D. L., 1979, Liquid formulas and protein requirements of nonhuman primates, in: *Primates in Nutritional Research* (K. C. Hayes, ed.), Academic Press, New York, pp. 39–57.

Ausman, L. M., and Hayes, K. C., 1974, Vitamin E deficiency anemia in Old and New World monkeys, *Am. J. Clin. Nutr.* **27**:1141–1151.

Ausman, L. M., Hayes, K. C., Lage, A., and Hegsted, D. M., 1970, Nursery care and growth of Old and New World infant monkeys, *Lab. Anim. Care* **20**:907–913.

Ausman, L. M., Hayes, K. C., and Hegsted, D. M., 1972, Protein deficiency and carbohydrate tolerance of the infant squirrel monkey (*Saimiri sciureus*), *J. Nutr.* **102**:1519–1528.

Ausman, L. M., Gallina, D. L., Hayes, K. C., and Hegsted, D. M., 1976, Hematological development of the infant squirrel monkey, *Folia Primatol.* **26**:292–300.

Ausman, L. M., Gallina, D. L., Camitta, B. M., Flath, L. C., and Hegsted, D. M., 1977, Acute erythroid hypoplasia in malnourished infant squirrel monkeys fed isolated soy protein, *Am. J. Clin. Nutr.* **30**:1713–1720.

Ausman, L. M., Gallina, D. L., Samonds, K. W., and Hegsted, D. M., 1979, Assessment of the efficiency of protein utilization in young squirrel and macaque monkeys, *Am. J. Clin. Nutr.* **32**:1813–1823.

Ausman, L. M., Rasmussen, L. M., and Gallina, D. L., 1981, Spontaneous obesity in maturing squirrel monkeys fed semipurified diets, *Am. J. Phsiol* **241**:R316–R321.

Ausman, L. M., Gallina, D. L., Hayes, K. C., and Hegsted, D. M., 1983, Comparative assessment of protein quality in infant cebus monkeys *Fed Proc.* **42**:540a.

Barney, G. H., Macapinlac, M. P., Pearson, W. M., and Darby, W. J., 1967, Parakeratosis of the tongue: A unique histopathologic lesion in the zinc-deficient squirrel monkey, *J. Nutr.* **93**:511–517.

Beischer, D. E., 1968, The squirrel monkey in aerospace medical research, in: *The Squirrel Monkey* (L. A. Rosenblum and R. W. Cooper, eds.), Academic Press, New York, pp. 347–364.

Blackwell, C. A., Manning, P. J., Hutchinson, T. C., and Fisk, S. K., 1974, Cranial hyperostosis of squirrel monkeys (*Saimiri sciureus*), *Lab. Anim. Sci.* **24**:541–544.

Brody, A., 1945, Basal energy and protein metabolism in relation to body weight in mature animals of different species, in: *Bioenergetics and Growth* (S. Brody, ed.), Reinhold, New York, pp. 352–403.

Brodie, D. A., and Marshall, R. W., 1963a, Gastric content of fasted primates: A survey, *Science* **141**:174–175.

Brodie, D. A., and Marshall, R. W., 1963b, Fasting gastric content of the squirrel monkey (*Saimiri sciurea*), *Am. J. Physiol.* **204**:681–685.

Buss, D. H., and Cooper, R. W., 1972, Composition of squirrel monkey milk, *Folia Primatol.* **17**:285–291.

Chadwick, R. W., Cranmer, M. F., and Peoples, A. J., 1971, Metabolic alterations in the squirrel monkey induced by DDT administration and ascorbic acid deficiency, *Toxicol. Appl. Pharmcol.* **20**:308–318.

Coe, C. L., and Rosenblum, L. A., 1978, Annual reproductive strategy of the squirrel monkey (*Saimiri sciureus*), *Folia Primatol.* **29**:19–42.

Corey, J. E., and Hayes, K. C., 1974, Effect of diet on hepatic and intestinal lipogenesis in squirrel, cebus and cynomolgus monkeys, *Atherosclerosis* **20**:405–416.

Corruccini, R. S., and Beecher, R. M., 1982, Occlusal variation related to soft diet in a nonhuman primate, *Science* **218**:74–76.

Coulson, R. A., Hernandex, T., and Herbert, J. D., 1977, Metabolic rate, enzyme kinetics *in vivo*, *Comp. Biochem. Physiol.* **56A**:251–262.

Davidson, I. W. F., and Blackwell, W. L., 1968, Changes in carbohydrate metabolism of squirrel monkeys with chromium dietary supplementation, *Proc. Soc. Exp. Biol. Med.* **127**:66–70.

Davidson, I. W. F., Lang, C. M., and Blackwell, W. L., 1967, Impairment of carbohydrate metabolism of the squirrel monkey, *Diabetes* **16**:395–401.

De la Iglesia, F. A., Porta, E. A., and Hartroft, W. S., 1967, Effects of dietary protein levels on the *Saimiri sciureus, Exp. Mol. Pathol.* **7:**182–195.

Delort, D., Chopard, M., and Tachon, J., 1976, The management of a squirrel monkey (*Saimiri sciureus*) unit, *Lab. Anim. Sci.* **26:**301–304.

Demaray, S. Y., Altman, N. H., and Ferrell, T. L., 1978, Suspected ascorbic acid deficiency in a colony of squirrel monkeys (*Saimiri sciureus*), *Lab. Anim. Sci.* **28:**457–460.

Dietschy, J. M., and Wilson, J. D., 1968, Cholesterol synthesis in the squirrel monkey. Relative rates of synthesis in various tissues and mechanisms of control, *J. Clin. Invest.* **47:**166–174.

Dietschy, J. M., and Wilson, J. D., 1970, Regulation of cholesterol metabolism (three parts), *N. Engl. J. Med.* **282:**1128–1249.

DuMond, F. V., 1968, The squirrel monkey in a seminatural environment, in: *The Squirrel Monkey* (L. A. Rosenblum and R. W. Cooper, eds.), Academic Press, New York, pp. 87–145.

Ellison, T., and Riddle, W. C., 1961, Commerical liquid diet for animals in behavioral studies, *J. Exp. Anal. Behav.* **4:**370.

ESPGCN (European Society for Paediatric Gastroenterology and Committee on Nutrition), 1977, Guidelines on infant nutrition. I. Recommendations for the composition of an adapted formula, *Acta Paediatr. Scand. Suppl.* 262.

Fiennes, R. N. T.-W., Sinclair, A. J., and Crawford, M. A., 1973, Essential fatty acid studies in primates. Linolenic acid requirements of capuchins, *J. Med. Primatol.* **2:**155–169.

Finn, M. G. C., Berrizbeitia, L. D., Colpoys, M., Moore-Ede, M. C., and Moore, F. D., 1982, Circadian timing of intravenous feeding: Day–night comparisons of nutrient effectiveness in the squirrel monkey, *J. Clin. Surg.* **1:**236–241.

Fooden, J., 1964, Stomach contents and gastro-intestinal proportions in wild-shot Guianan monkeys, *Am. J. Phys. Anthropol (N.S.)* **22:**227–231.

Gallina, D. L., and Ausman, L. M., 1979, Selected aspects of the metabolic behavior of the squirrel monkey, in: *Primates in Nutritional Research* (K. C. Hayes, ed.), Academic Press, New York, pp. 225–247.

Ganchrow, J., and Fisher, G. L., 1968, Two behavioral measures of the squirrel monkey's (*Saimiri sciureus*) taste for four concentrations of five sugars, *Psychol. Rep.* **22:**503–511.

Harris, R. S., 1970, Natural versus purified diets in research with nonhuman primates, in: *Feeding and Nutrition of Nonhuman Primates* (R. S. Harris, ed.), Academic Press, New York, pp. 251–262.

Hayes, K. C., 1979, Diet and atherosclerosis in nonhuman primates, in: *Primates in Nutritional Research* (K. C. Hayes, ed.), Academic Press, New York, pp. 181–198.

Hayes, K. C., Fay, G., Roach, A., and Stare, F. J., 1972, Breeding New World monkeys in a laboratory environment, in: *Breeding Primates* (W. I. B. Beveridge, ed.), Karger, Basel, pp. 122–129.

Hayes, K. C., Stephan, Z. I., and Sturman, J. A., 1980, Growth depression in taurine-depleted infant monkeys, *J. Nutr.* **110:**2058–2064.

Herndon, J. F., Greenburg, S. M., VanLoon, E. J., Kelleher, R. T., Cook, L., and Davidson, A., 1958, A liquid diet for animals in behavioral studies, *J. Exp. Anal. Behav.* **6:**249–252.

Hinkle, D. K., and Session, H. L., 1972, A method for hand rearing of *Saimiri sciureus, Lab. Anim. Sci.* **22:**207–209.

Hunt, R. D., Garcia, F. G., and Hegsted, D. M., 1967, A comparison of vitamin D_2 and D_3 in the New World primates. I. Production and regression of osteodystrophia fibrosa, *Lab. Anim. Care* **17:**222–234.

Hunt, R. D., Garcia, F. G., and Hegsted, D. M., 1969, Hypervitaminosis D in New World monkeys, *Am. J. Clin. Nutr.* **22:**358–366.

Illingworth, D. R., 1975, Metabolism of lipoproteins in nonhuman primates. Studies of the origin of low density lipoprotein apoprotein in the plasma of the squirrel monkey, *Biochim. Biophys. Acta* **388:**38–51.

Illingworth, D. R., Whipple, L. E., and Portman, O. W., 1975, Metabolism of lipoproteins in nonhuman primates: Reduced secretion of very low density lipoproteins in squirrel monkeys with diet-induced hypercholesterolemia, *Atherosclerosis* **22:**325–334.

Jones, D. C., Lofland, H. S., Clarkson, T. B., and St. Clair, R. W., 1975, Plasma cholesterol concentrations in squirrel monkeys as influenced by diet and phenotypes, *J. Food Sci.* **40:**2–7.

Kaack, B., Walker, L., and Brizzee, K. R., 1979, The growth and development of the squirrel monkey (*Saimiri sciureus*), *Growth* **43:**116–135.

Kaplan, J., 1974, Growth and behavior of surrogate-reared squirrel monkeys, *Dev. Psychobiol.* **7:**7–13.

Kaplan, J. N., 1977, Breeding and rearing squirrel monkeys (*Saimiri sciureus*) in captivity, *Lab. Anim. Sci.* **27:**557–567.

Kaplan, J. N., 1979, Growth and development of infant squirrel monkeys during the first six months of life, in: *Nursery Care of Nonhuman Primates* (G. C. Ruppenthal, ed.), Plenum Press, New York, pp. 153–164.

Kaplan, J. N., Hennessy, M. B., and Howd, R. A., 1982, Oral ethanol intake and levels of blood alcohol in the squirrel monkey, *Pharmacol. Biochem. Behav.* **17:**111–117.

Kerr, G. R., 1972, Nutritional requirements of subhuman primates, *Physiol. Rev.* **52:**415–467.

Kessler, M. J., 1980, Cephalhematomas due to suspected ascorbic acid deficiency in young squirrel monkeys (*Saimiri sciureus*), *J. Med. Primatol.* **9:**314–318.

Knapka, J. J., and Morin, M. L., 1979, Open formula natural ingredient diets for nonhuman primates, in: *Primates in Nutritional Research* (K. C. Hayes, ed.), Academic Press, New York, pp. 121–138.

Lang, C. M., 1966, Impaired glucose tolerance in the squirrel monkey (*Saimiri sciureus*), *Proc. Soc. Exp. Biol. Med.* **122:**84–85.

Lang, C. M., 1968, The laboratory care and clinical management of *Saimiri* (Squirrel monkeys), in: *The Squirrel Monkey* (L. A. Rosenblum and R. W. Cooper, eds.), Academic Press, New York, pp. 393–416.

Lang, C. M., 1970, Organoleptic and other characteristics of diet which influence acceptance by nonhuman primates, in: *Feeding and Nutrition of Nonhuman Primates* (R. S. Harris, ed.), Academic Press, New York, pp. 263–275.

Lehner, N. D. M., Bullock, B. C., Clarkson, T. B., and Lofland, H. B., 1967, Biological activities of vitamins D_2 and D_3 for growing squirrel monkeys, *Lab. Anim. Care* **17:**483–493.

Lehner, N. D. M., Bullock, B. C., and Clarkson, T. B., 1968, Ascorbic acid deficiency in the squirrel monkey, *Proc. Soc. Exp. Biol. Med.* **128:**512–514.

Lehner, L. D. M., Clarkson, T. B., and Lofland, H. B., 1971, The effect of insulin deficiency, hypothyroidism, and hypertension on atherosclerosis in the squirrel monkey, *Exp. Mol. Pathol.* **15:**230–244.

Lichtenstein, A. H., Nicolosi, R. J., and Hayes, K. C., 1980, Dietary fat and cholesterol effects on plasma lecithin:cholesterol acyltransferase activity in cebus and squirrel monkeys, *Atherosclerosis* **37:**603–616.

Lofland, H. B., Clarkson, T. B., St. Clair, R. W., and Lehner, H. D. M., 1972, Studies on the regulation of plasma cholesterol levels in squirrel monkeys of two genotypes, *J. Lipid Res.* **13:**39–47.

Long, J. O., and Cooper, R. W., 1968, Physical growth and dental eruption in captive-bred squirrel monkeys, *Saimiri sciureus* (Leticia, Colombia), in: *The Squirrel Monkey* (L. A. Rosenblum and R. W. Cooper, eds.), Academic Press, New York, pp. 193–205.

Macapinlac, M. P., Barney, G. H., Pearson, W. M., and Darby, W. J., 1967, Production of zinc-deficiency in the squirrel monkey (*Saimiri sciureus*)., *J. Nutr.* **93:**499–510.

Malinow, M. R., 1979, Important considerations for nonhuman primate models of diet-induced atherosclerosis, in: *Primates in Nutritional Research* (K. C. Hayes, ed.), Academic Press, New York, pp. 199–214.

Malinow, M. R., and Wagner, R., 1966, Oxygen uptake in squirrel monkeys (*Saimiri sciurea*), *Lab. Anim. Care* **16:**105–108.

Mann, G. V., and Crofford, O. B., 1970, Insulin levels in primates by immunoassay, *Science* **169:**1312–1313.

Manocha, S. L., 1979, Physical growth and brain development of captive-bred male and female squirrel monkeys, *Saimiri sciureus*, *Experientia* **35:**96–98.

Manocha, S. L., and Long, J., 1977, Experimental protein malnutrition during gestation and breeding performance of squirrel monkeys, *Saimiri sciureus*, *Primates* **18:**923–930.

Manocha, S. L., and Olkowski, Z. L., 1973, Experimental protein malnutrition in primates—Cytochemistry of the nervous system, *Am. J. Phys. Anthropol.* **38:**439–446.

Manocha, S. L., and Sharma, S. P., 1978, Nucleolar activity in the primate dorsal root ganglion cells associated with dietary protein malnutrition, *Acta Anat.* **100:**68–77.

Martin, G. D., Stanley, J. A., and Davidson, I. W. F., 1972, Corneal lesions in squirrel monkeys maintained on a low-chromium diet, *Invest. Ophthalmol.* **11:**153–158.

Middleton, C. C., and Rosal, J., 1972, Weights and measurements of normal squirrel monkeys (*Saimiri sciureus*), *Lab. Anim. Sci.* **22:**583–586.

Morello, A. M., and Nicolosi, R. J., 1981, Electrophoretic profiles of high density lipoprotein subclasses in primates, *Comp. Biochem. Physiol.* **69B:**291–294.

Muth, O. H., Weswig, P. H., Whanger, P. D., and Oldfield, J. E., 1971, Effect of feeding selenium-deficient ration to the subhuman primate (*Saimiri sciureus*), *Am. J. Vet. Res.* **32:**1603–1605.

National Research Council, 1978, *Nutrient Requirements of Nonhuman Primates*, National Academy of Sciences, Washington, D.C.

Navia, J. M., 1970, The preparation, use, and nutrient stability of nonhuman primate diets, in: *Feeding and Nutrition of Nonhuman Primates* (R. S. Harris, ed.), Academic Press, New York, pp. 277–297.

Nelson, C. A., Morris, M. D., and Geer, W. E., 1980, The distribution of high density lipoprotein subfractions in nonhuman primates, *Fed. Proc.* **39:**1718a.

Newberne, P. M., and Hayes, K. C., 1979, Semipurified diets for nonhuman primates, in: *Primates in Nutritional Research* (K. C. Hayes, ed.), Academic Press, New York, pp. 99–119.

Nicolosi, R. J., and Hayes, K. C., 1980, Composition of plasma and nascent very low density lipoprotein from perfused livers of hypercholesterolemic squirrel monkeys, *Lipids* **15:**549–554.

Nicolosi, R. J., and Hunt, R. D., 1979, Dietary allowances for nutrients in nonhuman primates, in: *Primates in Nutritional Research* (K. C. Hayes, ed.), Academic Press, New York, pp. 1–35.

Nicolosi, R. J., Hayes, K. C., el Lozy, M., and Herrera, M. G., 1977a, Hypercholesterolemia and triglyceride secretion rates in monkeys fed different dietary fats, *Lipids* **12:**936–940.

Nicolosi, R. J., Hojnacki, J. L., Llansa, N., and Hayes, K. C., 1977b, Diet and lipoprotein influence on primate atherosclerosis, *Proc. Soc. Exp. Biol. Med.* **156:**1–7.

Nicolosi, R. J., Yuan, R., and Rodger, R., 1982, Effect of longterm feeding of dietary fats containing polyunsaturated fatty acids and saturated fatty acids on high density lipoprotein metabolism in monkeys, *Fed. Proc.* **41:**2557a.

Portman, O. W., 1970, Nutritional requirements of nonhuman primates, in: *Feeding and Nutrition of Nonhuman Primates* (R. S. Harris, ed.), Academic Press, New York, pp. 87–115.

Portman, O. W., Alexander, M., Tanaka, N., and Soltys, P., 1976, The effects of dietary fat and cholesterol on the metabolism of plasma low density lipoprotein apoproteins in squirrel monkeys, *Biochim. Biophys. Acta* **450:**185–196.

Portman, O. W., Alexander, M., Tanaka, N., and Illingworth, D. R., 1977, Triacylglycerol and very low density lipoprotein secretion into plasma of squirrel monkeys, *Biochim. Biophys. Acta* **486:**470–482.

Portman, O. W., Alexander, M., and O'Malley, J. P., 1980a, Metabolism of free and esterified cholesterol and apoproteins of plasma low and high density lipoproteins, *Biochem. Biophys. Acta* **619:**545–558.

Portman, O. W., Alexander, M., Tanaka, N., and Osuga, T., 1980b, Relationship between cholesterol gallstones, biliary function and plasma lipoproteins in squirrel monkeys, *J. Lab. Clin. Med.* **96:**90–101.

Rasmussen, K. M., Thenen, S. W., and Hayes, K. C., 1979, Folacin deficiency and requirement in the squirrel monkey (*Saimiri sciureus*), *Am. J. Clin. Nutr.* **32:**2508–2518.

Rasmussen, K. M., Thenen, S. W., and Hayes, K. C., 1980, Effect of folic acid supplementation on pregnancy in the squirrel monkey (*Saimiri sciureus*), *J. Med. Primatol.* **9:**169–184.

Raymond, T. L., Lofland, H. B., and Clarkson, T. B., 1976, Cholesterol metabolism in squirrel monkeys. Analysis of long-term kinetic studies in plasma and body tissues, *Exp. Mol. Pathol.* **25:**344–351.

Rogers, Q. R., and Morris, J. G., 1980, Why does the cat require a high protein diet?, in: *Nutrition of the Dog and Cat* (R. S. Anderson, ed.), Pergamon Press, New York, pp. 45–66.

Rosenblum, L. A., 1968, Some aspects of female reproductive physiology in the squirrel monkey, in: *The Squirrel Monkey* (L. A. Rosenblum and R. W. Cooper, eds.), Academic Press, New York, pp. 147–169.

Rudel, L. L., and Lofland, H. B., 1976, Circulating lipoproteins in nonhuman primates, in: *Primates in Medicine*, Volume 9 (J. P. Strong, ed.), Karger, Basel, pp. 224–266.

Russo, A. R., Ausman, L. M., Gallina, D. L., and Hegsted, D. M., 1980, Developmental body composition of the squirrel monkey (*Saimiri sciureus*), *Growth* **44**:271–286.

Samonds, K. W., and Hegsted, D. M., 1978, Protein deficiency and energy restriction in young cebus monkeys, *Proc. Natl. Acad. Sci. USA* **75**:1600–1604.

Sharma, S. P., and Manocha, S. L., 1978, Experimental protein malnutrition in squirrel monkeys: Cytochemical interaction of nucleolus and cytoplasm in the anterior horn cells of the spinal cord of *Saimiri sciureus*, *Acta Anat.* **101**:19–32.

Shepherd, J., Packard, C. J., Patsch, J. R., Gotton, A. M., Jr., and Taunton, O. D., 1978, Effects of dietary polyunsaturated and saturated fat on the properties of high density lipoproteins and the metabolism of apolipoprotein A-I, *J. Clin. Invest.* **61**:1582–1592.

Smoake, J. A., Mulvey, P. F., Gerber, M., and Jones, L. G., 1977, Oxygen consumption and thyroid function in the squirrel monkey (*Saimiri sciureus*), *Lab. Anim. Sci.* **27**:655–659.

Srinivasan, R. R., Radhakrishnamurthy, B., Smith, C. S., Wold, R. H., and Berenson, G. S., 1976, Serum lipid and lipoprotein responses of six nonhuman primates species to dietary changes in cholesterol levels, *J. Nutr.* **106**:1757–1767.

Tashjian, A. H., Levine, L., Wilhelmi, A. E., and Parker, M. L., 1966, Rabbit, guinea pig, rat and human antibodies to human growth hormone: Immunological reactions with human and nonhuman primate growth hormones, *Endocrinology* **79**:615–623.

Thorington, R. W., Jr., 1968, Observations of squirrel monkeys in a Colombian forest, in: *The Squirrel Monkey* (L. A. Rosenblum and R. W. Cooper, eds.), Academic Press, New York, pp. 69–85.

Thorington, R. W., Jr., 1970, Feeding behavior of nonhuman primates in the wild, in: *Feeding and Nutrition of Nonhuman Primates* (R. S. Harris, ed.), Academic Press, New York, pp. 15–27.

Wang, A. C., Shuster, J., Epstein, A., and Fudenberg, H. H., 1968, Evolution of antigenic determinants of transferrin and other serum proteins in primates, *Biochem. Genet.* **1**:347–358.

Wen, C.-P., Hayes, K. C., and Gershoff, S., 1973, Effects of dietary supplementation of monosodium glutamate on infant monkeys, weanling rats, and suckling mice, *Am. J. Clin. Nutr.* **26**:803–813.

Westmoreland, N. P., and Hayes, K. C., 1974, Dietary carbohydrate influence on serum lipids in squirrel monkeys, *Fed. Proc.* **33**:718a.

Wilson, R. B., Martin, J. M., Kelly, H., and Newberne, P. M., 1971, Plasma and pancreatic insulin concentrations in adult squirrel and rhesus monkeys, *Diabetes* **20**:151–155.

15

Immunology and Pathology of the Squirrel Monkey

S. S. Kalter

1. INTRODUCTION

The squirrel monkey, as do other animals, reacts to an antigenic stimulus with an immunological response and, if the stimulus is sufficiently severe or not immunologically controlled, with pathology. It must be assumed, until information to the contrary is provided, that the immunological and pathological responses of all *Saimiri* would probably be similar. To avoid confusion in taxonomy, the squirrel monkey will be considered as a single species, i.e., *Saimiri sciureus,* without any attempt to differentiate the subspecies.

Specific immunological and pathological studies on nonhuman primates in general are few and frequently retrospective rather than designed. Although there are many biological similarities within a species, each animal may react individually, and generalizations may be misleading and confusing. Thus, we find on occasion that data provided in the literature may be difficult to interpret, if not actually inaccurate. In many instances, reports are anecdotal rather than the result of laboratory investigations. Studies may be based upon small numbers and these groups are frequently not under normal or natural conditions. Included in the study may be animals recently received in the laboratory that have been shipped and maintained under the most crowded and unhygienic conditions. The majority of monkeys and apes currently employed in biomedical research are the least standardized (by far) of all the experimental animals now in use (Kalter, 1980).

A brief reference to "ontogeny and phylogeny" and the role of immunology: No attempt will be made to referee the arguments concerning the development of the host's immune response or the relationship of one species to another. There is great variation in laboratory procedures as well as in host tissues selected for analysis in defining these parameters. The literature here is extensive and again, in many instances, contradictory and confusing. Clas-

S. S. Kalter • Southwest Foundation for Research and Education, San Antonio, Texas 78284.

sification schema have at times precipitated some of this confusion, but they also have provided a mechanism to separate as well as to define species. Thus, differences between monkeys and apes are well established, as is separation of monkeys into Old and New World species. It is generally agreed that the New World simians represent a completely different basic ancestral stock from the forms that gave rise to Old World species. Similarities between Old and New World species would be due to parallelisms and convergence and could influence interpretation of results. Differences in biological responses are also noted among individuals of the same species, principally due to geographic source, conditions of trapping and holding, and other variables.

2. SIMIAN IMMUNOLOGY

An animal's response to an antigen will be both humoral and cellular, with production and secretion of specific circulating antibody (immunoglobulins) by B lymphocytes ("bursa," bone marrow, Peyer's patches-derived lymphocytes), T cells (thymus-derived lymphocytes), and their subsets, which carry out a multitude of functions (cytotoxic, natural, helper, and suppressor cells) interferon response, complement, neutrophils, and macrophages. How these various components interact as part of the immune mechanism is not clear. That there are multiple interactions is recognized; the complexity of these interactions still confounds us. The involvement of T cells as modulators of the immune response by interactions with B cells has been well documented. Laboratory procedures for determining these functions are well described in the literature (Rose and Friedman, 1980).

Immunogenesis occurs early in the fetus. Species differences in responses to particular antigens do occur; however, the basic mechanisms are undoubtedly similar in most respects. The mammalian fetus is immunologically responsive and this responsiveness follows development of both B and T lymphocytes, but may be independent of the appearance of organized lymphoid tissue. Immunoglobulins have been found in the fetal rhesus monkey, and its immunological capability matches that of the adult animal (Silverstein *et al.*, 1970). This oversimplification of an extremely complex and varied phenomenon must be considered as a generalization, and the antigenic stimulation relative to gestation period as well as the particular host animal is highly important.

Variability in responsiveness may be illustrated by a comparison of the immunological competence of four simian species (including the squirrel monkey) exposed to the same experimental antigens (Harvey *et al.*, 1974). The baboon and squirrel monkey were found to be relatively intact immunologically and resembled the human. The cebus and marmoset species appeared to have abnormalities associated with susceptibilities to various replicating agents. The squirrel monkey was "immunologically competent," producing good antibody titers to microbial antigens and was resistant to the development of experimental infections and neoplasms. *In vitro* lymphocyte transformation to phytohemagglutinin and pokeweed mitogens was less in the squirrel monkey than that observed for both baboon and cebus monkey but better than that demonstrated by the marmoset. The ability to induce blastogenesis and mitosis in squirrel monkey peripheral leukocytes *in vitro* with a pokeweed mitogen was demonstrated by Srivastava *et al.* (1969).

Other immunological capabilities of the squirrel monkey have been described. A tissue antagonist to interferon was demonstrated in hair extracts of the squirrel monkey (Brown *et al.*, 1975). However, when squirrel monkey sera were examined for the presence of human isotypic and allotypic antigens, none was found (Gaarder and Natvig, 1974). These antigens were found to some extent among apes and Old World monkeys. Schmitt *et al.* (1965) previously noted the absence of human IgG antigen subclasses in squirrel monkey sera. Studies by Allen and Carstens (1968) and Carstens and Allen (1969) in squirrel monkeys indicate that total-body x-irradiation will induce an immunological depression, permitting overwhelming bacterial infections to occur.

Levy (1971), in discussing the use of nonhuman primates for studies of peridontal diseases, cites the work of Ranney and Zander (1970) in which it is shown that the squirrel monkey develops a gingival allergic reaction to crystalline albumen. A typical allergic response was observed: acute destructive periodontitis, which was followed on repeated application of antigen by chronic inflammation with plasma cell and lymphocyte infiltration of the gingiva. A resemblance to the human disease was noted.

It was recognized that α- and β-chain peptides of the squirrel monkey were similar to those of the human (R. L. Hill *et al.*, 1963). However, differences among the simians were also demonstrated when it was shown by Moor-Jankowski *et al.* (1964) and Moor-Jankowski and Wiener (1969, 1972) that variations in A-B-O blood groups existed, particularly between Old and New World species, and that other blood antigens of monkeys and apes show differences as well as parallelisms. Electrophoretic studies of simian myoglobins, as well as quantitative microcomplement fixation with serum albumin, clearly indicated the wide phylogenetic relationships among the Hominoidea, Cercopithecoidea, and Ceboidea, with the members of each family showing patterns similar to but distinct from the other families (Hudgins *et al.*, 1966; Sarich and Wilson, 1966; Ruffie, 1974). Wang *et al.* (1968) showed that the evolution of antigenic determinants on the transferrin molecule, as well as other primate serum proteins, was in good agreement with the taxonomic classification of primates, including the squirrel monkey. Numerous investigators subsequently used immunological procedures to demonstrate primate evolution and particularly the relationship of the nonhuman primate to the human. Damian and Greene (1972) provided one of the early reviews on nonhuman primate immunology, emphasizing that little was known relative to basic immunology. The majority of information available was obtained from Old World species. Of the New World animals studied, marmosets were generally the animal of choice. The authors pointed out, however, that "primates are not qualitatively vastly different from man in their immunological apparatus." These isolated reports on the immunological response of squirrel monkeys to immunogens is in keeping with this concept.

In this chapter, our discussion will be restricted to antigens derived from infectious agents rather than all immunogens that may be responsible for a host reaction. The various organisms capable of producing infection and/or disease in any given animal or species are varied, and successful infection is dependent upon many factors. An infectious agent that may not be pathogenic in one host may cause infection, disease, or even death in another. This "infectious" capability in turn is dependent upon the host's reactivity to the infecting agent.

Environmental and ecological factors associated with the animal's location may also influence infectivity. Specifically, the squirrel monkey acclimated to the South American geographic scene, when moved to a laboratory outside that area, faces not only a new environment and ecology in its new physical location (laboratory cage), but also a foreign microbial flora and fauna (humans, other laboratory animals, "wild" rodents, insects, etc.). The insult to the immune system must be tremendous, analogous to what occurs in a human from a rural or isolated area going off to school or to a large city for the first time. An entirely new and alien microflora is faced, often with dire results.

Defense against infection by an animal host is multifactoral and to be effective must be integrated and functioning. A breakdown in any of the components could lead to infection and subsequent disease. For example, complement is necessary as part of the host defense mechanism because it enhances phagocytosis, is involved in chemotaxis, and is essential for immune adherence, conglutination, and antibody-mediated cytotoxic and cytolytic activities. Alterations in complement amount or activity would reflect on the host's ability to mount a response to infection. A study of complement levels in baboons, squirrel monkeys, cebus monkeys, and cottontop marmosets demonstrated that the squirrel monkey had the highest levels (four times those of the human). Of the other species, the cebus monkey had complement levels similar to those of the human and the marmoset was the lowest (Rommel *et al.,* 1980). As will be noted below, this difference could reflect the ability of each of these species to reject (or accept) an invading organism.

There are various mechanical factors to combat the entry of infectious organisms into the host: skin and mucous membranes, tears, saliva, movement of cilia, and interference or direct antagonistic action by the normal microbial flora and fauna of the host. There are other nonimmunological host defenses that are also protective; their mechanism of action, however, has not been clearly defined. Susceptibility may be influenced by changes in nutrition, climatic condition, circadian rhythm, socialization, population density, light, caging and bedding (nesting boxes, etc.), handling, and microflora. While microflora are necessary for appropriate host metabolism, changes in intestinal organisms could result in clinical conditions evidenced by diarrhea and other constitutional symptomatology. Systemic defense mechanisms, humoral and cellular immunity, phagocytes, inflammatory reactions, and complement, all influenced by other host physiological capabilities, come into play to prevent the infecting organism from gaining a foothold in the host.

3. AGENTS OF DISEASE

In discussing the immunology and pathology of the squirrel monkey, it is necessary to provide some insight into those agents responsible for stimulating such responses. Sufficient data are available to suggest that the squirrel monkey can be infected with many of the same organisms (or their counterparts) that infect other primates. Melby and Altman (1974, esp. vols. II and III) provide comprehensive overviews of bacterial, mycotic, parasitic, rickettsial, and virological infections in laboratory animals.

Bacteria, fungi, parasites, rickettsiae, and viruses are all capable of stimulating the immune system as well as producing a pathological response in *Saimiri*. A series of reports appeared in the *Laboratory Primate Newsletter* (Vol. 2, Nos. 1, 3, and 4, 1963) extolling the virtues of the squirrel monkey as a laboratory animal. However, losses of these animals in early shipments exceeded 25%. Intestinal parasites, particularly *Strongyloides,* and blood parasites, microfilariae, appeared to be the main cause of death. Other parasitic infestations in *Saimiri* included, in addition to the above, *Trichuris* (whipworm), *Trichomonas,* and *Macracanthorhynchus hirudinaceus* (thorny-headed worm) in newly imported animals. These losses were reduced to less than 2–3% after treatment and handing procedures were modified. Included in the new therapy was administration of vitamins, antibiotics, anthelminthics, and fluids. Apparently other organisms were not observed or perhaps not investigated. Ruch (1967), in his volume on *Diseases of Laboratory Primates,* provides very little specific information on infectious agents of the squirrel monkey. Similarly, the volume on *The Squirrel Monkey* edited by Rosenblum and Cooper (1968) provides a chapter on "parasites" of *Saimiri* by Dunn (1968), who also emphasizes the lack of information on infectious agents of this animal. Other volumes on immunology and/or pathology of nonhuman primates include Fiennes (1967, 1972), Goldsmith and Moor-Jankowski (1969, 1972), Lapin and Yakovleva (1963), Whitney *et al.* (1973), Kalter and Hennesson (1980). First Intra-American Conference on Conservation and Utilization of American Nonhuman Primates in Biomedical Research, Pan American Health Organization, Sci. Pub. No. 317, 1976; Care and Diseases of the Research Monkey, N.Y. Acad. Sci., Vol. 85, 1960.

4. INFECTIOUS AGENTS, IMMUNOLOGICAL AND PATHOLOGICAL RESPONSES

4.1. Bacterial Infections

Bacterial infections are commonplace and generally not given serious consideration. Unfortunately, these organisms are capable of producing illness in spite of the availability of therapeutic agents. Several bacteria and their diseases are more prominent and should be considered carefully.

4.1.1. Tuberculosis

Most important, principally because of its resistance to the usual chemotherapeutic agents and its insidious pathogenesis, is *Myobacterium tuberculosis*. Mycobacterial infections, more frequent in Old World monkeys, do occur among New World monkeys. One of the early reports on tuberculosis indicates that the squirrel monkey is highly resistant to this disease (Gergen and Maclean, 1962). The lack of literature reports of tuberculosis in squirrel monkeys would tend to bear this out, although Fiennes (1963) reported the unexpected occurrence of nondisseminated disease in two squirrel monkeys at the London Zoo. Both animals apparently were infected with the human strain of *M. tuberculosis,* which causes an epizootic among Old World monkeys.

The finding that tuberculosis does occur in squirrel monkeys necessitated the inclusion of tuberculin testing in squirrel monkeys as part of the tuberculosis control management program (Chrisp *et al.* 1968; Hessler and Moreland, 1968; Leathers and Hamm, 1976; Mayhall *et al.,* 1981). Although the epidemiology associated with squirrel monkey tuberculosis is not clear, it appears that these species (as well as other New World monkeys) have a lower incidence of "natural" tuberculosis, but are susceptible. The majority of reported cases apparently have been the result of contact with Old World monkeys (*Macaca* sp.). It is not clear which strains of mycobacteria are infectious for New World species. As stated, variation in susceptibility to infectious agents is common among monkeys and apes. This is particularly true for tuberculosis, the human strain being recognized as most infectious for simians. Whether bovine and avian mycobacteria produce disease in simians is not clear. Soave *et al.* (1981) observed that an atypical mycobacterium may be responsible for producing positive tuberculin reactions, a reaction to be considered in making a diagnosis that may lead to the sacrifice of an animal. Tuberculosis in nonhuman primates has been well described, particularly in Old World species. The clinical disease in New World monkeys is similar, as are the pathological findings. The disease may be disseminated or may be limited to the thorax; microscopic examination of tissues will demonstrate granulomas as well as acid fast organisms that may be isolated by the usual laboratory procedures or seen in appropriately stained impression smears (Fiennes, 1972).

4.1.2. Gastroenteritis (Diarrhea)

Gastroenteritis (diarrhea) is probably one of the most common of disease entities noted in primates. The precise etiology of this syndrome is extremely varied, but various bacteria (as well as parasites and viruses) account for a large number of cases. Ruch (1967) provides an excellent review of diarrhea and dystentery in primates. Emphasis is placed on noninfectious causes of simian diarrhea: diet, emotional stress—anger, fright, but this should *not* be used as an excuse to avoid seeking an etiological agent. When attempting to determine the etiology of gastroenteritis, caution is urged in attributing etiology to one of the many agents comprising the normal flora of the intestinal tract. As in humans, the bacterial flora of the simian intestinal tract are extensive and variable. It is almost impossible to attribute pathogenicity to these organisms without detailed qualification. Given the appropriate set of circumstances, it is highly probable that any organism ("opportunist") can produce an infection and/or disease. Thus, such "intestinal" organism as *Proteus* spp,

Escherichia coli, and *Pseudomonas* spp. may cause disease with diarrhea as one of the manifestations (Fiennes, 1972).

Of the various enteropathogenic microorganisms, *Salmonella* spp. and *Shigella* spp. have received the most attention. The majority of information has been derived from *Macaca* spp. (rhesus) and apes (chimpanzees). Carpenter (1968) expresses concern for transmission of diseases from simians, particularly pets, to human contacts. In her studies on shigella in nonhuman primates, she observed the occurrence of *Shigella flexneri* (3a var. subserotype) in a squirrel monkey. Six human cases (children) resulted from contact with this single animal. A. Kaufmann (personal communication) indicated that of 36 *Saimiri sciureus* examined by rectal swab, none (0/36) showed the presence of *Shigella sonnei,* but 16/36 had a variety of *Salmonella* serotypes (*S. anatum, S. derby, S. newport, S. oranienburg, S. panama, S. poona, S. saint-paul, S. senftenberg, S. typhimurium,* and *S. waycross*). Similar findings were reported by Tokura and Katsubayashi (1973) in their study of quarantined monkeys.

These Enterobacteriaceae (nonmotile, Gram-negative rods) are all capable of producing diarrhea and dysentery; however, it is not clear if all the serotypes will cause disease in all simians. The existence and clinical capabilities of the various serotypes was reviewed by Ruch (1967). It becomes apparent, however, that disease, when it occurs, is similar (in most instances) in *all* primates. Further, the carrier state does occur, as in the human, with the animals appearing to be "normal" but able to transmit the organism to other animals, including humans.

4.1.3. Leptospirosis

Wolff *et al.* (1958) failed to find any evidence of leptospiral antibodies in two *Saimiri* sera tested. These negative findings were later confirmed in a serological survey of 12 squirrel monkeys (Minette, 1966). In that study, New World monkeys contrasted markedly with Old World monkeys insofar as there does not appear to be any evidence for natural leptospiral infections among the New World species, whereas antibody was frequently detected in Old World species. In an attempt to ascertain how simians react to leptospiral infection, Minette and Shaffer (1968) experimentally inoculated a number of Old and New World monkeys (including *S. Scuireus*) with various strains of leptospira. Marked variation in clinical response was observed, with the squirrel monkeys becoming acutely ill and occasionally dying as a result of infection.

Experimental infection of squirrel monkeys (Minette and Shaffer, 1968) with *Leptospira icterohemorrhagiae* resulted in an acute, severe leptospiral infection sometimes ending fatally. This was in marked contrast with the mild disease seen in rhesus and vervets. A pronounced icterus, but without significant hepatic lesions, a normal transaminase level, and an increased indirect bilirubin level were characteristic in squirrel monkeys. It was thought that hemolysis might be associated with the early development of jaundice in these animals. The liver showed no characteristic lesions, but sections of the kidneys indicated focal infiltrations of monocytic and histiocytic cells throughout the cortex and around the vessels. Microscopic examination showed masses of leptospira in the lumina of the renal tubules, sometimes extending to their lining cells, but not outside the basement membrane. Organisms could be readily recovered from blood. Lack of natural infection of *Saimiri* (as well as other New World monkeys) probably results from their arboreal habitat, which minimizes contact with leptospira-carrying rodents or contaminated materials.

4.1.4. Pasteurellosis

Greenstein *et al.* (1965) describe a severe outbreak of bacterial disease in which all but two of 29 squirrel monkeys died from infection with *Pasteurella multocida* following importation. It was suggested that the spread of the disease resulted from close contact and that the social behavior of these animals in the wild would exclude such contact. Clinically, the affected animals showed extreme weakness, unsteady head posture and gait, ocular and head nystagmus, edema of the eyelids, and a serosanguinous discharge from the ears. At necrospy, these findings were substantiated along with gross pathological findings of conjunctivitis with petechial hemorrhages, hemorrhages of the surface of the kidneys, cardiac enlargement, cloudy cerebrospinal fluid, hemorrhages in the back muscles, and petechiae of the serosal surface of the large intestines. Microscopic examination indicated an acute myocarditis with extension into the endocardium of the ventricles. This infiltration consisted of polymorphonuclear leukocytes with focal involvement of the myocardium along with degeneration of the myofibrils and loss of cross structures. A similar response was observed in the roof of the aorta involving the elastic laminae and the intima. There was a fibrous exudate on the epicardium. Focal areas of coagulative necrosis with polymorphonuclear leukocytes were observed in the liver. The outer medulla of the kidney also showed a polymorphonuclear infiltration along with abscess formation.

4.1.5. Syphilis

Experimental infection of squirrel monkeys with patient material (cerebrospinal fluid, aqueous humor) containing *Treponema pallidum* resulted in an infection similar to that seen in late human syphilis (J. L. Smith and Israel, 1968). Although the various serological tests on these animals were equivocal, one animal, about 1 year after inoculation, developed a lesion on the nose that was dark-field-positive and another animal showed the presence of spirochetes 3 months later in a skin biopsy of the inoculation site. Paracentesis of the anterior chamber of one animal's eyes 7 months postinoculation showed by dark-field examination that the aqueous humor contained spirochetes. No data were provided regarding the final outcome of these infections. The inoculated animals apparently were alive approximately 2 years after initiation of the study. It would appear that the squirrel monkey may be experimentally infected, but infection does not occur in nature. Sera from 12 squirrel monkeys were "weakly" positive when tested in the VDRL (Venereal Disease Research Laboratory) test and RPCF (Reiter protein complement fixation) test (Felsenfeld and Wolf, 1971). These findings may reflect cross-reactions with other organisms containing common antigens. The possibility of nonspecific serological reactivities must be considered as well.

4.1.6. Pneumonia (Klebsiella)

The squirrel monkey has been used for studies on the pathogenesis of *Klebsiella pneumoniae*. Henry *et al.* (1969) observed that *K. pneumoniae* given as an aerosol alone had little effect, but when animals were exposed to NO_2 (50 ppm) and *K. pneumoniae*, deaths occurred. Lesser concentrations of NO_2 did not have the same lethal effect. Similar results were obtained by Berendt (1978), who demonstrated that intratracheal administration of *K. pneumoniae* was usually fatal. These findings suggest that infection is not due to inhalation of the organisms. In a followup study on *K. pneumoniae*, Berendt *et al.* (1978) suggested that the squirrel monkey was a good model for the study of the human disease. Squirrel monkeys respond with a lobar pneumonia as well as the same form of clinical disease that occurs in humans, permitting study of the pathogenesis. It was noted that mice and rats may be appealing, particularly because of cost and size, but they do not provide the same model as does the monkey. In another study, these investigators report that kanamycin protected squirrel monkeys from *K. pneumoniae* infection (Berendt *et al.*, 1979).

Characteristic clinical findings include fever, anorexia, and coughing. Laboratory findings indicate a leukocytosis or leukopenia, bacteremia, and shedding of organisms into the pharynx. Mortality is high, with *K. pneumoniae* found at necropsy in a variety of tissues. Growth of the organisms occurs in the lungs and spreads to other tissues via the blood stream. Histopathologically, the disease is characterized by initial multifocal complications of the alveoli in the diaphragmatic lobe and involvement of major portions of the same lobe spreading to spaces, ducts, and a few respiratory bronchioles. An inflammatory exudate may develop in the larger bronchioles and bronchi, but not in the airways.

4.1.7. Pneumonia (Streptococcal)

Streptococcus pneumoniae in small doses produces clinical disease in squirrel monkeys following intratracheal challenge; larger doses are fatal (Berendt *et al.*, 1975). Lethality may be increased by the combined administration of *S. pneumoniae* and type A influenza virus (H3N2). The *S. pneumoniae* alone, administered intratracheally, produced a bacteremia; however, when given with influenza virus, a severe pneumonia resulted. Pathology included a patchy, red–gray lobar consolidation, which histologically included the following: loss of cilia in trachea and bronchi, minimal mucosa distortion, and minimal infiltration by polymorphonuclear neutrophils of the submucosa. Alveolar and bronchiolar epithelium was hyperplastic, with some alveolae containing an exudate of leukocytes. Peribronchiolar and perivascular foci of lymphocytes with some macrophages were also observed.

4.1.8. Mycoplasma

Mycoplasma have been isolated from *Saimiri,* suggesting susceptibility to this group of organisms (Davidson and Thomas, 1968; A. Hill, 1977; Koshimizu *et al.,* 1975; Taylor-Robinson *et al.,* 1971). A. Hill (1977) identified various *Mycoplasma* present in a variety of nonhuman primate species, noting *M. primatam* and *M. salivarium* in the squirrel monkey. Reported differences in *Mycoplasma* isolations as well as the failure to recover organisms must be considered in terms of contact with other animals, including humans. Apparently there are differences in susceptibility among monkey species, but more study is required to substantiate this suggestion.

4.1.9. Tularemia

Isolation of *Francisella tularensis* from the brain of an apparently naturally infected squirrel monkey with a presumed diagnoses of rabies has been reported (Emmons *et al.,* 1970). This finding has some significance for the epidemiology of infection. How a caged pet animal might acquire this type of infection could not be determined by the available data. Tularemia has been observed in other nonhuman primates (*Sanguinus nigricollis, Cercopithecus talapoin*) in a zoo as a result of contact with wild ground squirrels (Nayer *et al.,* 1979). Little information is available on the pathology induced in squirrel monkeys by this organism. Since the findings in both marmosets (*S. nigricollis*) and talapoin (*C. talapoin*) were essentially similar (Nayar *et al.,* 1979), we may assume that the squirrel monkey would react in like fashion. Splenomegaly and heptomegaly with a fibrous exudate and scattered caseatious necrosis were apparent. Various degrees of pneumonitis, glomerulitis, lymphadenitis, and enteritis may also be seen.

4.1.10. Tetanus

A major but infrequently reported problem in primates is the occurrence of tetanus. A clinical description of this disease in *S. sciureus* housed in outdoor corrals with soil-based floors has been provided by Kessler and Brown (1979). Diagnosis was based upon clinical signs, which simulated those seen in humans and other primates: trismus, extensor rigidity, and opisthotonus. According to these investigators, the case fatality was 100%. Gross findings included congestion of the major organs and atelectasis. A subacute bacterial dermatitis was most apparent, followed by chronic interstitial nephritis, mild glomerulonephritis, chronic inflammation of the adrenals, glossitis, and pulmonary edema.

4.2. Fungal Infections

The literature on mycotic infections of nonhuman primates is extremely limited. Al-Doory (1972a,b,c) reviewed mycoses as recognized in simians and described various infections in a variety of monkeys and apes. Dunn (1968) reported one study in *Saimiri* with *Cryptococcus neoformans* in which *Cryp-*

tococcus was found not to be lethal, but the organisms were found histologically in lungs (diffuse bronchopneumonia) of sacrificed animals. In considering fungal infections of primates, a report by Burge *et al.* (1979) is of interest. These investigators sampled various animal rooms, each housing dogs, rabbits, or monkeys (undesignated species), and noted that risk of fungus exposure in these rooms was no greater than that encountered in other types of interiors. It was noted, however, that fungi recovered in the animal rooms were different from those recovered from nonanimal rooms. Included among the agents recovered were, in descending frequency, *Penicillium, Graphium, Aspergillus, Sporobolomyces, Oidiodendron, Cladosporium cladosporioides, C. herbarum, Trichoderma, Scopulariopsis,* and *Sporotrichum,* as well as nonsporulating species and arthrospores.

Again it must be emphasized that the potential for infection of squirrel monkeys by all infectious fungi, particularly those affecting New World monkeys, must be considered even though references to specific occurrences may not have been recorded.

4.2.1. Histoplasmosis

A case of spontaneous histoplasmosis (*Histoplasma capsulatum*) was reported by Bergeland *et al.* (1970) in a squirrel monkey. This animal died after 3 days of illness characterized by anorexia, depression, and respiratory difficulties. At necropsy, a granulomatous pneumonia with diffuse, white, irregular nodules involving all lobes of the lung was seen. A serosanguinous pleural effusion was present as well. The pulmonary lesions were pyogranulomatous, with budding yeasts within macrophages seen in the affected areas. The causative agent could be readily seen when appropriately stained.

4.3. Parasitic Infections

Both ecto- and endoparasites are associated with all primates and have been recognized for many years (Hegner, 1935). Although extensive in distribution among the various simians, their general failure to be a major cause of fatalities or to be responsible for epidemics has relegated their significance to a secondary role. Nevertheless, parasites may be and often are harmful to the host. Their complex host relationships make them of interest in terms of understanding mechanisms of pathogenesis, particularly in their highly complicated immunological capabilities.

Broadly speaking, parasites have evolved a very carefully balanced relationship with hosts, in which the parasite is tolerated without injury to its host. This balance is not clinically apparent, but if disturbed may result in disease. At times, the host, although clinically well, may serve as a reservoir distributing infectious material to other hosts. What precipitates changes in this host–parasite balance is not clearly understood. It is recognized, however, that part of this balance is a coexistence with other microbial agents (normal flora and fauna). Disturbance of this coexistence will result in disease frequently more severe than that produced by only one organism. Unfavorable conditions, such as stress, travel, pregnancy, immunological suppression, crowded conditions, or malnutrition, may upset this balance. Newly imported animals are characterized by extensive parasitic infections. When these animals are maintained under proper husbandry conditions, a noticeable decrease in parasites is observed even without therapy.

The ability of any given parasite to produce infection and/or disease varies considerably, depending upon circumstances. Wild and captive animals differ markedly in their disease potential. Mixing of species or even different groups of the same species, especially newly imported animals, with stabilized animals must be avoided.

Hundreds of parasitic forms are recognized in nature; of these, perhaps 25–30 are thought to be indigenous to the nonhuman primate. Organisms generally considered as nonprimate in origin may also infect the primate. The natural host of many parasites is actually not known. Apparent host specificity for some parasitic forms occurs, but this phenomenon is rare.

Geographic and ecological factors also influence parasitism. Differences exist among the parasitic forms found in Old and New World monkeys. Some of these differences may reside in the different habits exhibited by these groups of monkeys. New World monkeys, being essential arboreal, may not have the same contact with parasitic forms found on the ground; this latter type of contact is common to Old World monkeys.

One of the earliest reports on parasites in squirrel monkeys was that of Middleton *et al.* (1964). In a group of 50 monkeys, thorny-headed worms (*Prosthenorchis elegans*) were observed in 33; abdominal cavity filaria (*Dipetalonema*) were seen in 17 animals; lungworms (*Filaroides barretoi*) in 16 cases; four animals were found with the nymphal stages of a pentastome; unidentified liver flukes were found in 17 animals; Sarcosporidia were seen in three animals; ten animals had circulating Microfilariae; one animal had a tapeworm (*Hymenolepiasis nana*); and one had the larval *Dermatobia hominis*. All animals had *Strongyloides* spp.

Table I provides a listing of recognized parasites in *Saimiri*. According to Dunn (1968), of the intestinal protozoa, only *Trichomonas* has thus far been observed in *Saimiri*. Other protozoan infections, however, include two species of *Plasmodium, Toxoplasma, Sarcocystis* (a genus of "unknown" taxonomic possition), and several species of *Trypanosoma*. An interesting observation was a case of encephalitozoonosis (*Encephalitozoon cuniculi*), apparently a congenital infection of the perinate, in a *Saimiri* less than 24 hr old (Anver *et al.*, 1979). A multifocal granulomatous encephalitis, with the organism present in the brain, similar to that seen in the rabbit was observed. The infant's dam was clinically normal. No lesions were present in the tongue, stomach, small intestine, colon, pancreas, trachea, lung, liver, heart, thyroid, and eye. There were congestion and focal hemorrhages in the adrenal cortex and thymus. This organism is common to laboratory rodents and has been reported in other animals. Its presence in primates is rare, with limited reports in both human and nonhuman primates.

Of the helminths (Cestodes), at least three tapeworms (*Hymenolepis* sp., *Atriotaenia megastoma*, and "*Spirometra*" = *Sparganum reptans*) have been

Table I
Parasites (and Anthropods) Reported in Saimiri sciureus[a]

Mastigophora (flagellates)
Trichomonas
Trypanosoma cruzi, T. hominis, T. saimiri, T. minasense
Sarcodina (amoebae)
Entamoeba histolytica
Sporozoa (malaria, etc.)
Plasmodium brasilianum, P. falciparum, P. simium
Sarcocystis
Toxoplasma gondii
Acanthocephala (thorny-headed worms)
Prosthenorchis elegans, P. spirula, P. sigmoides
Macracanthorhynchus hirudinaceus
Cestodes (tape worms)
Hymenolepis sp.
Atriotaenia megastoma
"Sparganum" (*Sparganum reptans*)
Trematodes (flukes)
Phaneropsolus orbicularis
Echinostoma
Schistosoma mansoni
Nematodes (round worms)
Strongyloides
Ancylostoma
Trichurus
Molineus
Trypanoxyuris
Longistriata
Vianna
Filaroides
Lobatorobius
Gongylonema
Dipetalonema
Tetrapetalonema
Arthropods (ticks, mites, flies, insects)
Demodex
Audycoptes
Fonsecalges
Mortelmansia
Pentastomes (tongue worms)
Porocephalus

[a] Adapted from Kuntz and Myers (1972) and Dunn (1968).

described along with a relatively large number of thorny-headed worms, flukes, and round worms.

In addition to the usual types of infection associated with parasites, an association between parasites (*Schistosoma haematobium*) and proliferative epithelial lesions of the urinary bladder of the squirrel monkey, as well as other nonhuman primates, has been suggested by Kuntz *et al.* (1972). These findings will be described in greater detail in Section 4.3.

Information on the immunological and pathological responses of squirrel monkeys to various parasitic infections is minimal. Even attempts to generalize by examining parasitic profiles in other species is made difficult by the failure of investigators to classify the parasitic forms correctly or even to define the particular simian species examined. Informative reports on parasitic infections of the nonhuman primate include Chitwood (1970), Dunn (1968), Fiennes (1972), Hunt *et al.* (1970a), Kuntz and Myers (1972), Levine (1970), Loew (1968), Neal (1970), Orihel (1970, 1971), Ruch (1967), Workshop on Nonhuman Primate Parasites (1970), and Worms (1967).

Mechanisms of inducing damage by parasites include: (1) pathogenic commensalism, (2) introduction of foreign body, (3) mechanical damage, (4) loss of blood, (5) liberation of toxins, (6) intracellular invasion and destruction, (7) utilization of host's nutrients, (8) hypersensitivity, (9) introduction of other pathogenic organisms, and (10) lowering host resistance to other diseases. The host response to these mechanisms may be one or more of the following: (1) no reaction—no lesion, (2) necrosis and inflammation, (3) secondary lesions, and (4) neoplasia. In the following, an attempt will be made to indicate how the squirrel monkey responds to these infections.

4.3.1. Protozoa

Of the protozoan infections of nonhuman primates, malaria has received extensive attention (Coatney *et al.*, 1971; Collins, 1974; Dunn, 1968; Voller, 1972). Dunn (1968) and Coatney *et al.* (1971) report the presence of *Plasmodium brasilianum* in *Saimiri* as a natural infection and experimental infection with *P. simium*. A survey of *P. brasilianum* infection in Peruvian and Colombian primates was reported by Dunn and Lambrecht (1963a). About 15% of the animals were found with *P. brasilianum*, but there was no evidence of *P. simian* infection. Splenectomy of the squirrel monkey did not elevate the low parasitemia. The infected animal spleens were grossly enlarged and revealed quantities of malarial pigment. Splenectomized squirrel monkeys were successfully infected with *P. vivax* (Deane *et al.*, 1966) and with *P. falciparum* (Young and Rossan, 1969).

Garnham *et al.* (1963) isolated a strain of *P. brasilianum* from a squirrel monkey and described the "11-day body" as containing vague clefts. A more detailed description of the exoerythrocytic stages of this plasmodium in the squirrel monkey is provided by the studies of Sodeman *et al.* (1969). Liver biopsies obtained at 7, 14, 18, and 21 days following intrahepatic and intravenous inoculation with sporozoites were examined for the exoerythrocytic (EE) stages. No EE bodies were observed at 7 days, but were seen and described in the biopsy material obtained at days 14, 18, and 21. Only limited information on the host response was provided, but enlarged cell nuclei with the presence of parasitized, double-nucleated liver cells, similar to those seen in the chimpanzee, were described. However, the presence of multinucleated liver cells in normal livers of these two species was noted. In a study with *P. vivax*, a patent infection was established in the squirrel monkey using an *Aotus trivirgatus*-adapted strain with serial trophozoite passages and parasitemia induced (Rossan *et al.*, 1972). Sporozoites from infected mosquitoes (*Anopheles albimanus*) similarly caused infections. Relapses were observed in monkeys with both trophozoite and sporozoite infections. However, the maximum parasite concentration achieved during relapses was less than that in primary infections, suggesting development of some immunity. Mortality (22%) in squirrel monkeys was lower than that observed in the owl monkeys (67%). Parasite concentrations exceeding 400,000/mm^3 and patent periods upward to 40 days with multiple attacks were recorded. Ultrastructural studies of *P. falciparum*-infected squirrel monkey red cells demonstrating membrane lesions similar to those produced by other strains in other hosts were done by Fremount and Rossan (1974). The highest concentration of schizonts was in spleen, bone marrow, and heart. Young *et al.* (1975) also indicate that *P. falciparum* can be directly adapted from the human to *A. trivirgatus* and then subsequently passaged in *Saimiri* (as well as *Cebus*). This report by Young *et al.* (1975) also provides a review of experimental monkey hosts to human plasmodia.

Leishmaniasis would be expected to occur in New World nonhuman primates; however, studies providing verification are not available. Baker (1972) indicates a natural infection, without clinical disease, of *Saguinus geoffoyi* with *Leishmania braziliensis*.

Trypanosomiasis (Chagas disease) is worldwide in distribution and several different primate species are affected. Dunn *et al.* (1963), in a survey of several hundred Peruvian and Colombian monkeys, detected trypanosomiasis in approximately 21% of the animals. Twenty-five of 82 squirrel monkeys (31%) were infected with *Trypanosama minasense* and/or *T. cruzi* (or *T. cruzi*-like) trypanosomes. There are a multitude of different names given to these parasites. Many are probably synonyms for the four or five major species: *T. cruzi, T. lambrechti, T. minasense, T. rangeli,* and *T. saimiri* (Baker, 1972; Dunn *et al.*, 1963). There does not appear to be any evidence for pathogenicity in the natural primate host, but in experimental animals the virulence of infection is varied. Concern for possible transmission to the human under special conditions apparently is justified, but specific information is not available. It is suggested, however, because *T. cruzi* is infectious for the human, that laboratory personnel using New World monkeys (the squirrel monkey specifically) should be aware of this potential danger. There are no indications that the other trypanosomes are infectious for primates.

Recently, it has been demonstrated that congenital transmission of *T. cruzi*, known to occur in the human, may also occur in the monkey. Eberhard and D'Alessandro (1982) described such an infection in a laboratory-born squirrel monkey. *Trypanosama cruzi* were found in blood smears of otherwise healthy-appearing mother and offspring.

There are several reports indicating *Toxoplasma* infections of primates, including the squirrel monkey. Ruch (1967), Nery-Guimaraes and Franken (1971), and Nery-Guimaraes *et al.* (1971) provide ample evidence of clinical disease in various species, and McKissick *et al.* (1968) describe an outbreak of toxoplasmosis among squirrel monkeys in the Philadelphia Zoo. Of 17 animals involved, nine died over a 22-month period. Clinical signs were variable and nonspecific, ranging from apparent good health to protracted diarrhea and other indications of gastrointestinal involvement. The clinical disease, when noted, lasted from 1 to 5 days. In addition to such findings as lethargy, emesis, anorexia, and dyspnea, two animals showed neurological signs (circling, incoordination, true convulsions). Grossly, lesions were also variable and nonspecific, with the most consistent findings observed in the lungs—diffuse congestion, edema, ecchymoses, hydrothorax; and pink, foamy exudate in bronchi, trachea, and external nares. Necrosis with or without an inflammatory cellular response was most characteristic. Both free and cystic forms of toxoplasma were observed in the necrotic areas as well as in normal tissues. Intracytoplasmic colonies were also seen in all these tissues. Free forms of the organism were *not* seen within a cell. Although not all animals showed the same extensive tissue involvement, lesions could be found in all tissues, including the brain. Ratcliffe and Worth (1951) had previously described the death of squirrel monkeys as a result of toxoplasmosis. In two *Saimiri* that died, necrotic lesions were seen in the heart, lungs, liver, spleen, and abdominal lymph nodes, and there was a hemorrhagic necrosis of the adrenals. Infiltration of monocytes characterized the acute lesions. A diffuse interstitial infiltration of monocytes and edema, with or without necrosis, was observed in the lungs. A similar pathology, due to *Toxoplasma* in a marmoset (*Oedipomidas oedipus*), was described by Benirschke and Richart (1960). Antibodies to *Toxoplasma* in monkeys were reported by Remington *et al.* (1965).

Sarcocystis, another member of the order Toxoplasmida, has been described in nonhuman primates, principally Old World species (*Macaca mulatta, Cercopithecus mitis, Erythrocebus patas*). Similarly, *Babesia* has also been reported in Old World primates. These protozoa appear to be nonpathogenic (or only mildly so) (Baker, 1972). However, in a survey of nonhuman primate tissues, 21% of wild-caught simians tested were infected (Karr and Wong, 1975). Included in this study were 18 squirrel monkeys, of which two (11%) were positive for an "unidentified species" of *Sarcocystis*. The cysts found in the squirrel monkeys were similar to *S. nesbitti*.

Protozoa of the intestinal tract are numerous and vary in their capacity to produce clinical disease. However, in spite of the many recognized genera of intestinal protozoa, little information has been derived from New World animals, particularly the squirrel monkey (Burrows, 1972; Dunn, 1968). Further, the trophozoites die rather rapidly and are often missed if the stool specimen is improperly handled or there is a delay in its examination. Most intestinal protozoa are considered nonpathogenic, but association with diarrhea has been suggested. Inasmuch as several of these protozoa are known to cause disease in humans, association of all intestinal protozoa in addition to those listed (Table I) with monkey diseases should be considered. *Entamoeba histolytica* is a well-known pathogen and responsible for clinical disease in a wide variety of monkeys and apes, causing severe diarrhea, anorexia, apathy, muscular weakness, blood and/or mucous in stool, vomiting, and death. Pathology is generally minimal, but may include shallow ulcerations in the colon and sloughing of the mucosa. Amoeba may penetrate the muscularis mucosae, and deaths have been reported in spider monkeys (Ruch, 1967).

4.3.2. Helminths (Worm Parasites)

Infections with these parasites appear to be more prevalent than those occurring with protozoa. Again, as seen in protozoan infections, little obvious disease occurs and mortality due to most helminths is rare (Kuntz, 1972; Myers, 1972; Orihel and Seibold, 1972; Schmidt, 1972; Yamashita, 1963). Because of the lack of specific information on the squirrel monkey, the following information is *per force* generalization based upon observations in primates. Variation in host response is to be expected. Helminths may be found throughout the animal body, but some tissue predilection may be demonstrated. Factors governing severity of infection are similar to those seen with the protozoan infections and depend upon the particular location of the parasite in the host, number of parasites present, migration capabilities, egg-laying capability, etc.

Of the Acanthocephala, *Prosthenorchis* (thorny-headed worms) have been seen in New World monkeys, where they may be found burrowing into or perforating the intestinal wall, giving rise to serious pathology, including death. Infection without clinical manifestations may also occur. *Prosthenorchis sigmoides* has been found in the squirrel monkey. Deaths are generally the result of a perforated intestine, with secondary infection producing a peritonitis. Acanthocephala is spread in animal colonies, particularly in zoos, via cockroaches (intermediate host) containing infective larva (cystacanth). Histopathology includes evidence of hemorrhage into the submucosa with or without an inflammatory response, depending upon perforation of the serosa. Secondary bacterial involvement results in an inflammatory response followed by a coagulative necrosis. Unoccupied lesions are common, indicating changes in position. Connective tissue proliferation occurs if the worm remains in place for any length of time. These nodules are smooth, hard, and pearly in appearance and may be easily differentiated from Peyer's patches. Perforation of the intestinal wall may occur, but this is probably due to bacterial activity. Intestinal obstruction occurs as a result of the presence of large numbers of worms (Schmidt, 1972). Other acanthocephalan infections (*P. elegans*) have been noted in New World monkeys, but the exact relationship of these to each host is obscure (Yamashita, 1963).

Cestode (tapeworm) infection of squirrel monkeys includes *Hymenolepis* and *Atiotaenia*. However, other cestodes have been reported in New World monkeys and their possible presence in squirrel monkeys should be kept in mind (Dunn, 1968; Kuntz, 1972; Myers, 1972). Cestode infections in nature are rarely of any clinical significance even when large numbers of parasites are present in the intestine. Under captive conditions, however, because of various factors generally associated with stress, active infection may occur. Pathology is related to attachment of the scolex to the intestinal wall, permitting secondary bacterial infection. Cestodes are transmitted via intermediate or carrier arthropod hosts, but may also be acquired by eating infected meat. The larval forms of diphyllobothriid cestodes (spargana) when invading muscle may give rise to pathological conditions as a direct result of migratory activities or secondary bacterial infection. Cestode infections of primates are considered as "deadend" because of the need for intermediate hosts as well as other biological characteristics.

Trematode (fluke) infections have been infrequently noted in nonhuman primates. Pathology is less than that seen with cestodes. Those flukes that are intestinal dwellers (*Phaneropsolus, Echinostoma*) are tolerated rather well, even when present in large numbers. Dicroeliids may also be found in the intestines, but more usually are found in bile and pancreatic ducts (Kuntz, 1972). Tissue invasion occurs by larval forms (*Diplostomid mesocercariae*) as a result of migration. Cosgrove (1966) describes in detail trematodes that are found in laboratory primates. Bostrom and Slaughter (1968) report on finding *Athesmia foxi* in liver of squirrel monkeys. Gross examination failed to detect the presence of lesions, which were noted microscopically. These parasites were located in the lumens of the interlobular bile ducts. Pathological lesions included cholangiectasis, partial biliary obstruction, and mild biliary stasis involving the bile canaliculi. Trematodes were not found in the gallbladder.

The Oriental lung fluke (*Paragonimus*) may cause pathology in the lung, where it becomes encysted. However, this fluke does not occur naturally in New World species. Similarly, *Schistosoma* is not indigenous to the New World, and, although it has been reported in the squirrel monkey (Swellengrebel and Rijpstra, 1965), this finding is considered to be accidental. Schistosome infections result in the presence of eggs in the tissues, where symptomatology and pathology is directly related to the intensity and duration of infection. Experimental infection of *S. sciureus,* among other nonhuman primates, was accomplished by Kuntz *et al.* (1971) in an attempt to find a satisfactory host for pathological studies of schistosome infections. Experimentally, the infection was found to cause fatalities after exposure to 500–1000 parasites. The schistosomes were moderately uniform in their distribution in the superior and inferior mesenteric venous systems. A number of the animals were observed to have worms and/or eggs associated with the genital and urinary systems, including the bladder. A small number of animals were found with macroscopic lesions in the bladder and one animal was found with hydronephrosis, hydroureter, and "cauliflower-like" growths in the bladder. In another study, these investigators (Kuntz *et al.*, 1972) describe production of tumors morphologically consistent with papillary transitional cell carcinoma of the bladder in experimental schistosome infections of squirrel monkeys, capuchins (*Cebus apella*), and talapoin (*Cercopithecus talapoin*).

Of the helminths found in primates, the majority are nematodes (roundworms). More than 100 nematodes are found in South American monkeys, of which ten genera have been described in *S. scuireus* (Dunn, 1968; Kuntz, 1972). An overview by Orihel and Seibold (1972) describes nematode infections of the bowel and tissues. Obvious disease is rare, but there are some exceptions, particularly when the intestinal wall is invaded by adult or larval forms. Brooks (1963) reports on the presence of *Trichuris* (whipworm) eggs in the squirrel monkey. This nematode has been found in both New and Old World monkeys. A report of champanzee infection suggests debilitating disease associated with anorexia and severe diarrhea (Thienpont *et al.*, 1962). Other Trichuridae are either not found in nonhuman primates or have a limited host and geographic distribution, or the available information is too scanty to clearly define their pathogenesis.

Hookworm (*Ancylostoma*) eggs have also been reported in the squirrel monkey (Brooks, 1963), but the identification of these eggs has been questioned (Dunn, 1968). Orihel and Seibold (1972) indicate that *Ancylostoma* and *Necator* are not commonly found in monkeys and apes and their presence in New World monkeys is "especially rare." Lungworms (*Pseudoliidae*) may be found with relative frequency in New World monkeys. Confusion exists in the precise terminology of the genera included herein (*Filaroides, Filariopsis*), although five species have been reported in primates. *Filaroides gordius* has been found in squirrel monkeys, and Orihel and Seibold (1972) report adult worms in the parenchyma, with sections of the parasite found in the alveoli, alveolar ducts, and respiratory bronchioles, usually without any inflammatory reaction. Clusters of leukocytes may be found adjacent to the parasites in the respiratory bronchioles, along with some macrophages in parasitized alveoli. Upon gross examination, the lungs appear normal except for the presence of small, elevated nodules seen just under the pleura.

Other helminths found in New World species include the trichostrongyles (Trichostrongylidae), which are found in *Saimiri* (*Molineus elegans, M. torulosus, Longistriata dubia*). These nematodes are very small parasites invading the mucosal surfaces of the intestinal tract and are considered nonpathogenic in their natural hosts even when present in large numbers. However, a report by Smetana and Orihel (1969) suggests a unique relationship in *M. speciosa*. *Nochtia nochti* were found in gastric tumors of this Old World monkey and tumors were never found in the absence of this parasite. Kuntz and Myers (1972) note the presence of *Vianna* in squirrel monkeys, but provide no information on this nematode.

Pinworms (Oxyuridae) have been found in primates, but only one, *Trypanoxyuris scelertus,* has been recovered from the squirrel monkey and is nonpathogenic. The human pinworm, *Enterobius vermicularis,* has been reported in several species of New World monkeys (Cameron, 1929), but these are considered not to be natural infections but to be derived from human contacts (Yamashita, 1963).

Spiruroidea are common nematodes found in the digestive tract and viscera of primates. Generally nonpathogenic, large numbers may produce serious consequences in a host animal. Foster and Johnson (1939) report an outbreak in laboratory *Cebus capucinus* in which large numbers of *Protospirura muricola* caused perforation of the stomach wall, and encapsulated abscesses were present in the viscera. *Gongylonema saimirisi* is found in the squirrel monkey and is present in the mucosal lining of the esophagus (Dunn, 1968). In other species, they have been observed in the bronchi, stomach, lip, tongue, and other tissues (Orihel and Seibold, 1972). Orihel and Seibold (1971) indicate the presence of another Spiruroidea, *Trichospirura leptostoma,* in squirrel monkeys. This parasite, present in several South American monkeys, is found in the pancreatic ducts. In *Callithrix jacchus* and *Aotus trivirgatus,* this worm causes a chronic pancreatitis with patchy, granulocytic infiltration of the interstitial tissue in relation to intralobular ducts.

Filariasis (Filariidae) is one of the more common infestations of New World monkeys, particularly Platyrrhines (Clark, 1931). These adult worms are tissue or somatic parasites, with filamentous worms found in the peritoneal cavity (where they may be found free) or mesenteries. Of the eight recognized species found in South American simians, all (with the exception of one species, *Parlitomosa zakii*) belong to two genera: *Dipetalonema* or *Tetrapetalonema* (Dunn, 1968; Dunn and Lambrecht, 1963b; Orihel and Seibold, 1972). However, Chalifoux *et al.* (1973) describe the characteristics of 11 species and indicate that six of these microfilarial types are present in *S. scuireus: Tetrapetalonema marmosetae* (50/95 positive), *Dipetalonema gracile* (48/95 positive), *P. zakii* (only one of 95 samples was positive), *D. caudispina* (8/95 positive), *T. panamensis* (50/95 positive, and "Microfilaria B" (3/95 positive). A total of 84 of the 95 animals were found with microfilariae (88%). These investigators indicate that differences in localization of the enzyme acid phosphatase provides a method for distinguishing these microfilariae in blood smears. Large worm populations will cause a fibrinous exudative reaction with adhesions in either the pleural or peritoneal cavities. The mesentery may be bound to the peritoneal wall by adhesions. Pleural infections may also result in adhesions. Recently, Eberhard (1982) examined the effect of some known antifilarial drugs on naturally occurring filarial infection in squirrel monkeys. Of the drugs used (naphuride sodium, thiacetarsamide sodium, diethylcarbamazine), diethylcarbamazine (50 mg/kg for 10 days) was effective in clearing the blood of *D. gracile, D. caudispina,* or *T. marmosetae* for at least 12–24 weeks following therapy.

Various Pentastomida have been reported in nonhuman primates, and their pathology has been reviewed (Cosgrove *et al., 1970). Encapsulated Porocephalus* nymphs were seen in tissues of two squirrel monkeys, but the lesions were not described. Fox *et al.* (1972) noted the encapsulated nymph (*Porocephalus clavatis*) in the brain of a squirrel monkey. Found in the occipital cortex, the nymph was viable without evidence of cuticles, fluids, or cellular infiltrate. Host response was minimal, with slight compression and replacement of the brain tissue by the cyst along with mild thickening of the pia-arachnoid. The inner lining of the cyst was periodic acid-Schiff-positive, smooth, and thin with an acellular membrane enclosing the nymph.

Numerous other parasitic forms have been described. They are not included in the above discussion principally for lack of specific information or because they have not been found in New World simian species. The role and distribution of various intermediate hosts in relationship to infection have not been elaborated in this discussion, but their involvement in the spread and maintenance of disease is paramount. Dunn (1963) provides such information for acanthocephalan and cestode infections of South American primates.

Dunn (1968), in commenting on information available on infections of primates, states "the gaps in knowledge are great for almost every parasite group." Although Dunn's comment was directed towards parasitic infection in the broad sense, information on helminths and protozoa is still incomplete for any given simian species. The comments of Wong (1970) are of interest to those attempting to examine necropsy material for the presence or absence of parasites. Of importance is a listing of nonhuman primate parasites that are found outside the gastrointestinal tract as well as a listing of those simian parasites that require dissection and a microscope for their detection. A model procedure for the satisfactory performance of this necropsy is provided.

4.4. Viral Infections

Viruses of nonhuman primates have probably received more attention than most other primate infectious agents. Two major findings account for this: (1) the recognition that nonhuman primates are parasitized with an entirely "new" viral flora than previously suspected, differing from other animal viruses (Table II), and (2) the fact that certain of these primate viruses are oncogenic.

It is important, however, to emphasize that the full significance of these findings is not understood. As with many of the parasitic infections, most of the viruses isolated from nonhuman primates have not been known to produce disease. Obviously, there are some marked exceptions to this. The two major simian virus groups, i.e., the adenoviruses and enteroviruses, are notably nonpathogenic; the herpesviruses have been recognized as the most pathogenic.

Following recognition that a nonhuman primate may carry a virus fatal to the human, i.e., *Herpesvirus simiae* (Sabin and Wright, 1934), a number of other simian viruses were described. Through the 1950s and 1960s, the isolation of large numbers of viral agents from nonhuman primates, principally Old World monkeys (*Macaca* spp. and *Cercopithecus aethiops*), was reported (Albano, 1957; Heberling and Cheever, 1960; Hoffert *et al.*, 1958; Hsiung and Melnick, 1958; Hull *et al.*, 1956, 1958; Hull and Minner, 1957; Kalter, 1960; Malherbe and Harwin, 1957; Malherbe *et al.*, 1963; Rustigian *et al.*, 1955). For the simian viruses isolated from New World monkeys, the squirrel monkey was the major contributor: *H. saimiri,* a herpes simplex-like virus (*H. tamarinus*), a cytomegalovirus, two adenoviruses, and two retroviruses (see below). A number of workshops reviewed the early virology of nonhuman primates as reported in *Laboratory Animal Science,* Vol. 21, Part II (1971) and Vol. 24, Part II (1974). Several major overviews provide detailed information on simian viruses (Fiennes, 1972; Hsiung, 1969; Hsiung *et al.*, 1964; Hull, 1968; Kalter, 1969, 1983). Problems in classification of these agents persist, and a WHO/FAO Simian Virus Working Team has proposed a numerical designation as a nomenclature schema for these viruses (Table II) (Kalter *et al.*, 1980).

In 1966, Melendez and co-workers described a herpesvirus isolated from normal squirrel monkey kidney cell cultures, which was designated as *Herpesvirus saimiri* (Melendez *et al.*, 1966). Subsequently, this virus was characterized (Melendez *et al.*, 1969a,b) and, more important, was demonstrated to produce malignant disease in owl monkeys (*Aotus trivirgatus*) and in marmosets (*Saguinus oedipus*) (Melendez *et al.*, 1971b). This discovery demonstrated for the first time the existence of a virus in a normal primate (*S. sciureus*) capable of inducing malignancy in another primate.

Primates had been used previously for "oncogenic" virus studies. A number of reports indicate the production of tumors in nonhuman primates, generally macaques and baboons, with Rous sarcoma virus. The pioneering efforts of Munroe in the U. S. and Zilber in the Soviet Union, along with their collaborators, in utilizing monkeys for such studies should be acknowledged (Munroe and Windle, 1963; Zilber *et al.*, 1965). An early study in squirrel monkeys was that of Masironi *et al.* (1968), in which the buccal mucosa was inoculated with the "Toolan H. Ad. # 1U tumor." The animals responded with formation

Table II
Simian Viruses: Suggested Nomenclature[a]

Current nomenclature			
Virus[c]	Strain	Species of origin	Proposed nomenclature[b]
Simian adenoviruses (Adenoviridae)			
SV1 (M1)	301	*Macaca fascicularis*	Adenovirus sim-1
SV11 (M5)	646776	*Macaca mulatta*	Adenovirus sim-2
SV15 (M4)	AP4398	*Macaca mulatta*	Adenovirus sim-3
SV17 (M6)	6630-IC	*Macaca mulatta*	Adenovirus sim-4
SV20 (M7)	P4	*Macaca mulatta*	Adenovirus sim-5
SV23 (M2)	74932WK	*Macaca mulatta*	Adenovirus sim-6
SV25 (M8)	8045-2WN	*Macaca mulatta*	Adenovirus sim-7
SV30	P5	*Macaca fascicularis*	Adenovirus sim-8
SV31	P6	*Macaca* sp.	Adenovirus sim-9
SV32 (M3)	P7	*Macaca* sp.	Adenovirus sim-10
SV33 (M10)	P10	*Macaca mulatta*	Adenovirus sim-11
SV34	A7644	*Macaca mulatta*	Adenovirus sim-12
SV36 (M11)	P9	*Macaca* sp.	Adenovirus sim-13
SV37	E4382	*Macaca mulatta*	Adenovirus sim-14
SV38	702707	*Macaca mulatta*	Adenovirus sim-15
SA7	C-8	*Cercopithecus aethiops*	Adenovirus sim-16
SA17	B105	*Cercopithecus aethiops*	Adenovirus sim-17
SA18	C626	*Cercopithecus aethiops*	Adenovirus sim-18
AA153	AA153	*Papio cynocephalus*	Adenovirus sim-19
V340	V340	*Cercopithecus aethiops*	Adenovirus sim-20
C-1	Bertha	*Pan satyrus*	Adenovirus sim-21
Pan5	CV-23	*Pan satyrus*	Adenovirus sim-22
Pan6	CV-32	*Pan satyrus*	Adenovirus sim-23
Pan7	CV-33	*Pan satyrus*	Adenovirus sim-24
Simian picornaviruses (Picornaviridae)[b]			
SV2	2383	*Macaca mulatta*	Enterovirus S-1
SV6	1631	*Macaca mulatta*	Enterovirus S-2
SV16	2450-SD	*Macaca mulatta*	Enterovirus S-3
SV18	2481-B2	*Macaca mulatta*	Enterovirus S-4
SV19	P2	*Macaca fascicularis*	Enterovirus S-5
SV26	3163	*Macaca mulatta*	Enterovirus S-6
SV28	9128	*Macaca mulatta*	Enterovirus S-7
SV35	A7987	*Macaca mulatta*	Enterovirus S-8
SV42	P1	*Macaca fascicularis*	Enterovirus S-9
SV43	P12	*Macaca fascicularis*	Enterovirus S-10
SV44	P13	*Macaca mulatta*	Enterovirus S-11
SV45	P14	*Macaca fascicularis*	Enterovirus S-12
SV46	P15	*Macaca* sp.	Enterovirus S-13
SV47	P16	*Macaca fascicularis*	Enterovirus S-14
SV49	P19	*Macaca mulatta*	Enterovirus S-15
SA4	L79C2	*Cercopithecus aethiops*	Enterovirus S-16
SA5	B165	*Cercopithecus aethiops*	Enterovirus S-17
A13	A13	*Papio cynocephalus*	Enterovirus S-18
Simian herpesviruses (Herpesviridae)[d]			
Herpesvirus simiae	E-2490	*Macaca* sp.	Herpesvirus S-1
SA6	ST3835-7	*Cercopithecus aethiops*	Herpesvirus S-2
SA8	B264	*Cercopithecus aethiops*	Herpesvirus S-3

Table II. (continued)

Current nomenclature			
Virus[c]	Strain	Species of origin	Proposed nomenclature[b]
Herpesvirus tamarinus (*Herpesvirus platyrrhinae* type 1)	MV-5-4PSL	*Saimiri sciureus*	Herpesvirus S-4
Herpesvirus saimiri	S295C	*Saimiri sciureus*	Herpesvirus S-5
SMV	TA4-4646	*Ateles geoffroyi*	Herpesvirus S-6

[a] Adapted from Kalter *et al.* (1980). A similar sequential numerical system is suggested for the remaining simian virus families as they became defined. See text for references on reviews of simian viruses.

[b] The vernacular name will continue: simian adenovirus 1, 2, etc.; the formal taxon, if approved by the ICTV, will be adenovirus sim-1, sim-2, etc.

[c] SV27 (M9) and SV39 have been excluded from this group because of their close antigenic relationship to SV31 and SV23, respectively. Squirrel monkey isolates SqM-1 and SqM-2 have not been sufficiently characterized to be included here. See text concerning antigenic relationships between SV28, SV4, and SA4.

[d] Approximately 30 additional simian herpesviruses will be added to this list after final characterization. The vernacular names, i.e., simian herpes viruses 1, 2, etc., with the current nomenclature in parentheses, are suggested for *Herpesvirus simiae*, SA6, etc.; the proposed nomenclature, if approved by the ICTV, is the formal taxon.

of granulation tissue and fibroblast proliferation around groups of tumor cells at the site of inoculation. Additional pathology included infiltration of small lymphocytes and other blood and reticuloendothelial cells. Rabin and Cooper (1971) included the squirrel monkey in their experiments on tumor induction by Rous sarcoma virus. Numerous reviews provide the historical background on these efforts and should be consulted (Deinhardt, 1980; Fiennes, 1972; Heberling, 1983; Kalter, 1983; Kalter *et al.*, 1979; Rabin, 1971, 1983).

Several investigators have found that other viruses—feline sarcoma virus, SV40, woolly monkey tumor virus, Yaba-pox virus, and Yaba-like disease (Tana-pox) virus—are oncogenic in simians. In several of these experiments, the squirrel monkey was found to be susceptible.

Feline sarcoma virus produced tumors in various age groups of squirrel monkeys. These undifferentiated sarcomas generally regressed within a short period of time (Rabin *et al.*, 1971, 1972). "Successful" tumor production was also achieved in squirrel monkeys with the woolly monkey tumor virus. This virus produced nodules in two of five animals; however, it is questionable whether these could be classified as sarcomas (Rabin, 1971; Theilen *et al.*, 1973). Ambrus *et al.* (1963) were unsuccessful in an attempt to stimulate Yaba virus tumor growths in primates (including squirrel monkeys) by use of various carcinogenic chemicals as well as radiation.

In spite of numerous reports on the presence of viruses in nonhuman primates, it should be reemphasized that the majority of these studies reflect information collected on captive animals rather than on animals in their "natural" environment. Probably the only primate studied to any great extent in its natural environment is the baboon (Kalter, 1973).

Several different simian viruses have achieved considerable prominence either as pathogenic agents (herpes B, Marburg virus) or perhaps as models for laboratory studies on tumorigenicity (*H. saimiri*) and transformation (SV40). As a group, however, the simian herpesviruses have undoubtedly received the most attention.

4.4.1. Herpesviruses (Herpesviridae)

Three subfamilies constitute the Herpesviridae: Alphaherpesvirinae (herpes simplex virus group as well as varicellz-zoster virus and viruses of various animal species), Betaherpesvirinae (cytomegalovirus group), and Gammaherpesvirinae (lymphoproliferative virus group). Simian herpesviruses are represented within all of these subfamilies (Table III). Included here are viruses that are counterparts of all the human herpesvirus subfamilies as well as of other animal herpesviruses. No attempt will be made to review the classification or biological characteristics of the simian herpesviruses, as these have been extensively documented in the literature (Ablashi *et al.*, 1976, 1979; Barahona *et al.*, 1974; Brack, 1977; Deinhardt, 1973; Hull, 1968; Hull *et al.*, 1972; Hunt and Melendez, 1972; Kalter, 1982; Kalter *et al.*, 1980; Kaplan, 1973; Matthews, 1982; McCarthy and Tosolini, 1975; Melendez *et al.*, 1971a; Nahmias and Dowdle, 1968; Neubauer and Rabin, 1979; Rabin, 1983; Rapp, 1974; Roizman, 1974; Zur Hausen, 1975). The proceedings of a recent symposium should be consulted for information on herpesviruses in general (Bonneau and Hennessen, 1981). Nonetheless, the universal nature of this group of agents as well as their ability to be associated with a wide variety of clinical manifestations warrant comment.

Herpesviruses as a group undoubtedly are among the most widely distributed of viruses. Every animal species apparently is or will be infected with one or more of the herpesviruses. Although a herpesvirus is generally either specific or characteristic for each particular host animal species, crossing of species barriers occurs, with development of infection, disease, and fatalities. Thus, in terms of colony management and husbandry, the transmission of herpesviruses within the framework of animal–human interrelationships is of concern and importance. In addition to the practical aspects of herpesvirus infections, their involvement in such a diverse spectrum of disease offers investigators the potential for a multitude of laboratory investigations.

The observed clinical diseases and cellular invasiveness of the herpesviruses may be grouped into the following categories [adapted from McCarthy and Tosolini (1975)].

4.4.1a. Generalized (Possibly Neurological) Disease. Initially all herpesvirus infection caused by viruses in this category, regardless of the host species, is characterized by the occurrence of spontaneous localized herpetic vesicular or ulcerative lesion(s), usually restricted to the mucous membranes. Extensive variation may occur in primate species, depending upon the infecting herpesvirus and host species. Transient oral (labial or lingual) lesions, conjunctivitis, and/or cutaneous lesions may be noted. These will heal without scarring, but may develop into ulcers prior to healing. Localized adenopathy may or may

Table III
Herpesviruses of Nonhuman Primate[a]

Family	Vernacular name	Species of origin	Proposed name
Alphaherpesvirinae	B Virus (herpes B)	*Macaca* sp.	Cercopithecid herpesvirus 1 (B), Herpesvirus S-1, *Herpesvirus simiae*
	SA8	*C. aethiops, P. cynocephalus*	Herpesvirus S-3
	Herpesvirus tamarinus (marmoset herpesvirus, herpesvirus-T)	*S. sciureus*	*Herpesvirus platyrrhinae* type 1, Herpesvirus S-4
	SMHV (spider monkey herpesvirus)	*A. geoffroyi*	Herpesvirus S-6
Betaherpesvirinae	SA6	*C. aethiops*	Herpesvirus S-2
	SA15	*C. aethiops*	—
	AGM-CMV (African green monkey CMV)	*C. aethiops*	—
	LVMV (Liverpool vervet monkey virus, LVP)	*C. aethiops*	—
	PMHV (patas monkey herpesvirus)	*E. patas*	—
	RhM-CMV (rhesus monkey CMV)	*M. mulatta*	—
	MLM (Medical Lake *Macaca* herpesvirus)	*M. nemestrina*	—
	LAHM (leukocyte-associated herpesvirus)	*M. mulatta*	—
	THV (tree shrew herpesvirus)	*T. glis*	—
	OM-CMV (owl monkey CMV)	*A. trivirgatus*	—
	Other CMV (MV) isolates[b]	—	—
Gammaherpesvirinae	*H. saimiri* (HVS)	*S. sciureus*	Herpesvirus S-5
	H. ateles (HVA)[c] (spider monkey herpesvirus)	*A. geoffroyi*	—
	H. saguinus	*S. oedipus*	—
	AT-46	*A. geoffroyi*	—
	H. aotus, types 1, 2, 3 (owl monkey virus)	*A. trivirgatus*	—
	H. papio	*P. hamadryas*	—
	H. pan	*P. troglodytes*	—
	H. pongo	*P. pygmaeus*	—
	H. gorilla	*G. gorilla*	—
	H. macaca	*M. fascicularis*	—

[a] Various other herpesviruses have been recovered from monkeys and apes. Their exact relationship to the simian is unknown; several are human viruses.

[b] A number of herpesviruses resembling human CMV have been recovered from different monkeys and apes. Additional studies are required for appropriate classification.

[c] At least three types are recognized; it is not clear to which subfamily each type belongs. Type 1 is probably an Alphaherpesvirinae, the other two types are probably Gammaherpesvirinae.

Table IV
Presence in Squirrel Monkey of Antibody to Herpesviruses[a]

H. hominis	*H. tamarinus*	*H. simiae*	SA8
0/117	64/99	4/91	0/99

[a] The numerator is the number of serologically positive squirrel monkeys; the denominator is the number of squirrel monkeys examined. SA8 is a virus antigenically closely related to herpes simplex and B virus. [From Kalter and Heberling (1972).]

not be detected. The mucocutaneous lesions resulting from infection with natural herpesvirus in any given host are indistinguishable from those seen in humans following herpes simplex virus infection. The viruses included here are classified as Alphaherpesvirinae.

Depending upon the particular simian species involved, the other clinical manifestations that may develop are extremely protean and noncharacteristic. Any or all tissues may be involved, frequently leading to a fatal disease—involvement of the digestive tract, respiratory disease with pneumonia, lymphadenopathies, adrenal necrosis, genital involvement, and finally neurological complications.

Neurological disease, although uncommon, is probably the greatest cause for concern. Little information is available on nonhuman primate herpetic neurological disease. Although involvement of the central nervous system of primates has long been recognized following herpes infections, much of our information is derived from disease in the human with herpes simplex virus. As a generalization, an animal herpesvirus infecting an alien species will cause a disease dissimilar from that seen in the natural host. Under such circumstances, this disease frequently involves the central nervous system. Thus, it is somewhat difficult to define specifically the clinical disease for any given species. Sabin (1934) was able to infect rhesus monkeys with B virus and demonstrated that central nervous system signs did not occur after peripheral inoculation. Intracerebral inoculation, however, was usually fatal.

Primate herpesviruses associated with generalized or neurological disease include the recognized prevalent viruses herpes simplex (*H. hominis*), B virus (*H. simiae*), herpes-T (*H. tamarinus*), *H. ateles, H. aotus,* and others. These herpesviruses grow well in cell cultures and are cytopathic, with cell-free virus detected in the culture fluid. Neurological disease resulting from infection with any of the viruses in this group has not been described for the squirrel monkey.

Because of the concern associated with viral tumorigenesis, particularly with the herpesviruses, an attempt was made to develop a model of herpes simplex-induced cervical cancer (see Section 4.4.1d for discussion on herpesvirus tumorigenesis). Squirrel monkeys (among other simians) could not be infected with this virus (London *et al.*, 1971; Nahmias *et al.*, 1971); however, baboons (*Papio cynocephalus*), marmosets (*Saguinus oedipus*), and cebus monkeys (*Cebus albifrons*) were, although with different clinical responses (Kalter *et al.*, 1972a). Antibody to several of these herpesviruses is observed in the squirrel monkey (Table IV) (Kalter and Heberling, 1972).

The failure to find antibody to *H. hominis* may be due to the possible lethality of herpes simplex virus for squirrel monkeys and the lack of survivors. Although four animals demonstrated B virus antibody, this contrasts with the herpes simplex findings. One would expect B virus to have the same lethal effect as herpes simplex on this species. The explanation for finding these four B virus antibody-positive animals eludes us at this time, but the antibody response may be due to a common, unrelated virus antigen. The B virus is naturally found in *Macaca* spp. and would not be expected in New World species. This would lend credence to the suggestion that the four seropositive animals reflect a response to a common antigen on another virus. The negative serology for SA8, a virus antigenically closely related to herpes simplex and B virus, may be similarly explained, i.e., lethality of virus. It is also conceivable that the squirrel monkey may never have encountered SA8, an Old World primate (baboon and African green monkey) virus.

Herpes-T virus, one of the first of the New World cytocidal herpesviruses to be described (*H. tamarinus*, marmoset virus, *H. platyrrhinae*), was isolated during several outbreaks of a highly fatal disease in marmosets (W. R. Holmes *et al.*, 1964; Melendez *et al.*, 1966, 1971a; Melnick *et al.*, 1964). It was then demonstrated that the marmoset was not the natural host for this virus. Significant numbers of *Saimiri* sp., *Ateles* sp., and *Cebus albifrons* were found to have high antibody titers to this virus (A. W. Holmes *et al.*, 1966), suggesting that herpes-T is indigenous to one or all three of these species. Melendez *et al.* (1966, 1971a) also reported that herpes-T exists as a latent infection in the squirrel monkey. Confirmation of this finding was reported by Daniel *et al.* (1967) with isolation of herpes-T from naturally infected squirrel monkeys. Additional studies strengthened the suggestion that the squirrel monkey was the natural host, with the virus carried as a latent infection. Disease in the squirrel monkey is analogous to herpes simplex infection in the human (King *et al.*, 1967).

Pathology in the experimentally infected marmoset is one of rhinitis and pneumonitis. At necropsy following intradermal inoculation, gross findings are restricted to bilateral adrenal hemorrhage. Extensive pneumonitis has been observed after intranasal inoculation. Feeding virus causes inflammation and necrosis of the small intestine. Upon histological examination, all animals show various degrees of necrosis, particularly in the liver, spleen, and adrenals. Eosinophilic intranuclear inclusions are seen in and around the areas of necrosis. These lesions resemble those occurring in the human after generalized herpes simplex infections. In the squirrel monkey, which appears healthy, the labial lesions are similar to those in humans, as are the necrotic plaques on tongue and palate. Grossly, the lesions consist of ulcerations of the mucosa and adjacent skin covered by crusts of dried serum. When the scars are removed, the excoriated tissue is swollen and red. Lip lesions are characterized by necrosis of the mucosa and adjacent epidermis. The necrotic tissue contains large numbers of polymorphonuclear leukocytes, chromatin debris, and secondary invasive bacteria. Adjacent tissue indicates ballooning degeneration of the epithelial cells, which may contain intranuclear inclusion bodies characteristic of herpesviruses. It is not known if the squirrel monkey succumbs to herpes-T infection as can the human following herpes simplex infection.

4.4.1b. Exanthematous Disease. Herpesviruses producing this clinical syndrome in nonhuman primates are either *H. varicellae,* the causative agent of human chickenpox (varicella-zoster), or an antigenically closely related virus. These and similar animal viruses will probably be included among the Alphaherpesvirinae. Several isolates from different primate species have been described. All are cell-associated and the virus is not found in the culture fluids.

The nonhuman primates from which viruses in this group have been isolated include chimpanzees (McClure and Keeling, 1971), gorillas (White *et al.,* 1972), and orangutans (White *et al.,* 1972); also included are the Liverpool Vervet Virus (LVV) isolated by Clarkson *et al.* (1967) from *C. aethiops,* Patas monkey herpesvirus (McCarthy *et al.,* 1968), Delta herpesvirus (Riopelle *et al.,* 1971), also from a patas monkey, and Macaque vesicular disease virus, from a stumptail macaque (Blakely *et al.,* 1973). Only limited information is available on these viruses, but it appears that they are all related. Clinically, the disease produced in all the animals is similar, consisting of a vesicular rash associated with a generalized infection. Hemorrhage and tissue ulceration occur, with intranuclear inclusions observed in involved cells. Respiratory disease is common and is associated with pneumonia, hemorrhagic necrosis, and death. Although "chicken pox" has not been reported in the squirrel monkey, it may be assumed that this same clinical condition would result from infection with any of these agents.

4.4.1c. Cytomegalovirus-like Disease (Betaherpesvirinae). The full extent of cytomegalovirus infection in primates is unknown and only recently has been recognized and appreciated as a potential clinical problem. Because these viruses are highly host-specific, it is probable that each animal species has its own complement of one or more cytomegaloviruses. These may be "readily" recovered from tissue, saliva, and/or urine and are usually *not* associated with clinical disease in the immunologically normal host. Humans subjected to chemotherapy (immunosuppresion) frequently develop a cytomegalic pneumonia. This type of disease has not been reported in nonhuman primates. However, as will be noted below, immunosuppression has resulted in the recovery of lymphotropic viruses. The relatedness of the various cytomegalovirus isolates is not clear, but certain isolates from different simian species are antigenically related.

Cytomegalovirus isolates include chimpanzee cytomegalovirus (Vogel and Pinkerton, 1955), SA6 (vervet) (Malherbe and Harwin, 1957), SA15 (vervet) (Malherbe *et al.,* 1963), vervet cytomegalovirus (Black *et al.,* 1963), rhesus monkey cytomegalovirus (Asher *et al.,* 1969), tarsier cytomegalovirus (A. A. Smith and McNulty, 1969), and squirrel monkey cytomegalovirus (Rangan and Chaiban, 1980).

One aspect of cytomegalovirus infection that has been studied in the squirrel monkey concerns congenital infection, a problem recognized in the human. Ordy *et al.* (1981) noted that inoculation of the squirrel monkey with a cytomegalovirus during gestation often results in stillbirths; survivors develop specific IgM antibody. It was also noted that reflexes and neuromuscular responses were slow and poorly coordinated in the newborn. Other deficiencies, involving eyesight, learning, and growth, were also apparent.

4.4.1d. Lymphoproliferative Disease (Gammaherpesvirinae). Recent recognition of herpesviruses with the potential to be oncogenic has received considerable attention. Included are such viruses as EBV (Epstein–Barr virus), *H. saimiri, H. ateles, H. papio, H. pongo, H. pan, H. gorilla,* cynomolgus herpesviruses, AGM-EBV (from *C. aethiops*), and tree shrew herpesviruses. Several of these viruses are closely related to EBV. The Old World isolates have the capacity to transform B lymphocytes *in vitro:* the New World isolates transform T cells *in vitro* (B- or T-cell tropic).

Wright *et al.* (1976), using a rosette-enrichment technique, were able to demonstrate that the squirrel monkey herpesvirus (*H. saimiri*) (HVS) could be isolated only from the T lymphocytes. More recently, it has been suggested that some variability may exist among the strains of HVS isolated from squirrel monkeys; these differences were not related to the geographic origin of the host animal (Desrosiers and Falk, 1982). The New World monkey T-cell herpesviruses are biologically similar to the B-cell herpesviruses (Old World), but are clearly distinct (Ablashi *et al.*, 1979; Barahona *et al.*, 1974; Deinhardt, 1973; Deinhardt *et al.*, 1974; Falk, 1974; Fleckenstein, 1979; Fleckenstein and Mulder, 1980; Rabin, 1983).

Herpesvirus saimiri, as reported by Melendez *et al.* (1966, 1968, 1969a,b), is considered to be the first oncogenic nonhuman primate virus to be isolated from New World monkeys. Previously, *H. tamarinus* (herpes-T, marmoset virus, *H. platyrrhinae*) had been recovered from outbreaks of a highly fatal disease in marmosets (A. W. Holmes *et al.*, 1964; Melendez *et al.*, 1966; Melnick *et al.*, 1964) and was thought to be a marmoset virus. As described above, however, herpes-T, now recognized as an indigenous virus of the squirrel monkey, is a cytocidal herpesvirus, is probably an Alphaherpesvirinae, and is not to be confused with *H. saimiri,* an oncogenic herpesvirus (Gammaherpesvirinae).

HVS is one of the most extensively studied of primate herpesviruses. This virus is readily isolated from kidney cultures, whole blood, or peripheral leukocytes of the squirrel monkey, in which the natural disease resembles that seen in the human following infection with herpes simplex. Squirrel monkeys have circulating neutralizing antibody, which may be found in 100% of adult animals. Experimental inoculation of this virus into adult squirrel monkeys fails to induce any evidence of disease. In various other species of New World monkeys (*Saguinus oedipus, S. fuscicollis, S. nigricollis, Aotus trivirgatus, Ateles geoffroyi*), a malignant lymphoma follows *H. saimiri* inoculation (Hunt *et al.*, 1970b). Similarly, the rabbit (*Oryctolagus cuniculus*) also will develop a malignant lymphoma following experimental inoculation with HVS. In certain instances, and for reasons not entirely understood, a reticuloproliferative disease, lymphocytosis, or lymphocytic hyperplasia may develop in marmosets, owl monkeys, and spider monkeys. Although HVS will induce a malignant lymphoma when inoculated into the spider monkey (*A. geoffroyi*), *H. ateles,* an oncogenic herpesvirus isolated from this species, does *not* produce any evidence of malignancy when inoculated into the squirrel monkey. Lymphoma in cottontop marmosets and owl monkeys is histologically characterized by a marked reticulum-cell (immature lymphocytes) invasion with associated cell replacement in a variety of tissues: liver, spleen, kidney, lymph nodes, adrenal, and lungs. Blood changes are characterized by leukocytosis, lymphocytosis, the presence of immature lymphocytes, lymphoblasts, and occasional reticulum cells. All animals developed this uniformly fatal malignancy (Wolfe *et al.*, 1971).

Melendez *et al.* (1971b) were also able to demonstrate that HVS in the owl monkey causes lymphocytic leukemia in association with malignant lymphoma. The disease is characterized by lymphocytic hyperplasia of the lymph nodes as well as an increase in lymphocytes in hepatic sinusoids. White cell counts in excess of $90{,}000/mm^3$, primarily consisting of lymphocytes as well as polymorphocytes and lymphoblasts, developed prior to death.

Falk *et al.* (1972) studied the epidemiology of the natural infection of *H. saimiri* in the squirrel monkey and noted that latent infections were the result of horizontal transmission of HVS during the first year of life. Animals less than 6 months of age did not yield virus and were antibody-negative. At 1 year of age, virus could be isolated from approximately half the animals studied, with about 40% of the animals now seropositive. Colony animals over 2 years of age were "all" seropositive.

No antigenic cross-reactions are observed between HVS and herpes-T, *H. ateles,* herpes simplex, EBV, Marek disease virus, and a herpesvirus isolated from a rabbit lymphoma. Experimentally induced HVS lymphoproliferative disease, particularly in marmosets, resembles in many respects the lymphoproliferative disease in susceptible primates as a result of EBV infection (Deinhardt *et al.*, 1973). In both instances, free virus or intracellular virus particles are not readily demonstrated in normal or tumor tissues at various stages of the disease. Epstein (1972) and Ablashi and Pearson (1974) suggest that HVS infection of the squirrel monkey may serve as a model for the study of the role of EBV in human neoplasia.

Although the squirrel monkey was not included in the study, a survey of simian sera for antibody to EBV failed to demonstrate this antibody in the New World monkeys (*S. oedipus, A. palliata*) examined (Kalter *et al.*, 1972b). Dunkel *et al.* (1972), in their EBV survey of New World monkeys, which included squirrel monkeys, also failed to detect seropositives.

Preliminary information indicated that HVS was *not* shed in oropharyngeal secretions such as saliva, nor was the virus cell-free. A study by Falk *et al.* (1973) clarified this question by demonstrating that, contrary to previous findings, not only is virus shed in pharyngeal washings of healthy animals, but degenerating salivary gland cells release infectious virus. Further substantiation of virus shedding may be found in the report of Barahona *et al.* (1975). HVS was naturally transmitted from squirrel monkeys to an owl monkey housed in the same cage. One may speculate on this mechanism as a possible method for transmission of oncogenic viruses and development of tumors in humans and other animals.

Pertinent to understanding oncogenesis is the question of how viruses relate to this problem. Are viruses directly responsible for induction of the tumor, or do they interact with another virus; for example, a type C endogenous virus (or other factors)? It has previously been demonstrated that avian leukosis virus (RAV-2) infection of chicken embryo fibroblast cultures interacts with Marek disease herpesvirus (Frankel and Groupe, 1971). The studies of Neubauer *et al.* (1974) demonstrated that HVS-infected lymphoid cells could be stimulated to express virus; however, these stimulated cultures showed no evidence of type C virus. This virus expression is similar to that observed in EBV-infected Burkitts lymphoma cells. Electron microscopic examination of mink lung cells previously inoculated with a squirrel monkey throat swab revealed the simultaneous infection of a single cell with the squirrel monkey retrovirus (see Section 4.4.7) and a herpesvirus that was neutralized by anti-HVS serum. These findings indicate that a type D retrovirus and a herpesvirus will proliferate in the same cell. The significance of this finding is not known, but could be another example of interaction between two viruses in the process of tumorigenesis (G. C. Smith *et al.*, 1978). It was also demonstrated that baboons inoculated with HVS showed increased cellular and humoral immunity to purified whole baboon virus as measured by lymphocyte transformation and radioimmunoprecipitation (Kalter *et al.*, 1978a).

Another aspect of HVS infection of squirrel monkeys is associated with their response to primary infection. As in the human following infection with herpes simplex, little if any clinical disease is noted. The antibody response will influence the initial infection; however, this humoral antibody is not capable of eliminating the latent infection. Although virus has been reported in the oropharynx (Falk *et al.*, 1973), the exact relationship of antibody to virus shedding is not clear. Martin and Allen (1975) noted that altering the immune response of the natural host does change the course of infection. HVS could be isolated from juvenile squirrel monkeys only in the absence of antibody. Immunosuppression permitted the expression of the virus because of its inhibition of antibody development. Thus, the expression of the virus is related to the duration of the infection and the status of the antibody. When antibody and other immune factors decline, virus shedding occurs. There is immunological control of experimental virus-induced tumors in primates. In RNA virus infections, both humoral and cellular responses are detected; however, if the animal is immunologically deficient, tumor growth is accelerated and regressing tumors recur. Oncogenic herpesvirus (DNA virus) infections, which cause minor or no disease in the natural host (although many of these animals are clinically infected with HVS), may induce lymphomas or lymphoblastic leukemias in susceptible primate species. The immune response of the natural host is greater than that of the susceptible host and is associated with the antibody response to the early antigen (EA) (Deinhardt *et al.*, 1975; Pearson and Davis, 1976). Klein *et al.* (1973) examined the antibody reactions to HVS in experimentally infected squirrel, marmoset, and owl monkeys. These studies showed that healthy squirrel monkeys had antibodies to late antigens (LA) but not to EA. Squirrel monkeys reared in isolation, and antibody-negative, responded with LA and EA antibodies following HVS inoculation. These animals shed virus that caused infection in uninoculated cage-mate squirrel monkeys, who in turn responded with development of LA and EA antibodies. Squirrel monkeys developed antibody within 2–3 weeks, whereas the other species did not respond until 28–80 days postinoculation. This response may be partly responsible for the resistance of squirrel monkeys to the oncogenic effect of HVS.

Natural HVS infection of the squirrel monkey is similar in pathogenesis to EBV in humans, although quite distinct virologically. An interesting corollary of this relationship was reported by Shope and Miller (1973). These investigators noted that squirrel monkeys given a series of autologous EBV-converted lymphoblasts developed the heterophile antigen of infectious mononucleosis on the cell membrane of transformed cells as well as EBV-specific antibodies. More recently, Pizza *et al.* (1982) demonstrated that, although natural antibody to HVS was present in squirrel monkeys, there was no evidence of cell-mediated immunity (CMI). Immunization with HVS did not alter CMI in this species, but did in the rhesus monkey, which originally had neither natural antibody nor CMI. In this context, passive immunization can protect against tumor development due to HVS infection (Laufs and Steinke, 1975). In a comparison of lymphocyte responsiveness to infection with HBS by cebus and marmoset monkeys with that of squirrel monkeys, it was found that the squirrel and cebus monkeys are comparable in their ability to respond to mitogens and to HVS antigen. The marmosets receiving HVS developed typical lymphoproliferative disease characteristic of this virus, indicating a depressed immunological responsiveness (Harvey *et al.*, 1974; Kalter *et al.*, 1976).

Investigations of herpesvirus infections have helped provide an insight into the host response to viral neoplasia. Although the entire picture is not clear, it would appear that an immunologically competent host will reject sarcomas resulting from RNA virus transformation of fibroblastic cells. Lymphomas or leukemias resulting from herpesviruses are usually fatal because of the immunosuppresive capabilities of this virus group. In the natural host, neoplasia does not occur, because of the rapid immunological response to the infection. If tumors do develop in the natural host, this could be considered an immunological accident (Deinhardt *et al.*, 1975).

Any discussion of the host response to herpesvirus infection cannot be complete without some reference to herpesvirus vaccination as well as some insight into the antigenic structure of herpesviruses, particularly HVS. Attempts to develop herpesvirus vaccines are not new. A herpesvirus vaccine against Marek disease has been successful (Biggs *et al.*, 1972; Churchill *et al.*, 1969). Vaccines against herpes simplex virus have been discussed (Hilleman, 1976; Wise *et al.*, 1977). Other vaccines against one or another of the herpesviruses have also been proposed, including one against HVS (Laufs, 1974; Laufs and Steinke, 1975), but the consensus is that herpesvirus vaccination raises a number of still unresolved problems. There is no question regarding the ability of various vaccines to raise antibody levels, but, as is evident to those with recurrent herpes simplex (cold-sore) infections, antibody level is obviously not the answer. Numerous studies are currently in progress attempting to use subunit antigenic preparations as vaccines. The success of these as vaccines is still to be determined. Such preparations have been described for most of the herpesviruses, including HVS (Randall and Honess, 1980, 1982).

4.4.2. *Adenoviruses (Adenoviridae)*

These viruses, counterparts of the human adenoviruses, have the same biological characteristics as, but are antigenically distinct from, human adenoviruses. For a description of their classification and characteristics, see reports by Kalter *et al.* (1980), Norrby *et al.* (1976), Wigand *et al.* (1982), and Willimzik *et al.* (1981). Heberling (1972) provides a detailed report on simian adeno viruses and their sources, including two squirrel monkey isolates (SqM-1, SqM-2).

Adenoviruses are readily isolated and like herpesviruses are present in all monkeys and apes thus far examined. Squirrel monkey isolates frequently have been recovered from throat and fecal specimens as well as from latent infections of kidney cells. Virus excretion is most frequent in newly captured animals, particularly upon arrival in the laboratory. Antibody to SqM-1 is found in virtually 100% of tested squirrel monkeys. Although this virus is apparently highly infectious, pathogenicity is unknown. A number of the human and simian adenoviruses, however, are oncogenic in rodents (Heberling, 1972; Hull, 1968). Ad 12 (the only adenovirus studied for this capability) produces retinoblastomas when inoculated into the eye of the newborn baboon (Mukai *et al.*, 1980). It is not known if any of the simian adenoviruses inoculated under similar conditions would produce retinoblastomas. Limited studies for the presence of adenovirus antibody in squirrel monkey sera to an adenovirus group antigen (complement-fixation test) failed to show any seropositives, suggesting lack of infection with these viruses (Kalter and Heberling, 1971) These findings contrast with surveys using SqM-1 antigen. As complement-fixing antibody disappears rather rapidly, this may explain the apparent discrepancy.

4.4.3. *Arboviruses (Togaviridae, Bunyaviridae)*

The arthropod-borne viruses are included in two families: Togaviridae and Bunyaviridae. The Togaviridae consist of the *Alphavirus* (arbovirus group A), *Flavivirus* (arbovirus group B), rubella virus (*Rubivirus*), and *Pestivirus* of pigs and cattle. The latter two genera are *not* arthropod-borne. Included in the Bunyaviridae family are four genera: *Bunyavirus* (Bunyamwera supergroup), *Phlebovirus* (sandyfly fever group), *Nairovirus* (Nairobi sheep disease), and *Uukuvirus* (Uukuniemi). One would suspect, considering the geographic proximity of these viruses and simians, that arboviruses would be one of the more commonly isolated viruses of nonhuman primates. With few exceptions, however, little is known about arbovirus infections of nonhuman primates (Felsenfeld, 1972). Of the vast number of recognized arboviruses, only a few have been reported in nonhuman primates. Several viruses have been reported in one or another monkey; for example, Kyasanur Forest disease in *Macaca* spp., but not in other simians. See Section 4.11.1 for other arbovirus infections.

Yellow fever (*Flavivirus;* arbovirus group B) has been one of the more extensively studied arboviruses. *Urban* yellow fever involves humans and mosquitoes without an animal reservoir. *Sylvatic* (forest) yellow fever only incidentally involves the human and is primarily a disease of nonhuman primates, both in the Old and New World. The squirrel monkey is among the many species of monkeys involved in sylvatic yellow fever. Fatalities are frequent, usually because of liver damage. Antibody is found in serum obtained from adult animals, indicating that not all infected animals succumb. Davis (1930) successfully produced the disease in squirrel monkeys by inoculation with blood of affected individuals or by infected mosquito bites. Lesions occurring in these animals resembled those seen in the human. Pathologically, this included liver (orange or yellowish in color) necrosis and fatty peripheries, which, upon microscopic examination, showed completely disorganized structure with infiltration of mononuclear and polymorphonuclear cells. Blood spaces were congested, with evidence of petechial hemorrhages, and areas of fatty infiltration were apparent. Kidneys showed injection of blood vessels and suggestions of cloudy swelling and a few early casts. At times the intima of the aorta was involved. Similarly, Bates (1944) was able to transmit the disease to the squirrel monkey when it was reported that this species is more susceptible to yellow fever virus than is the white mouse. Clinically, the response in *Saimiri* could be categorized as "mild, moderate, or severe." Passage through these animals enhances pathogenicity. The liver pathology generally was not as intense as occurs in the human nor as described by Davis (1930) in his studies. Virulence of the virus must be variable, because in Davis' study (1930) a high mortality rate (80%) was observed. This does not seem to be what occurs in nature, because the squirrel monkey population is rather high in yellow fever endemic areas. In a study to determine the efficacy of sequential immunization (live 17 D yellow fever, Langat and dengue 2 viruses) against group B antigens, Price *et al.* (1974) observed that the squirrel monkey was protected against challenges with dengue 1 and St. Louis encephalitis viruses as determined by the absence of viremia and lack of an anamnestic antibody response. "Natural" antibody to yellow fever but not to Ilheus, St. Louis encephalitis, or Bussuquara viruses had previously been reported by Bensabath *et al.* (1966).

Other arboviruses studied included dengue viruses in the squirrel monkey without clinically apparent disease (Sabin, 1959). Nathanson and Harrington (1966) reported that the squirrel monkey developed a viremia with Langat virus as well as with Japanese encephalitis virus. Naturally occurring HI antibody to dengue 2, Langat, West Nile, Japanese encephalitis, Ilheus, and yellow fever viruses was not detected in the seven squirrel monkeys examined. However, high antibody titers developed to both Langat and Japanese encephalitis viruses following experimental infection. In a serological survey done in this laboratory (Kalter and Heberling, 1971), complement-fixing antibody was not found to Eastern, Western, or St. Louis encephalitis viruses nor to Colorado tick virus in squirrel monkeys. In another study, HI antibody was not detected to Bunyamwera, Buttonwillow, and dengue 1 and 2 viruses, but was found to Uganda, LaCrosse, Trivitatus, Sindbis, Chickungunya, Semliki forest, and Turlock viruses (Kalter and Heberling, 1972).

Because of antigenic crossing among many of these arboviruses, the exact significance of the findings must be carefully considered. It should be noted that experimental inoculation of monkeys often fails to elicit disease or measurable antibody. Each simian species, as well as each individual animal, may react differently. For a review of arbovirus infections in other New World monkeys see Dunn (1968).

A number of attempts have been made to simulate the known human transmission of rubella virus to an experimental animal *in utero*. The squirrel monkey has been used, but demonstrates no evidence of transplacental infection (Schiff and Sever, 1966). Those studies should be repeated, as there is evidence for successful transmission of rubella as well as other viruses (including a number of Togaviridae) to the animal fetus.

4.4.4. *Picornaviruses (Picornaviridae: Enterovirus, Rhinovirus, Cardiovirus, Apthovirus)*

The original simian picornaviruses were recovered from *Macaca* sp., African green monkeys, and baboons, and only limited information is available on New World species. Counterparts of human rhinoviruses have not been reported in nonhuman primates. Rhinovirus infections of chimpanzees have been limited to human strains (Dick and Dick, 1968). A *Cardiovirus* (encephalomyocarditis) infection of primates, including the squirrel monkey, has been reported (Roca-Garcia and San Martin-Barberi, 1957).

Antibody surveys of squirrel monkey sera for evidence of infection with selected picornaviruses failed to show any detectable antibody to coxsackievirus A20, echoviruses types 3, 4, 7, 11, 12, and 13, or SV45 (Kalter and Heberling, 1972). As noted, encephalomyocarditis (EMC), a *Cardiovirus*, had been isolated from a dead squirrel monkey that may have had contact with experimentally inoculated mice. Regardless of the source, the isolation of virus from various tissues of this animal suggests susceptibility (Roca-Garcia and San Martin-Barberi, 1957). Necropsy findings on this animal showed congestion of lungs along with pleuritic fluid. Evidence of a myocarditis with pulmonary edema was observed microscopically. Virus could be isolated from lungs, liver, spleen, and kidney. Recently this virus was suspected of killing 104 squirrel monkeys and three capuchin monkeys. Various laboratory studies, however, failed to confirm this suspicion (Kessler *et al.*, 1982).

Burch *et al.* (1982) demonstrated the susceptibility of the squirrel monkey to coxsackievirus B4 infection. Intravenous or intraperitoneal inoculation resulted in the development of mitral valve and aortic valve lesions. The squirrel monkey developed viremia and seroconverted following inoculation with foot-and-mouth virus types O, A, and C. Although no clinical disease was observed, nor could horizontal transmission be demonstrated, virus was found in the oropharynx (Schudel *et al.*, 1982).

An interesting aspect of picornavirus studies in nonhuman primates is the failure to appreciate fully the usefulness of various species of monkeys, particularly in the development of the poliomyelitis vaccine and its testing. Although it was recognized early in the program that a wide range in susceptibilities existed among the different species of monkeys and apes, only a small number were actually evaluated. Among those successfully used, albeit to a very limited extent, was the squirrel monkey. The evolution of the poliovirus vaccine is well known, as is the use of nonhuman primates in its development

(Paul, 1971). As early as 1908 the susceptibility of nonhuman primates (chimpanzee, baboon, rhesus) to poliovirus was recognized (Landsteiner and Popper, 1908, 1909). Involvement of various simian species in the study of poliovirus pathogenesis and particularly isolation, cultivation, and characterization of the virus types is a hallmark in virology. However, although some comparative studies on species susceptibility to polioviruses were undertaken, by and large the choice of experimental animal was based on availability. As pointed out by Paul (1971), of the monkeys and apes investigated for susceptibility, it was clearly shown that Old World species were more susceptible than were New World animals. This led to the choice of the rhesus monkey for many of the poliovirus studies. Actually, the infection in chimpanzees resembles most closely the human situation; however, limited availability of chimpanzees as well as the recognition that rhesus monkey kidney cells were suitable for cultivation of the virus led to the popularity of this latter species. Although other studies showed that various New World monkey kidney cells were susceptible to poliovirus infection, availability of macaques suggested continued use of that species.

4.4.5. Poxviruses (Poxviridae)

Poxviruses of vertebrates are in the subfamily Chordopoxvirinae, containing six genera: *Orthopoxvirus, Parapoxvirus, Avipoxvirus, Capripoxvirus, Leporipoxvirus,* and *Suipoxvirus*. The host range of these viruses in nature is extremely narrow, usually limited to a single animal species. However, crossing of species does occur and is well recognized. Primates are usually involved only with the *Orthopoxvirus* and *Parapoxvirus* genera. Monkeypox virus is considered to be a poxvirus of monkeys and is closely related antigenically to the variola–vaccinia virus complex (*Orthopoxvirus*). The reservoir species for monkey pox and its epidemiological relationship to the monkey are unknown. Tanapox and Yaba monkey tumor pox viruses, which are serologically related to each other but not to the other poxviruses, have not been allocated to specific genera and their relationship to the monkey is also not clear.

Of greatest concern to all is the status of human smallpox. Because smallpox has been technically eliminated from the human population (Henderson, 1975), the role of monkeys as a possible nonhuman primate reservoir for this disease becomes increasingly important (Arita and Henderson, 1976). Monkeypox virus and resulting human cases have principally been restricted to Africa. Monkeypox virus and poxvirus infections in nonhuman primates have been reviewed (Arita and Henderson, 1976; Cho and Wenner, 1973; Fenner, 1979; Hahon, 1961).

Hahon (1961) in his review indicates that the squirrel monkey, among many other simian hosts, is susceptible to variola virus. Inhalation of this virus results in a generalized exanthem with fever. In this review, Hahon questions (1) the extent of simian involvement, and the need for serological surveys to determine this, and (2) the potential for an infected monkey to serve as a source of human infection. Although only limited studies were reported in New World monkeys (Hahon, 1961), the clinical manifestations with vaccinia and alastrim in all test animals were essentially the same, i.e., a localized lesion. Some species differences were noted, including encephalitis (rarely), fever, and death. It was emphasized that differences in susceptibility exist among the species and that New World monkeys may be more susceptible than are Old World animals. The histopathology of the individual lesion induced was in most ways similar, although generalized infection usually did not occur with vaccinia or with alastrim. Intracytoplasmic inclusion bodies (Guarnieri bodies) were observed in the cells of the lesion. Monkeypox occasionally induced eosinophilic intranuclear inclusion bodies, but never concurrently in a cell with an intracytoplasmic inclusion body.

Cross-protection studies indicated that a number of factors were involved in the host response. Cross-immunity was strong, but not necessarily completely reciprocal. Vaccinia virus protected against both viruses and variola was completely protective against itself, but protection varied from partial to complete against vaccinia. Variability in other strains was noted as well. It would appear that dose of inoculum was important, as were strains employed, species of monkey used, and site of inoculation.

Both New and Old World simian sera have been examined for antibody to various poxviruses (Arita *et al.*, 1972; Breman *et al.*, 1977; Kalter and Heberling, 1971, 1972; Marennikova *et al.*, 1975; McConnell *et al.*, 1962; Noble, 1970; Sehgal and Ray, 1974). The results are difficult to evaluate, but current thinking suggests that while poxvirus infection occurs in nature, these infections do not constitute a threat of smallpox in humans. As indicated, interpretation of serology results is still in need of critical evaluation. Among the New World monkeys, squirrel monkeys were found with antibody to vaccinia-monkeypox viruses (Kalter and Heberling, 1972). A fatal case of monkeypox in a squirrel monkey was reported by Peters (1966), confirming the possibility of poxvirus seropositives. Typical pox symptoms and virus isolation were noted.

Yaba virus, first described by Bearcroft and Jamieson (1958), is another poxvirus responsible for the production of skin lesions (tumors) as well as lesions on internal organs. These tumors were classified by Bearcroft and Jamieson (1958) as histiocytomas. In general, tumors developed only on Asiatic monkeys (*Macaca* sp.) and not on African or New World monkeys (including squirrel monkeys) (Ambrus *et al.*, 1963). These histiocytomas may be characterized by perivascular collars of plasma cells, lymphocytes, histiocytes, and some eosinophilic leukocytes. Tumors occur about 2 weeks after inoculation in the dermis and subcutaneous tissues. The proliferating cells are elongated, with spindle-shaped nuclei with scanty cytoplasm, and closely resemble fibroblasts. Cytoplasmic (basophilic staining) inclusions appeared early (third day) but are relatively inconspicuous until the third week and thereafter. Tumor degeneration occurs without zones of necrosis except in relation to ulceration, with little residual scarring of the skin. The presence of multinucleated cells appears to be the end stage of incompletely divided cells with many large inclusions present (Sproul *et al.*, 1963).

Squirrel monkeys can be infected with Yaba virus (Akers and Sheek, 1975). In this study, illness became apparent 2–4 days postinfection, with tumors developing approximately 2 weeks following signs of illness. Virus could be recovered from plasma and white cells by inoculating CF-1 cells. Another poxvirus, originally referred to as OrTeCa (Oregon Texas California) virus or Yaba-like virus, 1211 virus, was described by several groups of investigators following outbreaks of a benign epidermal pox disease in various colonies of macaques (Crandell *et al.*, 1969; Espana, 1971; Hall and NcNulty, 1967). The disease had first occurred in humans in 1957 and 1962 along the Tana River in Kenya (Downie *et al.*, 1971). Nicholas (1970) examined squirrel monkeys, among other uninfected simian species, for antibody to this virus, with negative results.

Monkeypox virus has not been described in New World monkeys, nor is there any evidence for its natural existence in simians derived from Central and South America except for the outbreak occurring in the Blijdorp Zoo (Peters, 1966). In this outbreak, one of three squirrel monkeys, along with gorillas, chimpanzees, a gibbon, and four owl-faced monkeys (*Cercopithecus hamlyni*), did become infected. The squirrel monkey demonstrated multiple vesicles and virus was isolated from the lesions. Monkeypox was first described by von Magnus *et al.* (1959) in cynomolgus monkeys after an occurrence of this disease in 1958. Shortly thereafter, similar outbreaks were reported in a number of *Macaca* colonies (rhesus, cynomolgus). As indicated above, the human outbreaks occurring in Africa are presumed to be monkey-derived. The relationship of the macaque monkeypox virus to African monkeypox strains, and the possible epidemiological considerations, have never been determined.

The immunological relationships of these viruses were studied by Nicholas (1970). She reported that OrTeCa (Tanapox virus) poxvirus shared antigens with Yaba virus, but no cross-protection was observed with vaccinia and monkeypox viruses.

Molluscum contagiosum, another poxvirus, once considered to be specific for the human, has been reported in chimpanzees, but no other nonhuman primate (Douglas *et al.*, 1967). The disease is nonfatal and the pox appear as papular (papillomatous) elevations.

4.4.6. *Influenza Virus (Orthomyoxviridae)*

These viruses are influenza virus A, B, and possibly C. Of these, influenza A infects a wide variety of animal species, including the nonhuman primate. Influenza B strains are primarily infectious for the human, although there is some evidence for animal infections. Little information concerning susceptibility is available on animal infections with influenza C. Influenza A viruses have been reported to occur naturally in various animal and avian species. Few attempts have been made to ascertain occurrence of these viruses in monkeys and apes. It has been demonstrated that nonhuman primates (chimpanzees, baboons) do become naturally infected with influenza viruses and respond with clinical disease as well as antibody development (Kalter and Heberling, 1978).

A number of simian species, principally macaques, have been used for influenza studies (Saslaw and Carlisle, 1969); however, use of the squirrel monkey for study of these viruses is relatively recent. Serological studies for influenza A and B showed that a number of squirrel monkeys possessed antibodies. Of interest was the finding of a progressive increase in number of seropositives to the newer strains. Thus, the greatest number of seropositives were for the A_2 strain, with fewer for the older PR8 and FM1 strains. Only one of four colonies of squirrel monkeys was found with antibody to influenza B (Kalter and Heberling, 1971).

Squirrel monkeys immunosuppressed with cyclophosphamide developed an encephalitis following intracerebral inoculation with either of the neurotropic strains NWS or WSN. Sham-inoculated, immunosuppressed animals or animals given virus alone did not develop clinical symptoms (Miyoshi *et al.*, 1971). In another study, Henry *et al.* (1970) observed that influenza A/PR/8 virus was able to kill squirrel monkeys if followed by 10 ppm NO_2. Less NO_2 (5 ppm) resulted in a decreased mortality. This latter level of NO_2 did not influence the development of HI antibody, but apparently slowed the initial production of serum-neutralizing (SN) antibody. However, after 133 days, no significant differences were observed in the production of HI, SN antibody, or other physiological values by the test or control animals (Fenters *et al.*, 1971).

Berendt (1974) was able to infect rhesus monkeys with the Aichi strain of type 2 influenza virus. The animals responded with virus shedding and seroconversion. Rechallenge resulted in no virus replication, but a serological anamnestic reaction occurred. As a followup of studies in the rhesus, Berendt *et al.* (1975) examined the response of squirrel monkeys to influenza virus and *S. pneumoniae* infections. The intratracheal inoculation of influenza type A (Aichi/2/68) or *S. pneumoniae* produced clinical illness. However, when the two organisms were given sequentially, a severe disease resulted. As this infection was reproducible, it was suggested that the squirrel monkey could be used in the study of sequential infections with other bacteria. Infection with influenza A/New Jersey/8/76 (Swine-H_{SW}1N1) was also acomplished in squirrel monkeys (Berendt and Hall, 1977). The disease was similar to that seen with other types of influenza A viruses. Stephen *et al.* (1977, 1978) then noted that the squirrel monkey may be used to determine the efficacy of aerosol therapy. Ribavirin therapy significantly reduced the clinical signs (sneezing, coughing, nasal discharge) of monkeys when given 6 hr after exposure to the virus. Amantadine treatment that started either 24 hr before or 48 hr after intratracheal inoculation of virus caused a significant decrease in days of virus shedding as well as in severity of illness.

An interesting aspect of this study was the finding that there was a lack of correlation between HI antibody titer following vaccination and protection. Vaccinated animals given either whole virus or split virus, however, had significantly less illness than unvaccinated controls. Anamnestic reactions were observed only in those squirrel monkeys receiving the whole vaccine (Berendt and Scott, 1977). Apparently, cell-mediated immunological factors are associated with influenza infection, as these factors could be observed in the squirrel monkeys in this study. All the monkeys vaccinated with the whole virus had an immunological recognition and memory mediated by T lymphocytes for the whole antigen, but not for the split antigen.

In a similar study, Murphy *et al.* (1980) compared the responses of humans and three species of New World monkeys (owl, cebus, squirrel) to strains of influenza type A viruses. The squirrel monkeys, like humans, responded to the three strains of test viruses with clinical disease and virus shedding. The other two simian species remained clinically well, although they became infected. In another study by this group (Murphy *et al.*, 1982), an attempt was made to compare in squirrel monkeys the replication and virulence of avian strains with those of human type A influenza strains. The ten avian type A influenza viruses exhibited a spectrum of virulence and replication. Two avian viruses resembled the human virus, whereas five of the avian virus strains were clearly attenuated. As a consequence of both these studies, it was suggested that use of the squirrel monkey permits investigation into the genetic basis of virulence as well as providing a mechanism to determine the genes involved in the restriction of replication observed with attenuated viruses. It was also suggested that this type of study could permit the development of a live attenuated avian influenza virus for use in humans.

Influenza pathology in the squirrel monkey is characterized by extensive consolidation of the lung. Microscopically, there is a loss of cilia in the bronchi and trachea, some distortion of the mucosa, and minimal infiltration of polymorphonuclear neutrophils of the submucosa. Hyperplasia of the alveolar and bronchiolar epithelium may be observed, and in some areas the alveoli adjacent to bronchioles contain an exudate of macrophages, neutrophils, and lymphocytes in a pale eosinophilic homogeneous material. Peribronchiolar and perivascular foci of lymphocytes with small numbers of macrophages may be seen. Deaths are usually associated with a bacterial etiology.

4.4.7. Retroviruses (Retroviridae) [RNA Tumor (Oncoviruses) and Related Viruses]

Three subfamilies, Oncovirinae (RNA tumor virus group, type C oncovirus group, type B oncovirus group), Spumavirinae (foamyvirus group), and Lentivirinae (Maedi/Visna group), make up this virus family. Interest has been centered principally around the Oncovirinae because of their association with tumor development as well as other disease processes. However, some retroviruses have thus far not been found to be pathogenic. There are two oncovirus genera, type B and C oncovirus groups, and one proposed genus, type D retrovirus group. Spumaviruses (foamyviruses) have been reported in various animal species (cattle, rats, human and nonhuman primates). The Maedi/Visna group has thus far been reported only as natural infections in sheep and will not be discussed. For detailed discussion of all these well-described Retroviridae pertinent to primates, see Deinhardt (1980), Fine and Schochetman (1978), Gallo and Wong-Staal (1980), Heberling (1983), Kalter *et al.* (1979), Matthews (1982), and Rangan and Gallagher (1979).

The retroviruses are transmitted vertically as well as horizontally. As endogenous viruses, they are genomic and inherited as Mendelian genes. Types B and C are distinguished by an acentric central nucleoid in type B and a concentric central nucleoid in type C. Both type- and group-specific determinants are recognized; enveloped glycoproteins are type-specific, whereas the major internal *gag* group-specific protein determinants define the subgenera. All retroviruses initiate replication by means of a reverse transcriptase, which transcribes virus RNA to DNA. Type C oncoviruses have been seen in the placentas of most primates, including the squirrel monkey. Isolation of these agents, however, has been limited to only a few species. No isolations of this virus group have been made from the squirrel monkey. Type B oncoviruses have not been reported in nonhuman primates, but type D has been recovered from primates, including squirrel monkeys (Heberling *et al.*, 1977).

In addition to the studies with DNA viruses, principally herpesviruses, RNA viruses such as Rous sarcoma virus (an avian type C oncovirus) have been inoculated into a variety of simians. Rabin and Cooper (1971) demonstrated that squirrel monkeys responded with tumors following Rous sarcoma virus inoculation. Solid fibrosarcomas developed at the sites of inoculation (9/12 animals) after an average latent period of 37.2 days. There was no indication of metastasis in spite of the large tumor size. Although one animal died, regression occurred in most instances. Older animals were refractory even after treatment with cyclophosphamide. However, a 5-year-old male treated with anti-squirrel monkey thymocyte globulin developed a tumor 35 days postinoculation. Virus particles could be seen in tumor tissue by electron microscopy but could not be reisolated. In another study, Rabin *et al.* (1972) reported that the newborn squirrel monkey develops undifferentiated sarcomas 14–15 days after inoculation with the ST strain of feline sarcoma virus (ST-FeSV). The latent period in older animals was longer (45 days) and all tumors regressed within 1–2 weeks. The cells were rounded and occasionally spindle-shaped and had prominent nuclei with moderate amounts of fairly granular cytoplasm. The cell pattern was one of closely crowded cells with imperceptible cell boundaries. Giant cells were present throughout the tumor mass and many mitotic figures were observed. Neutrophils could be seen among the neoplastic cells.

A type C simian virus, isolated from a woolly monkey (simian sarcoma virus) with a fibrosarcoma (Theilen *et al.*, 1971), was inoculated into four species of nonhuman primates, including squirrel monkeys (Theilen *et al.*, 1973). Of the four newborns (less than 6 days old) and two fetuses (inoculated 11 and 15 days prior to birth), two animals developed nodules 37 and 67 days postinoculation. The C-type virus was not seen in these tissues, nor in cell cultures derived from these tissues. In contrast with the squirrel monkeys, marmosets (*Saguinus nigricollis, S. fuscicollis*) developed fibromas or well-differentiated fibrosarcomas. The C-type virus was found in these tumors (Deinhardt *et al.*, 1970).

The observation and isolation of an endogenous type C virus from baboon (Benveniste *et al.*, 1974; Kalter and Heberling, 1974; Kalter *et al.*, 1973a, 1975) and other primate placentas (Kalter *et al.*, 1973b, 1975), including placenta from the squirrel monkey (Kalter *et al.*, 1979), caused a marked increase in studies related to the role of viruses in primate oncology. Numerous reports and reviews attest to the emphasis placed on this possible relationship. Both DNA (see above) and RNA viruses are now suggested as candidate oncogenic agents. Unfortunately, after years of extensive study, this elusive relationship defies a positive conclusion. Isolation from primates of type C virus or close antigenic relatives has been limited to the baboon and related species. As indicated, however, the morphological presence of this virus has now been confirmed in tissues of practically all mammals (Deinhardt, 1980). Attempts to produce metastatic tumors in a variety of animal hosts with type C virus have thus far been essentially unsuccessful. Pseudotypes of this virus produced by cocultivation of BaEV (baboon endogenous virus)-infected canine thymus cells (Cf2Th) with a nonproducer human osteosarcoma cell line carrying the Kirsten murine osteosarcoma virus genome (NP/KHOS) did produce metastatic disease in beagle dogs, chimpanzees, cynomolgus monkeys, and marmosets, but not cebus or squirrel monkeys (Kalter and Heberling, 1976; Kalter *et al.*, 1978a,b). Recently, however, a type C retrovirus was isolated from a human tumor in two separate laboratories, one in the U. S. (Poiesz *et al.*, 1980) and one in Japan (Hinuma *et al.*, 1981). At this time, no relationship to other primate retroviruses has been demonstrated.

A type D oncornavirus was observed and isolated from squirrel monkey fetal lung tissue (squirrel monkey retrovirus, SMRV) of a stillborn, full-term male animal as well as from squirrel monkey placenta (Heberling *et al.*, 1977; G. C. Smith *et al.*, 1977). This virus, an endogenous virus of the squirrel monkey (Colcher *et al.*, 1977a), is morphologically similar to the Mason–Pfizer monkey virus (MPMV), an Old World (rhesus) monkey isolate. SMRV is antigenically distinct, but does share some common antigenic determinants with MPMV (Colcher *et al.*, 1977b). It may be of interest to note that this squirrel monkey virus shares gene sequences with cellular DNA of normal New World skunks (*Mephitis mephitis* and *Spilogale putorius*). This is considered to be another example of the evolutionary divergence of primate virus genes between vertebrate species that are only remotely phylogentically related (Benveniste *et al.*, 1978).

In an attempt to determine the occurrence of humoral and cell-mediated immune responses to type D retroviruses, Massey *et al.* (1978) observed that 10% of the squirrel monkeys tested had antibody to MPMV and 46% showed a blastogenic response to SMRV. Neither virus induced a blastogenic response with human or nonprimate lymphocytes. A word of caution in interpreting host–virus immune responses is expressed by Barbacid *et al.* (1980). When attempting to ascertain the immune response of humans to purified oncovirus glycoproteins, the reaction was shown to be due to cell-specified rather than virus encoded antigenic determinants and was not the result of oncovirus infection.

Foamyviruses (*Spumavirinae*) are also included among the Retroviridae. These viruses have many of the morphological and biological characteristics of the retroviruses, but have not been shown to cause any *in vivo* pathology. Although recovered from several mammalian species, including primates (squirrel monkeys), they are associated with persistent infections despite a host immune response. Recognition of their presence is by unmasking infection in cell culture with the formation of foamy vacuolated syncytia. There are at least nine primate serotypes (Hooks and Gibbs, 1975), with type 4 originally isolated from the squirrel monkey. Antibody to this type, but not to the other serotypes, has been reported in the squirrel monkey (Johnston, 1971; Kalter and Heberling, 1972).

4.4.8. *Papovaviruses (Papovaviridae)*

This family is composed of two virus genera, *Papillomavirus* and *Polyomavirus*. The family name is derived by combining the first two letters of the genera plus the first two letters of vacuolating virus (the early name for SV40). These viruses, as their names imply, are associated with tumors, although two human isolates, BK and JC, may not be oncogenic. A number of the polyomaviruses have been isolated from nonhuman primates (Matthews, 1982). Most notable of these isolates is SV40, which has been administered, inadvertently incorporated in the poliovirus vaccines, to millions of individuals, with no apparent illness thus far.

Early studies with SV40 indicated that the kidney cells of squirrel monkeys failed to show any lytic activity after 35 days of observation (Ushijima *et al.*, 1966). Attempts by Koller and Olson (1972) to produce cutaneous growths in squirrel monkeys with bovine, human, and equine papilloma viruses as well as with deer fibroma viruses were negative.

4.4.9. *Parainfluenza (Paramyxoviridae)*

Several of the Paramyxoviridae have been reported in nonhuman primates (parainfluenza 5, measles, mumps, respiratory syncytial virus) however, specific infections of the squirrel monkey with this virus family are few. Two paramyxoviruses (SV5, SV41) have been described as simian in origin (Hull, 1968); however, the presence of SV5 in dogs and humans as well as the limited occurrence of SV41 has raised questions regarding the origin as well as relationship of these two viruses to nonhuman primates. Neither of these two viruses has been reported in the squirrel monkey or in other New World simian species. Of the human viruses, measles (rubeola) has been notorious in its ability to cause highly lethal infections of both Old and New World monkey species. Infection of squirrel monkeys with this virus has not been reported. The disease in monkeys is very much like that seen in humans. Similarly, mumps has occasionally been seen in simians, but has not been studied to any significant degree (Hull, 1968).

Respiratory syncytial virus (RS) has been experimentally studied in squirrel monkeys (Belshe *et al.*, 1977). Although this virus was originally isolated from a chimpanzee with coryza (Morris *et al.*, 1956), RS virus is now recognized to be a human agent. The squirrel monkey, as well as rhesus, cebus, and chimpanzee, has been readily infected with RS virus as determined by seroconversion and virus shedding 3–7 days postinoculation. However, only the chimpanzee develops signs of clinical infection. In addition, the chimpanzee sheds a greater quantity of virus than do monkeys.

4.4.10. *Hepatitis*

Hepatitis is a clinical entity found in humans and other animal species following infection by a number of different viruses (as well as other microbial and chemical agents) involving the liver. Generally, however, hepatitis is considered to be the result of infection with hepatitis virus A (HAV), B (HBV), or nonA-nonB (NANB). The latter group (NANB) consists of at least two viruses. Hepatitis in humans is considered as an example of a virus disease in which nonhuman primates have played a significant role in furthering our understanding of the disease as well as in its final resolution. Recently Lucke and Bennett (1982) reported on an outbreak of fatal hepatitis in marmosets (*Callimico goeldii, Callithrix argentata melanura,* and *Saguinus oedipus*) due to an agent other than *H. tamarinus* or HAV. This would suggest the existence of still another hepatitis virus infectious for nonhuman primates. The lesions seen in the liver of these animals strongly resembled those of human viral hepatitis (HAV) in marmosets.

The various hepatitis viruses are unrelated but do have certain biological and clinical capabilities in common. HAV is now considered to be a picornavirus and is so classified. HBV and NANB are not included among any of the recognized virus families. For early studies on attempts to transmit human hepatitis to nonhuman primates, see Deinhardt (1970, 1976), Smetana (1972), and Anon (1975). Squirrel monkeys (among many other New and Old World primates) have been inoculated with human infectious materials with negative results. Some squirrel monkeys have been reported to be "Australia antigen"-positive, but were not tested for antibody. The finding that chimpanzees and marmosets are susceptible to HAV and HBV and the chimpanzee susceptible to HBV resulted in concerted studies on hepatitis with these species. That other species may be susceptible to HAV and HBV was indicated by a recent serological survey of nonhuman primate sera for HAV, HBs antibody, and HBsAg. Of 20 squirrel monkey sera examined, none were found with anti-HAV nor were any found with HBgAg. However, 5.0% of the squirrel monkeys had anti-HBs (Eichberg and Kalter, 1980). Other species were also found with hepatitis antibody.

Current "successful" studies, principally with HAV, in various nonhuman primates would suggest renewed efforts in species that are available and nonendangered. It is apparent that studies in the chimpanzee will be restricted by availability, to say nothing of cost. Success in transmission of HAV to the owl monkey (*Aotus trivirgatus*), with demonstration of virus in the liver as well as seroconversion, would substantiate this suggestion (Lemon *et al.*, 1982).

4.4.11. Hemorrhagic Diseases

This is another clinical entity associated with different viruses, of which Marburg disease in monkeys is probably the most infamous. Certain of the viruses responsible for some hemorrhagic diseases are classified, e.g., Lassa virus is an *Arenavirus* (Arenaviridae). Natural infection of monkeys with Lassa virus has not been reported. A rodent host is suspected and its antigenic relationship to such arenaviruses of the Tacaribe complex (natural hosts are rodents) as Junin virus (Argentine hemorrhagic fever) and Machupo virus (Bolivian hemorrhagic fever) strengthen this suspicion. Other hemorrhagic diseases include dengue, which is a group B arbovirus (*Flavivirus*) transmitted by mosquitoes, and tick-borne *Flavivirus* hemorrhagic diseases, including Omsk hemorrhagic fever and Kyasunur Forest disease. Crimean-Congo hemorrhagic fever viruses are in the family Bunyaviridae (*Nairovirus*), and are primarily tick-borne diseases. Rift Valley fever is another member of the Bunyaviridae (*Phlebovirus*). Korean hemorrhagic fever is a rodent-borne disease related to other diseases with a renal syndrome, a disease found in the Soviet Union, and "nephropathia epidemica," a disease found in Finland. These are probably all capable of infecting nonhuman primates, but there are no data to substantiate this suggestion.

Data on Marburg/Ebola viruses are too scanty for appropriate classification of these viruses into virus families. Diagnosis of these diseases is frequently accomplished by immunofluorescence with patient or animal serum added to slides containing Congo-Rift Valley-Ebola-Lassa-Marburg virus antigens, hence the acronym "CRELM." Natural infection of monkeys with the majority of the above viruses has been infrequently reported (Kyasanur forest, dengue, and Marburg). It is highly probable that the nonhuman primate is only indirectly related to these diseases and is not a natural host. A word of caution in interpreting the data on Marburg virus infections: Although several laboratories have reported antibody to this virus in a number of Old World monkey species (none in New World species), it is still not clear whether or not the serological reaction is specific (Kafuko *et al.*, 1970; Kalter, 1971; Kalter *et al.*, 1969). Other investigations have not been able to confirm these findings (Malherbe and Strickland-Cholmley, 1971; Slenczka *et al.*, 1971). Further studies are indicated, but the virulence of Marburg virus has caused investigators to be reluctant to enter into such studies.

The epidemiology of Marburg disease is also obscure; although this disease clearly was transmitted to humans by the African green monkey-vervet (*Cercopithecus aethiops*), this infectious relationship apparently was limited to the original outbreak in 1967 (Martini and Siegert, 1971; Siegert, 1972). Subsequent occurrences appeared not to involve monkeys, although this aspect has not been clearly determined. Because of the finding in both natural infection (vervets) and experimental infection of several monkeys species that infection led to 100% fatalities, nonhuman primates are not considered to be the natural reservoir.

Zlotnik (1969) provides data on the pathology of Marburg virus following experimental infection of squirrel, rhesus, and vervet monkeys as well as hamsters and guinea pigs. The macroscopic lesions were similar irrespective of the species and route of inoculation and were similar to those observed in the human disease. Most striking was finding severe degeneration and necrosis of lymphoid tissue, spleen, and liver, including large accumulations of cellular and nuclear debris. Basophilic bodies were generally found in cells surrounding the necrotic centers; however, it is not clear if these reflect debris or intracytoplasmic inclusion bodies. Experimental infection of vervet, rhesus, and squirrel monkeys was also reported by Simpson (1969). Again the clinical disease was similar in all animals regardless of the species and was uniformly fatal. Following an incubation period of 2–13 days (depending upon virus dose), during which time the animals appeared to be healthy, rapid deterioration occurred. Dosage of virus was not important in terms of lethality, nor was the route of inoculation. Viremia was observed throughout the febrile period. Death was preceded by clinical signs occurring for 1–2 days: anorexia, apathy, weight loss, petechial skin rash. The terminal stage was characterized by difficulty in breathing, coarse rales in thorax, and diarrhea. Blood collected by venipuncture failed to clot. Virus could be isolated from blood, urine, and saliva. Control (uninoculated) animals in the same room became ill and died.

Studies on primate sera done shortly after the original outbreak in humans provided confusing results (Slenczka *et al.*, 1971). Several laboratories screened primate sera for antibody to the Marburg agent (Kafuko *et al.*, 1970; Kalter *et al.*, 1969; Kalter, 1971). A large number of serum samples from African and Asian primates were positive in these studies. None of the New World monkey sera, including that from the squirrel monkey, were found with antibody. Other investigators (Malherbe and Strickland-Cholmley, 1971; Slenczka *et al.*, 1971) were unable to confirm these findings. More recently, studies from two laboratories have again shown Marburg virus antibody to be present in Old World monkey sera (Johnson *et al.*, 1981; Kalter *et al.*, in preparation). Thus the question of simian reservoirs remains unanswered, and the specificity of reaction also remains undetermined.

A hemorrhagic disease of monkeys involving *Macaca* spp. has been recognized in several primate centers. Initial studies on this disease came from the Soviet Union and the U. S. (Allen *et al.*, 1968; Lapin and Shevtsova, 1971; Lapin *et al.*, 1967; Palmer *et al.*, 1968; Shevtsova, 1967; Tauraso *et al.*, 1968a,b, 1971). An overview of the outbreaks as reported from various primate facilities was provided by Tauraso *et al.* (1971). Although highly lethal for macaques, this particular hemorrhagic disease is evidently restricted to this host species. Extensive serological studies have failed to find antibody to this virus in any other simian host, including the squirrel monkey. It is of some importance that outbreaks have not been reported in recent years.

Squirrel monkeys were inoculated with Lassa virus in an attempt to develop information on the pathogenesis of this disease (Walker *et al.*, 1975a,b). The incubation period following intramuscular inoculation was 8–18 days. Virus was isolated from virtually every tissue of sacrificed animals or those dying of infection. Lymph nodes, livers, and kidneys appeared to be target tissues. Persistent viremia developed, with antibody not detected in an animal sacrificed 28 days postinoculation. Pathological observations included necrosis in spleen and lymph nodes; myocarditis; acute arteritis; renal tubular necrosis and regeneration; hepatocytic regeneration; chronic inflammation of choroid plexus, ependyma, and meninges; and cerebral perivascular cuffing. It was suggested that the pathology showed a certain similarity with other arenavirus infections. These results in many respects resemble the disease in humans (Buckley and Casals, 1978).

4.4.12. *Slow and Exotic Virus Diseases (Subacute Spongiform Viral Encephalopathies)*

Included here are those agents of scrappie, transmissible mink encephalopathy, Kuru, Creutzfeldt–Jakob, and "other" diseases. These agents remain unclassified, principally because information on biological characteristics is inadequate. Nonhuman primates, however, have played a major role in the development of our present information. Other neurological and obscure diseases are now suspected of being viral in etiology as a result of these findings, and slow virus diseases may be caused by conventional viruses. The early history of animal inoculation in attempts to demonstrate a transmissible agent in various chronic neurological diseases is reviewed by Brody *et al.* (1967), Gibbs (1967), Gibbs *et al.* (1969), and Halonen *et al.* (1973).

The term "slow virus" was first introduced by Sigurdsson (1954; Sigurdsson *et al.*, 1957) in describing chronic diseases of Icelandic sheep. However, persistent virus infections had been recognized for some time, principally as a result of the studies by Traub (1935) on lymphocytic choriomeningitis in mice. Gajdusek *et al.* (1966, 1967) and Gajdusek (1978) reported on the transmission of kuru to young, immature chimpanzees by inoculation of patient brain material. Other animals, including squirrel monkeys, were also used in unsuccessful attempts to isolate an agent (Gajdusek *et al.*, 1967). In this early report, the similarities between kuru and scrapie were indicated and experimental infection of squirrel monkeys by inoculation with scrapie-infected mouse brain was also attempted unsuccessfully. The pathology of kuru in the chimpanzee and its similarity to scrapie were described in several reports (Beck and Daniel, 1969; Beck *et al.*, 1966). These reports suggest similarities with Creutzfeldt–Jakob disease and indicate that this disease might also be transmissible (Gibbs *et al.*, 1968). Most striking of the pathological findings in the chimpanzee was the spongy degeneration of the gray matter associated with an intense proliferation and hypertrophy of astrocytes. Demyelination was not a basic feature of the disease, nor was there any prominent evidence of an inflammatory response.

Other investigators attempting to determine the host range of transmissible mink encephalopathy (TME) inoculated a young rhesus monkey with mink brain. Although the animal remained asymptomatic for 33 months, upon sacrifice a polioencephalopathy with vacuolization of the gray matter similar to TME in the mink was observed (Marsh *et al.*, 1969). This led to additional studies using second-passage rhesus monkey brain material in the squirrel monkey (Eckroade *et al.*, 1970). Eleven months after inoculation, a fatal spongiform encephalopathy similar to that seen in the mink developed in three of four inoculated animals. Gross pathology was not apparent; however, microscopic examination revealed a severe spongiform polioencephalopathy involving most of the gray matter. In the spinal cord, a similar pathological response was observed only in the posterior horns. Loss of nerve cells was not conspicuous. In the cerebellum, vacuolization was limited to the granular layer of the cortex with focal involvement of the subjacent white matter of the folia. Areas of intense vacuolization also showed an extensive proliferation and hypertrophy of astrocytes. Again, as in kuru, demyelinization and inflammatory infiltrates were not observed. Morphological changes occurred prior to the advent of clinical signs. A triad of pathological changes characterize spongiform encephalopathies: neuron vacuolation occurs early in the disease, neuronal loss probably develops with clinical signs, and proliferation of fibrous astrocytes also occurs with the clinical disease. Additional studies by this group (Grabow *et al.*, 1973) on the squirrel monkey provide data on EEG findings following TME infection. The data indicate that TME in the squirrel monkey is similar to Creutzfeldt–Jakob disease in humans. For EEG patterns in the squirrel monkey, among other New and Old World simians, see Jurko *et al.* (1974).

Kuru and Creutzfeldt–Jakob disease were subsequently transmitted to New World monkeys—spider (*Ateles geoffroyi*), capuchin (*Cebus* sp.), squirrel (*S. scuireus*), and woolly (*Lagothrix lagothrica*)—as reported by Gibbs and Gajduzek (1973). It was suggested, as a result of the successful transmission of both natural and experimental diseases to the squirrel monkey, that the squirrel monkey could provide "a new laboratory model, more convenient than the chimpanzee, for the study of these diseases." The pathology of disease in the various positive simians is the same as the neuropathology in human patients with Creutzfeldt–Jakob disease: generalized neuronal loss, marked astrogliosis, a vacuolization of neurons, and a resulting status spongiosus of gray matter. An early comparison of patterns of slow virus diseases in various animal species was provided by Leader and Hurvitz (1972) and Fuccillo *et al.* (1974). The host range of subacute spongiform virus encephalopathies is provided by Gibbs and Gajdusek (1973). Confirmation of these findings in the squirrel monkey may be seen in reports by Zlotnik *et al.* (1974, 1976) and Cathala *et al.* (1974).

Gajdusek *et al.* (1974) reported on the successful production of a subacute spongiform encephalopathy in both chimpanzee and squirrel monkey following inoculation of human brain tissue from a patient dying with progressive dementia associated with malignant atrophic arteritis of Kohlmeier and Degos (papulosis atrophicans maligna). The clinical signs in the squirrel monkey consisted of ataxia, marked tremors of hands and legs, and incoordination. This was followed by intensification of the neurological signs, lethargy, and terminal paralysis. Histopathological findings consisted of a severe spongiform encephalopathy involving the striatum, gliosis, and moderate astrocytosis.

Schilders disease, which is a demyelinating disease resembling multiple sclerosis, was studied in the squirrel monkey by Elizan *et al.* (1974). Although membrane-bound, dense, core particles were seen in the human material by electron microscopy, no disease developed after 8 months in three inoculated squirrel, spider, or capuchin monkeys.

Gibbs *et al.* (1980) demonstrated that the viruses of kuru, Creutzfeldt–Jakob disease, and scrapie could induce disease in squirrel monkeys following feeding of brain, kidney, and spleen of infected animals. The disease was identical in all respects to that observed following intracerebral injection of these agents.

The occurrence of heterogeneic autoantibodies (autoantibodies that react with neurofilament proteins of neuron cultures prepared from several species of mammalian hosts) in sera of various primates, laboratory rodents, and sheep following experimental infection with kuru, Creutzfeldt–Jakob disease, and scrapie (natural infection in sheep) has been reported by Aoki *et al.* (1982). The overall incidence of autoantibodies in the experimentally infected animals was fairly low (high in humans with Creutzfeldt–Jakob disease or kuru and in sheep with natural scrapie). It was suggested that autoantibodies against neurofilament proteins cannot serve as clinical markers of experimental transmission of these diseases in animals. Only one of 20 squirrel monkeys inoculated with kuru reacted and none of 46 squirrel monkeys inoculated with Creutzfeldt–Jakob disease or of ten inoculated with scrapie reacted in the immunofluorescent test used to determine the incidence of autoantibodies in animal sera.

A brief remark concerning the host immune response to slow virus infection may provide the reader with some idea of the magnitude of the problem when attempting to generalize data acquired from model systems. There is great variation in susceptibility among species to infectious agents and in host immune response, particularly to these central nervous system infections. As pointed out by ter Meulen and Hall (1978), in Visna, the immune reaction is not impaired and may contribute to the development of the lesion; a humoral hyperimmune response against the agent of subacute sclerosing panencephalitis is found in the presence of a functional cell-mediated immune system; in progressive multifocal leukoencephalopathy, on the other hand, the disease develops only in an immunologically compromised host.

Also, in brain tissue of animals, including the squirrel monkey (Beal *et al.*, 1977), the pathological significance of the presence of cytoplasmic inclusions in neurons of both the dorsal horn and the supraoptic nucleus is unexplained. Other inclusions have similarly been reported in simians, although the squirrel monkeys examined were negative (Fussell and Roberts, 1979). The inclusion bodies in ureters of macaques are also of unknown relationship to disease.

Many questions still need to be answered in our attempt to understand slow virus infections. Probably most perplexing are: (1) How does the virus persist in the host tissues in defiance of the immune (and other) protective mechanisms? (2) What activates this latent infection? (3) Why does the ensuing disease follow a slow progressive clinical course? (4) How and when might a conventional virus become a slow virus?

Summarizing the various experimental efforts on slow viruses in nonhuman primates is made difficult by the voluminous amounts of material available, much of which is repetitious. Some suggested readings include *Advances in Neurology* (1975), Gajdusek (1978), Gajdusek *et al.* (1965), Hotchin (1971), and ter Meulen and Hall (1978).

4.4.13. Miscellaneous Viruses

Other viral agents have been reported in the squirrel monkey either as the result of "natural" infection or as a result of attempts to develop animal models.

Lassa virus (*Arenavirus*) has been discussed in Section 4.4.11. Another *Arenavirus*, lymphocytic choriomeningitis (LCM) virus, has been extensively studied in other simian species, but not the squirrel monkey. Although not recognized as a natural disease of nonhuman primates, the presence of LCM virus in mouse (the natural host) excreta would strongly suggest that this disease could occur in any simian species that has contact with LCM-carrying rodents. Antibody to LCM was not found in 40 squirrel monkey sera tested for this virus in the CF test; however, such antibody was found in many other simian sera (Kalter and Heberling, 1971).

Of the Rhabdoviridae, rabies (*Lyssavirus*) infection of nonhuman primates does occur, but has not been studied in any detail. Squirrel monkeys have been reported as having rabies (Richardson and Humphrey, 1971). The presence of virus was determined by FA, mouse inoculation, and negri bodies. The disease in simians appears to be similar to that seen in humans.

Several other viruses have been reported in one or another nonhuman primate either as "natural" or experimental infections, including vesicular stomatitis, reoviruses, rotaviruses, and various simian viruses. These have been to our knowledge not seen in squirrel monkeys. It is probable that squirrel monkeys can be infected with one or more of these.

5. *COMMENTS AND PERSPECTIVES*

The squirrel monkey is one of the more popular of nonhuman primates used in biomedical research and has played a leading role in the progress made against disease. Squirrel monkeys comprise approximately half of the 10,000 New World monkeys estimated to be used annually in the U. S. (Held, 1982). However, despite its popularity, and in common with most of the other nonhuman primates used in the laboratory, the squirrel monkey suffers from a lack of information regarding infectious agents occurring in nature or as a result of laboratory infection and regarding its immunological and pathological response to these infections. In reviewing primate literature, it is evident that there are major gaps in our knowledge, and even misinformation, regarding infectious agents. With more than 200 recognized simian species (Napier and Napier, 1967), the full extent of susceptibilities and variation in response among these species to infectious agents has yet to be determined (National Research Council, 1975).

Held (1981) makes a strong plea for the importance of laboratory animals in the successful execution of many health-related programs and notes the need for information on infectious diseases. Held (1981) also emphasized that reciprocal benefits are derived by animal populations from such studies. There is abundant evidence that infectious agents are responsible for numerous deaths in natural populations. Detailed examinations of the various species used in research, both in their natural habitat and in captivity, are still required to provide a more complete basic pattern of normal or baseline values for experimental hosts. J. C. Holmes (1982) recognized this problem and emphasized that "infectious disease agents may cause considerable mortality in animal populations." In addition, Holmes raised the question regarding whether disease agents produce "selective mortality" and whether such mortality is additive or compensating and suggested the need for field studies to resolve this quiestion. Chamove *et al.* (1979) made an interesting comparison of primate disease categories (clinical) with popularity, birth, death, and illness rates. Of the 33 captive simian species surveyed, the squirrel monkey, although one of the more popular of species used (listed third), ranked 15th with regard to disease morbidity and 25th in mortality.

A certain amount of baseline information has been provided on the squirrel monkey by Rosenblum and Cooper (1968) as well as by subsequent studies. However, as emphasized by Dunn (1968), the information on infectious agents is extremely limited. This lack of information principally reflects the need for research on this subject. In contemplating the collection of such data, however, several factors must be considered. Obviously, it is important to differentiate between animals under natural conditions and those in captivity. Inherent in such studies, regardless of animal location, are all the contrasting peripheral ecological influences that may exert an effect on the findings such as the nature of contact with other animals (population density, mixing of species), nutritional aspects, factors relating to geography (temperature, humidity), arboreal or terrestrial habit, and conditions of cageing. For example, Kaack *et al.* (1979) compare Bolivian and Columbian subspecies, noting apparent likenesses and differences between the two groups. Squirrel monkeys are widely distributed in Central and South America and the influence of this distribution on baseline parameters has yet to be established. This is true of other species as well. The effect of these factors on the health of the animal has been demonstrated in the baboon (Kalter, 1973). Similar ideas have been suggested by Held (1981, 1982).

When choosing the squirrel monkey for infectious disease studies, consideration should be given to the following. The squirrel monkey is host to an oncogenic herpesvirus (among other herpesviruses) and at least two adenoviruses, as well as type C and type D retroviruses. These retroviruses (and probably other viruses as well) have all the characteristics of the oncogenic retroviruses, but thus far this potential has not been demonstrated. The oncogenicity of *H. saimiri* has been well documented. On the positive side, we have a relatively small animal that is still available and about which a comparatively good deal of data have been collected. Although certain viruses are present, *H. simiae* (B virus), which is present in most macaques, is not found in the squirrel monkey. Simian viruses as a group have not been found to any great extent in New World monkeys. With few exceptions, the recognized simian viruses (principally adenoviruses and enteroviruses) have been derived from Old World (Asian, African) species. Perhaps this reflects the lack of detailed study on New World species, inasmuch as these species, as demonstrated, do have indigenous viral agents. This lack of information may also be indicative of the failure to report many recognized infections. Do these infections exist unrecognized or does this lack of information indicate a "natural" resistance to these agents and their absence?

In conclusion, an attempt has been made to provide current information on infectious diseases of the squirrel monkey as well as its immunological and pathological responses. In developing this information, it is our contention that every effort should be made to ascertain the precise nature and source of these organisms in order to develop methods for their elimination either by vaccination, therapy, or improved husbandry. Continued laboratory investigations should provide new methods to most effectively institute preventive and curative medicine. It is now apparent that the presence of many of these disease incitants reflects contact with other host species. Good husbandry practices will do much to control the spread of these infectious agents.

Provisions must be made when removing these species from their natural environment to that of the laboratory or zoo to reduce activation of latent infections as well as to reduce exposure of the animal to alien microorganisms. This will help curtail loss of life of these desired species as well as lessen any undesirable transfer of indigenous agents to other animal populations, including the human. It is evident that much needs to be learned concerning infectious agents of the squirrel monkey and its response. We know very little about diseases occurring naturally in the geographic area that serves as a source of this species. It is strongly suggested that every effort be made to ameliorate this lack of knowledge.

A number of viruses have not been discussed, as they have not been found to cause disease in squirrel monkeys. These viruses, along with many bacteria, parasites, rickettsiae, and fungi, similarly have not been described because of a lack of information. However, their potential to cause disease should not be ignored and must be given consideration when attempting to make a differential diagnosis. Obviously their ability to cause an immune response as well as pathology will need to be examined.

REFERENCES

Ablashi, D. V., and Pearson, G. R., 1974, Animal models: *Herpesvirus saimiri*, a nonhuman primate model for herpesvirus-associated neoplasia of man, *Cancer Res.* **34:**1232–1236.

Ablashi, D. V., Easton, J. M., and Guegan, J. H., 1976, Herpesviruses and cancer in man and subhuman primates, *Biomedicine* **24:**286–305.

Ablashi, D. V., Gerber, P., and Easton, J., 1979, Oncogenic herpesviruses of nonhuman primates, *Comp. Immunol. Microbiol. Infect. Dis.* **2:**229–241.

Akers, J. E., and Sheek, M. R., 1975, Serology and pathology of Yaba tumor virus in squirrel monkeys, *Abstr. Annu. Meet. Am. Soc. Microbiol.* **75:**278 (RT 15).

Albano, A., 1957, Agenti citopatogeni isolati da scimmie (*M. rhesus* e *M. cynomolgus*), *Bull. Ist. Sieroterap. Milan* **36:**564–573.

Al-Doory, Y., 1972a, Superficial mycoses, in: *Pathology of Simian Primates*, Part II (R. N. T. W. Fiennes, ed.), Karger, Basel, pp. 206–213.

Al-Doory, Y., 1972b, Systemic mycoses, in: *Pathology of Simian Primates*, Part II, (R. N. T. W. Riennes, ed.), Karger, Basel, pp. 224–242.

Al-Doory, Y., 1972c, Intermediate mycoses, in: *Pathology of Simian Primates*, Part II (R. N. T. W. Fiennes, ed.), Karger, Basel, pp. 214–223.

Allen, J. R., and Carstens, L. A., 1968, Response of *Saimiri sciureus* monkeys to total body X-irradiation: Clinical signs and pathologic changes, *Am. J. Vet. Res.* **29:**2179–2188.

Allen, A. M., Palmer, A. E., Tauraso, N. M., and Shelokov, A., 1968, Simian hemorrhagic fever. II. Studies in pathology, *Am. J. Trop. Med.* **17:**413–421.

Ambrus, J. L., Feltz, E. T., Grace, J. T., Jr., and Owens, G., 1963, A virus-induced tumor in primates, *Natl. Cancer Inst. Monogr.* **10:**447–458.

Anon, 1975, Viral hepatitis, Report of a WHO meeting, *WHO Tech. Rep. Ser.* **570:**1–51.

Anver, M. R., King, N. W., and Hunt, R. D., 1972, Congenital encephalitozoonosis in a squirrel monkey (*Saimiri sciureus*), *Vet. Pathol.* **9:**475–480.

Aoki, T., Gibbs, C. J., Jr., Sotelo, J., and Gajdusek, D. C., 1982, Heterogeneic autoantibody against neurofilament protein in the sera of animals with experimental kuru and Creutzfeldt–Jakob disease and natural scrapie infection, *Infect. Immunol.* **38:**316–324.

Arita, I., and Henderson, D. A., 1976, Monkeypox and whitepox viruses in West and Central Africa, *Bull. WHO* **53:**347–353.

Arita, I., Gispen, R., Kalter, S. S., Teong Wah, L., Marennikova, S. S., Netter, R., and Tagaya, I., 1972, Outbreaks of monkeypox and serological surveys in nonhuman primates, *Bull. WHO* **46:**625–631.

Asher, D. M., Gibbs, C. J., Jr., and Lang, D. J., 1969, Rhesus monkey cytomegalovirus: Persistent asymptomatic viruria, *Bacteriol. Proc.* **69:**191(V269).

Baker, J. R., 1972, Protozoa of tissues and blood (other than the Haemosporina), in: *Pathology of Simian Primates,* Part II (R. N. T. W. Fiennes, ed.), Karger, Basel, pp. 29–56.

Barahona, H., Melendez, L. V., and Melnick, J. L., 1974, A compendium of herpesviruses isolated from non-human primates, *Intervirology* **3:**175–192.

Barahona, H., Melendez, L. V., Hunt, R. D., Forbes, M., Fraser, C. E. O., and Daniel, M. D., 1975, Experimental horizontal transmission of *Herpesvirus saimiri* from squirrel monkeys to an owl monkey, *J. Infect. Dis.* **132:**694–697.

Barbacid, M., Bolognesi, D., and Aaronson, S. A., 1980, Humans have antibodies capable of recognizing oncoviral glycoproteins: Demonstration that these antibodies are formed in response to cellular modification of glycoproteins rather than as a consequence of exposure to virus, *Proc. Natl. Acad. Sci. USA* **77:**1617–1621.

Bates, M., 1944, The saimiri monkey as an experimental host for the virus of yellow fever, *Am. J. Trop. Med.* **24:**83–89.

Beal, J. A., Cooper, M. H., and LuQui, I. J., 1977, Neuronal cytoplasmic inclusions in the dorsal horn and supraoptic nucleus of the squirrel monkey, *Saimiri sciureus, Cell Tissue Res.* **176:**37–46.

Bearcroft, W. G. C., and Jamieson, M. F., 1958, An outbreak of subcutaneous tumours in rhesus monkeys, *Nature* **182:**195–196.

Beck, E., and Daniel, P. M., 1969, Degenerative diseases of the central nervous system transmissible to experimental animals, *Postgrad. Med. J.* **45:**361–370.

Beck, E., Daniel, P. M., Alpers, M., Gajdusek, D. C., and Gibbs, C. J., Jr., 1966, Experimental "kuru" in chimpanzees. A pathological report, *Lancet* **2:**1056–1059.

Belshe, R. B., Richardson, L. S., London, W. T., Sly, D. L., Lorfeld, J. H., Camargo, E., Prevar, D. A., and Chanock, R. M., 1977, Experimental respiratory syncytial virus infection of four species of primates, *J. Med. Virol.* **1:**157–162.

Benirschke, K., and Richart, R., 1960, Spontaneous acute toxoplasmosis in a marmoset monkey, *Am. J. Trop. Med. Hyg.* **9:**269–273.

Bensabath, G., Shope, R. E., De Andrade, A. H. P., and De Souza, A. P., 1966, Recuperacion de virus amarilico, procedente de un mono centilena, en las cercanias de Belem, Brasil, *Bol. Of. Sanit. Panam.* **60:**187–192.

Benveniste, R., Sherr, C. J., Lieber, M., Callahan, R., Todaro, G., and Kalter, S., 1974, Characterization of baboon type C virus, *Abstr. Annu. Meet. Am. Soc. Microbiol.* **74:**233(V195).

Benveniste, R. E., Wilson, G. L., and Todaro, G. J., 1978, Skunks have gene sequences in their cellular DNA related to squirrel monkey retrovirus: Transmission between species of a New World primate endogenous type D retrovirus, *Biochem. Biophys. Res. Commun.* **81:**1363–1369.

Berendt, R. F., 1974, Simian model for the evaluation of immunity to influenza, *Infect. Immunol.* **9:**101–105.

Berendt, R. F., 1978, Relationship of method of administration to respiratory virulence of *Klebsiella pneumoniae* for mice and squirrel monkeys, *Infect. Immunol.* **20:**581–583.

Berendt, R. F., and Hall, W. C., 1977, Reaction of squirrel monkeys to intratracheal inoculation with influenza/A/New Jersey/76 (Swine) virus, *Infect. Immunol.* **16:**476–479.

Berendt, R. F., and Scott, G. H., 1977, Evaluation of commercially prepared vaccines for experimentally induced type A/New Jersey/8/76 influenza virus infections in mice and squirrel monkeys, *J. Infect. Dis.* **136**(Suppl.):S712–S717.

Berendt, R. F., Long, G. G., and Walker, J. S., 1975, Influenza alone and in sequence with pneumonia due to *Streptococcus pneumoniae* in the squirrel monkey, *J. Infect. Dis.* **132:**689–693.

Berendt, R. F., Knutsen, G. L., and Powanda, M. C., 1978, Nonhuman primate model for the study of respiratory *Klebsiella pneumoniae* infection, *Infect. Immunol.* **22:**275–281.

Berendt, R. F., Schneider, M. A., Young, H. W., and Frola, F. R., 1979, Protection against *Klebsiella pneumoniae* respiratory tract infection of mice and squirrel monkeys given kanamycin by aerosol and injection, *Am. J. Vet. Res.* **40:**1231–1235.

Bergeland, M. E., Barnes, D. M., and Kaplan, W., 1970, Spontaneous histoplasmosis in a squirrel monkey, *Primate Zoonosis Surveillance,* Report no. 1, U. S. Department of Health, Education and Welfare, Center for Disease Control, Atlanta, Georgia, p. 10.

Biggs, P. M., Jackson, C. A. W., Bell, R. A., Lancaster, F. M., and Milne, B. S., 1972, A vaccination study with an attenuated Marek's disease virus, in: *Oncogenesis and Herpesvirus* (P. M. Biggs, G. de-The, and L. N. Payne, eds.), International Agency for Research on Cancer (IARC Sci. Pub. no. 2), Lyon, pp. 139–146.

Black, P. H., Hartley, J. W., and Rowe, W. P., 1963, Isolation of a cytomegalovirus from African green monkey, *Proc. Soc. Exp. Biol. Med.* **112:**601–605.

Blakely, G. A., Lourie, B., Morton, W. G., Evans, H. H., and Kaufmann, A. F., 1973, A varicella-like disease in macaque monkeys, *J. Infect. Dis.* **127:**617–625.

Bonneau, M., and Hennessen, W., 1981, *XVIIth International Congress on Herpes Virus of Man and Animal. Standardization of Immunological Procedures.* Volume 52. *Developments in Biological Standardization,* Karger, Basel.

Bostrom, R. E., and Slaughter, L. J., 1968, Trematode (*Athesmia foxi*) infection in two squirrel monkeys (*Saimiri sciureus*), *Lab. Anim. Care* **18:**493–495.

Brack, M., 1977, Morphological and epidemiological aspects of simian herpesvirus infections, *Schriftenr. Versuchstierkd.* **5:**1–63.

Breman, J. G., Bernadou, J., and Nakano, J. H., 1977, Poxvirus in West African nonhuman primates: Serological survey results, *Bull. WHO* **55:**605–612.

Brody, J. A., Henle, W., and Koprowski, H., eds., 1967, Chronic infectious neuropathic agents (CHINA) and other slow virus infections, *Curr. Top. Microbiol. Immunol.* **40:**1–74.

Brooks, B. A., 1963, More notes on *Saimiri sciureus, Lab. Primate Newsl.* **2:**3–4.

Brown, P., Sarragne, M., Chang, C., Gibbs, C. J., Jr., and Gajdusek, D. C., 1975, Tissue antagonists of interferon in primate brain, in: *Effects of Interferon on Cells, Viruses, and the Immune System* (A. Geraldes, ed), Academic Press, London, pp. 337–343.

Buckley, S. M., and Casals, J., 1978, Pathobiology of Lassa fever, *Int. Rev. Exp. Pathol.* **18:**97–136.

Burch, G. E., Chu, K.-C., and Soike, K. F., 1982, Experimental coxsackievirus B4 valvulitis in squirrel monkeys, *Exp. Mol. Pathol.* **36:**414–422.

Burge, H. A., Solomon, W. R., and Williams, P., 1979, Fungus exposure risks associated with animal care units, *J. Allergy Clin. Immunol.* **64:**29–31.

Burrows, R. B., 1972, Protozoa of the intestinal tract, in: *Pathology of Simian Primates,* Part II (R. N. T. W. Fiennes, ed.), Karger, Basel, pp. 2–28.

Cameron, T. W. M., 1929, The species of *Enterobius* Leach, in primates, *J. Helminthol.* **7:**161–182.

Carpenter, K. P., 1968, Human and simian shigellosis, in: *Some Diseases of Animals Communicable to Main in Britain* (O. Graham-Jones, ed.), Pergamon Press, Oxford, pp. 35–46.

Carstens, L. A., and Allen, J. R., 1969, Response of *Saimiri sciureus* monkeys to total body X-irradiation: Electron microscopic changes in the liver, *Am. J. Vet. Res.* **30:**623–635.

Cathala, F., Hauw, J. J., and Escourolle, R., 1974, Maladie de Creutzfeldt–Jakob et kuru experimentaux chez le singe ecureuil (*Saimiri scieureus*), in: *Immunopathologie du System Nerveux* (Rapport 4, Proc. INSERM Symp., Hospital de la Salpetriere, Paris-Grignon), pp. 13–18.

Chalifoux, L. V., Hunt, R. D., Garcia, F. G., Sehgal, P. K., and Comiskey, J. R., 1973, Filariasis in New World monkeys: Histochemical differentiation of circulating microfilariae, *Lab. Anim. Sci.* **23:**211–220.

Chamove, A., Cameron, G., and Nash, V., 1979, Primate disease and breeding rates, *Lab. Anim.* **13:**313–316.

Chitwood, M., 1970, Comparative relationships of some parasites of man and Old and New World subhuman primates, *Lab. Anim. Care* **20:**389–394.

Cho, C. T., and Wenner, H. A., 1973, Monkeypox virus, *Bacteriol. Rev.* **37:**1–18.

Chrisp, C. E., Cohen, B. J., Ringler, D. H., and Abrams, G. D., 1968, Tuberculosis in a squirrel monkey (*Saimiri sciureus*), *J. Am. Vet. Med. Assoc.* **153:**918–922.

Churchill, A. E., Payne, L. N., and Chubb, R. C., 1969, Immunization against Marek's disease using a live attenuated virus, *Nature* **221**:744–747.

Clark, H. C., 1931, Progress in the survey for blood parasites of the wild monkeys of Panama, *Am. J. Trop. Med.* **11**:11–20.

Clarkson, M. J., Thorpe, E., and McCarthy, K., 1967, A virus disease of captive vervet monkeys (*Cercopithecus aethiops*) caused by a new herpesvirus, *Arch. Ges. Virusforsch.* **22**:219–234.

Coatney, G. R., Collins, W. E., Warren, M., and Contacos, P. G., 1971, *The Primate Malarias*, U. S. Government Printing Office, Washington, D.C.

Colcher, D., Heberling, R. L., Kalter, S. S., and Schlom, J., 1977a, Squirrel monkey retrovirus: An endogenous virus of a New World primate, *J. Virol.* **23**:294–301.

Colcher, D., Teramoto, Y. A., and Schlom, J., 1977b, Interspecies radioimmunoassay for the major structural proteins of primate type-D retroviruses (squirrel monkey retrovirus/Mason–Pfizer monkey virus/viral proteins/endogenous primate viruses/RNA tumor viruses), *Proc. Natl. Acad. Sci. USA* **74**:5739–5743.

Collins, W. E., 1974, Primate malarias, *Adv. Vet. Sci.* **18**:1–23.

Cosgrove, G. E., 1966, The trematodes of laboratory primates, *Lab. Anim. Care* **16**:23–39.

Cosgrove, G. E., Nelson, B. M., and Self, J. T., 1970, The pathology of pentastomid infection in primates, *Lab. Anim. Care* **20**:354–360.

Crandell, R. A., Casey, H. W., and Brumlow, W. B., 1969, Studies of a newly recognized poxvirus of monkeys, *J. Infect. Dis.* **119**:80–88.

Damian, R. T., and Greene, N. D., 1972, The immunology of non-human primates, in: *Pathology of Simian Primates*, Part I (R. N. T. W. Fiennes, ed.), Karger, Basel, pp. 342–379.

Daniel, M. D., Karpas, A., Melendez, L. V., King, N. W., and Hunt, R. D., 1967, Isolation of herpes-T virus from a spontaneous disease in squirrel monkeys (*Saimiri sciureus*), *Arch. Virusforsch.* **22**:324–331.

Davidson, M., and Thomas, L., 1968, Mycoplasma in primates, *Bacteriol. Proc.* **68**:79(M79).

Davis, N. C., 1930, The transmission of yellow fever. Experiments with the "woolly monkey" (*Lagothrix lagotriche* Humboldt), the "spider monkey" (*Ateleus ater* F. Curier), and the "squirrel monkey" (*Saimiri scireus* Linnaeus), *J. Exp. Med.* **51**:703–720.

Deane, L. M., Ferreira Neto, J., and Silveira, I. P. S., 1966, Experimental infection of a splenectomized squirrel-monkey, *Saimiri sciureus*, with *Plasmodium vivax*, *Trans. R. Soc. Trop. Med. Hyg.* **60**:811–812.

Deinhardt, F., 1970, Hepatitis in subhuman primates and the hazards to man, in: *International Symposium on Infections and Immunosuppression in Subhuman Primates* (H. Balner and W. I. B. Beveridge, eds.), Munksgaard, Copenhagen, pp. 55–63.

Deinhardt, F., 1973, *Herpesvirus saimiri*, in: *The Herpesviruses* (A. S. Kaplan, ed.), Academic Press, New York, pp. 595–625.

Deinhardt, F., 1976, Hepatitis in primates, *Adv. Virus Res.* **20**:113–157.

Deinhardt, F., 1980, Biology of primate retroviruses, in: *Viral Oncology* (G. Klein, ed.), Raven Press, New York, pp. 357–398.

Deinhardt, F., Wolfe, L. G., Theilen, G. H., and Snyder, S. P., 1970, ST-feline fibrosarcoma virus: Induction of tumors in marmoset monkeys, *Science* **167**:881.

Deinhardt, F., Falk, L., Marczynska, B., Shramek, G., and Wolfe, L., 1973, *Herpesvirus saimiri:* A simian counterpart of Epstein–Barr virus of man, *Bibl. Haematol.* **39**:416–427.

Deinhardt, F. W., Falk, L. A., and Wolfe, L. G., 1974, Simian herpesviruses and neoplasia, *Adv. Cancer Res.* **19**:167–205.

Deinhardt, F., Wolfe, L., Falk, L., Johnson, T., Johnson, D., and Massey, R., 1975, Immunological control of virus-induced tumors in primates, in: *Comparative Leukemia Research, 1973, Leukemogenesis* (Y. Ito and R. M. Dutcher, eds.), University of Tokyo Press, Tokyo, pp. 639–648.

Desrosiers, R. C., and Falk, L. A., 1982, *Herpesvirus saimiri* strain variability, *J. Virol.* **43**:352–356.

Dick, E. C., and Dick, C. R., 1968, A subclinical outbreak of human rhinovirus 31 infection in chimpanzees, *Am. J. Epidemiol.* **88**:267–272.

Douglas, J. D., Tanner, K. N., Prine, J. R., Van Riper, D. C., and Derwelis, S. K., 1967, Molluscum contagiosum in chimpanzees, *J. Am. Vet. Med. Assoc.* **151**:901–904.

Downie, A. W., Taylor-Robinson, C. H., Caunt, A. E., Nelson, G. S., Manson-Bahr, P. E. C., and Matthews, T. C. H., 1971, Tanapox: A new disease caused by a pox virus, *Br. Med. J.* **1**:363–368.

Dunkel, V. C., Pry, T. W., Henle, G., and Henle, W., 1972, Immunofluorescence tests for antibodies to Epstein–Barr virus with sera of lower primates, *J. Natl. Cancer Inst.* **49**:435–440.

Dunn, F. L., 1963, Acanthocephalans and cestodes of South American monkeys and marmosets, *J. Parasitol.* **49**:717–722.

Dunn, F. L., 1968, The parasites of *Saimiri:* In the context of platyrrhine parasitism, in: *The Squirrel Monkey* (L. A. Rosenblum and R. W. Cooper, eds.), Academic Press, New York, pp. 31–68.

Dunn, F. L., and Lambrecht, F. L., 1963a, The hosts of *Plasmodium brasilianum* Gonder and von Berenberg-Gossler, 1908, *J. Parasitol.* **49**:316–319.

Dunn, F. L., and Lambrecht, F. L., 1963b, On some filarial parasites of South American primates, with a description of *Tetrapetalonema tamarinae* n. sp. from the Peruvian tamarin marmoset, *Tamarinus nigricollis* (Spix, 1823), *J. Helminthol.* **37**:261–286.

Dunn, F. L., Lambrecht, F. L., and du Plessis, R., 1963, Trypanosomes of South American monkeys and marmosets, *Am. J. Trop. Med. Hyg.* **12**:524–534.

Eberhard, M. L., 1982, Chemotherapy of filariasis in squirrel monkeys (*Saimiri sciureus*), *Lab. Anim. Sci.* **32**:397–400.

Eberhard, M., and D'Alessandro, A., 1982, Congenital *Trypanasoma cruzi* infection in a laboratory-born squirrel monkey, *Saimiri sciureus, Am. J. Trop. Med. Hyg.* **31**:931–933.

Eckroade, R. J., Zu Rhein, G. M., Marsh, R. F., and Hanson, R. P., 1970, Transmissible mink encephalopathy: Experimental transmission to the squirrel monkey, *Science* **169**:1088–1090.

Eichberg, J. W., and Kalter, S. S., 1980, Hepatitis A and B: Serologic survey of human and nonhuman primate sera, *Lab. Anim. Sci.* **30**:541–543.

Elizan, T. S., Schwartz, J., Kott, E., and Pedersen, B., 1974, Schilder's disease: Virology studies concerning dense core particles, *Neurology* **24**:428–430.

Emmons, R. W., Woodie, J. D., Taylor, M. S., and Nygaard, G. S., 1970, Tularemia in a pet squirrel monkey (*Saimiri sciureus*), *Lab. Anim. Care* **20**:1149–1153.

Epstein, M. A., 1972, Burkitt's lymphoma and *Herpesvirus saimiri* lymphoma: Comparative aspects, *J. Natl. Cancer Inst.* **49**:213–217.

Espana, C., 1971, A pox disease of monkeys transmissible to man, in: *Medical Primatology, 1970* (E. I. Goldsmith and J. Moor-Jankowski, eds.), Karger, Basel, pp. 694–708.

Falk, L. A., 1974, Oncogenic DNA viruses of nonhuman primates: A review, *Lab. Anim. Sci.* **24**:182–192.

Falk, L., Wolfe, L., and Deinhardt, F., 1972, Epidemiology of *Herpesvirus saimiri* infection in squirrel monkeys, in: *Medical Primatology,* Part III (E. I. Goldsmith and J. Moor-Jankowski, eds.) Karger, Basel, pp. 151–158.

Falk, L. A., Nigida, S., Deinhardt, F., Cooper, R. W., and Hernandez-Camacho, J. I., 1973, Oral excretion of *Herpesvirus saimiri* in captive squirrel monkeys and incidence of infection in feral squirrel monkeys, *J. Natl. Cancer Inst.* **51**:1987–1989.

Felsenfeld, A. D., 1972, The Arboviruses, in: *Pathology of Simian Primates,* Part II (R. N. T. W. Fiennes, ed.), Karger, Basel, pp. 523–536.

Felsenfeld, O., and Wolf, R. H., 1971, Serological reactions with treponemal antigens in nonhuman primates and the natural history of treponematosis in man, *Folia Primatol.* **16**:294–305.

Fenner, F., 1979, Portraits of viruses: The poxviruses, *Intervirology* **11**:137–157.

Fenters, J. D., Ehrlich, R., Findlay, J., Spangler, J., and Tolkacz, V., 1971, Serologic response in squirrel monkeys exposed to nitrogen dioxide and influenza virus, *Am. Rev. Respir. Dis.* **104**:448–451.

Fiennes, R. N., 1963, Incidence of TB in squirrel monkeys and marmosets, *Lab. Primate Newsl.* **4**:10.

Fiennes, R., 1967, *Zoonoses of Primates,* Cornell University Press, Ithaca.

Fiennes, R. N. T. W., ed., 1972, *Pathology of Simian Primates,* Volumes I and II, Karger, Basel.

Fine, D., and Schochetman, G., 1978, Type D primate retrovirus: A review, *Cancer Res.* **38**:3123–3139.

Fleckenstein, B., 1979, Oncogenic herpesviruses of non-human primates, *Biochim. Biophys. Acta* **560**:301–342.

Fleckenstein, B., and Mulder, C., 1980, Molecular biological aspects of *Herpesvirus saimiri* and *Herpesvirus ateles*, in: *Viral Oncology* (G. Klein, ed.), Raven Press, New York, pp. 799–812.

Foster, A. O., and Johnson, C. M., 1939, A preliminary note on the identity, lifecycle, and pathogenicity of an important nematode parasite of captive monkeys, *Am. J. Trop. Med.* **19:**265–277.

Fox, J. G., Diaz, J. R., and Barth, R. A., 1972, Nymphal *Porocephalus clavatus* in the brain of a squirrel monkey, *Saimiri sciureus, Lab. Anim. Sci.* **22:**908–910.

Frankel, J. W., and Groupe, V., 1971, Interactions between Marek's disease herpesvirus and avian leucosis virus in tissue culture, *Nature New Biol.* **234:**125–126.

Fremount, H. N., and Rossan, R. N., 1974, The sites of sequestration of the Uganda–Palo Alto strain of *Plasmodium falciparum*-infected red blood cells in the squirrel monkey, *Saimiri sciureus, J. Parasitol.* **60:**534–536.

Fuccillo, D. A., Kurent, J. E., and Sever, J. L., 1974, Slow virus diseases, *Annu. Rev. Microbiol.* **28:**231–264.

Fussell, E. N., and Roberts, J. A., 1979, Cytoplasmic inclusion bodies in the monkey ureter, *Vet. Pathol.* **16:**127–129.

Gaarder, P. I., and Natvig, J. B., 1974, Distribution of isotypic and allotypic human IgG antigens in non-human primates, *J. Immunol.* **113:**635–644.

Gajdusek, D. C., 1978, Slow infections with unconventional viruses, *Harvey Lect.* **72:**283–353.

Gajdusek, D. C., Gibbs, C. J., Jr., and Alpers, M., eds., 1965, *Workshop and Symposium on Slow, Latent, and Temperate Virus Infections, National Institutes of Health, 1964*, U. S. Government Printing Office, Washington, D.C.

Gajdusek, D. C., Gibbs, C. J., Jr., and Alpers, M., 1966, Experimental transmission of a kuru-like syndrome to chimpanzees, *Nature* **209:**794–796.

Gajdusek, D. C., Gibbs, C. J., Jr., and Alpers, M., 1967, Transmission and passage of experimental "kuru" to chimpanzees, *Science* **155:**212–214.

Gajdusek, D. C., Gibbs, C. J., Jr., Earle, K., Dammin, G. J., Schoene, W. C., and Tyler, H. R., 1974, Transmission of subacute spongiform encephalopathy to the chimpanzee and squirrel monkey from a patient with papulosis atrophicans maligna of Kohlmeier-Degos, in: *Neurology* (A. Busirana, J. M. Espadaler, and E. H. Burrows, eds.), American Elsevier, New York, pp. 391–392.

Gallo, R. C., and Wong-Staal, F., 1980, Molecular biology in primate retroviruses, in: *Viral Oncology* (G. Klein, ed.), Raven Press, New York, pp. 399–432.

Garnham, P. C. C., Baker, J. R., and Nesbitt, P. E., 1963, Transmission of *Plasmodium brasilianum* by sporozoites, and the discovery of an exoerythrocytic schizont in the monkey liver, *Parasitologia* **5:**5–9.

Gergen, J. A., and Maclean, P. D., 1962, *A Sterotaxic Atlas of the Squirrel Monkeys Brain*, U. S. Government Printing Office, Washington, D.C.

Gibbs, C. J., Jr., 1967, Search for infectious etiology in chronic and subacute degenerative diseases of the central nervous system, *Curr. Top. Microbiol. Immunol.* **40:**44–58.

Gibbs, C. J., Jr., and Gajdusek, D. C., 1973, Experimental subacute spongiform virus encephalopathies in primates and other laboratory animals, *Science* **182:**67–68.

Gibbs, C. J., Jr., Gajdusek, D. C., Asher, D. M., Alpers, M. P., Beck, E., Daniel, P. M., and Matthews, W. B., 1968, Creutzfeldt–Jakob disease (spongiform encephalopathy): Transmission to the chimpanzee, *Science* **161:**388–389.

Gibbs, C. J., Jr., Gajdusek, D. C., and Alpers, M. P., 1969, Attempts to transmit subacute and chronic neurological diseases to animals, *Int. Arch. Allergy Appl. Immunol.* **36**(Suppl.):519–552.

Gibbs, C. J., Jr., Amyx, H. L., Bacote, A., Masters, C. L., and Gajdusek, D. C., 1980, Oral transmission of kuru, Creutzfeldt–Jakob disease, and scrapie to nonhuman primates, *J. Infect. Dis.* **142:**205–208.

Goldsmith, E. I., and Moor-Jankowski, J., eds., 1969, *Experimental Medicine and Surgery in Primates*, New York Academy of Science, Vol. 162.

Goldsmith, E. I., and Moor-Jankowski, J., eds., 1972, *Conference on Experimental Medicine and Surgery in Primates, 3rd, Lyon, 1972;* Part 3: *Infectious Diseases, Oncology, Pharmacology and Toxicology, Cardiovascular Studies*, Karger, Basel.

Grabow, J. D., ZuRhein, G. M., Eckroade, R. J., Zollman, P. E., and Hanson, R. P., 1973, Transmissible mink encephalopathy agent in squirrel monkeys. Serial electroencephalographic, clinical and pathologic studies, *Neurology* **23**:820–832.

Greenstein, E. T., Doty, R. W., and Lowy, K., 1965, An outbreak of a fulminating infectious disease in the squirrel monkey, *Saimiri sciureus*, *Lab. Anim. Care* **15**:74–80.

Hahon, N., 1961, Smallpox and related poxvirus infections in the simian host, *Bacteriol. Rev.* **25**:459–476.

Hall, A. S., and McNulty, W. P., Jr., 1967, A contagious pox disease in monkeys, *J. Am. Vet. Med. Assoc.* **151**:833–838.

Halonen, P., Panelius, M., and Salmi, A., eds., 1973, Slow virus infections of the central nervous system, *Ann. Clin. Res.* **5**:253–355.

Harvey, J. S., Jr., Felsburg, P. J., Heberling, R. L., Kniker, W. T., and Kalter, S. S., 1974, Immunological competence in non-human primates: Differences observed in four species, *Clin. Exp. Immunol.* **16**:267–277.

Heberling, R. L., 1972, The simian adenoviruses, in: *Pathology of Primates*, Part II (R. N. T. W. Fiennes, ed.), Karger, Basel, pp. 572–591.

Heberling, R. L., 1983, Nonhuman primates in viral oncology, in: *Viral and Immunological Diseases in Nonhuman Primates* (S. S. Kalter, ed.), A. R. Liss, New York, pp. 91–110.

Heberling, R. L., and Cheever, F. S., 1960, Enteric viruses of monkeys, *Ann. N. Y. Acad. Sci.* **85**:942–950.

Heberling, R. L., Barker, S. T., Kalter, S. S., Smith, G. C., and Helmke, R. J., 1977, Oncornavirus: Isolation from a squirrel monkey (*Saimiri sciureus*) lung culture, *Science* **195**:289–292.

Hegner, R., 1935, Intestinal protozoa from Panama monkeys, *J. Parasitol.* **21**:60–61.

Held, J. R., 1981, Requirements for laboratory animals in health programmes, *Bull. WHO* **59**:513–517.

Held, J. R., 1982, Conservation and management of nonhuman primates, *Microbiologica* **5**:161–169.

Henderson, D. A., 1975, Smallpox eradication—The final battle, *J. Clin. Pathol.* **28**:843–849.

Henry, M. C., Ehrlich, R., and Blair, W. H., 1969, Effect of nitrogen dioxide on resistance of squirrel monkeys to *Klebsiella pneumoniae* infection, *Arch. Environ. Health* **18**:580–587.

Henry, M. C., Findlay, J., Spangler, J., and Ehrlich, R., 1970, Chronic toxicity of NO_2 in squirrel monkeys. III. Effect on resistance to bacterial and viral infection, *Arch. Environ. Health* **20**:566–570.

Hessler, J. R., and Moreland, A. F., 1968, Pulmonary tuberculosis in a squirrel monkey (*Saimiri sciureus*), *J. Am. Vet. Med. Assoc.* **153**:923–927.

Hill, A., 1977, The isolation of mycoplasmas from non-human primates, *Vet. Rec.* **101**:117.

Hill, R. L., Buettner-Janusch, J., and Buettner-Janusch, V., 1963, Evolution of hemoglobin in primates, *Proc. Natl. Acad. Sci. USA* **50**:885–893.

Hilleman, M. R., 1976, Herpes simplex vaccines, *Cancer Res.* **36**:857–858.

Hinuma, T., Nagata, K., Hanaoka, M., Nakai, M., Matsumoto, T., Kinoshito, K., Kirakawa, S., and Miyoshi, I., 1981, Adult T-cell leukemia: Antigen in an ATL cell line and detection of antibodies to the antigen in human sera, *Proc. Natl. Acad. Sci. USA* **78**:6478–6480.

Hoffert, W. R., Bates, M. E., and Cheever, F. S., 1958, Study of enteric viruses of simian origin, *Am. J. Hyg.* **68**:15–30.

Holmes, A. W., Caldwell, R. G., Dedmon, R. E., and Dienhardt, F., 1964, Isolation and characterization of a new herpes virus, *J. Immunol.* **92**:602–610.

Holmes, A. W., Devine, J. A., Nowakowski, E., and Deinhardt, F., 1966, The epidemiology of a herpes virus infection of New World monkeys, *J. Immunol.* **96**:668–671.

Holmes, J. C., 1982, Impact of infectious disease agents on the population growth and geographical distribution of animals, in: *Population Biology of Infectious Diseases* (R. M. Anderson and R. M. May, eds.), Springer-Verlag, New York, pp. 37–51.

Hooks, J. J., and Gibbs, C. J., Jr., 1975, The foamy viruses, *Bacteriol. Rev.* **39**:169–185.

Hotchin, J., 1971, Persistent and slow virus infections, *Monogr. Virol.* **3**:1–211.

Hsiung, G. D., 1969, Recent advances in the study of simian viruses, *Ann. N. Y. Acad. Sci.* **162**:483–498.

Hsiung, G. D., and Melnick, J. L., 1958, Orphan viruses of man and animals, *Ann. N. Y. Acad. Sci.* **70**:342–361.

Hsiung, G.-D., Black, F. L., and Henderson, J. R., 1964, Susceptibility of primates to viruses in relation to taxonomic classification, in: *Evoutionary and Genetic Biology of Primates,* Volume II (J. Buettner-Janusch, ed.), Academic Press, New York, pp. 1–23.

Hudgins, P. C., Whorton, C. M., Tomoyoshi, T., and Riopelle, A. J., 1966, Comparison of the molecular structure of myoglobin of fourteen primate species, *Nature* **212:**693–695.

Hull, R. N., 1968, The simian viruses, *Virol. Monogr.* **2:**1–66.

Hull, R. N., and Minner, J. R., 1957, New viral agents recovered from tissue cultures of monkey kidney cells. II. Problems of isolation and identification, *Ann. N. Y. Acad. Sci.* **67:**413–423.

Hull, R. N., Minner, J. R., and Mascoli, C. C., 1958, New viral agents recovered from tissue cultures of monkey kidney cells. III. Recovery of additional agents both from cultures of monkey tissues and directly from tissues and excreta, *Am. J. Hyg.* **68:**31–44.

Hull, R. N., Dwyer, A. C., Holmes, A. W., Nowakowski, E., Deinhardt, F., Lennette, E. H., and Emmons, R. W., 1972, Recovery and characterization of a new simian herpesvirus from a fatally infected spider monkey, *Pan Am. Health Or. Sci. Publ.* **235:**137–144.

Hull, R. N., Minner, J. R., and Smith, J. W. , 1956, New viral agents recovered from tissue cultures of monkey kidney cells. I. Origin and properties of cytopathogenic agents S. V. 1, S. V. 2, S. V. 4, S. V. 5, S. V. 6, S. V. 11, S. V. 12, and S. V. 15, *Am. J. Hyg.* **63:**204–215.

Hunt, R. D., Jones, T. C., and Williamson, M., 1970a, Mechanisms of parasitic damage and the host response, *Lab. Anim. Care* **20:**345–353.

Hunt, R. D., Melendez, L. V., King, N. W., Gilmore, C. E., Daniel, M. D., Williamson, M. E., and Jones, T. C., 1970b, Morphology of a disease with features of malignant lymphoma in marmosets and owl monkeys inoculated with *Herpesvirus saimiri, J. Natl. Cancer Inst.* **44:**447–465.

Hunt, R. D., and Melendez, L. V., 1972, Clinical, epidemiologic, and pathologic features of cytocidal and oncogenic herpesvirus in South American monkeys, *J. Natl. Cancer Inst.* **49:**261–271.

Johnson, B. K., Gitau, L. G., Gichogo, A., Tukei, P. M., Else, J. G., Suleman, M. A., and Kimani, R., 1981, Marburg and Ebola virus antibodies in Kenyan primates, *Lancet* **1:**1420–1421.

Johnston, P. B., 1971, Taxonomic features of seven serotypes of simian and ape foamy viruses, *Infect. Immunol.* **3:**793–799.

Jurko, M. F., Giurintano, L. P., Giurintano, S. L., and Andy, O. J., 1974, Spontaneous awake EEG patterns in three lines of primate evolution, *Behav. Biol.* **10:**377–384.

Kaack, B., Walker, L., and Brizzee, K. R., 1979, The growth and development of the squirrel monkey (*Saimiri sciureus*), *Growth* **43:**116–135.

Kafuko, G. W., Henderson, B. E., Williams, M. C., and Kissling, R. E., 1970, Investigations in Uganda relating to the Marburg agent, in: *International Symposium on Infections and Immunosuppression in Subhuman Primates* (H. Balner and W. T. B. Beveridge, eds.), Munksgaard, Copenhagen, pp. 45–53.

Kalter, S. S., 1960, Animal "orphan" enteroviruses, *Bull. WHO* **22:**319–337.

Kalter, S. S., 1969, Nonhuman primates in viral research, *Ann. N. Y. Acad. Sci.* **162:**499–528.

Kalter, S. S., 1971, A serological survey of primate sera for antibody to the Marburg virus, in: *Marburg Virus Disease* (G. A. Martini and R. Siegert, eds.), Springer-Verlag, New York, pp. 177–187.

Kalter, S. S., 1973, The Baboon. Microbiology, clinical chemistry and some hematological aspects, *Primates Med.* **8:**1–171.

Kalter, S. S., 1980, Standardized laboratory animals, their need. Why the nonhuman primate, *Dev. Biol. Stand.* **45:**3–11.

Kalter, S. S., 1982, Herpesviruses of nonhuman primates: Their significance, *Microbiologica* **5:**149–159.

Kalter, S. S., 1983, Primate viruses their significance, in: *Viral and Immunological Diseases in Nonhuman Primates* (S. S. Kalter, ed.,), A. R. Liss, New York, pp. 67–90.

Kalter, S. S., and Heberling, R. L., 1971, Comparative virology of primates, *Bacteriol. Rev.* **35:**310–364.

Kalter, S. S., and Heberling, R. L., 1972, Serologic evidence of viral infection in South American monkeys, *J. Natl. Cancer Inst.* **49:**251–259.

Kalter, S. S., and Heberling, R. L., 1974, Isolation of C-type viruses from baboon (*Papio cynocephalus*) placental tissue, *Abstr. Annu. Meet. Am. Soc. Microbiol.* **74:**233(V194).

Kalter, S. S., and Heberling, R. L., 1976, Primate endogenous viruses: Their role in oncogenesis and as biohazards, *Dev. Biol. Stand.* **37**:219–228.

Kalter, S. S., and Heberling, R. L., 1978, Serologic response of primates to influenza viruses, *Proc. Soc. Exp. Biol.* **159**:414–417.

Kalter, S. S. and Hennessen, W., eds., 1980, The standardization of animals to improve biomedical research, in: *Developments in Biological Standardization. The Standardization of Animals to Improve Biomedical Research,* Vol. 45, Karger, Basel.

Kalter, S. S., Ratner, J. J., and Heberling, R. L., 1969, Antibodies in primates to the Marburg virus, *Proc. Soc. Exp. Biol. Med.* **130**:10–12.

Kalter, S. S., Felsburg, P. J., Heberling, R. L., Nahmias, A. J., and Brack, M., 1972a, Experimental *Herpesvirus hominis* type 2 infection in nonhuman primates, *Proc. Soc. Exp. Biol. Med.* **139**:964–968.

Kalter, S. S., Heberling, R. L., and Ratner, J. J., 1972b, EBV antibody in sera of non-human primates, *Nature* **238**:353–354.

Kalter, S. S., Helmke, R. J., Panigel, M., Heberling, R. L., Felsburg, P. J., and Axelrod, L. R., 1973a, Observations of apparent C-type particles in baboon (*Papio cynocephalus*) placentas, *Science* **179**:1332–1333.

Kalter, S. S., Helmke, R. J., Heberling, R. L., Panigel, M., Fowler, A. K., Strickland, J. E., and Hellman, A., 1973b, C-Type particles in normal human placentas, *J. Natl. Cancer Inst.* **50**:1081–1084.

Kalter, S. S., Heberling, R. L., Hellman, A., Todaro, G. J., and Panigel, M., 1975, C-Type particles in baboon placneta, *Proc. R. Soc. Med.* **68**:135–140.

Kalter, S. S., Kniker, W. T., Harvey, J. S., Jr., Felsburg, P. J., and Heberling, R. L., 1976, Lymphocyte responsiveness in squirrel, cebus and marmoset monkeys infected with *Herpesvirus saimiri, Bibl. Haematol.* **43**:384–387.

Kalter, S. S., Eichberg, J. W., and Heberling, R. L., 1978a, Possible induction of baboon endogenous type C virus following *H. saimiri* inoculation, in: *International Congress for Virology, 4th, The Hague, The Netherlands, 1978, International Virology IV, Abstracts,* Centre for Agricultural Publishing and Documentation, Wageningen, The Netherlands, p. 373 (W26/3).

Kalter, S. S., Heberling, R. L., Eichberg, J. W., Boenig, D. M., and Cummins, L. B., 1978b, Pathologic and immunologic response to the baboon endogenous virus and its pseudotype, in: *Advances in Comparative Leukemia Research, 1977* (P. Bentvelzen, J. Hilgers, and D. S. Yohn, eds.), Elsevier North-Holland, Amsterdam, pp. 100–102.

Kalter, S. S., Hellman, A., and Gruber, J., eds., 1979, *Primates and Human Cancer,* U. S. Government Printing Office, Washington, D. C.

Kalter, S. S., Ablashi, D., Espana, C., Heberling, R. L., Hull, R. N., Lennette, E. H., Malherbe, H. H., McConnell, S., and Yohn, D. S., 1980, Simian virus nomenclature, 1980, *Intervirology* **13**:317–330.

Kalter, S. S., Rodriguez, A. R., and Heberling, R. L., in preparation, "CRELM" antibodies in nonhuman primates.

Kaplan, A. S., ed., 1973, *The Herpesviruses,* Academic Press, New York.

Karr, S. L., Jr. and Wong, M. M., 1975, A survey of *Sarcocystis* in nonhuman primates, *Lab. Anim. Sci.* **25**:641–645.

Kessler, M. J., and Brown, R. J., 1979, Clinical description of tetanus in squirrel monkeys (*Saimiri sciureus*), *Lab. Anim. Sci.* **29**:240–242.

Kessler, M. J., Brown, R. J., Kalter, S. S., and Altman, N. H., 1982, A fatal epizootic of undetermined etiology in New World monkeys, *Am. J. Primatol.* **3**:253–261.

King, N. W., Hunt, R. D., Daniel, M. D., and Melendez, L. V., 1967, Overt Herpes-T infection in squirrel monkeys (*Saimiri sciureus*), *Lab. Anim. Care* **17**:413–423.

Klein, G., Pearson, G., Rabson, A., Ablashi, D. V., Falk, L., Wolfe, L., Deinhardt, F., and Rabin, H., 1973, Antibody reactions to *Herpesvirus saimiri* (HVS)-induced early and late antigens (EA and LA) in HVS-injected squirrel, marmoset and owl monkeys, *Int. J. Cancer* **12**:270–289.

Koller, L. D., and Olson, C., 1972, Attempted transmission of warts from man, cattle, and horses and of deer fibroma, to selected hosts, *J. Invest. Dermatol.* **58**:366–368.

Koshimizu, K., Magaribuchi, T., Yamamoto, K., and Ogata, M., 1975, Characterization of mycoplasmas isolated from imported nonhuman primates, *Exp. Anim.* **24**:95–101.

Kuntz, R. E., 1972, Trematodes of the intestinal tract and biliary passages, in: *Pathology of Simian Primates* Part II (R. N. T. W. Fiennes, ed.), Karger, Basel, pp. 104–123.

Kuntz, R. E., and Myers, B. J., 1972, Parasites of South American primates, *Int. Zoo Yearb.* **12**:61–68.

Kuntz, R. E., Myers, B., Huang, T. C., and Moore, J. A., 1971, Use of nonhuman primates in experimental *Schistosomiasis haematobia*, in: *Proceedings of the Third International Congress of Primatology*, Volume 2 (J. Biegert and W. Leutenegger, eds.), Karger, Basel, pp. 162–172.

Kuntz, R. E., Cheever, A. W., and Myers, B. J. 1972, Proliferative epithelial lesions of the urinary bladder of nonhuman primates infected with *Schistosoma haematobium*, *J. Natl. Cancer Inst.* **48**:223–235.

Landsteiner, K., and Popper, E., 1908, Mikroscopishe Praparate von einem menslichen und zwei Afftenruchenmarken, *Wien. Klin. Wochenschr.* **21**:1830.

Landsteiner, K., and Popper, E., 1909, Uebertragung der Poliomyelitis acuta auf Affen, *Z. Immunitaetsforsch. Exp. Ther.* **2**:377–390.

Lapin, B. A., and Shevtsova, Z. V., 1971, On the identity of two simian hemorrhagic fever virus strains (Sukhumi and NIH), *Z. Versuchstierkd.* **13**:21–23.

Lapin, B. A., and Yakovleva, L. A., 1963, *Comparative Pathology in Monkeys*, C. C. Thomas, Springfield, Illinois.

Lapin, B. A., Pekerman, S. M., Yakovleva, L. A., Dzhikidze, E. K., Shevtsova, Z. V., Kuksova, M. I., Danko, L. V., Krilova, R. I., Akbroit, E. Ya, and Agraba, V. Z., 1967, Gemorragicheskaia likhoradka obez'ien, *Vopr. Virusol.* **12**:168–173 [in Russian].

Laufs, R., 1974, Immunisation of marmoset monkeys with a killed oncogenic herpesvirus, *Nature* **249**:571–572.

Laufs, R., and Steinke, H., 1975, Passive immunisation of marmoset monkeys against neoplasia induced by a herpesvirus, *Nature* **255**:226–228.

Leader, R. W., and Hurvitz, A. I., 1972, Interspecies patterns of slow virus diseases, *Annu. Rev. Med.* **23**:191–200.

Leathers, C. W., and Hamm, T. E., Jr., 1976, Naturally occurring tuberculosis in a squirrel monkey and a cebus monkey, *J. Am. Vet. Med. Assoc.* **169**:909–911.

Lemon, S. M., LeDuc, J. W., Binn, L. N., Escajadillo, A., and Ishak, K. G., 1982, Transmission of hepatitis A virus among recently captured Panamanian owl monkeys, *J. Med. Virol.* **10**:25–36.

Levine, N. D., 1970, Protozoan parasites of nonhuman primates as zoonotic agents, *Lab. Anim. Care* **20**:377–382.

Levy, B. M., 1971, The nonhuman primate as an analogue for the study of peridontal disease, *J. Dent. Res.* **50**:246–253.

Loew, F. M., 1968, A review of some helminths of laboratory animals, *Cornell Vet.* **58**:408–421.

London, W. T., Catalano, L. W., Jr., Nahmias, A. J., Fuccillo, D. A., and Sever, J. L., 1971, Genital *Herpesvirus hominis* Type 2 infection of monkeys, *Obstet. Gynecol.* **37**:501–509.

Lucke, V. M., and Bennett, A. M., 1982, An outbreak of hepatitis in marmosets in a zoological collection, *Lab. Anim.* **16**:73–77.

Malherbe, H., and Harwin, R., 1957, Seven viruses isolated from the vervet monkey, *Br. J. Exp. Pathol.* **38**:539–541.

Malherbe, H., and Strickland-Cholmley, M., 1971, Studies on the Marburg virus, in: *Marburg Virus Disease* (G. A. Martini, and R. Siegert, eds.), Springer-Verlag, New York, pp. 188–194.

Malherbe, H., Harwin, R., and Ulrich, M., 1963, The cytopathic effects of vervet monkey viruses, *S. Afr. Med. J.* **37**:407–411.

Marennikova, S. S., Shelukhina, E. M., Shenkman, L. S., Maltseva, N. N., and Matsevich, G. R., 1975, The results of examinations of wildlife monkeys for the presence of antismallpox antibody and viruses of the smallpox group, *Vopr. Virusol.* **3**:321–328 [in Russian].

Marsh, R. F., Burger, D., Eckroade, R., Za Rhein, G. M., and Hanson, R. P., 1969, A preliminary report on the experimental host range of the transmissible mink encephalopathy agent, *J. Infect. Dis.* **120**:713–719.

Martin, L. N., and Allen, W. P., 1975, Response to primary infection with *Herpesvirus saimiri* in immunosuppressed juvenile and newborn squirrel monkeys, *Infect. Immunol.* **12**:528–535.

Martini, G. A., and Siegert, R., eds., 1971, *Marburg Virus Disease,* Springer-Verlag, New York.

Masironi, R., Godlewski, H. G., and Bourne, G. H., 1968, Experiments on heterotransplantation of the Toolan H. Ad. # 1U tumor to monkeys. I. The cheek pouch of *Macaca rhesus* and buccal mucosa of *Saimiri sciureus, Neoplasma (Bratisl.)* **15:**145–156.

Massey, R. J., Arthur, L. O., and Fine, D. L., 1978, Spontaneously occurring humoral and cell-mediated immune responses to type-D retroviruses in Old and New World primates, *Abstr. Annu. Meet. Am. Soc. Microbiol.* **78:**240 (S165).

Matthews, R. E. F., 1982, Classification and nomenclature of viruses, *Intervirology* **17:**1–199.

Mayhall, C. G., Lamb, V. A., and Coleman, P. H., 1981, Infection in rhesus (*Macaca mulatta*) and squirrel (*Saimiri sciureus*) monkeys due to *Mycobacterium tuberculosis* phage type B, *J. Med. Primatol.* **10:**302–311.

McCarthy, K., and Tosolini, F. A., 1975, A review of primate herpes viruses, *Proc. R. Soc. Med.* **68:**145–150.

McCarthy, K., Thorpe, E., Laursen, A. C., Heymann, C. S., and Beale, A. J., 1968, Exanthematous disease in patas monkeys caused by a herpes virus, *Lancet* **2:**856–857.

McClure, H. M., and Keeling, M. E., 1971, Viral diseases noted in the Yerkes Primate Center Colony, *Lab. Anim. Sci.* **21:**1002–1010.

McConnell, S. J., Herman, Y. F., Mattson, D. E., and Erickson, L., 1962, Monkey pox disease in irradiated cynomologous monkeys, *Nature* **195:**1128–1129.

McKissick, G. E., Ratcliff, H. L., and Koestner, A., 1968, Enzootic toxoplasmosis in caged squirrel monkeys *Saimiri sciureus, Pathol. Vet.* **5:**538–560.

Melby, E. C., Jr., and Altman, N. H., eds., 1974, *Handbook of Laboratory Animal Science,* Volumes I, II, III, CRC Press, Cleveland, Ohio.

Meldrum, B. S., and Marsden, C. D., eds., 1975, *Advances in Neurology,* Vol. 10, Raven Press, New York.

Melendez, L. V., Hunt, R. D., Garcia, F. G., and Trum, B. F., 1966, A latent herpes-T infection in *Saimiri sciureus* (squirrel monkey), *Symp. Zool. Soc. Lond.* **17:**393–397.

Melendez, L. V., Daniel, M. D., Hunt, R. D., and Garcia, F. G., 1968, An apparently new herpesvirus from primary kidney cultures of the squirrel monkey (*Saimiri sciureus*), *Lab. Anim. Care* **18:**374–381.

Melendez, L. V., Daniel, M. D., Garcia, F. G., Fraser, C. E. O., Hunt, R. D., and King, N. W., 1969a, *Herpesvirus Saimiri.* I. Further characterization studies of a new virus from the squirrel monkey, *Lab. Anim. Care* **19**(3):372–377.

Melendez, L. V., Hunt, R. D., Daniel, M. D., Garcia, F. G., and Fraser, C. E. O., 1969b, *Herpesvirus saimiri.* II. Experimentally induced malignant lymphoma in primates, *Lab. Anim. Care* **19:**378–386.

Melendez, L. V., Daniel, M. D., Barahona, H. H., Fraser, C. E. O., Hunt, R. D., and Garcia, F. G., 1971a, New herpesviruses from South American monkeys. Preliminary report, *Lab. Anim. Sci.* **21:**1050–1054.

Melendez, L. V., Hunt, R. D., Daniel, M. D., Blake, B. J., and Garcia, F. G., 1971b, Acute lymphocytic leukemia in owl monkeys inoculated with *Herpesvirus saimiri, Science* **171:**1161–1163.

Melnick, J. L., Midulla, M., Wimberly, I., Barrera-Oro, J. G., and Levy, B. M., 1964, A new member of the herpesvirus group isolated from South American marmosets, *J. Immunol.* **92:**596–601.

Middleton, C. C., Clarkson, T. B., and Garner, F. M., 1964, Parasites of squirrel monkeys (*Saimiri sciureus*), *Lab. Anim. Care* **14:**335(52).

Minette, H. P., 1966, Leptospirosis in primates other than man, *Am. J. Trop. Med. Hyg.* **15:**190–198.

Minette, H. P., and Shaffer, M. F., 1968, Experimental leptospirosis in monkeys, *Am. J. Trop. Med. Hyg.* **17:**202–212.

Miyoshi, K., Gamboa, E. T., Harter, D. H., Wolf, A., and Hsu, K. C., 1971, Influenza virus encephalitis in squirrel monkeys receiving immunosuppressive therapy, *J. Immunol.* **106:**1119–1121.

Moor-Jankowski, J., and Wiener, A. S., 1969, Blood group antigens in primate animals and their relation to human blood groups, *Primates Med.* **3:**64–77.

Moor-Jankowski, J., and Wiener, A. S., 1972, Red cell antigens of primates, in: *Pathology of Simian Primates,* Part 1, (R. N. T. W. Fiennes, ed.), Karger, Basel, pp. 270–317.

Moor-Jankowski, J., Wiener, A. S., and Rogers, C. M., 1964, Human blood group factors in nonhuman primates, *Nature* **202:**663–665.

Morris, J. A., Blount, R. E., Jr., and Savage, R. E., 1956, Recovery of cytopathogenic agent from chimpanzees with coryza, *Proc. Soc. Exp. Biol. Med.* **92:**544–549.

Mukai, N., Kalter, S. S., Cummins, L. B., Matthews, V. A., Nishida, T., and Nakajima, T., 1980, Retinal tumor induced in the baboon by human adenovirus 12, *Science* **210:**1023–1025.

Munroe, J. S., and Windle, W. F., 1963, Tumors induced in primates by chicken sarcoma virus, *Science* **140:**1415–1416.

Murphy, B. R., Sly, D. L., Hosier, N. T., London, W. T., and Chanock, R. M., 1980, Evaluation of three strains of influenza A virus in humans and in owl, cebus, and squirrel monkeys, *Infect. Immunol.* **28:**688–691.

Murphy, B. R., Hinshaw, V. S., Sly, D. L., London, W. T., Hosier, N. T., Wood, F. T., Webster, R. G., and Chanock, R. M., 1982, Virulence of avian influenza A viruses for squirrel monkeys, *Infect. Immunol.* **37:**1119–1128.

Myers, B. J., 1972, Echinococcosis, coenurosis, cysticercosis, sparganosis, etc., in: *Pathology of Simian Primates,* Part II (R. N. T. W. Fiennes, ed.), Karger, Basel, pp. 124–143.

Nahmias, A. J., and Dowdle, W. R., 1968, Antigenic and biologic differences in *Herpesvirus hominis, Progr. Med. Virol.* **10:**110–159.

Nahmias, A. J., London, W. T., Catalano, L. W., Fuccillo, D. A., Sever, J. L., and Graham, C., 1971, Genital *Herpesvirus hominis* type 2 infection: An experimental model in cebus monkeys, *Science* **171:**297–298.

Napier, J. R., and Napier, P. H., 1967, *A Handbook in Living Primates,* Academic Press, New York.

Nathanson, H., and Harrington, B., 1966, Experimental infection of monkeys with Langat virus. I. Comparison of viremia following peripheral inoculation of Langat and Japanese encephalitis viruses, *Am. J. Epidemiol.* **84:**541–556.

National Research Council. Institute of Laboratory Animal Resources. Committee on Conservation of Nonhuman Primates, 1975, *Nonhuman Primates: Usage and Availability for Biomedical Programs,* National Academy of Sciences, Washington, D. C.

Nayar, G. P. S., Crawshaw, G. J., and Neufeld, J. L., 1979, Tularemia in a group of nonhuman primates, *J. Am. Vet. Med. Assoc.* **175:**962–963.

Neal, R. A., 1970, Intestinal protozoa of simians as a handling hazard, *Lab. Anim. Handb.* **4:**77–81.

Nery-Guimaraes, F., and Franken, A. J., 1971, Toxoplasmose em primatas nao humanos. II. Tentativas de infeccoes experimentais em *Macacca mulatta, Cebus apella* e *Callithrix jacchus;* e pesquisa de anticorpos em varias especies de platyrrhinus, *Mem. Inst. Oswaldo Cruz Rio de J.* **69:**97–111.

Nery-Guimaraes, F., Franken, A. J., and Chagas, W. A., 1971, Toxoplasmose em primatas nao humanos. I. Infeccoes naturais em *"Macacca mulatta"* e *"Cebus apella," Mem. Inst. Oswaldo Cruz Rio de J.* **69:**77–87.

Neubauer, R. H., and Rabin, H., 1979, Herpesvirus-induced lymphomas: Immunodepression and disease, in: *Naturally Occurring Biological Immunosuppressive Factors and Their Relationship to Disease,* CRC Press, Boca Raton, Florida, pp. 203–231.

Neubauer, R. H., Wallen, W. C., and Rabin, H., 1974, Stimulation of *Herpesvirus saimiri* expression in the absence of evidence for type C virus activation in marmoset lymphoid cell line, *J. Virol.* **14:**745–750.

Nicholas, A. H., 1970, A poxvirus of primates. II. Immunology, *J. Natl. Cancer Inst.* **45:**907–914.

Noble, J., Jr., 1970, A study of New and Old World monkeys to determine the likelihood of a simian reservoir of smallpox, *Bull. WHO* **42:**509–514.

Norrby, E., Bartha, A., Boulanger, P., Dreizin, R. S., Ginsberg, H. S., Kalter, S. S., Kawamura, H., Rowe, W. P., Russell, W. C., Schlesinger, R. W., and Wigand, r., 1976, Adenoviridae, *Intervirology* **7:**117–125.

Ordy, J. M., Rangan, S. R. S., Wolf, R. H., Knight, C., and Dunlap, W. P., 1981, Congenital cytomegalovirus effects on postnatal neurological development of squirrel monkey (*Saimiri sciureus*) offspring, *Exp. Neurol.* **74:**728–747.

Orihel, T. C., 1970, The helminth parasites of nonhuman primates and man, *Lab. Anim. Care* **20**:395–401.

Orihel, T. C., 1971, Primates as models for parasitological research, in: *Medical Primatology, 1970* (E. I. Goldsmith and J. Moor-Jankowski, eds.), Karger, Basel, pp. 772–782.

Orihel, T. C., and Seibold, H. R., 1971, Trichospirurosis in South American monkeys, *J. Parasitol.* **57**:1366–1368.

Orihel, T. C., and Seibold, H. R., 1972, Nematodes of the bowel and tissues, in: *Pathology of Simian Primates,* Part II (R. N. T. W. Fiennes, ed.), Karger, Basel, pp. 76–103.

Palmer, A. E., Allen, A. M., Tauraso, N. M., and Shelokov, A., 1968, Simian hemorrhagic fever. I. Clinical and epizootiologic aspects of an outbreak among quarantined monkeys, *Am. J. Trop. Med. Hyg.* **17**:404–412.

Paul, J. R., 1971, *A History of Poliomyelitis,* Yale University Press, New Haven.

Pearson, G. R., and Davis, S., 1976, Immune response of monkeys to lymphotrophic herpesvirus antigens *Cancer Res.* **36**:688–691.

Peters, J. C., 1966, A monkeypox-enzooty in the Blijidorp Zoo, *Tijdschr. Diergeneeskd.* **91**:387–391.

Pizza, G., Levine, P. H., Ablashi, D. V., Armstrong, G., Bengali, Z., and Cannon, G. B., 1982, Variations in the immune response to *Herpesvirus saimiri* in squirrel and rhesus monkeys, *Comp. Immunol. Microbiol. Infect. Dis.* **5**:437–446.

Poiesz, B. L., Ruschetti, F. W., Gozdar, A. F., Bunn, P. A., Minna, J. D., and Gallo, R. C., 1980, Detection and isolation of type C retrovirus particles from fresh and cultured lymphocytes of a patient with cutaneous T-cell lymphoma, *Proc. Natl. Acad. Sci. USA* **77**:7415–7419.

Price, W. H., Casals, J., and O'Leary, W., 1974, Studies on the sequential immunization against group B arboviruses in squirrel monkeys, cynomolgus monkeys, rhesus monkeys, and chimpanzees, *Am. J. Trop. Med. Hyg.* **23**:118–130.

Rabin, H., 1971, Assay and pathogenesis of oncogenic viruses in nonhuman primates, *Lab. Anim. Sci.* **21**:1032–1049.

Rabin, H., 1983, Lymphotropic herpesviruses of nonhuman primates, in: *Viral and Immunological Diseases in Nonhuman Primates* (S. S. Kalter, ed.), A. R. Liss, New York, pp. 111–133.

Rabin, H., and Cooper, R. W., 1971, Tumor production in squirrel monkeys (*Saimiri sciureus*) by Rous sarcoma virus, *Lab. Anim. Sci.* **21**:705–711.

Rabin, H., Theilen, G., and Cooper, R., 1971, Sarcomas induced in squirrel monkeys by feline sarcoma virus, *Bacteriol. Proc.* **71**:193(V154).

Rabin, H., Theilen, G. H., Sarma, P. S., Dungworth, D. L., Nelson-Rees, W. A., and Cooper, R. W., 1972, Tumor induction in squirrel monkeys by the ST strain of feline sarcoma virus, *J. Natl. Cancer Inst.* **49**:441–450.

Randall, R. E., and Honess, R. W., 1980, Proteins of *Herpesvirus saimiri;* Identification of two virus polypeptides released into the culture medium of productively infected cells, *J. Gen. Virol.* **51**:445–449.

Randall, R. E., and Honess, R. W., 1982, Proteins specified by *Herpesvirus saimiri:* Purification and properties of a single polypeptide which elicits virus-neutralizing antibody, *J. Gen. Virol.* **58**:149–161.

Rangan, S. R. S., and Chaiban, J., 1980, Isolation and characterization of a cytomegalovirus from the salivary gland of a squirrel monkey (*Saimiri sciureus*), *Lab. Anim. Sci.* **30**:532–540.

Rangan, S. R. S., and Gallagher, R. E., 1979, Tumors and viruses in nonhuman primates, *Adv. Virus Res.* **24**:1–123.

Ranney, R. R., and Zander, H. A., 1970, Allergic periodontal disease in sensitized squirrel monkeys, *J. Periodontol.* **41**:12–21.

Rapp, F., 1974, Herpesviruses and cancer, *Adv. Cancer Res.* **19**:265–302.

Ratcliffe, H. L., and Worth, C. B., 1951, Toxoplasmosis of captive wild birds and mammals, *Am. J. Pathol.* **27**:655–667.

Remington, J. S., Soave, O. A., and Davis, J., 1965, A serological survey in three species of monkeys for antibodies to toxoplasma, *Am. J. Trop. Med. Hyg.* **14**:724–726.

Richardson, J. H., and Humphrey, G. L., 1971, Rabies in imported nonhuman primates, *Lab. Anim. Sci.* **21**:1083.

Riopelle, A. J., Ayres, J. P., Seibold, H. R., and Wolf, R. H., 1971, Studies of primate diseases at the Delta Center, in: *Medical Primatology, 1970* (E. I. Goldsmith and J. Moor-Jankowski, eds.), Karger, Basel, pp. 826–838.

Roca-Garcia, M., and San Martin-Barberi, C., 1957, The isolation of encephalomycarditis virus from *Aotus* monkeys, *Am. J. Trop. Med. Hyg.* **6**:840–852.

Roizman, B., 1974, Herpesviruses, latency and cancer: A biochemical approach, *J. Reticuloendothel. Soc.* **15**:312–321.

Rommel, F. A., Bendure, D. W., and Kalter, S. S., 1980, Hemolytic complement in nonhuman primates, *Lab. Anim. Sci.* **30**:1026–1029.

Rose, N. R., and Friedman, H., eds., 1980, *Manual of Clinical Immunology*, 2d ed., American Society for Microbiology, Washington, D. C.

Rosenblum, L. A., and Cooper, R. W., eds., 1968, *The Squirrel Monkey*, Academic Press, New York.

Rossan, R. N., Baerg, D. C., and Young, M. D., 1972, Characterization of *Plasmodium vivax* infections in *Saimiri sciureus* (squirrel monkeys), *Proc. Helminthol. Soc. Wash.* **39**:24–28.

Ruch, T. C., 1967, *Diseases of Laboratory Primates*, Saunders, Philadelphia.

Ruffie, J., 1974, Immunogenetics of primates, *Adv. Behav. Biol.* **9**:177–219.

Rustigian, R., Johnston, P., and Reihart, H., 1955, Infection of monkey kidney tissue cultures with virus-like agents, *Proc. Soc. Exp. Biol. Med.* **88**:8–16.

Sabin, A. B., 1934, Studies on the B virus. III. The experimental disease in *Macacus rhesus* monkeys, *Br. J. Exp. Pathol.* **XV**:321–334.

Sabin, A. B., 1959, Dengue, in: *Viral and Rickettsial Infections of Man*, 3rd ed. (T. M. Rivers and F. L. Horsfall, eds.), Lippincott, Philadelphia, pp. 361–373.

Sabin, A.B., and Wright, A. M., 1934, Acute ascending myelitis following a monkey bite, with the isolation of a virus capable of reproducing the disease, *J. Exp. Med.* **9**:115–136.

Sarich, V. M., and Wilson, A. C., 1966, Quantitative immunochemistry and the evolution of primate albumins: Micro-complement fixation, *Science* **154**:1563–1566.

Saslaw, S., and Carlisle, H. N., 1969, Use of subhuman primates in experimental infections, *Ann. N. Y. Acad. Sci.* **162**:568–586.

Schiff, G. M., and Sever, J. L., 1966, Rubella: Recent laboratory and clinical advances, *Progr. Med. Virol.* **8**:30–61.

Schmidt, G. D., 1972, Acanthocephala of captive primates, in: *Pathology of Simian Primates*, Part II (R. N. T. W. Fiennes), Karger, Basel, pp. 144–156.

Schmitt, J., Krupe, M., and Deicher, H., 1965, Untersuchungen uber Immunoglobulin-Ermerkmale (Gm- und InV-Gruppen) bei subhumanen Primaten, *Humangenetik* **1**:571–573.

Schudel, A. A., Sadir, A. M., Etcheverrigaray, M. E., Samus, S., Colilla, O., and Rivenson, S., 1982, Susceptibility of squirrel and cebus monkeys to foot-and-mouth disease virus, *J. Med. Primatol.* **11**:126–132.

Sehgal, C. L., and Ray, S. N., 1974, Survey of rhesus monkeys (*Macaca mulatta*) for haemagglutination-inhibition antibody against vaccinia/variola and monkey pox visuses, *J. Commun. Dis.* **6**:233–235.

Shevtsova, Z. V., 1967, Izuchenie etiologii gemorragicheskoi likhoradki obez'ian, *Vopr. Virusol.* **12**:47–51 [in Russian].

Shope, T., and Miller, G., 1973, Epstein–Barr virus. Heterophile responses in squirrel monkeys inoculated with virus-transmormed autologous leucocytes, *J. Exp. Med.* **137**:140–147.

Siegert, R., 1972, Marburg virus, *Virol. Monogr.* **11**:97–153.

Sigurdsson, B., 1954, Rida, a chronic encephalitis of sheep with general remarks on infections wihich develop slowly and some of their special characteristics, *Br. Vet. J.* **110**:341–354.

Sigurdsson, B., Palsson, P., and Grimsson, H., 1957, Visna, a demyelinating transmissible disease of sheep, *J. Neuropathol. Exp. Neurol.* **16**:389–403.

Silverstein, A. M., Prendergast, R. A., and Parshall, C. J., Jr., 1970, Cellular kinetics of the antibody response by the fetal rhesus monkey, *J. Immunol.* **104**:269–271.

Simpson, D. I. H., 1969, Marburg agent disease: In monkeys, *Trans. R. Soc. Trop. Med. Hyg.* **63**:303–309.

Slenczka, W., Wolff, G., and Siegert, R., 1971, A critical study of monkey sera for the presence of antibody against the Marburg virus, *Am. J. Epidemiol.* **93**:496–505.

Smetana, H. F., 1972, Infectious hepatitis, in: *Pathology of Simian Primates*, Part II (R. N. T. W. Fiennes, ed.), Karger, Basel, pp. 663–683.

Smetana, H. F., and Orihel, T. C., 1969, Gastric papillomata in *Macaca speciosa* induced by *Nochtia nochti* (Nematoda: Trichostrongyloidea), *J. Parasitol.* **55**:349–351.

Smith, A. A., and McNulty, W. P., Jr., 1969, Salivary gland inclusion disease in the tarsier, *Lab. Anim. Care* **19**:479–481.

Smith, G. C., Heberling, R. L., Helmke, R. J., Barker, S. T., and Kalter, S. S., 1977, Oncornavirus-like particles in squirrel monkey (*Saimiri sciureus*) placenta and placenta culture, *J. Natl. Cancer Inst.* **59**:975–979.

Smith, G. C., Barker, S. T., and Kalter, S. S., 1978, Coexistance of squirrel monkey retrovirus and *Herpesvirus saimiri* in a squirrel monkey throat swab culture, *Abst. Annu. Meet. Am. Soc. Microbiol.* **78**:276(S389).

Smith, J. L., and Israel, C. W., 1968, Recovery of spirochaetes in the monkey by passive transfer from human late sero-negative syphilis, *Br. J. Vener. Dis.* **44**:109–115.

Soave, O., Jackson, S., and Ghumman, J. S., 1981, Atypical mycobacteria as the probable cause of positive tuberculin reactions in squirrel monkeys (*Saimiri sciureus*), *Lab. Anim. Sci.* **31**:295–296.

Sodeman, T. M., Held, J. R., Contacos, P. G., Jumper, J. R., and Smith, C. S., 1969, Studies of the exoerythrocytic stages of simian malaria. IV. *Plasmodium brasilianum*, *J. Parasitol.* **55**:963–970.

Sproul, E. E., Metzgar, R. S., and Grace, J. T., Jr., 1963, The pathogenesis of Yaba virus-induced histiocytomas in primates, *Cancer Res.* **23**:671–675.

Srivastava, P. K., Srivastava, A. K., and Lucas, F. V., 1969, Squirrel monkey (*Saimiri sciureus*) leucocyte culture: Pokeweed mitogen and human AB serum, important adjuvants, *Stain. Technol.* **44**:205–207.

Stephen, E. L., Walker, J. S., Dominik, J. W., Young, H. W., and Berendt, R. F., 1977, Aerosol therapy of influenza infections of mice and primates with rimantadine, ribavirin, and related compounds, *Ann. N. Y. Acad. Sci.* **284**:264–271.

Stephen, E. L., Scott, G. H., and Berendt, R. F., 1978, Experimental infection of mice and squirrel monkeys using swine influenza virus: Effect of amantadine, rimantadine, and ribavirin, *Fed. Proc.* **37**:379 (Abstract).

Swellengrebel, N. H., and Rijpstra, A. C., 1965, Lateral-spined schistosome ova in the intestine of a squirrel monkey from Surinam, *Trop. Geogr. Med.* **17**:80–84.

Tauraso, N. M., Shelokov, A., Allen, A. M., Palmer, A. E., and Aulisio, C. G., 1968a, Epizootic of simian haemorrhagic fever, *Nature* **218**:876–877.

Tauraso, N. M., Shelokov, A., Palmer, A. E., and Allen, A. M., 1968b, Simian hemorrhagic fever. III. Isolation and characterization of a viral agent, *Am. J. Trop. Med. Hyg.* **17**:422–431.

Tauraso, N. M., Kalter, S. S., Ratner, J. J., and Heberling, R. L., 1971, Simian hemorrhagic fever, in: *Medical Primatology, 1970*, (E. I. Goldsmith and J. Moor-Jankowski, eds.), Karger, Basel, New York, pp. 660–670.

Taylor-Robinson, D., Martin-Bourgon, C., Watanabe, T., and Addey, J. P., 1971, Isolation of T-mycoplasmas from dogs and squirrel monkeys: Biological and serological comparison with those isolated from man and cattle, *J. Gen. Microbiol.* **68**:97–107.

Ter Meulen, V., and Hall, W. W., 1978, Slow virus infections of the nervous system: Virological, immunological and pathogenetic considerations, *J. Gen. Virol.* **41**:1–25.

Theilen, G. H., Gould, D., Fowler, M., and Dungworth, D. L., 1971, C-Type virus in tumor tissue of a woolly monkey (*Lagothrix* spp.) with fibrosarcoma, *J. Natl. Cancer Inst.* **47**:881–889.

Theilen, G. H., Wolfe, L. G., Rabin, H., Deinhardt, F., Dungworth, D. L., Fowler, M. E., Gould, D., and Cooper, R., 1973, Biological studies in four species of nonhuman primates with simian sarcoma virus (*Lagothrix*), *Bibl. Haematol.* **39**:251–257.

Thienpont, D. D., Mortelmans, J., and Vercruysse, J., 1962, Contribution a l'étude de la Trichuriose du Chimpanzé et de son traitement avec la Methyridine, *Ann. Soc. Belg. Med. Trop.* **42**:211–218.

Tokura, H., and Katsubayashi, K., 1973, Experiences in quarantine of imported monkeys, *Exp. Anim.* **22**(Suppl.):443–446.

Traub, E., 1935, A filterable virus recovered from white mice, *Science* **81**:298–299.

Ushijima, R. N., Shininger, F. S., and Gardner, C. E., 1966, Susceptibility of cultured renal cells from different species of subhuman primates to simian virus 40, *Proc. Soc. Exp. Biol. Med.* **122**:673–675.

Vogel, F. S., and Pinkerton, H., 1955, Spontaneous salivary gland virus disease in chimpanzees, *Arch. Pathol.* **60**:281–285.

Voller, A., 1972, Plasmodium and hepatocystis, in: *Pathology of Simian Primates,* Part II (R. N. T. N. Fiennes, ed.), Karger, Basel, pp. 57–73.

Von Magnus, P., Andersen, E. K., Petersen, K. B., and Birch-Andersen, A., 1959, A pox-like disease in cynomolgus monkeys, *Acta Pathol. Microbiol. Scand.* **46**:156–176.

Walker, D. H., Wulff, H., Lange, J. V., and Murphy, F. A., 1975a, Comparative pathology of Lassa virus infection in monkeys, guinea-pigs, and *Mastomys natalensis, Bull. WHO* **52**:523–534.

Walker, D. H., Wulff, H., and Murphy, F. A., 1975b, Experimental Lassa virus infection in the squirrel monkey, *Am. J. Pathol.* **80**:261–278.

Wang, A. C., Shuster, J., Epstein, A., and Fudenberg, H. H., 1968, Evolution of antigenic determinants of transferrin and other serum proteins in primates, *Biochem. Genet.* **1**:347–358.

White, R. J., Simmons, L., and Wilson, R. B., 1972, Chickenpox in young anthropoid apes: Clinical and laboratory findings, *J. Am. Vet. Med. Assoc.* **161**:690–692.

Whitney, R. A., Jr., Johnson, D. J., and Cole, W. C., 1973, *Laboratory Primate Handbook,* Academic Press, New York.

Wigand, R., Bartha, A., Dreizin, R. S., Esche, H., Ginsberg, H. S., Green, M., Hierholzer, J. C., Kalter, S. S., McFerran, J. B., Pettersson, U., Russell, W. C., and Wadell, G., 1982, Adenoviridae: Second Report, *Intervirology* **18**:169–176.

Willimzik, H.-F., Kalter, S. S., Lester, T. L., and Wigand, R., 1981, Immunological relationship among adenoviruses of humans, simians, and nonprimates as determined by the neutralization test, *Intervirology* **15**:28–36.

Wise, T. G., Pavan, P. R., and Enis, F. A., 1977, Herpes simplex virus vaccines, *J. Infect. Dis.* **136**:706–711.

Wolfe, L. G., Falk, L. A., and Deinhardt, F., 1971, Oncogenicity of *Herpesvirus saimiri* in marmoset monkeys, *J. Natl. Cancer Inst.* **47**(5):1145–1162.

Wolff, J. W., Collier, W. A., Bool, P. H., and Bohlander, H., 1958, Investigations on the occurrence of leptospirosis in Surinam, *Trop. Geogr. Med.* **10**:341–346.

Wong, M. M., 1970, Procedure in laboratory examination of primates with special reference to necropsy technics, *Lab. Anim. Care* **20**:337–341.

Workshop on Nonhuman Primate Parasites, 1st, Southwest Foundation for Research and Education, San Antonio, Texas, 1970, Nonhuman primate parasites, *Lab. Anim. Care* **20**:319–409.

Worms, M. J., 1967, Parasites in newly imported animals, *J. Inst. Anim. Tech.* **18**:39–47.

Wright, J., Falk, L. A., Collins, D., and Deinhardt, F., 1976, Mononuclear cell fraction carrying *Herpesvirus saimiri* in persistently infected squirrel monkeys, *J. Natl. Cancer Inst.* **57**:959–962.

Yamashita, J., 1963, Ecological relationships between parasites and primates. I. Helminth parasites and primates, *Primates* **4**:1–96.

Young, M. D., and Rossan, R. N., 1969, *Plasmodium falciparum* induced in the squirrel monkey, *Saimiri sciureus, Trans. R. Soc. Trop. Med. Hyg.* **63**:686–687.

Young, M. D., Baerg, D. C., and Rossan, R. N., 1975, Parasitological review. Experimental monkey hosts for human plasmodia, *Exp. Parasitol.* **38**:136–152.

Zilber, L. A., Lapin, B. A., and Adgighytov, F. I., 1965, Pathogenicity of Rous sarcoma virus for monkeys, *Nature* **205**:1123–1124.

Zlotnik, I., 1969, Marburg agent disease: Pathology, *Trans. R. Soc. Trop. Med. Hyg.* **63**:310–323.

Zlotnik, I., Grant, D. P., Dayan, A. D., and Earl, C. J., 1974, Transmission of Creutzfeldt–Jakob disease from man to squirrel monkey, *Lancet* **2**:435–438.

Zlotnik, I., Grant, D. P., Dayan, A. D., Earl, C. J., Illis, L. S., and Weller, R. O., 1976, Further observations on experimental transmission of Creutzfeldt–Jakob disease from man to squirrel and spider monkeys, *Neuropathol. Appl. Neurobiol.* **2**:125–130.

Zur Hausen, H., 1975, Oncogenic herpes viruses, *Biochim. Biophys. Acta* **417**:25–53.

16

Medical Care and Management of the Squirrel Monkey

Christian R. Abee

1. INTRODUCTION

The use of the squirrel monkey in biomedical research has helped to improve our understanding of many diseases and other disorders of human beings in the past 20 years. Most of this research has been and continues to be carried out using animals that are captured as juveniles or adults from the wild. The introduction and long-term maintenance of these animals in the laboratory requires an understanding of the husbandry and medical care needs of this species. With improvements in facility design, husbandry techniques, and equipment for medical care, morbidity and mortality among laboratory-housed squirrel monkeys have been greatly reduced. An excellent chapter on the care and management of the squirrel monkey was written by Lang (1968) and is a valuable source of additional information on the care of the squirrel monkey.

This chapter is devoted to the development of an understanding of those factors that must be considered when using squirrel monkeys in research. Since most studies involve the use of groups or colonies of squirrel monkeys, care and management will be discussed from the perspective of the colony rather than the individual.

2. COLONY HUSBANDRY

Perhaps the single most important but often overlooked aspect of the care of laboratory primates is daily husbandry. Close attention to the basic needs of the colony will often prevent serious problems from occurring. Daily observations made during the course of cage cleaning and feeding provide many useful clues to the condition of individual animals and the colony as a whole.

Christian R. Abee • Department of Comparative Medicine, University of South Alabama College of Medicine, Mobile, Alabama 36688.

The choice of functional housing designed to facilitate an effective husbandry program will minimize stress and ultimately contribute to the success of the research program.

2.1. Housing and Cage Requirements

Squirrel monkeys are housed in a variety of ways. Appropriate housing requires thoughtful consideration of many factors. Physical comfort of the animals is important in reducing stress and promoting the well-being of colony members. The requirements of the research to be carried out must be considered. A research project that involves only occasional manipulation of animals can be done with a large group cage or pen, which will provide free movement of animals within the cage. However, studies that require daily or constant manipulation may be done best in an individual animal cage arrangement.

The Institute for Laboratory Animal Resources (ILAR) has established guidelines (National Institutes of Health, 1978) that should be considered in the selection of a cage system. These are:

1. The system should be designed to promote physical comfort of the animals being maintained.
2. The system should be compatable with normal growth and development and prevention of disease.
3. The system should be designed to facilitate sanitary and service maintenance.
4. The system should meet the investigators' research needs.

ILAR has also established cage space recommendations for squirrel monkeys, which are classified as group I, small primates weighing less than 1 kg. It is recommended that they be provided a minimum of 1.6 ft^2 of living area and a minimum of 20 inches of cage height. The National Institutes of Health requires grant and contract recipients to comply with these guidelines.

2.1.1. Types of Cage Systems

A variety of single-animal cages are available from commercial sources. Most single cages used today are constructed of stainless steel welded wire (Fig. 1). These cages can be suspended on racks mounted on casters or hung on the wall using wall-mounted frames made for that purpose. There are advantages to each arrangement that should be considered in deciding whether to use rack- or wall-mounted cages. Rack-mounted cages offer mobility and can be quickly moved via castors if necessary. The rack and cages can be washed as a unit in a rack washer. The wall-mounted cages, on the other hand, use space more efficiently, thus maximizing the number of cages that can be placed within a room.

There are several common group or "gang" housing cage systems used for squirrel monkeys. The most common is the "run"-style cage, which is a rectangular cage with dimensions 3–6 ft wide by 8–20 ft long. The height of the cage varies, but is usually sufficient for an attendant to stand upright while

Fig. 1. A cage rack mounted on casters containing four stainless steel individual cages is easily moved and cleaned due to its portability. (Courtesy of Allentown Caging Equipment Company, Allentown, New Jersey.)

cleaning inside it. Squirrel monkey groups ranging from 5 to 20 animals can be housed in this type of cage. This cage is often used for housing breeding colonies. The run-type cages are constructed of a variety of materials, from chainlink fencing to prefabricated aluminum paneling (Fig. 2). These cages cannot be easily moved and therefore must be hand cleaned. Cages of this type

Fig. 2. "Run"-type group cage for squirrel monkeys may be constructed of prefabricated aluminum panels bolted to the floor. (Courtesy of Primate Research Laboratory of the University of South Alabama, College of Medicine; photo by W. E. Beckham, Jr.)

can be used in an indoor environmentally controlled area or outdoors. If run-type cages are to be used outdoors, arrangements must be made to protect the animals from temperature extremes. A corral-type cage system has also been used to house groups of squirrel monkeys. Corrals are outdoor pens usually constructed of wire and sheet metal. The enclosure may or may not have a wire ceiling, depending upon whether the wall can be climbed by the animals within the cage. The floor of the corral is usually gravel or earth. Corral-type pens usually provide a large amount of living area relative to the number of animals being housed.

The advantages of outdoor cages versus indoor group and single cages must be carefully considered prior to acquiring a colony of squirrel monkeys for a research project. The outdoor systems are generally less costly to operate, since ventilation and air conditioning are unnecessary. The labor required to maintain animals in corral-type pens with gravel or natural grass floors is much less, since cleaning and sanitation requirements are reduced. However, animals housed in outdoor pens are less accessible for manipulation and are more difficult to examine for detection of medical problems.

The indoor cage systems provide for much better environmental control and the animals are more easily observed and manipulated. For most research uses, one of the indoor cage systems will be the easiest way to provide for the needs of the squirrel monkey and the research to be carried out. Some facilities have incorporated the advantages of both the indoor and outdoor cage systems by providing outdoor runs or group cages that can be enclosed during inclement weather. These indoor/outdoor systems provide many of the advantages of each system.

There are two basic types of watering devices used for squirrel monkeys. These are bottles with sipper tubes and automatic watering systems, which consist of low-pressure water pipes equipped with watering valves. Squirrel monkeys unaccustomed to these devices must be shown how they operate, to avoid dehydration. This is most often a problem with newly imported animals.

Water bottles with sipper tubes are usually mastered quickly after a technician has manipulated the bottle so that some water drips from the tube. Automatic watering devices are sometimes more difficult, since the animals are unable to see the water. Squirrel monkeys will quickly adapt to using these devices if the valves are made to leak slowly for the first 24 hr after new animals are moved into the cage. Automatic watering valves must be checked regularly to make certain that they are functioning properly.

A variety of feeders may be attached to cages. Since group-housed animals may fight over feed, it is a good practice to provide sufficient feed to allow *ad libitum* feeding throughout the daylight hours. A second feeder separately located from the first will help to assure that less dominant group members are not prevented from eating.

Perches are an important part of the cage, since they provide a surface for sitting and sleeping. A variety of materials have been used as perches, from wood to steel pipe. In large cages, an effort should be made to place perches at several levels within the living area for climbing and jumping from one perch to another. It is best to construct perches from materials that are not highly thermoconductive, since such materials tend to dissipate body heat (Moreland, 1972). We have found that polyvinyl chloride pipe of the type used by plumbers can be cut and joined to meet the requirements of almost any cage configuration. Plastic pipe is easily cleaned and is less thermoconductive than metal perches.

Squirrel monkeys are almost entirely arboreal in their natural habitat. They lack ischeal callosities and therefore are prone to the development of sores if they are not provided with suitable structures on which to climb and perch. Squirrel monkeys prefer a flat, shelf-type surface for sleeping, but animals using such perches frequently develop pressure ulcers on the dorsal aspect of the tail (Clewe, 1965). By using large-diameter plastic pipe (1.5 inch), a highly

desirable perch can be provided. These perches have a broad surface yet are sufficiently contoured to avoid tail sores.

2.1.2. Facility Design and Environment

The design of a facility for housing squirrel monkeys must be responsive to the basic needs of the species and the type of research to be carried out. Many of the design features of a multispecies vivarial facility as described in the National Institutes of Health (1978) are directly applicable for housing the squirrel monkey. In this chapter, design considerations of specific importance to the squirrel monkey will be described.

Squirrel monkeys do not tolerate low temperatures well. Facilities for housing this species must be designed to prevent temperature extremes. In general, the smaller the cage in which the animal is housed, the greater the importance of temperature control. Squirrel monkeys in individual cages are more prone to cold and heat stress than animals housed in larger group enclosures. Animals housed in individual cages do very well with environmental temperatures of 75–80°F. Squirrel monkeys that are debilitated due to disease or other causes seem to require somewhat higher temperatures, 78–85°F. Although the majority of problems associated with environmental temperature extremes are related to cold stress, squirrel monkeys can experience heat exhaustion with extreme elevations of temperature and humidity. Hickey and Kelly (1972) reported four cases of heat exhaustion resulting in death in a colony of 37 squirrel monkeys housed in a room where a mechanical failure resulted in room temperatures of 100°F and 100% humidity for 2 hr. Although this report describes heat exhaustion due to mechanical failure, it provides an excellent description of the clinical signs of a disease that could be encountered with some outdoor housing systems. The clinical signs included high-pitched screaming, attempts at copulation, panting, depression, and death.

Relative humidity is another important environmental concern. The percentage of moisture held in air depends upon the temperature of the air. Therefore, humidity is relative to temperature. A room in which outside air initially at 25°F is warmed to 75°F without humidification may be uncomfortable even though the temperature is 75°F. The warmed air has a much greater capacity for holding moisture, resulting in rapid evaporation from skin and mucous membranes. Squirrel monkeys housed under conditions of low relative humidity frequently develop crusts around the nose and dry rales as a result of drying and irritation of nasal and respiratory membranes. Animals housed in areas with low relative humidity are also prone to chilling. This chilling effect is especially apparent in squirrel monkeys maintained in stainless steel or aluminum cages, since contact with the metal surfaces further aggravates the problem by conducting heat away from the body. Relative humidity should be maintained between 40 and 60% for squirrel monkeys housed within the temperature ranges described in this chapter.

The air conditioning and heating system should be designed to provide approximately 10–15 air changes per hour. Supply air should be 100% fresh air makeup; however, the ever-increasing cost of utilities may necessitate consideration of a recirculating system. If a recirculating ventilation system is used, high-efficiency particulate air (HEPA) filtration prior to recirculation should be designed into the system. The rationale for using fresh air or HEPA filtered air involves two considerations. First, very strong odors in poorly ventilated housing areas can be reduced. Second, clean air is an aid in promoting good health by minimizing the potential for spread of pathogens. Air from squirrel monkey housing areas should not be recirculated to areas of human habitation. Static air pressures in rooms housing squirrel monkeys should be negative to adjacent halls or rooms. This will further aid in the containment of odors and help to provide isolation between animal housing and other work areas.

2.1.3. Cleaning and Sanitation

The maintenance of squirrel monkeys in cages limits their living space to a relatively small area. This results in concentration of feces and urine within the animals' immediate living area. Many pathogens are shed in large numbers in feces. Therefore, the daily cleaning of cages and contact surfaces is an important part of a preventive medicine program. Accumulations of feces and wasted feed attract vermin, which can also spread disease. A stringent cleaning and sanitation program is an effective means of reducing cockroach and rodent infestation.

Cleaning procedures vary, depending upon the type of cage system in use. The outdoor corral-type cage requires a minimum of cleaning. Occasional raking of the gravel floor and routine sanitation of feeders is usually adequate for a healthy outdoor colony. Indoor cage systems must be cleaned more regularly. Rack-mounted stainless steel and aluminim cages are usually constructed with wire mesh or bar floors, which allow urine and feces to fall through the cage floor to a catch pan located under the cage. Catch pans should be cleaned once or twice each day. The entire cage should be changed out, cleaned, and sanitized once each week. These cages are best cleaned in a rack washer that has both a wash and high-temperature rinse cycle. Rinse temperatures must reach 180°F (82.2°C) for a sufficient period to eliminate coliform organisms. Disinfectant detergents can be used along with or in lieu of high-temperature washing, provided that these chemicals are completely rinsed from the cages. Detergent disinfectants should have tuberculocidal activity.

Frequently, accessory items that come in contact with squirrel monkeys are overlooked during cleaning. Catch gloves are often soiled with feces and urine. These gloves are sometimes moved from one housing area to another by technicians unaware that such a practice may spread disease. Catch gloves should be cleaned and disinfected regularly. It is also wise to permanently place catch gloves in each area of use with instructions that gloves from one area are not be used in other areas. This will help reduce the possible spread of pathogens between housing areas. Other items that should be regularly cleaned and disinfected include water bottles, sipper tubes, treatment tables, surrogates used for hand-rearing infants, automatic watering valves, catch nets, and any other items that come in direct contact with squirrel monkeys.

Another aspect of the sanitation program is personal hygiene. Technicians and other personnel working with squirrel monkeys should be provided with clean protective clothing. When working directly with squirrel monkeys, personnel should wear a surgical mask or respirator and gloves. Persons developing positive tuberculin reactions should not work with squirrel monkeys until a sufficient period of treatment has elapsed to prevent the possible transmission of the disease to the colony. This is best determined through consultation with the physician providing treatment.

Staff eating areas should be physically separated from animal housing or work areas. Eating, drinking, and smoking should be prohibited in areas where squirrel monkeys are maintained or manipulated.

2.2. Colony Records

Many colony management decisions will depend upon information obtained from colony records. The discovery of tuberculosis in an animal within the colony will necessitate identification of all animals that may have come in contact with that animal. In a breeding colony, the selection of a male to be placed with a group of females may require knowledge of the pedigrees of the animals involved. This is particularly important if inbreeding is to be prevented or if the purpose of breeding is to foster a heritable trait. The database for the colony should be designed to provide adequate information for making both management and research decisions.

The availability of relatively inexpensive microcomputers capable of maintaining, sorting, and merging large data files is rapidly changing biomedical research. This "computer revolution" has made possible the development of comprehensive animal record databases previously unattainable without a mainframe-based computer center. Progress in the field of computers and programs for use with various brands of computers is changing so rapidly that anything written in a text will be rapidly outdated within a few months. Therefore, only the types of records that should be stored and the fundamental design of the database will be discussed.

Research and animal husbandry data are stored by the computer in fielded data files. These files are sets of lines or records of data stored so that all lines are divided the same way. This concept can best be perceived by imagining a long, continuous stream of paper with columns of information on it. Each file can be thought of as a stream of paper with the columns of data being fields and lines of data being individual records.

The computer software used to enter and manipulate data must be a database manager that allows the creation and maintenance of files. This should include sorting, merging, editing, and transformation of files along with other logical operations. Software should include a report-writing facility that allows the output of data stored in any number of files in a prespecified fashion. The database should be designed with prewritten statistical subroutines or with the capability of transferring data sets into files for statistical computation.

The following is a listing of files that should be included in a database for management of a large colony of squirrel monkeys. The files presented here would be subject to modification and additions, depending upon the type of colony.

1. A *Basic Animal History File* contains all basic nonvariable information on each animal in the colony. This includes birth information if the animal was born in the colony, source of the animal if it was acquired from an importer or other laboratory, sex, and disposition.
2. A *Cage History File* is a record of locations of all animals in the colony according to the date they were moved to a given location. The file should be designed so that each time an animal is moved, a record is made of the new location. When a location history is requested, the file must be sortable by animal identification number and cage dates. This will provide a means for tracing an animal's location chronologically for the entire period it has resided in the colony.
3. A *Medical Procedures File* contains records of every treatment received by each animal within the colony. These records should include identifying information about the animal, the date of treatment, and the nature of the problem requiring treatment. This file is very useful in obtaining epidemiological data about the colony.
4. A *Tuberculin Test History File* stores the dates and results of tuberculin tests carried out on each animal within the colony.
5. A *Research Use File* contains a listing and dates of experiments in which each animal has been used. An experimental accession number and a protocol describing each experiment should be referenced in this file. This file provides an important check procedure to ensure that animals selected for a particular study have not been used in a previous study that would preclude their use.
6. A *Body Weight History File* contains animal identification numbers and body weights obtained at routine intervals and when animals are manipulated. Body weights should be obtained whenever an animal is anesthetized. This file is particularly useful in following seasonal fluctuations in body weight and in estimating the onset of illness in animals that develop chronic or degenerative diseases.
7. A *Necropsy File* should contain all information gathered at necropsy. Consideration should be given to the use of standard descriptions, such as those provided in Coté (1979). This file should be designed to allow sorting of deaths by etiological or morphological diagnosis, origin, age, and sex.

Other files can be created, depending on the special needs within the colony. An example would be the development of a pregnancy palpation file for breeding females and a conception history file keyed to breeding males to help determine breeding performance of both males and females.

3. PRINCIPLES OF MEDICAL MANAGEMENT

3.1. Preventive Medicine

An effective preventive medicine program is the foundation from which a healthy, successful squirrel monkey colony is built. Many factors contribute to a preventive medicine program. These include sanitation procedures, design of the facility, personal hygiene, quarantine and isolation procedures, and clinical care. Squirrel monkeys are stoical animals that do not show clinical signs of disease until they are seriously sick. In clinical terms, this means that animals identified as sick are often critically ill by the time they are discovered. Therefore, prevention of disease rather than treatment should be the primary focus of the colony medical care program.

3.1.1. Quarantine

The quarantine program is the first step in identifying and eliminating diseases in newly acquired squirrel monkeys. There is frequently a temptation to reduce or attenuate quarantine by investigators eager to begin their research. However, this can be a very costly shortcut. It is much better to eliminate disease and stabilize new animals prior to beginning a project than to use animals of questionable health status. The quarantine of new animals is also a very effective means of preventing the introduction of disease into an established colony.

Quarantine should be carried out in an area designed for this purpose and separate from other animal housing areas. Squirrel monkeys should be quarantined in small groups or in individual cages. Once a quarantine group is formed, it must be regarded as a single unit throughout the quarantine period. A standard quarantine period for squirrel monkeys should span at least 60 days. However, if an infectious disease is discovered, this period should be lengthened to 90 days. If one animal within a quarantine group develops an infectious disease, no animal from that group should be released until the disease has been successfully treated and eliminated. This is particularly important in the management of a group in which tuberculosis has been diagnosed. Animals should not be released from a quarantine group with tuberculosis until the entire group has completed at least an additional 90 days of quarantine from the last documented occurrence of the disease. Squirrel monkeys should never be quarantined or housed with other species since each may be a health hazard to the other.

Table I lists a standard quarantine protocol for squirrel monkeys. This protocol may be changed, depending on the origin of the animals to be quarantined and their ultimate research use.

3.1.2. Isolation

Isolation of squirrel monkeys is an important part of the preventive medicine program after quarantine is completed. It should not be assumed that squirrel monkeys successfully completing quarantine can be safely housed in the same room with other primate species. A large percentage of squirrel monkeys harbor latent or subclinical infections that can result in fatal disease in other primate species (Hunt, 1970; Kalter and Heberling, 1976; Rangan *et al.*, 1977).

Table I
Standard Squirrel Monkey Quarantine Procedures

Quarantine procedure	Frequency
Physical examination and treatment by veterinarian	On arrival, as necessary during, and at completion of quarantine
Complete blood counts	Three-week intervals
Blood chemistries	Three-week intervals
Blood films for detection of malaria	Twice
Fecal analyses for parasites	Twice or as needed
Stool culture for enteric pathogens	Twice or as needed
Tuberculin test	On arrival and every 2 weeks

Animals found in group cages or general colony housing areas with clinical signs of disease should be removed and isolated for diagnosis and treatment. The isolation of diseased animals from the colony is an effective way to limit the spread of infectious diseases.

3.1.3. Daily Observations

The daily observation of all squirrel monkeys within the colony is an important husbandry procedure. Technicians should be trained to take this part of their duties very seriously. An overlooked animal during the morning observations could die before being discovered. Daily observations should be carried out at least twice and preferably three times a day.

The first observation should be made before cages are cleaned in the morning. The floor of the cage should be observed for signs of blood, aborted and stillborn fetuses (during the birth season), and animals sitting or lying on the floor of the cage. An estimation of the amount of feed consumed should also be made at this time. If water bottles are being used, the amount of water consumed should also be estimated.

The second observation should be made during or immediately after feeding. Animals that do not appear interested in eating should be observed for signs of injury or disease.

In a breeding colony, an evening observation is often very helpful. This observation is made before the lights are turned off for the night. Observations at this time are often useful in determining which females are nearing parturition. The beginning of labor and abnormal fetal presentations may be observed at this time.

3.1.4. Routine Colony Health Evaluation

Periodic medical examinations of all squirrel monkeys within the colony should be carried out by a veterinarian. This should be done at 3- to 6-month intervals. The following procedures should be carried out at routine intervals: tuberculin test, physical examination, dental examination (teeth should be cleaned if necessary), body weight determination, identification marking as necessary.

3.2. Therapeutic Approaches and Restraint

The determination of appropriate therapy requires the use of diagnostic resources. These resources should include facilities for clinical pathology, diagnostic microbiology, and radiology. Necropsy and histopathology laboratories are also an essential part of the medical care program. Diagnoses made at necropsy will help provide the necessary information for management of remaining squirrel monkeys within an affected group.

The basic approach to treatment of squirrel monkeys should be aimed at providing adequate care with a minimum of manipulation. The choice of drugs and routes of administration should be made with the understanding that the stress associated with treatment can worsen the condition being treated. If the treatment to be used requires frequent chemical restraint, a judgment must be made between the potential benefits of the treatment and the stress of manipulation. Such decisions require experience and an understanding of the research in which the animals are being used.

3.2.1. Intensive Care

An intensive care system for squirrel monkeys can be made by modifying an intensive care cage of the type commercially available for dogs and cats. These support systems are designed with a chamber in which oxygen, relative humidity, and temperature can be controlled independently from the room in which the unit is located. The most frequently used settings in my laboratory for oxygen, relative humidity, and temperature are 40%, 60%, and 82°F, respectively. These parameters may be modified to meet specific clinical requirements. The intensive care chamber is also very useful for maintaining orphaned infant squirrel monkeys during the first weeks of life.

3.2.2. Physical Restraint

Squirrel monkeys can be captured with leather gloves similar to those used by welders. Care must be taken to handle animals gently to avoid accidental injury during the catching process. Nets have also been used to catch squirrel monkeys; however, the need for such devices is very limited and extreme care must be taken to avoid injury to animals being caught.

Several restraint devices have been described for squirrel monkeys. Inexpensive devices for restraining squirrel monkeys in dorsal recumbency have been described (Clarkson *et al.*, 1968; Kuehl and Dukelow, 1974). Restraint chairs have also been devised (Adair, 1969; Dost *et al.*, 1972). The use of these restraint devices requires careful judgment, since unnecessarily prolonged whole-body restraint may not be the most humane method for achieving the goals of the research being done.

Several other restraint devices can be made to partially restrict movement. An "Elizabethean" collar can be made from a "doughnut"-shaped piece of stiff cardboard (Fig. 3). This collar provides an effective way to prevent the animal from self-mutiliation. When fitting the animal with the collar, care must be taken to assure a proper fit around the throat. Squirrel monkeys quickly learn to eat and drink with the collar in place. Plaster casts can also be used to prevent animals from mutilating wounds of the arms, legs, or abdomen. Casts are also useful for immobilizing broken bones. The cast should extend beyond the joints proximal and distal to the fracture. A frequently encountered problem in group-housed squirrel monkeys is tail-chewing. A tail guard made from a 1-ml plastic syringe cover placed over the end of the tail is an excellent way to protect the tail until it has healed.

3.2.3. Chemical Restraint and Anesthesia

Chemical restraint for squirrel monkeys is usually accomplished with ketamine HCl. Greenstein (1975) reported that dosages of 13 mg/kg were adequate for restraint and dosages above 25 mg/kg produced surgical anesthesia. Deaths did not occur at dosages below 325 mg/kg. This indicates a wide margin of safety of ketamine in the squirrel monkey. I have also combined ketamine HCl (14 mg/kg) with xylazine HCl (1.2 mg/kg) to increase muscle relaxation and extend the period of anesthesia. This drug combination appears to be a safe and reliable method for anesthesia in this species.

Anesthesia can be induced with barbiturates and gas anesthetics if necessary. Pentobarbital sodium at a dosage of 25 mg/kg administered intravenously will produce anesthesia for 1–3 hr. This drug has a narrow margin of safety and should be used with great caution in the squirrel monkey. Recovery from barbiturate anesthesia is prolonged and may occasionally span 24 hr. Grimm *et al.* (1969) reported that Halothane® (Ayerst Laboratories, New York), a gas anesthetic, produced a predictable anesthetic state. Recovery following anesthesia was approximately 45 min.

3.3. Methods of Treatment

The route of administration of drugs and other therapeutic agents must be carefully considered. Since the squirrel monkey is a small primate, intravenous therapy requires considerable technical skill and a knowledge of the anatomical sites for administration. Judgment must be exercised in deciding the route for most effective administration. There are situations in which treatment methods may be chosen because they are less stressful to the animal than other methods.

3.3.1. Intravenous Administration

There are six commonly used sites for intravenous administration. These are the paired lateral tail veins, saphenous veins, and femoral veins. The lateral tail veins are located approximately 45° either side of the dorsal midline of the tail. These vessels run the entire length of the tail and are especially useful for

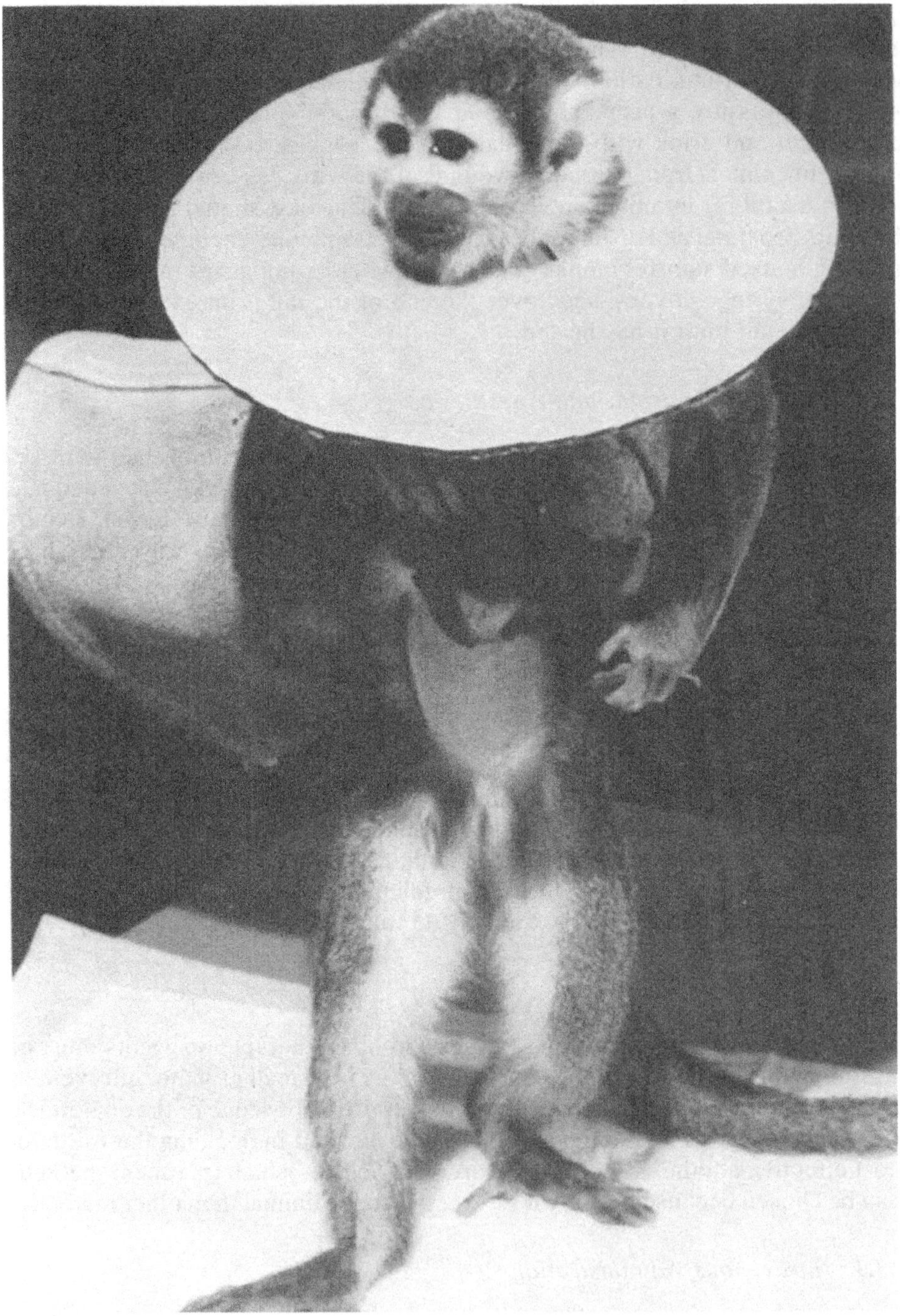

Fig. 3. An "Elizabethean" collar made from stiff cardboard may be used to restrict the squirrel monkey from biting at wounds on its body or touching its head and neck.

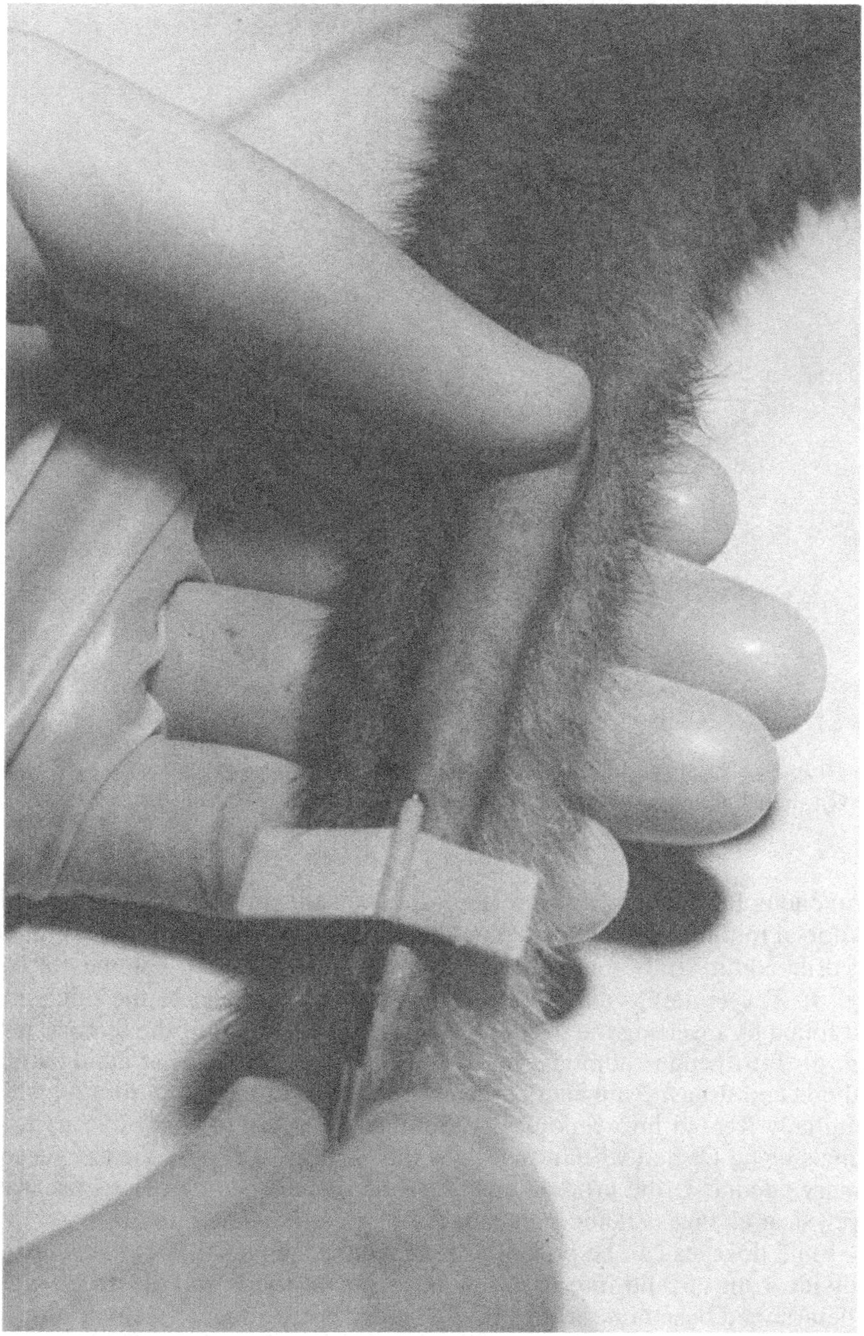

Fig. 4. A butterfly infusion set with a 25-gauge needle is used for intravenous administration via the lateral tail vein. (Photo by W. E. Beckham, Jr.)

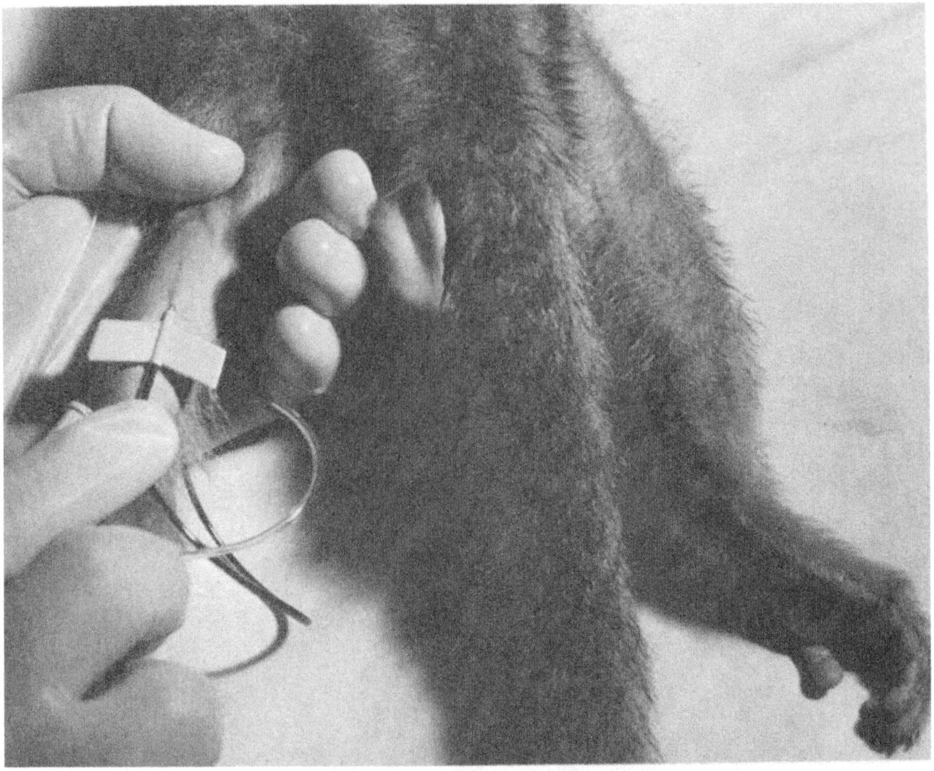

Fig. 5. The saphenous vein located on the dorsal midline of the leg below the knee is used for intravenous administration of drugs and fluids. (Photo by W. E. Beckham, Jr.)

intravenous fluid administration (Fig. 4). The saphenous veins are located on the dorsal midline of each leg below the knee. These vessels are used for fluid and drug administration and can be used for drawing small volumes of blood (Fig. 5). The femoral vein is located on the medial aspect of the thigh and is best found by inserting the needle just medial to the pulse of the femoral artery (Fig. 6). Intravenous administration is the most rapid route of administration of fluids and drugs. Animals that are comatose or rapidly deteriorating should be initially treated intravenously. Care must be taken to avoid overhydrating animals being treated with intravenous fluids, since it is possible to cause pulmonary edema. Urine production should be monitored and close observation of physical changes should be made during fluid administration.

Fluid dosages can be precisely measured using pediatric IV sets designed to divide 1 ml of fluid into 60 drops. Dosages of fluids and electrolytes must be determined based on the size of the animal, percentage of dehydration, and the clinical state. Basic rules for fluid administration in small animals are described by Short (1980). Butterfly infusion sets with 23 or 25 gauge needles are used for venipuncture. The infusion set is secured with adhesive tape for prolonged fluid administration.

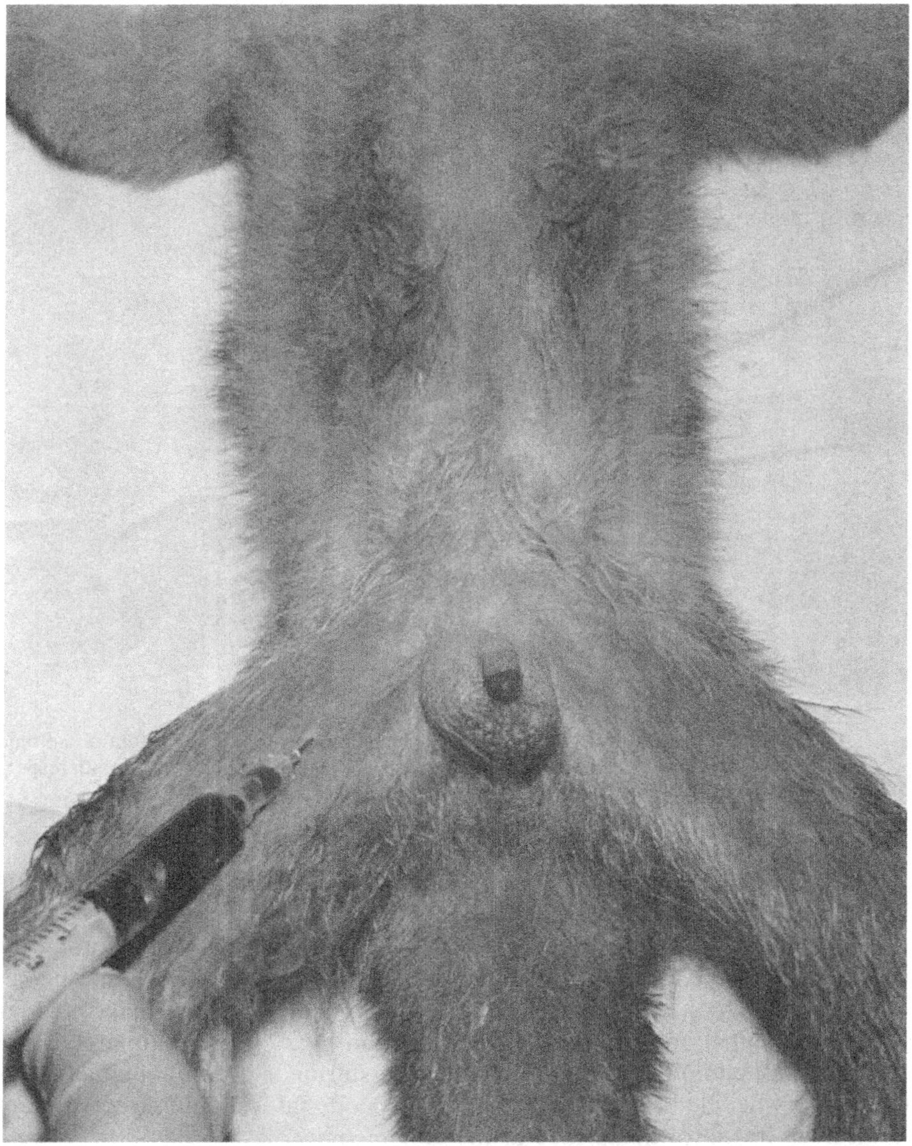

Fig. 6. The femoral vein is used for obtaining blood samples from squirrel monkeys. The vein is located medial to the femoral artery on the medial aspect of the thigh. Skilled personnel can obtain blood samples quickly and without the aid of anesthetics. (Photo by W. E. Beckham Jr.)

3.3.2. Intraperitoneal Administration

Intraperitoneal (IP) injection of fluids and electrolytes is the next most rapidly absorbed route of administration. The IP infusion of fluids can be used when the animal has become so hypotensive that venipuncture is no longer

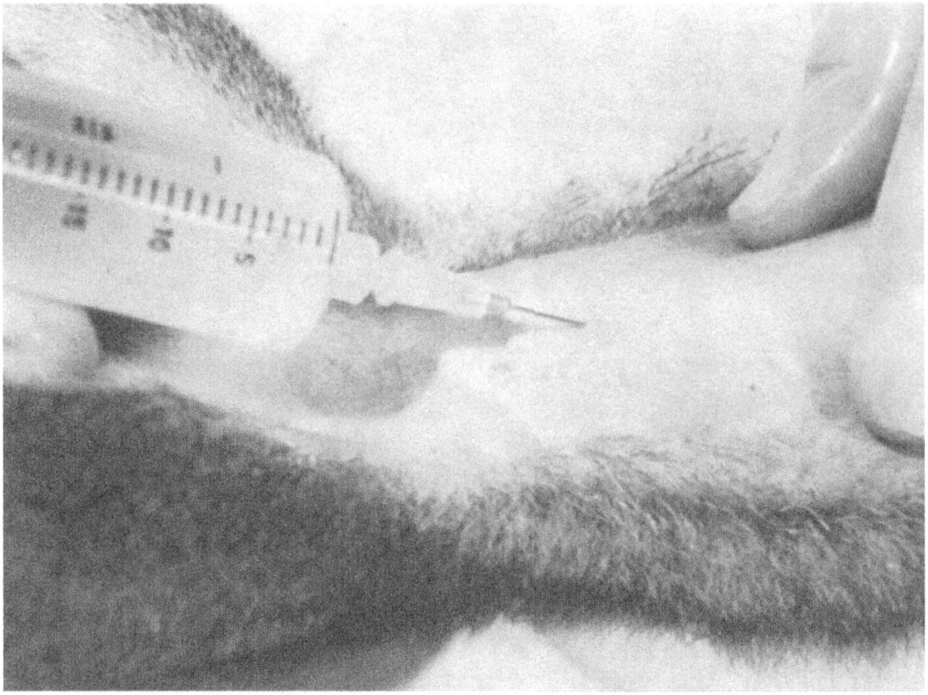

Fig. 7. Intraperitoneal administration of drugs and fluids may be used for infusion of isotonic fluid and electrolyte solutions. The needle is directed cranially and at an oblique angle with respect to the abdominal surface to avoid puncture of organs.

possible. A severely dehydrated squirrel monkey will begin to respond to fluids administered IP in approximately 45 min.

Fluids and electrolytes administered by IP infusion must be isotonic to prevent further dehydration of the animal. The site of injection should be selected to minimize the possibility of accidental puncture of the urinary bladder or other abdominal organs. The site illustrated in Fig. 7 is approximately 1 cm lateral to the abdominal midline and 1 cm posterior to the umbilicus. A 22-gauge, 1-inch needle is inserted through the abdominal wall and directed cranially at a 60° angle to the surface of the abdomen.

3.3.3. Subcutaneous Administration

Fluids and electrolytes administered subcutaneously are absorbed much more slowly than IV and IP routes. Several hours are required for absorption, making this approach to fluid therapy excellent for overnight treatment. Ideally, overnight treatment should be done intravenously, but that is often not practical for squirrel monkeys. Therefore, the patient can be treated initially intravenously. Subcutaneous treatment can be used to provide a continuous source of fluid and electrolytes for several hours thereafter. Only isotonic fluids should be given subcutaneously.

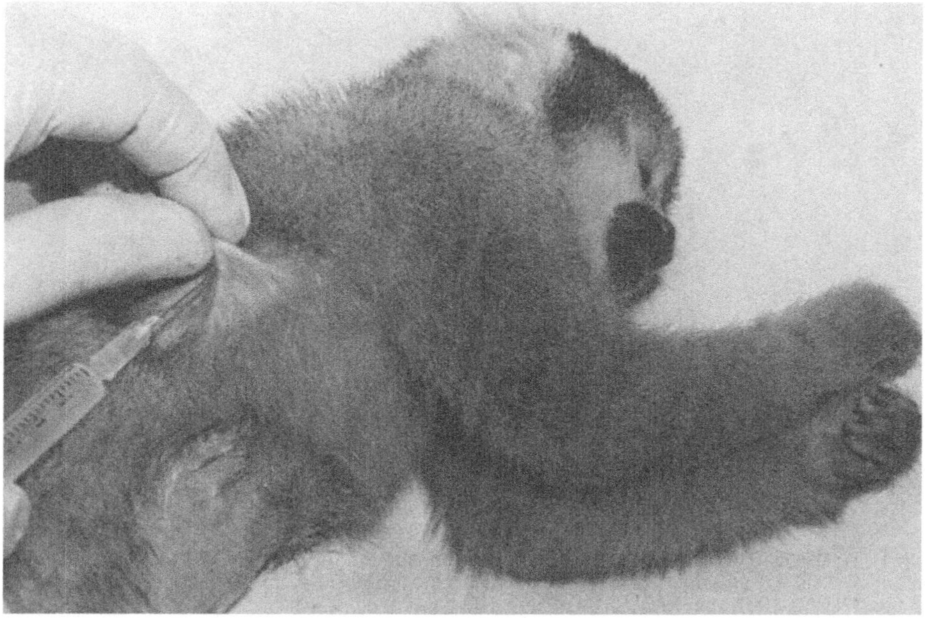

Fig. 8. Subcutaneous injections are made in the fold of skin created by pinching the skin between the thumb and index finger. (Photo by W. E. Beckham, Jr.)

The skin of the axillary or scapular region is pinched between the thumb and index finger. The needle (22 gauge) is passed under the skin in the fold of skin created (Fig. 8). Fluid volumes ranging between 10 and 20 ml can be given. Antibiotics and other drugs also can be administered subcutaneously. Label instructions on drug vials describe appropriate routes of administration.

3.3.4. Intramuscular Administration

Intramuscular injection (IM) is a recommended route of administration for many drugs, particularly antibiotics. The squirrel monkey does not have an abundance of large muscle masses suitable for deep IM injection. The best sites are the caudal and lateral muscles of the mid thigh. Care must be taken to insert the needle obliquely to avoid the femur, which is relatively superficial. The IM administration of drugs that irritate tissue should be divided among multiple sites. Long-term IM treatment with drug volumes that exceed 0.5 ml should be avoided due to the limited number of suitable sites of injection. Fluids and electrolytes should not be given IM because the volume displacement could cause excessive trauma of muscle tissue.

3.3.5. Oral Administration

Oral treatment is an excellent way to administer fluids, electrolytes, and drugs to squirrel monkeys. Some squirrel monkeys can be taught to take pal-

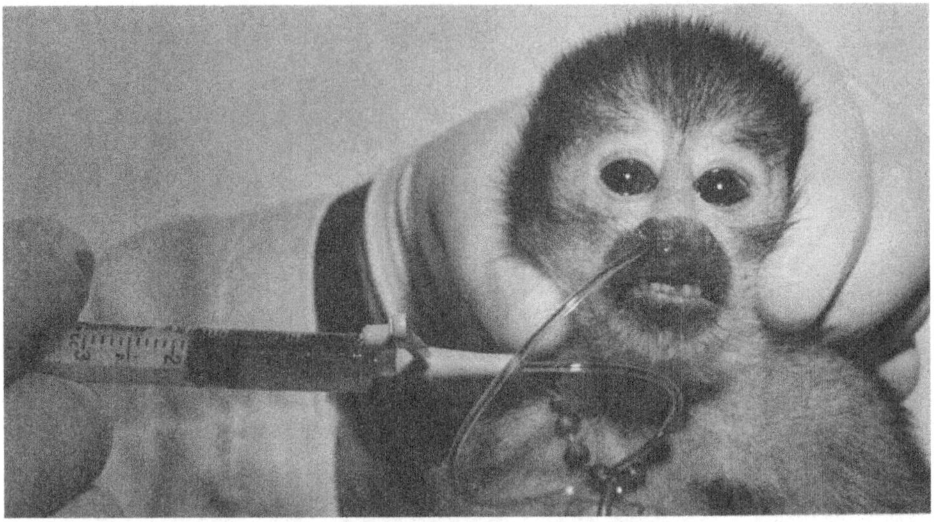

Fig. 9. Nasogastric intubation is useful for oral administration when the squirrel monkey is unwilling to drink voluntarily. The animal does not require anesthesia for this procedure. (Photo by W. E. Beckham, Jr.)

atable compounds voluntarily. There are a number of fruit-flavored pediatric antibiotic preparations available that can be used in squirrel monkeys. Drugs can also be mixed with fruit juices to increase their acceptance.

Less palatable or larger volumes can be administered by nasogastric intubation (Fig. 9). This is done by passing a #5 French infant feeding tube through the naris beyond the nasopharynx and into the stomach by way of the esophagus. The end of the feeding tube should be moistened with a water-soluble lubricant before inserting to facilitate passage of the tube. Squirrel monkeys have a laterally directed first segment of the nostril; therefore, the tube must be directed medially as it is inserted, then directed toward the back of the head. The animal usually swallows as the tube passes into the esophagus. Gagging, coughing, and labored breathing often indicate that the tube has accidentally passed into the trachea. To assure proper placement, an attempt should be made to aspirate gastric contents up the feeding tube by placing a vacuum on the tube with a syringe. If there is continued uncertainty whether the tube is in the esophagus, a small volume of sterile normal saline can be slowly administered; coughing or moist chest sounds indicate that the tube should be removed and reinserted.

4. *DISEASES OF THE SQUIRREL MONKEY*

Many natural and experimental diseases have been reported in the past 30 years. Fiennes (1972) provides a comprehensive review of diseases of squirrel monkeys and other primates. This chapter will focus on the natural diseases and disorders that are likely to be encountered in a large colony.

4.1. Bacterial Diseases

4.1.1. Tuberculosis

The bacterial disease causing the most concern to primatologists is tuberculosis. Quarantine programs for primates are designed primarily for management of this disease. The incidence of tuberculosis in the squirrel monkey appears to be much lower than in the Old World primate species. With this in mind, it is not surprising that many quarantine programs for squirrel monkeys are 60 days compared to 90 days for other primate species.

The lower incidence of tuberculosis should not leave the impression that this disease is an unimportant consideration in the medical management of the squirrel monkey. Tuberculosis has been reported as a fatal disease of newly imported squirrel monkeys (Chrisp *et al.*, 1968). Mayhall *et al.* (1981) reported that five squirrel monkeys contracted tuberculosis after completing quarantine. These animals appeared to have been infected following exposure to rhesus monkeys that had the disease. The attack rate among exposed squirrel monkeys was 100%. The first two squirrel monkeys were found dead and the remaining three were euthanatized and subsequently found to have lesions even though they had negative tuberculin tests. Leathers and Hamm (1976) reported a similar incident in which a squirrel monkey housed in the same room with a tuberculous rhesus monkey developed disseminated disease. Most of the reported cases of tuberculosis in squirrel monkeys describe what appears to be a rapidly progressing disseminated form of the disease.

Medical management of tuberculosis in squirrel monkeys must be directed toward prevention through quarantine and isolation procedures, and eradication of infected animals. Squirrel monkeys with tuberculosis should be euthanatized and strick isolation measures should be instituted for all remaining animals. Isolation of exposed animals should span at least 90 days, with tuberculin testing every 2 weeks. Chest X rays should be obtained for all suspect animals.

Tuberculin testing is done by injecting 0.1 ml of Tuberculin, Mammalian, Human isolates (Jensen-Salsbery, Division of Burroughs Wellcome Company, Kansas City, Missouri), intradermally into skin of the upper eyelid. The test is read at 24, 48, and 72 hr. Other types of tuberculin, such as the purified protein derivative (PPD) preparations, are unreliable in nonhuman primates and should not be used.

A positive reaction is characterized by varying degrees of erythema, swelling, and induration. Animals with extensive disease may have no reaction to tuberculin during the final stages of the disease; this is referred to as an anergic response. An additional complicating factor in squirrel monkeys is the relatively thick skin of the upper eyelid. This may obscure a mild positive reaction. The interpretation of tuberculin tests requires experience and is a process in which many factors must be considered. The appearance of the eyelid, medical and research history of the animals being tested, and the number of animals at risk must all be evaluated in the decision. From the perspective of management of a large colony, it is better to euthanatize an animal with a questionable reaction than to risk other animals.

4.1.2. Other Bacterial Infections

Bacterial infections in primates generally result in pneumonia, enteritis, and septicemias. The use of antibiotics should be decided by identification of the pathogenic agent and antibiotic sensitivity testing. However, seriously debilitated animals must first be stabilized before specific treatment can be instituted. Supportive treatment with fluids and electrolytes and maintenance of body temperature are an important adjunct to antibiotic treatment. Intensive care chambers, commercially available for small-animal veterinary practice, are an effective way to provide oxygen therapy and the humidity and temperature controls helpful in reducing dyspnea and maintenance of body temperature homeostasis.

Squirrel monkeys may be predisposed to some bacterial diseases due to the cage system or husbandry program being used. Kessler and Brown (1979) described the clinical signs of tetanus that occurred in 19 squirrel monkeys housed in outdoor corrals with gravel-covered earthen floors. The initial signs of the disease were characterized by stiffness in gait, progressing to extensor rigidity and opisthotonus. The case fatality rate was 100%. Vaccination with tetanus toxoid appeared to prevent new cases. Since *Clostridium tetanae* is a soil-borne organism, the cause of this epidemic was thought to be the result of wound contamination from the earthen floors in the corrals.

Yersiniosis (*Y. pseudotuberculosis*) was found by *postmortem* examination in six squirrel monkeys housed in outdoor pens (Buhles *et al.*, 1981). Clinical signs were characterized by weakness, depression, diarrhea, and abortion. The primary lesion observed was multifocal, ulcerative enteritis. The causitive agent was also cultured from wild rodents (thought to be the source of the disease) trapped in the cage area.

Pneumonias caused by a number of bacteria have been observed in squirrel monkeys. Pneumococcus, staphylococcus, and several types of streptococcus have been cultured from the lungs of animals with bacterial pneumonias. We have also identified aeromonas hydrophila bacteremias from cultures of heart blood obtained at necropsy from some squirrel monkeys in our laboratory.

The best approach to the management of bacterial diseases in a colony of squirrel monkeys is prevention rather than treatment. This is addressed by maintenance of a high level of environmental sanitation and a responsive husbandry program. In a stable colony, bacterial diseases usually occur as sporadic isolated cases. Treatment is most effective early in the course of infection and group treatment of cage mates should be considered. Table II provides a listing of antibiotics and routes of administration that may be used.

Table II
Antibiotics Useful in Squirrel Monkeys

Drug	Route of administration	Dosage	Regimen
Gentamycin sulfate injection	Intramuscular or subcutaneous	1–2 mg/kg	Every 8 hr for at least 5 days
Tetracyline HCl	Intramuscular	15 mg/kg	Every 12 hr for 7 days
Cephapirin sodium	Intramuscular or intravenous	20 mg/kg	Every 6 hr for 4 days
Penicillin G potassium	Intravenous, intramuscular, or subcutaneous	20,000 U/kg	Every 8 hr, variable time course
Flocillin® (penicillin G benzathine and penicillin G procaine)	Subcutaneous	15,000 U penicillin G benzathine and 15,000 U penicillin G procaine	Every 48 hr for 6 days
Septra® (trimethoprim and sulfamthoxazole)[a]	Oral	4 mg trimethoprim and 20 mg sulfamethoxazole/kg	Every 12 hr for 10 days
Cephradine suspension	Oral	25 mg/kg	Every 12 hr for 5 days
Chloramphenicol palmitate suspension	Oral	50 mg/kg	Every 6 hr, variable time course
Chloramphenical opthalmic ointment	Topical	—	Once or twice each day
Neomycin and isoflupredine acetate ointment	Topical	—	Daily
Nitrofurazone ointment	Topical	—	Daily

[a] Should not be used in pregnant squirrel monkeys since this drug may interfere with folic acid metabolism.

4.2. Viral Infections

4.2.1. Herpesvirus Saimiri and Herpes-T

Viral infections of greatest concern in the management of squirrel monkeys in the laboratory are *Herpesvirus saimiri* (HVS) and Herpes-T. These viruses occur primarily as latent infections in the squirrel monkey, but both cause fatal diseases in other primate species (Falk *et al.*, 1972a; King *et al.*, 1967; Falk *et al.*, 1973; Melendez *et al.*, 1970).

HVS is indigenous in squirrel monkeys and does not appear to produce any disease in the natural host. However, malignant lymphoma and leukemia following experimental infections in eight primate species (Falk *et al.*, 1972b; Rangan *et al.*, 1977) have been reported. The naturally occuring disease has been described only once (Hunt *et al.*, 1973), but the potential for accidental cross-infection to susceptible species must be considered when designing a management program for the squirrel monkey.

Herpes-T was first described as the etiological agent of a fatal systemic disease in marmosets (Melnick *et al.*, 1964). It has also been reported to cause a similar disease in owl monkeys (Hunt and Melendez, 1966). Characteristic lesions in marmosets include multiple necrotic foci in parenchymous organs and ulceration of mucous membranes. Ulcers and necrotic foci appear to spread and coalesce as the disease progresses, culminating in coma and death. Herpes-T has been reported as a latent infection (Melendez *et al.*, 1966) and as a pathogen (Daniel *et al.*, 1967; King *et al.*, 1967) in the squirrel monkey. The disease in squirrel monkeys is characterized by labial and oral ulcers resembling cold sores in human beings. Since Herpes-T can be shed from mucous membranes of apparently healthy squirrel monkeys, the potential for natural infections in susceptible species is a serious management concern.

The high incidence of these viruses in clinically healthy squirrel monkeys necessitates careful consideration of housing and handling procedures in institutions maintaining multiple New World primate species. Squirrel monkeys should not be housed in the same room with marmosets and owl monkeys. Needles and syringes used in squirrel monkeys should never be shared with other primate species. Squirrel monkeys with labial or oral ulcers should be isolated, since Herpes-T can be shed in large numbers during this stage of the infection.

4.2.2. Other Viral Infections

One outbreak of measles (rubeola) has been reported that resulted in a case fatality rate of approximately 10%. The clinical signs of the disease included sneezing, serous rhinitis, coughing, and maculopapular rash (Abee *et al.*, 1973). Microscopically, syncytia were observed in the dermis and bronchiolar epithelium. Deaths resulted primarily from interstitial pneumonitis and secondary bacterial pneumonia.

Recently, the squirrel monkey has been shown to develop clinical signs of disease following experimental infection with three human influenza A virus strains (Murphy *et al.*, 1980). This suggests the possibility that influenza may be passed to squirrel monkeys by personnel with the disease. Personnel with influenza or other infectious diseases should not work with squirrel monkeys during the infectious stage of the disease.

4.3. Parasitic Diseases

Parasitic diseases are usually most common and severe in newly imported squirrel monkeys. Following treatment for parasitic infestations, a good colony husbandry program will help in limiting the possibility of future clinical cases.

4.3.1. Protozoal Parasites

There are two protozoal parasites of concern in the management of squirrel monkeys. *Plasmodium brasilianum,* the most common agent of malaria in squirrel monkeys, appears to have a variable incidence. The infection is characterized by mild to moderate hemolytic anemia, listlessness, and splenomegaly. The infection is spread by a mosquito vector.

Trypanisoma cruzi is a hemoflagellate occasionally found in the squirrel monkey. The incidence appears low in squirrel monkeys; however, the importance of this parasite lies in the potential for transmission to laboratory personnel. *T. cruzi* causes Chagas disease in man. The disease is transmitted by an insect vector or by accidental inoculation with blood containing the parasite (Flynn, 1973a).

Each squirrel monkey entering the colony should be examined for the presence of these parasites. Diagnosis is made by identifying the organism in blood or organ smears or histological sections. Control is based on elimination of the insect vector. Treatment of malaria consists of intramuscular administration of chloroquine, 2.5 mg/kg of body weight for 7 days, followed by oral administration of primaquine, 0.75 mg/kg of body weight for 14 days (Flynn, 1973b). There is no reported treatment for *T. cruzi* in the squirrel monkey.

4.3.2. Helminth Parasites

Most helminth parasites are treated and eliminated from animals during quarantine. Table III lists helminths that are frequently found in newly imported squirrel monkeys and the recommended treatment.

Of the parasites listed in Table III, *Prosthenorchis elegans* is of most clinical concern. Clinical signs of infestation consist of diarrhea, abdominal distention, progressive debilitation, dehydration, and death. I have observed parasitized squirrel monkeys eating large amounts of feed followed by diarrhea and steady weight loss. This syndrome is apparently the result of hyperperistalsis and malabsorption. Lesions are primarily associated with the site of attachment of the proboscis to the intestinal mucosa. Circular white nodules containing thick, white purulent material are observed from the serosal surface of the intestine. These nodules are abscesses formed around the site of attachment of the parasite on the luminal surface. These abscesses occasionally

Table III
Common Helminth Parasites of the Squirrel Monkey

Parasite	Type	Location	Treatment	Reference
Athesmia foxi	Trematode	Bile ducts	None	Bostrom and Slaughter (1968)
Strongyloides cebus	Nematode	Small intestine	Thiabendazole	Bingham and Rabstein (1964)
Dipetalonema sp. and *Tetrapetalonema*	Nematode	Peritoneal cavity	Diethylcarbamazine	Eberhard (1982)
Prosthenorchis elegans	Acanthocephalad	Ileum, cecum	Carbon tetrachloride	Mackenzie (1979)

rupture, resulting in purulent peritonitis. Mechanical obstruction of the intestine may also occur with severe infestations.

Diagnosis is made by a combination of clinical signs and examination of feces for ova and direct observation. Morin *et al.* (1980) compared the diagnostic accuracy of flexible fiberoptic proctoscopy, formalin-ether fecal sedimentation, and sodium nitrate fecal flotation. Proctoscopy was clearly the best of the three methods evaluated for diagnosis of *Prosthenorchis* infestations.

Anthelmintics used for treatment of other helminth parasites are not effective for *Prosthenorchis*. Mackenzie (1979) reported that oral administration of carbon tetrachloride at a dosage of 0.5 ml/kg of body weight was an effective treatment with minimal side effects. However, the efficiency and long-term safety of this treatment have not been established.

Prosthenorchis infestations are of clinical concern mostly in newly imported squirrel monkeys. Those animals that survive the first months in the laboratory appear to lose the infestation. This may reflect a rather short life span of the parasite in this species.

The best approach to management of *Prosthenorchis* infestations is to provide supportive treatment for newly imported parasitized animals; efforts to prevent reinfection are important also. A cockroach intermediate host is required to complete the life cycle; therefore, an effective vermin control program will limit the possibility of reinfection and spread within the laboratory. Squirrel monkeys consider cockroaches a culinary delicacy; the observation of insect parts in the fecal collection pan should alert laboratory personnel to the potential hazzard.

All helminth infestations should be diagnosed and treated appropriately during quarantine. Periodic fecal examinations should be done to determine the prevalence of parasites within the colony.

4.4. Deficiency Diseases

Vitamin deficiency diseases occur rarely in squirrel monkeys fed commercial diets. However, the use of experimental diets or accidental feeding of outdated diet may result in vitamin deficiencies. Nutritional requirements of the squirrel monkey are reviewed in Hayes (1979) and Harris (1970). These texts are particularly useful when designing an experimental diet for squirrel monkeys.

Experimentally induced vitamin C deficiency in the squirrel monkey was described by Lehner *et al.* (1968). Initial signs of disease were the development of large, fluctuant swellings over the parietal bones of the head and anemia. These signs were first observed after 3 months of vitamin C deficiency. A number of reports of cephalhematomas related to suspected vitamin C deficiency have been published (Blackwell *et al.*, 1974; Demaray *et al.*, 1978; Kessler, 1980). These spontaneous cases of hypovitaminosis C occurred in squirrel monkeys fed commercial monkey chow diets. Therefore, it is important to make certain that freshly milled monkey chow is fed to squirrel monkeys. This can be determined by checking milling dates which are printed on each sack of

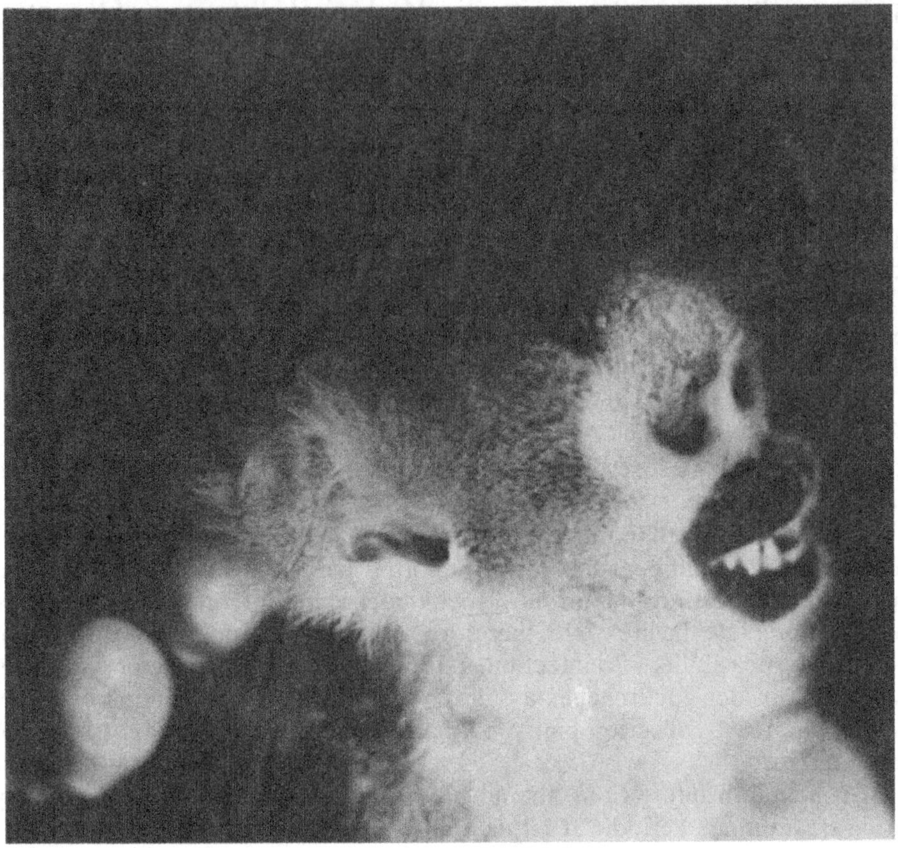

Fig. 10. Cephalhematomas are an early clinical sign of vitamin C deficiency disease. This squirrel monkey had a fluctuant swelling of the head entirely the result of periosteal hemorrhage of the cranial bones.

commercial monkey chow. Also, soaking feed biscuits in water should be avoided, since vitamin C is rapidly destroyed by moisture.

Cephalhematomas are the result of subperiosteal hemorrhage, which separates the periosteum from cranial bones. Figure 10 is a photograph of a young squirrel monkey with a large cephalhematoma extending over the entire cranium. Subsequently, hyperostosis occurs due to the formation of new bone by the periosteum. Severe deformity of the skull may result (Fig. 11). Vitamin C treatment of animals with cephalhematomas will rapidly reverse the signs of disease; however, hyperostosis is not reversible.

Rickets and osteomalacia in squirrel monkeys were first described by Bullock and Bowen (1966). Osteomalacia is the result of chronic vitamin D deficiency in the adult and is characterized by generalized demineralization of bone, deformed long bones, and spontaneous fractures. Rickets is the form of the disease that occurs in young, growing animals. Lehner *et al.* (1967) compared the biological activities of vitamins D_2 and D_3 in squirrel monkeys 6–12 months of age. After 2 months of feeding vitamin D_2 or diet devoid of vitamin D,

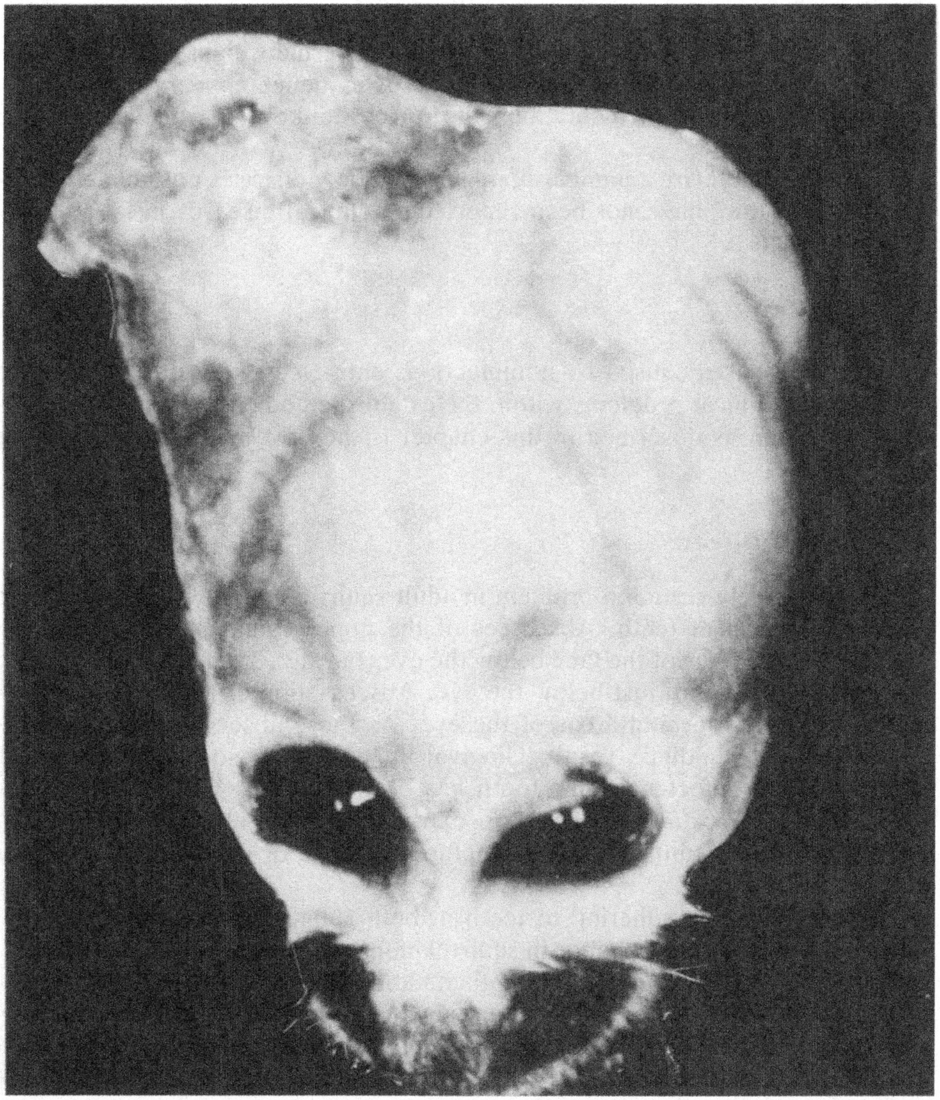

Fig. 11. Hyperostosis of the skull caused by vitamin C deficiency is observed subsequent to cephalhematomas and results in permanent distortion of the skull.

radiographs revealed thickened epiphyseal plates and frayed, widened metaphyses characteristic of rickets. Squirrel monkeys fed a diet containing vitamin D_3 did not develop lesions. It is now generally accepted that laboratory-housed New World primates, including squirrel monkeys, must be provided vitamin D_3 to prevent the development of rickets and osteomalacia.

Folic acid deficiency in pregnant squirrel monkeys leading to pregnancy wasteage was first suggested by Rosenblum (1972). He found that squirrel monkeys with stillbirths had hemograms characteristic of megaloblastic anemia of pregnancy, which resembled folic acid deficiency observed in pregnant women.

Rasmussen *et al.* (1980) reported that folic acid supplementation of pregnant squirrel monkeys resulted in improvements in maternal weight gain, increased birth weights, and improved hemograms. This evidence appears sufficient to warrant folic acid supplementation of squirrel monkeys, particularly during gestation.

Although there are a number of other deficiency diseases possible in squirrel monkeys, they have not been reported as clinical diseases in laboratory-housed animals.

4.5. Other Diseases

Those diseases caused by noninfectious entities are the most commonly encountered clinical problems within the established colony. A routine observation program as described in this chapter is the best method for detecting these diseases.

4.5.1. Dental Disease

An extremely common problem in adult squirrel monkeys is severe wear and abscessation of teeth. Abscesses of the upper canine teeth are usually detected as swellings of the face below the eye (Fig. 12). If left untreated, these abscesses will rupture just below the eye. Abscessation of molars usually extend into the infraorbital areas of the eye. As these abscesses enlarge, exophthalmos and blindness result. Gingival hyperplasia associated with malar abscess and an infected canine tooth has been reported (Williams and Whitenack, 1975). Treatment consists in extraction of the abscessed tooth and irrigation of the fistulus tract with tincture of iodine or other disinfectant solutions.

Dental deposits adhering to teeth in both subgingival and supragingival areas are a common occurrence in squirrel monkeys (Chase and Cooper, 1968). Accumulations of calculus cause inflammation and necrosis of the gingiva. Cleaning of the teeth at 6-month intervals is an important preventive medical procedure. Occasionally, severe bacterial infections occur subsequent to cleaning. It is a good practice to pretreat squirrel monkeys with broad-spectrum antibiotics prior to cleaning teeth.

4.5.2. Renal Disease

Renal diseases consisting of renal lithiasis, pyelonephritis, and glomerulonephritis have been reported as common necropsy findings in squirrel monkeys of the Columbian–Brazilian subspecies (Stills and Bullock, 1981). Renal mineralization of varying degree was also observed in 32 of 38 consecutive necropsies of adult squirrel monkeys. The primary diagnosis at necropsy in one-third of these animals was glomerulonephritis.

Four types of glomerulonephritis have been described in the squirrel monkey by Stills and Bullock (1981). These include focal sclerosing glomerulonephritis, membranoproliferative glomerulonephritis, membranous glomeru-

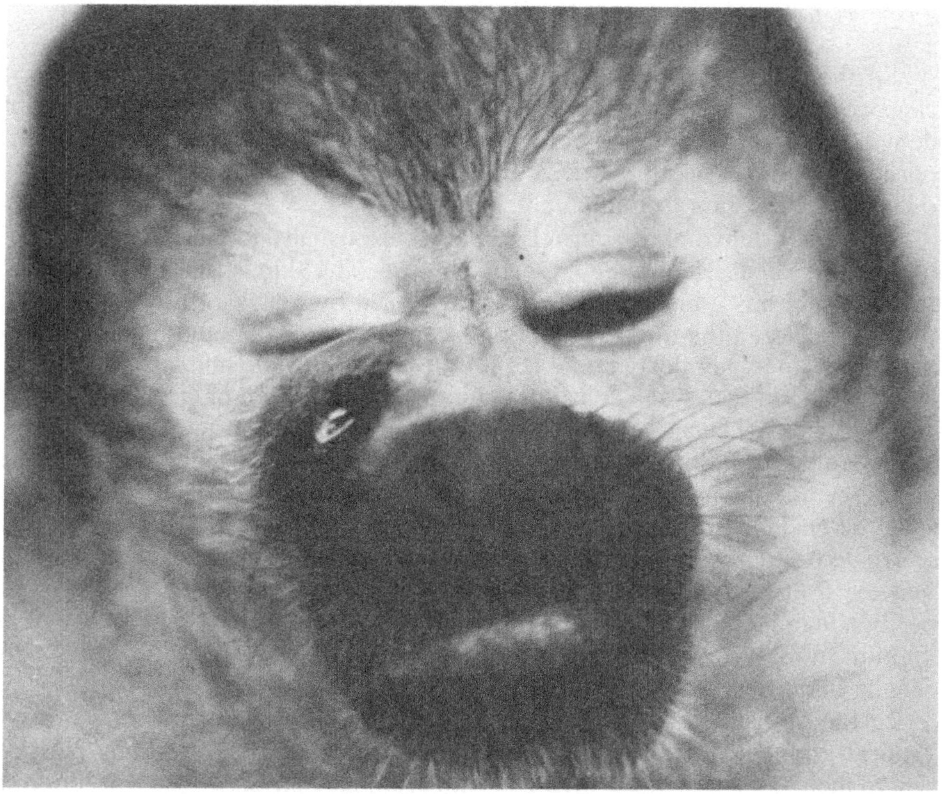

Fig. 12. Swelling of the face below the eye and malar abscess are frequently caused by an abscessed canine tooth.

lonephritis, and a combination of membranoproliferative and membranous glomerulonephritis. The cause of renal disease in squirrel monkeys is not known; however, chronic parasitic infections such as malaria and genetic factors have been postulated. The incidence of glomerulonephritis in other colonies and subspecies has not been determined. Although I have observed renal lesions in the Bolivian subspecies, it appears that the Columbian–Brazilian subspecies has the greatest prevalence of renal disease.

Management and care of squirrel monkeys with renal disease consists in diagnosis of affected animals and supportive treatment. Diagnosis is based on elevations of blood urea nitrogen (BUN) and serum creatinine and examination of urine for the presence of protein. Animals with renal dysfunction should never be deprived of water, since increased water intake is an important compensatory mechanism for these animals. Appropriate antibiotic therapy is indicated when bacteria are cultured from urine. Since the cause of this disease in squirrel monkeys is unknown, no method for preventing renal disease has been proposed.

4.5.3. Dehydration

Although dehydration is usually observed secondary to other diseases, it can occur as a primary disease in the squirrel monkey. Common causes of dehydration are malfunctions of automatic watering devices, deprivation imposed on animals being ostracized by cage mates, and the effects of stress related to separation from a parental group.

Severely dehydrated squirrel monkeys are usually found on the floor of the cage markedly depressed or semicomatose. The skin is tightly fixed to the subcutis and has a plastic texture characterized by a lack of resilience when pinched between the fingers. Fluid replacement therapy should be started immediately since any delay could result in the death of the animal.

The treatment of choice is intravenous administration of sterile fluid electrolyte solutions such as lactated Ringer's solution and normal saline. Circulatory collapse is sometimes a complications intravenous therapy. In these cases, percutaneous venipuncture may be difficult or impossible. A decision must be made as to whether a venous cutdown procedure is to be done or an alternate route of fluid administration chosen. A venous cutdown requires skill and experience to be successful, whereas intraperitoneal fluids can be more easily administered. Response to intravenous fluids is usually observed within 15 min; intraperitoneal fluids are more slowly absorbed, requiring 30–60 min for the initial response to treatment.

After initial treatment, additional fluids should be given more slowly. Squirrel monkeys are very difficult to treat intravenously over long periods due to their small size and difficulties with prolonged physical restraint. I prefer to first treat dehydrated squirrel monkeys with intravenous fluids, followed by subcutaneous fluid administration. Fluids given under the skin are absorbed slowly to meet body requirements over a period of several hours. Subcutaneous fluids provide added safety over other routes of administration since there is very little danger of fluid overdose.

The administration of fluids and electrolytes is a clinical procedure requiring considerable basic understanding of fluid therapy and an understanding of the special problems posed by the squirrel monkey. The small size of these animals and their high metabolic rate must be remembered when fluid therapy is chosen. A moderately dehydrated adult squirrel monkey can be given between 15 and 50 ml of fluid intravenously over a 1-hr period followed by 15–20 ml divided among several subcutaneous sites. The animal should be observed carefully for urine production and the chest should be auscultated to determine whether pulmonary edema has resulted from fluid treatment. A pediatric IV set that divides each milliliter of fluid into 60 drops is very useful for accurately determining the rate and amount of fluid given. Fluids should be warmed to body temperature before administration.

4.5.4. Hypoglycemia

Squirrel monkeys occasionally develop hypoglycemia as a primary disorder. The basal metabolic rate of approximately 107 kcal/kg per 24 hr (Malinow

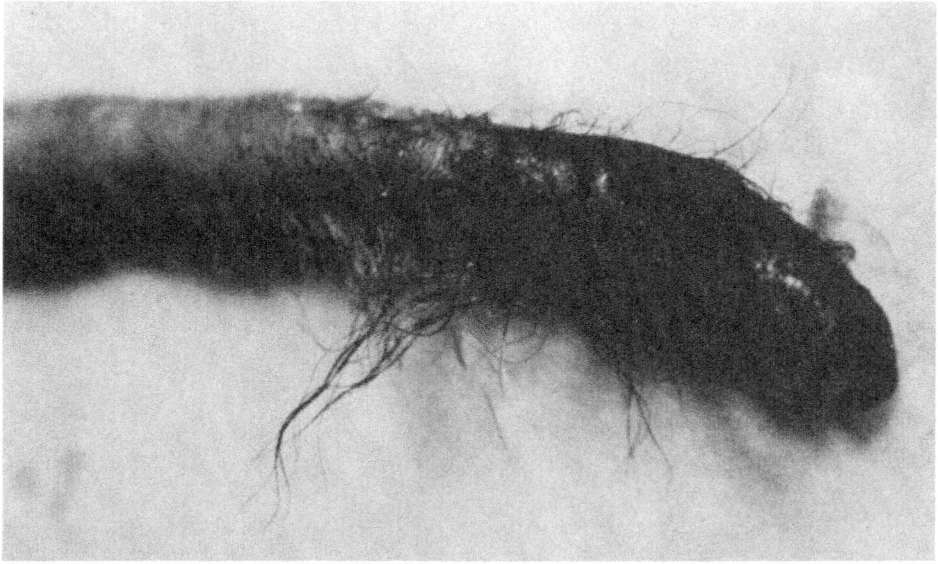

Fig. 13. Gangrene of the end of the tail is a common complication following infection of a fight wound.

and Wagner, 1966) combined with the low percentage of body adipose stores (Gallina and Ausman, 1979) predispose the squirrel monkey to hypoglycemic episodes. This condition is often observed in juveniles subjected to stress of separation from their parental group, accidental deprivation of feed, or complications during weaning.

Characteristic signs of hypoglycemia include lethargy, weakness, and disorientation. Severely hypoglycemic animals frequently move in slow motion, resembling exhausted wind-up toy soldiers. Treatment consists in intravenous administration of 5% dextrose in water at a dosage of 5–8 ml/100 g body weight. This treatment results in an amazing reversal of clinical signs within 30 min after beginning treatment.

4.5.5. Trauma

Trauma is the most common medical problem requiring treatment in large colonies of group-housed squirrel monkeys. Injuries result from a number of causes, which include fighting, cage injuries, and handling and restraint accidents.

Wounds from fighting are of several types. Tail chewing may be self-inflicted, or one or two animals within a cage may be biting the tails of the other cage members. Once a tail wound is infected, the irritation is often sufficient to cause the injured animal to continue to pick and chew at the wound. The blood supply to the distal segment of the tail is rather poor, which can lead to secondary infections and gangrene (Fig. 13). Occasionally, I have observed animals with abscessed teeth or other painful medical problems mutilate their

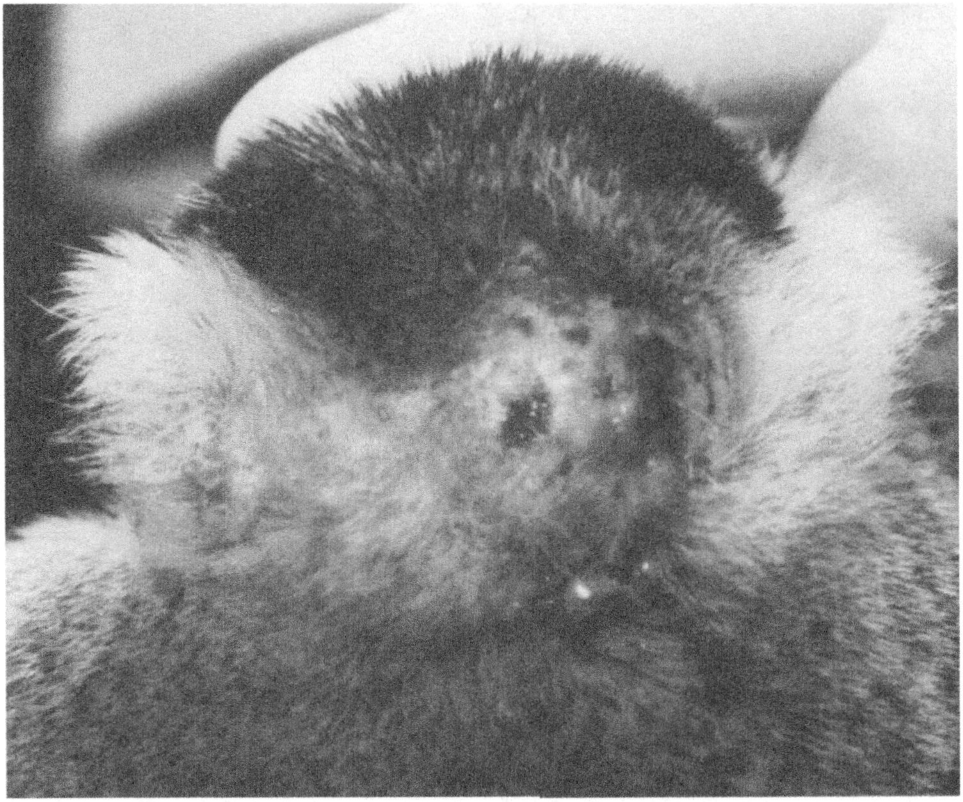

Fig. 14. A freshly inflicted "crush" wound to the back of the head may be followed by necrosis, sloughing of the skin, and exposure of the cranial bones.

own tails in an attempt to alleviate the primary source of pain. Routine physical examination should always include examination of the tail for injuries. Treatment of tail wounds often requires partial amputation of the tail. Tail guards made from plastic syringe covers or similar items may help to prevent new injuries while the tail is healing.

A second type of wound observed in group-housed animals is the "crush" wounds frequently found over the elbows, knees, and back of the head (Fig. 14). These wounds are usually inflicted by females and result from repeated biting at the same site, crushing the tissue and causing the affected area to become necrotic. Upon close examination, small puncture wounds are also found. This wound type is often not detected until severe undermining infections have resulted. Treatment consists in trimming necrotic tissue, cleaning the wound, and applying a strong disinfectant. These wounds must be allowed to drain for proper healing. The use of absorbant dressings to draw the infection from the wound is often effective. Severe wounds of the arm or leg with extensive necrosis of muscle and nerves may require amputation. Crush wounds to the back of the head may result in necrosis of the scalp and exposure of the

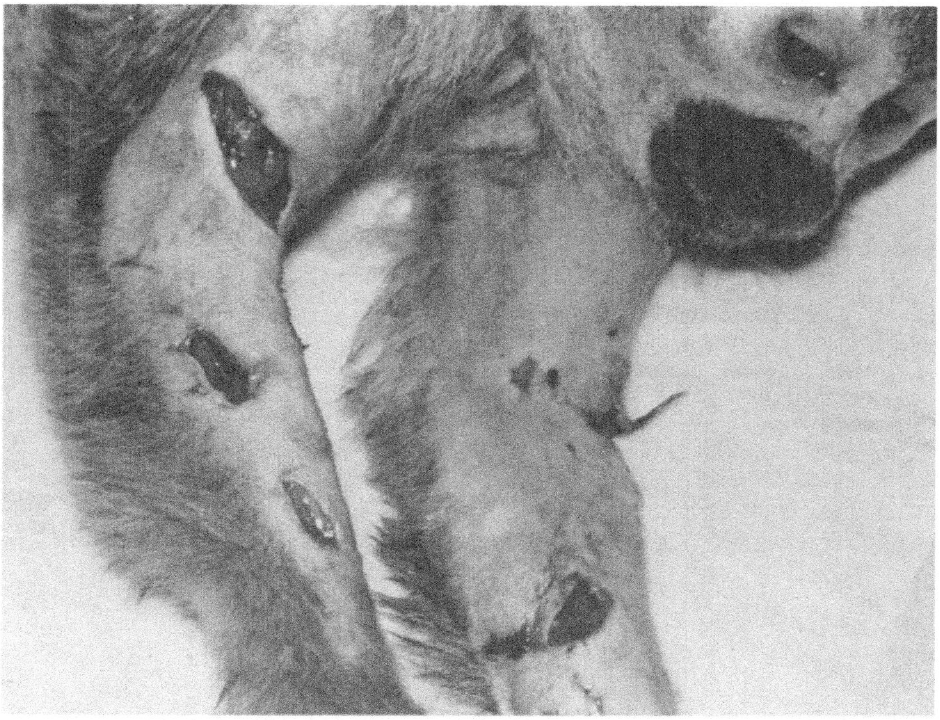

Fig. 15. Deep lacerations of the skin and underlying muscles are usually inflicted by adult male squirrel monkeys.

cranium. These wounds should be cleaned and debraided every 2–3 days. This will allow gradual healing of the skin over the exposed portion of the skull.

Deep lacerations of the thighs, shoulders, forearm, and upper back are usually caused by adult males (Fig. 15). These wounds are easily observed after finding blood in the cage or on the wounded animal. Fresh lacerations should be cleaned thoroughly, trimmed, and sutured with monofiliament stainless steel or other suitable suture material.

Skull fractures of infant squirrel monkeys are a common cause of neonatal mortality in some breeding colonies (Lee *et al.*, 1980). These injuries seem to occur most often while the infant is being carried on the mother's chest during nursing. When the mother jumps to the cement floor, her chin and chest often hit the floor. If the infant happens to be clinging to the chest when the mother jumps, the infant's head hits the floor. There is very little that can be done for the infant after the trauma has occurred; therefore, a preventive medical approach is necessary. We have greatly reduced infant head trauma by installing artificial turf carpets or flexible vinyl mats in our breeding cages. These mats serve to cushion the floor surface of the cage. Artificial turf can be cleaned easily and autoclaved as necessary.

Pressure ulcers on the dorsal base of the tail occur frequently in squirrel monkeys. These sores may be caused by the type of perch used (Clewe and

Table IV
Hematological Determinations in the Bolivian Squirrel Monkey[a]

	N	Mean	Standard deviation	Minimum value	Maximum value	Standard error of mean
Red cell count (10^6/mm^3)	397	6.26	1.06	3.3	10.7	0.05
Hemoglobin (g/dl)	375	11.72	1.70	1.8	17.5	0.09
Hematocrit (%)	397	35.26	4.56	2.4	48.9	0.23
MCV (μ^3)[b]	317	59.15	2.39	52.0	68.0	0.13
MCH (pg)[b]	317	19.59	1.07	17.3	31.0	0.06
MCHC (g/dl)[b]	317	33.19	0.90	30.2	36.5	0.05
White cell count (10^3/mm^3)	397	7.12	3.46	2.6	43.5	0.17
Segmented neutrophils (%)	348	52.95	13.20	6.0	88.0	0.71
Band cells (%)	296	0.22	0.55	0.0	4.0	0.03
Lymphocytes (%)	396	43.19	12.79	7.0	86.0	0.64
Monocytes (%)	337	0.77	2.19	0.0	37.0	0.12
Eosinophils (%)	366	2.25	2.46	0.0	14.0	0.13
Basophils (%)	281	0.06	0.25	0.0	2.0	0.02

[a] From the Primate Research Laboratory, University of South Alabama, College of Medicine.
[b] Abbreviations: Mean corpuscular volume (MCV); mean corpuscular homoglobin (MCH); mean corpuscular hemoglobin concentration (MCHC).

Table V
Biochemical Determinations in the Bolivian Squirrel Monkey[a]

	N	Mean	Standard deviation	Minimum value	Maximum value	Standard error of mean
Iron (μg/dl)	331	144.21	55.01	38.0	435.0	3.02
Sodium (meq/liter)	343	144.16	4.34	123.0	154.0	0.23
Potassium (meq/liter)	345	4.47	1.13	2.2	10.0	0.06
Chloride (meq/liter)	346	110.06	6.31	92.0	124.0	0.34
Carbon dioxide content (meq/liter)	326	19.92	4.54	9.0	32.0	0.25
Protein (g/dl)	330	6.56	0.47	4.4	8.1	0.03
Albumin (g/dl)	330	3.80	0.27	2.9	4.5	0.01
Calcium (mg/dl)	343	9.24	0.70	4.9	10.9	0.04
Inorganic phosphorus (mg/dl)	344	3.40	1.50	0.5	22.6	0.08
Alkaline phosphatase (U/l)	342	221.01	86.85	65.0	514.0	4.70
Bilirubin (mg/dl)	344	0.32	0.25	0.0	1.8	0.01
Lactic dehydrogenase (U/l)	330	254.92	131.75	53.0	810.0	7.25
Cholesterol (mg/dl)	330	139.87	32.22	32.0	362.0	1.77
Glucose (mg/dl)	342	85.09	45.93	21.0	760.0	2.48
Uric acid (mg/dl)	330	0.89	0.82	0.1	14.0	0.05
Blood urea nitrogen (mg/dl)	344	27.49	7.92	10.0	55.0	0.43
SGPT (MIU/ml)	324	108.73	114.76	4.0	1612.0	6.38

[a] From the Primate Research Laboratory, University of South Alabama, College of Medicine.

Table VI
Hematological Determinations in the Colombian–Brazilian Squirrel Monkey[a]

Constituent	Method of determination	Number of subjects	Mean	Standard deviation	Observed range
Red blood cells ($10^6/mm^3$)	Coulter counter	67	8.3	0.6	7.1–10.9
Hemoglobin (g %)	Cyanmet hemoglobin	67	14.8	0.8	12.9–17.0
Packed cell volume (%)	Microhematocrit	67	46.9	2.4	43.0–56.0
Mean corpuscular volume (μ^3)	—	67	57.1	3.8	41.4–62.7
Mean corpuscular hemoglobin (pg)	—	67	18.1	1.1	13.9–20.1
Mean corpuscular hemoglobin concentration (%)	—	67	31.7	1.1	29.2–34.8
White blood cells (WBC) ($10^3/mm^3$)	Coulter counter	67	8.2	2.4	3.4–14.8
Lymphocytes (per 100 WBC)	—	67	51.9	13.7	19.0–82.0
Neutrophils (per 100 WBC)	—	67	41.2	13.8	13.0–79.0
Band cells (per 100 WBC)	—	67	0.1	0.4	0.0–2.0
Eosinophils (per 100 WBC)	—	67	5.3	5.2	0.0–22.0
Monocytes (per 100 WBC)	—	67	1.2	1.3	0.0–6.0
Basophils (per 100 WBC)	—	67	0.3	0.7	0.0–4.0

[a] From New (1968).

Table VII
Biochemical Determinations in the Colombian–Brazilian Squirrel Monkey[a]

Constituent	Method of determination	Number of subjects	Mean	Standard deviation	Observed range
Sodium (meq/liter)	Flame photometry	67	159.7	6.1	144.0–173.0
Potassium (meq/liter)	Flame photometry	67	5.6	1.2	3.5–9.5
Calcium (meq/liter)	Stanbio, AZO Cal	67	5.1	0.3	4.2–5.8
Chloride (meq/liter)	Schales and Schales	48	113.6	5.2	103.0–118.0
Inorganic phosphorus (mg/100 ml)	Fiske-subarrow	67	5.3	1.4	3.2–9.2
Alkaline phosphate (mmole units/liter)	Bessy–Lowery–Brock	66	21.8	10.7	6.0–49.0
Glutamic oxalacetic transaminase (U/ml)	Reitman–Frankel, 37°C	67	138.1	61.7	56.0–384.0
Glutamic pyruvic transaminase (U/ml)	Reitman–Frankel, 37°C	64	117.2	74.9	38.0–470.0
Lactic dehydrogenase (U/ml)	Sigma Tech. Bull. #500	32	381.9	110.9	230.0–760.0
Bilirubin, total (mg/100 ml)	Evelyn Malloy	38	0.2	0.3	0.0–1.9
Blood urea nitrogen (mg/100 ml)	Diacetyl monoxime	67	20.7	6.2	11.4–42.2
Nonprotein nitrogen (mg/100 ml)	Micro-Kjeldahl	32	28.8	8.6	12.0–49.6
Cholesterol (mg/100 ml)	Liebermann–Burchard	67	199.1	34.2	116.0–272.0
Glucose (mg/100 ml)	Somogyi	67	72.3	19.7	35.0–148.0
Uric acid (mg/100 ml)	Caraway	67	1.0	0.4	0.2–2.1

[a] From New (1968).

Duvall, 1966) or may be the result of debilitation from some other undetected medical problem. Squirrel monkeys lack ischeal callosities and are not anatomically equipped for sitting. Therefore, animals that are weak or reluctant to move will sit on their tails. These wounds should be cleaned and disinfected. The perch arrangement should be examined to determine whether the perches are contributing to the cause. A complete medical examination of the animal should be carried out to determine whether other medical problems are present.

5. NORMATIVE DATA

Tables IV–VII provide normal hematological and serum chemistry values for the squirrel monkey. The subspecies from which values were derived is presented with each table. The number of observations used to calculate the mean for each parameter is also provided. These normal values are very useful in assessing the clinical condition of sick animals and in determining appropriate therapeutic intervention.

ACKNOWLEDGMENTS. I am pleased to acknowledge the support of NIH grant RR01254, from which a portion of my experience with squirrel monkeys was derived. I am also indebted to Jane L. Martin for her efforts in the preparation of this chapter.

REFERENCES

Abee, C. R., Martin, P. D., Lehner, N. D. M., and Falk, L. E., Jr., 1973, Clinical and pathologic characteristics of measles in squirrel monkeys (*Saimiri sciureus*), abstract presented to American Association for Laboratory Animal Science.

Adair, E. R., 1969, An effective restraining chair for small primates, *Lab. Primate Newsl.* **8**(1):3–6.

Bingham, G. A., and Rabstein, M. M., 1964, A study of the effectiveness of Thiabendazole in the rhesus monkey, *Lab. Anim. Care* **14**(5):357–365.

Blackwell, C. A., Manning, P. J., Hutchinson, T. C., and Fisk, S. K., 1974, Cranial hyperostosis of squirrel monkeys (*Saimiri sciureus*), *Lab. Anim. Sci.* **24**(3):541–544.

Bostrom, R. E., and Slaughter, L. J., 1968, Trematode (*Athesmia foxi*) infection in two squirrel monkeys (*Saimiri sciureus*), *Lab. Anim. Care* **18**(4):493–495.

Buhles, W. C., Vanderlip, J. E., Russell, S. W., and Alexander, N. L., 1981, *Yersinia pseudotuberculosis* infection: Study of an epizootic in squirrel monkeys, *J. Clin. Microbiol.* **13**(3):519–525.

Bullock, B. C., and Bowen, J. A., 1966, Rickets and osteomalacia in squirrel monkeys (*Saimiri sciureus*), *Fed. Proc.* **25**:533.

Chase, J. E., and Cooper, R. W., 1968, Dental deposits and their control: An important aspect of the preventative medical program for laboratory primates, *Lab. Anim. Care* **18**(2):186–191.

Chrisp, C. E., Cohen, B. J., Ringler, D. H., and Abrams, G. D., 1968, Tuberculosis in a squirrel monkey (*Saimiri sciureus*), *J. Am. Vet. Med. Assoc.* **153**(7):918–922.

Clarkson, T. B., Bullock, B. C., Lehner, N. D. M., and Feldner, M. A., 1968, The squirrel monkey as a laboratory animal, in: *Animal Models for Biomedical Research*, National Academy of Sciences, Washington, D. C.

Clewe, T. H., 1965, Squirrel monkey perching habits, *Lab. Primate Newsl.* **4**(1):7.

Clewe, T. H., and Duvall, W., 1966, Sore tails in squirrel monkeys produced by perches, *Lab. Primate Newsl.* **5**(4):8.

Coté, R. A., ed., 1979, *Systematized Nomenclature of Medicine*, College of American Pathologists, Skokie, Illinois.

Daniel, M. D., Karpas, A., Melendez, L. V., King, N. W., and Hunt, R. D., 1967, Isolation of herpes-T virus from a spontaneous disease in squirrel monkeys (*Saimiri sciureus*), *Arch. Ges. Virusforsch.* **22**:324–331.

Demaray, S. Y., Altman, N. H., and Ferrell, T. L., 1978, Suspected ascorbic acid deficiency in a colony of squirrel monkeys (*Saimiri sciureus*), *Lab. Anim. Sci.* **28**(4):457–460.

Dost, F. N., Johnson, D. E., and Wang, C. H., 1972, A restraint system for squirrel monkeys, *Lab. Anim. Sci.* **22**(6):893–897.

Eberhard, M. L., 1982, Chemotherapy of filariasis in squirrel monkeys (*Saimiri sciureus*), *Lab. Anim. Sci.* **32**(4):397–400.

Falk, L., Wolfe, L., and Deinhardt, F., 1972a, Epidemiology of *Herpesvirus Saimiri* infection in squirrel monkeys, in: *Medical Primatology 1972, Part III: Infectious Diseases, Oncology, Pharmacology and Toxicology, Cardiovascular Studies* (E. I. Goldsmith and J. Moor-Jankowski, eds.), Karger, Basel, pp. 151–158.

Falk, L. A., Wolfe, L. G., and Deinhardt, F., 1972b, Ioslation of *Herpesvirus saimiri* from blood of squirrel monkeys (*Saimiri sciureus*), *J. Natl. Cancer Inst.* **48**(5):1499–1505.

Falk, L. W., Nigida, S., Deinhardt, F., Cooper, R. W., and Hernandez-Camacho, J. I., 1973, Brief communication: Oral excretion of *Herpesvirus saimiri* in captive squirrel monkeys and incidence of infection in feral squirrel monkeys, *J. Natl. Cancer Inst.* **51**(6):1987–1989.

Fiennes, R. N. T. W., ed., 1972, *Pathology of Simian Primates*, S. Karger, Basel.

Flynn, R. J., 1973a, Flagellates, in: *Parasites of Laboratory Animals*, Iowa State University Press, Ames, Iowa, pp. 6–7.

Flynn, R. J., 1973b, Sporozoans and neosporans, in: *Parasites of Laboratory Animals*, Iowa State University Press, Ames, Iowa, p. 69.

Gallina, D. L., and Ausman, L. M., 1979, Selected aspects of the metabolic behavior of the squirrel mokey, in: *Primates in Nutritional Research* (K. C. Hayes, ed.), Academic Press, New York, pp. 225–247.

Greenstein, D. T., 1975, Ketamine HCl, a dissociative anesthetic for squirrel monkeys (*Saimiri sciureus*), *Lab. Anim. Sci.* **25**(6):774–777.

Grimm, R. J., Savić, M., Peterson, P. F., and Griffith, J. S., 1969, Halothane anesthesia for squirrel monkeys used in neurophysiology studies, *Lab. Primate News.* **8**(1):7–16.

Harris, R. S., ed., 1970, *Nutrition of Non-Human Primates*, Academic Press, New York.

Hayes, K. C., ed., 1979, *Primates in Nutritional Research*, Academic Press, New York.

Hickey, T. E., and Kelly, W. A., 1972, Heatstroke in a colony of squirrel monkeys, *J. Am. Vet. Med. Assoc.* **161**(6):700–702.

Hunt, R. D., 1970, Nonhuman primates of different species should not be housed together, *Lab. Anim. Care* **20**(5):1007–1008.

Hunt, R. D., and Melendez, L. V., 1966, Spontaneous herpes-T infection in the owl monkey (*Aotus trivirgatus*), *Pathol. Vet.* **3**:1–26.

Hunt, R. D., Garcia, F. G., Barahona, H. H., King, N. W., Fraser, C. E., and Melendez, L. V., 1973, Spontaneous *Herpesvirus saimiri* lymphoma in an owl monkey, *J. Infect. Dis.* **127**(6):723–725.

Kalter, S. S., and Heberling, R. L., 1976, Virology, in: *Handbook of Laboratory Animal Science, Vol. III.* (E. C. Melby, Jr. and N. H. Altman, eds.), CRC Press, Cleveland, Ohio, pp. 609–611.

Kessler, M. J., 1980, Cephalhematomas due to suspected ascorbic acid deficiency in young squirrel monkeys (*Saimiri sciureus*), *J. Med. Primatol.* **9**:314–318.

Kessler, M. J., and Brown, R. J., 1979, Clinical description of tetanus in squirrel monkeys (*Saimiri sciureus*), *Lab. Anim. Sci.* **29**(2):240–242.

King, N. W., Hunt, R. D., Daniel, M. D., and Melendez, L. V., 1967, Overt herpes-T infection in squirrel monkeys (*Saimiri sciureus*), *Lab. Anim. Care* **17**(4):413–423.

Kuehl, T. J., and Dukelow, W. R., 1974, A restraint device for electroejaculation of squirrel monkeys (*Saimiri sciureus*), *Lab. Anim. Sci.* **24**(2):364–366.

Lang, C. M., 1968, The laboratory care and clinical management of *Saimiri* (squirrel monkeys), in: *The Squirrel Monkey* (L. A. Rosenblum and R. W. Cooper, eds.), Academic Press, New York, pp. 393–416.

Leathers, C. W., and Hamm, T. E., 1976, Naturally occurring tuberculosis in a squirrel monkey and a cebus monkey, *J. Am. Med. Assoc.* **169**(9):909–911.

Lee, K. J., Hughes, H. C., and Lang, C. M., 1980, Skull fractures in infant squirrel monkeys (*Saimiri sciureus*), *Lab. Anim. Sci.* **30**(6):1006–1008.

Lehner, N. D. M., Bullock, B. C., Clarkson, T. B., and Lofland, H. B., 1967, Biological activities of vitamins D_2 and D_3 for growing squirrel monkeys, *Lab. Anim. Care* **17**(5):483–493.

Lehner, N. D. M., Bullock, B. C., and Clarkson, T. B., 1968, Ascorbic acid deficiency in the squirrel monkey, *Proc. Soc. Exp. Biol. Med.* **128**:512–514.

Mackenzie, P. S., 1979, Pathogenicity, identification and treatment of *Prosthenorchis elegans* infestation in squirrel monkeys (*Saimiri sciureus*), *Primate Supply* **4**(1):5–7.

Malinow, M. R., and Wagner, R., 1966, Oxygen uptake in squirrel monkeys (*Saimiri sciureus*), *Lab. Anim. Care* **16**:105–108.

Mayhall, C. G., Lamb, V. A., and Coleman, P. H., 1981, Infection in rhesus (*Macaca mulatta*) and squirrel (*Saimiri sciureus*) monkeys due to mycobacterium tuberculosis phage type B—Outbreak in a primate colony, *J. Med. Primatol.* **10**:302–311.

Melendez, L. V., Hunt, R. D., Garcia, F. G., and Trum, B. F., 1966, A latent herpes T infection in *Saimiri sciuresu* (squirrel monkey), in: *Some Recent Developments in Comparative Medicine* (R. N. T. W. Fiennes, ed.), Academic Press, London, pp. 393–397.

Melendez, L. V., Hunt, R. D., Daniel, M. D., Fraser, C. E. O., Garcia, F. G., and Williamson, M. E., 1970, Lethal reticuloproliferative disease induced in *Cebus albifrons* monkeys by *Herpesvirus saimiri, Int. J. Cancer* **6**:431–435.

Melnick, J. L., Midulla, M., Wimberly, I., Barrera-Ora, J. G., and Levy, B. M., 1964, A new member of the *Herpesvirus* group isolated from South American marmosets, *J. Immunol.* **92**:596–601.

Mooreland, A. F., 1972, Design and management of a primate research facility for an academic institution located in an area of mild climate, in: *Medical Primatology*, Part 1 (E. I. Goldsmith and J. Moor-Jankowski, eds.), Karger, Basel, pp. 130–137.

Morin, M. L., Renquist, D. M., Johnson, D. K., and Strumpf, I. J., 1980, Flexible fiberoptic protoscopy compared with fecal examination techniques for diagnosis of *Prosthenorchis* infection in squirrel monkeys (*Saimiri sciureus*), *Lab. Anim. Sci.* **30**(6):1009–1011.

Murphy, B. R., Sly, D. L., Hosier, N. T., London, W. T., and Chanock, R. M., 1980, Evaluation of three strains of influenza A virus in humans and in owl, cebus, and squirrel monkeys, *Infect. Immunol.* **28**(3):688–691.

National Institutes of Health, 1978, *Guide for Care and Use of Laboratory Animals*, (NIH publication 80-23, rev. 1978), National Institutes of Health.

New. A. E., 1968, Base-line blood determinations of the squirrel monkey (*Saimiri sciureus*), in: *The Squirrel Monkey* (L. A. Rosenblum and R. W. Cooper, eds.), Academic Press, New York, pp. 417–419.

Rangan, S. R. S., Martin, L. N., Enright, F. M., and Abee, C. R., 1977, *Herpes virus Saimiri* induced lymphoproliferative disease in howler monkeys, *J. Natl. Cancer Inst.* **59**(1):165–171.

Rasmussen, K. M., Thenen, S. W., and Hayes, K. C., 1980, Effect of folic acid supplementation on pregnancy in the squirrel monkey (*Saimiri sciureus*), *J. Med. Primatol.* **9**:169–184.

Rosenblum, L. A., 1972, Reproduction of squirrel monkeys in the laboratory, in: *Breeding Primates*, Karger, Basel, pp. 130–143.

Short, C. E., 1980, Fluid and electrolyte therapy, in: *Current Veterinary Therapy VII* (R. W. Kirk, ed.), W. B. Saunders, Philadelphia, pp. 49–53.

Stills, H. F., and Bullock, B. C., 1981, Renal disease in squirrel monkeys (*Saimiri sciureus*), *Vet. Pathol.* **18**:38–44.

Williams, C. S. F., and Whitenack, D. L., 1975, Gingival hyperplasia and malar abscesses in a squirrel monkey, *Veterinary Med.* **70**:1094–1095.

Index

Callicebus, 1, 57, 89, 259

Made in the USA
Monee, IL
13 September 2026

62976351R00289